PHYSICS RESEARCH AND TECHNOLOGY

FUTURE RELATIVITY, GRAVITATION, COSMOLOGY

Physics Research and Technology

Additional books in this series can be found on Nova's website
under the Series tab.

Additional e-books in this series can be found on Nova's website
under the e-book tab.

PHYSICS RESEARCH AND TECHNOLOGY

FUTURE RELATIVITY, GRAVITATION, COSMOLOGY

VALERIY V. DVOEGLAZOV
M. DE G. CALDERA CABRAL
J. A. CÁZARES MONTES
J. L. QUINTANAR GONZÁLEZ
EDITORS

nova
science publishers
www.novapublishers.com

DOI: https://doi.org/10.52305/RLIT5885

NOTICE TO THE READER

Library of Congress Cataloging-in-Publication Data

ISBN: 979-8-88697-455-3

Published by Nova Science Publishers, Inc. † New York

Contents

EDITORIAL INTRODUCTION

This book may be considered as continuation of books of our Book Series "Contemporary Fundamental Physics" published by us in the past years. The thematic issue "Future Relativity, Gravitation, Cosmology" contains articles related to problems of modern physics. It is dedicated to various significant dates of the 20th century. The book includes the Editorial Introduction and fifteen chapters written by well-known researchers. V. V. Dvoeglazov presents his earlier paper, as always, dedicated to the fundamental problems of high-energy physics. His recent research has been presented at the XII and XIII Workshop (2017 and 2019), at the XI and XII Schools (2016 and 2018) of the DGFM of the Sociedad Mexicana de Física, at the LXII Congreso Nacional de Física (2019), and at many International Conferences in Italy, Czechia, Brasil etc.

M. Kirchbach and C. Compean consider conformal symmetry in quantum chromodynamics (QCD). However, they found its relation to the dS_4, the parametrization was previously used by Kadyshevsky *et al.* and Yamaleev, see for example [1]. The phenomenon of charge confinement predicted by the dS_4 special relativity is well known from the theory of strong interactions, the Quantum Chromodynamics (QCD) [2], "in which the fundamental matter degrees of freedom, the quarks, are endowed with so called color charges". "Free quantum motions of quark-anti-quark dipoles on S_3, when perturbed by the potential generated by a gluon-anti-gluon dipole, give rise to an instantaneous interaction capable of explaining a large amount of data". Next, we find the paper of R. Dahm who insists on a geometrical approach. The paper tells for itself since the first words. His approach goes back to the J. Plücker approach [3]. I. Krouglov presents again his view on non-linear electrodynamics. He presents many plots to prove his model. However, "it was demonstrated that at $|\mathbf{B}| \geq |\mathbf{E}|$ the principles of causality and unitarity take place. In his model the singularity of the electric field at the center of charges is absent and the maximum electric field in the origin is $E(0) = 1/\sqrt{\beta}$." In the mathematical paper by R. Yamaleev "the second order polynomial relationship between energy, mass and momentum, the mass-shell equation, is considered as the characteristic polynomial of the second-order ordinary differential equation describing the damped oscillator model". M. Land continues with the Stueckelberg-Horwitz-Piron theory [4]. "Historically, SHP electrodynamics was constructed by generalizing the O(3,1)- symmetric action for a particle in Maxwell theory by adding τ - dependence to the vector potential, along with a new scalar potential". "Stueckelberg-Horwitz-Piron (SHP) theory is a framework for posing classical and quantum relativistic physics in canonical form with an external parameter of evolution τ". Non-Riemannian generalization of the standard Born-Infeld (BI) Lagrangian was recently introduced and analyzed from the theory of gravitation with dynamical torsion [5]. The great idea in the Debergh et al. paper is "the assumption of positive and negative masses reacting in the same way must be forgotten and replaced by a bimetric model giving rise to two field equations (generalizing the Einstein one) with the corresponding new interaction rules". They proposed the unitary time reversal operator. "Group theory clearly shows that a relativistic equation can be associated with a Poincare group which is no longer restricted and, therefore, which predicts the existence of negative mass (and energy) particles". H. Múnera presents an interesting historical review of the MMMM experiments since 1881. "The crucial 1881 and 1887 experiments never were null... Morozov claims that 'there are no reasons for the energy of the empty space being equal to zero'". His argument is that "the dark energy is nothing but the energy of the empty space".

H. V. Grushevskaya and N. G. Krylova paper from the Belarusian State University contains many useful references. As given in the Abstract, "while within the minimal Standard Model the electroweak phase transition is considered to be of the second order, the first-order electroweak phase transitions proposed in the back- ground of beyond Standard Models may solve some cosmological problems". In the Kisel et al. paper "starting from general the Gelfand-Yaglom approach a new 20-component wave equation for spin-1/2 fermion, which is characterized by three mass parameters, is derived". Ivashkevich et al. consider relationships between Fradkin and Pauli - Fierz models. Finally, we publish the controversial article by Lukaszyk on the new approach to black holes.

Acknowledgments. In fact, this book continues the Book Series "Relativity, Gravitation, Cosmology". I am grateful to Mrs. N. Columbus-Gotsiridze, the president of the Nova Science Publishers (Hauppauge, NY, USA) and staff for their assistance in the publication of this book.

Next, the appearance of the arxiv.org in 1991 (xxx.lanl.gov in that time) was like a breath of freedom. It helped much in distribution of preprints, scientific ideas, and in the academic contacts. As a reader probably knows that I have 86 papers in arxiv.org (in the primary submission categories hep-th, hep-ph, quant-ph, math- ph) and several more with the physics primary category; 52 articles in the mainstream journals with impact index, more articles in proceedings and local journals; 29 plenary invited talks at the international conferences (and many contributed talks); 22 editorial books and special issues (and several more invited chapters). However, the moderators of arxiv.org began to suppress and misclassify even my submissions. (What kind of secret wisdom did I discover?) Since 2014 the current Board members are: Eberhard Bodenschatz, (Max-Planck-Institut fr Dynamik und Selbstorganisation) Ralf Bundschuh (Chair, Ohio State University), Paul Ginsparg (Cornell University), Daniel Gottesman (Perimeter Institute for Theoretical Physics), Steven Gottlieb (Indiana University), Joe Halpern (Cornell University) Chris Myers (Cornell University), Greg Kuperberg (University of California, Davis), Shude Mao (Tsinghua University), David Morrison (University of California, Santa Barbara), Jennifer Ross (University of Massachusetts, Amherst), Yuri Tschinkel (Courant Institute of Mathematical Sciences), Karen Vogtmann (University of Warwick). The arXiv Physics Advisory Committee is: Andrew Connolly (Washington): astrophysics; Jacques Distler (Texas): high-energy theory; Paul Fendley (Virginia): high- energy and condensed-matter theory; Paul Ginsparg (Cornell): high-energy theory; Daniel Gottesman (Perimeter): quantum physics (chair); Dong Lai (Cornell): astrophysics; Michael Lawler (SUNY Binghamton): condensed-matter physics; Brian Maple (UCSD): condensed-matter experiment; Bruno Nachtergaele (UC Davis): mathematical physics. Unfortunately, many scientists complained to me concerning the unjustified actions of the arxiv founder (Dr. Ginsparc), moderators and administrators.

And, again about the newest history as in Refs [6, 7]. The administration of the Zacatecas University was/is: I.Q. Armando Silva Cháirez (Secretario General, UAZ, 2008-2012; Rector, UAZ, 2012- 2016), Dr. Antonio Guzmán Fernández (Rector, UAZ, 2016-2020), Dr. Edmundo Guerrero Sifuentes (Secretario Administrativo, Sept. 2016 - Jun. 2017) [8], MSc. I. L. Torres Villa (CIG-UAZ), Lic. Felipe Correa García (Planeación, PROFEXCE), Dr. J. de J. Araiza Ibarra (Director, UAF-UAZ, 2008-2012; Responsible of the Area de Ciencias Básicas, 2012-2016, Director Interim, 2019-2020), Dr. Sinhué Lizandro Hinojosa Ruiz (Responsible of the Maestría, 2006- 2008; Responsible of the Licenciatura, 2008-2012; Director, UAF- UAZ, 2012-2016), Dr. Felipe Román Puch Ceballos (Responsible of the Maestría, 2008-2014; Responsible of the Licenciatura, 2014- 2016; Director, UAF-UAZ, 2016-2019), MSc. Armando de la Cruz Arteaga, Dra. Sandra García Guerrero, Lic. Abraham (Defensoría Universitaria, UAZ), Mtro. Miguel Sánchez Salas (Coordinación de Personal, UAZ), M. Trinidad Romo Guerrero (Proyectos Federales), Dr. Manuel Reta Hernez (CGIP-UAZ), R. Ivanov Tzonchev (ex-lider of the "Cuerpo Académico de Optica"). Nevertheless, 15 ex-professors of the UAF-UAZ (including its founders) and a few others quitted and formed new Academic Unit, the LUMAT.

I am very grateful to the Publisher, to our authors, referees and friends.

REFERENCES

[1] V. G. Kadyshevsky et al., *Phys. Part. Nucl.* 2 (1972) 69.
[2] Kerson Huang, *Quarks, Leptons and Gauge Fields* (World Scientific, Singapore, 1982).
[3] J. Plücker, *Neue Geometrie des Raumes* [*New Geometry of Space*]. Ed A. Clebsch and F. Klein (Leipzig: B. G. Teubner, 1868/1869)
[4] E. Stueckelberg, *Helv. Phys. Acta* 14 (1941) 321322; E. Stueckelberg, *Helv. Phys. Acta* 14 (1941) 588594; L. P. Horwitz, *Relativistic Quantum Mechanics* (Dordrecht, Netherlands: Springer, 2015).
[5] D. J. Cirilo-Lombardo, *Int. J. Theor. Phys.* 50 (2011) 3621.
[6] V. V. Dvoeglazov and A. Enciso Muñoz, Hadronic J. Suppl. 17 (2002) – *Proceedings of the Zacatecas School on Theoretical Physics*; V. V. Dvoeglazov, Einstein and Hilbert: *Dark Matter*. (Nova Science Pubs, NY, 2011); V. V. Dvoeglazov, Einstein and Others: *Unification*. (Nova Science Pubs, NY, 2014); *Relativity, Gravitation, Cosmology: Foundations*. (Nova Science Pubs, NY, 2016); *Relativity, Gravitation, Cosmology: Beyond Foundations*. (Nova Science Pubs, NY, 2019).
[7] https://zacatecasonline.com.mx/noticias/universidad/26227-eligen-tribunal-universitario.html.
[8] https://ljz.mx/2020/08/29/rectoria-de-la-buaz-no-realizara-accion-alguna-en-contra-de-involucrados-en-la-estafa-maestra/.

Prof. Dr. Valeriy V. Dvoeglazov
Profesor-Investigador Titular "C"
Universidad Autónoma de Zacatecas
Apartado Postal 636, Suc. 3
Zacatecas 98061 Zac., México

December 2018-June 2019-September 2020.

In: Future Relativity, Gravitation, Cosmology
Editors: V. V. Dvoeglazov et al.
ISBN: 979-8-88697-455-3

Chapter 1

deSitter Special Relativity as a Possible Reason for Conformal Symmetry and Confinement in QCD

Mariana Kirchbach[1],[*] ***and Cliffor B. Compean***[2],[†]
[1]Instituto de Física, UASLP, Zona Universitaria,
San Luis Potosí, S. L. P., México
[2] Facultad de Ciencias, UASLP, Privadas del Pedregal,
San Luis Potosí, S. L. P., México

Dedicated to the 60th birthday of Prof. Valeri V. Dvoeglazov.
Our wish, an easy going journey through his successful life,
full of valuable goals, achievements, and memories.

Abstract

Conformal symmetry and color confinement in the infrared regime of QCD are interpreted by means of a conjectured deSitter dS_4 geometry of the internal space-time of hadrons, an assumption inspired by the hypothesis on deSitter special relativity. Within such a scenario, the interactions involving the virtual gluon- and constituent quark degrees of freedom of hadrons are deduced from the Green functions of Laplace operators on the dS_4 geodesics. Then the conformal symmetry of QCD emerges as a direct consequence of the conformal symmetry of the dS_4 space-time, while the color confinement, understood as colorlessness of hadrons, appears as a consequence of the inevitable charge neutrality of the unique closed space-like manifold, the three dimensional hyper-sphere S^3, on whose geodesics the hadron's constituents are conjectured to reside when near rest frame. Mesons are now modelled as quarkish color-anticolor dipoles, whose free quantum motions on the aforementioned S^3 geodesics are perturbed by a potential generated by a gluon–anti-gluon color dipole. The potential predicted presents itself as the color charge analogue to the "curved" Coulomb potential, i.e. to the electric potential that defines a consistent electrostatic theory on a hyper-spherical surface. The advantage of this method is that it allows to establish a direct relationship of the potential parameters to the fundamental constants of QCD.

[*]Corresponding Author's Email: mariana@ifisica.uaslp.mx.
[†]Corresponding Author's Email: cliffor7@gmail.com.

We apply the model to the description of the spectra of the a_1 and f_1 mesons, and the pion electric charge form factor, finding fair agreement with data.

PACS 03.30.+p (Special relativity), 11.30.-j (Symmetry in theory of fields and particles), 12.38.Aw (Quark confinement)

Keywords: deSitter special relativity, conformal symmetry, color confinement, unflavored mesons, strong coupling constant

"Think geometrically, prove algebraically",
John Torrence Tate Jr.
"...verify experimentally!"
Galileo

1. Introduction

Theoretical models in physics are frequently developed with the aim to explain particular observational data. Such approaches, termed to as "bottom-up", obviously depend on the data set chosen as the point of departure. It quite might happen that at different energy scales, different models can be adequate. For example, at low-, up to pretty high energies, the electron charge in the Quantum Electrodynamics (QED) can be considered as a constant, but in very strong electromagnetic fields, generated in the interiors of stars, and in collisions, it can become dependent on the transferred momentum. Particles interacting via the so called "strong" interactions, the hadrons, are described in analogy to QED by a properly designed fundamental gauge theory, the Quantum Chromodynamics (QCD), named in this way in reference to the hypothesis that the hadron constituents, the quarks, are endowed with three different degrees of freedom, conditionally termed to as red-, blue-, and green "color charges". All strong interacting systems observed so far have been found to be color-neutral, a phenomenon known as color-non-observability, or, color confinement. The phenomenon has been built into the fundamental gauge theory of strong interactions through a special choice of the gauge group, $SU(3)_c$, which first prescribes the number of "colored" gauge fields, the gluons, to be eight, and then "locks" all the color degrees of freedom, through limiting the irreducible representations to those of the factor group with respect to the central charges, $SU(3)_c/Z_3$. The space-time wave equations following from QCD are differential equations of high order and difficult to tackle. For the immediate analyzes of the rapidly accumulating data on hadrons concerning their masses, decay modes, collision probabilities (scattering cross sections), compiled in [1], a model has been developed, the "constituent" quark model which is based on second order differential equations of the Sturm-Liouville type, equivalent to employing potentials in the quantum mechanical wave equations of Schrödinger-, Klein-Gordon-, or Dirac. Historically, as point of departure for the traditional quark model [2] have served the hadrons of the lowest masses, the pseudo-scalar mesons η, π (pion), and K (kaon), on the one side, and the spin-$1/2$ baryons, the nucleon N, the Σ, and Ξ hyperons, as well as their spin-$3/2$ excitations Δ, Σ^*, and Ξ^*, supplemented by the Ω-Hyperon, on the other. The potentials of conventional use employed so far within this framework have been predominantly power functions of relative distances. As a rule, the quark models can not be directly linked to the gauge theory of strong interactions, possibly with the exception of the MIT bag model [3], and versions of it [4]. To the best of our knowledge, none of the conventional constituent quark models has provided hints on possible reasons behind the confinement, nor enabled extraction from data on spectra of the value of the fundamental QCD coupling. In a recent work of ours [5], the choice of the data set to be explained by a "bottom-up" quark model has been shifted from the lowest to the highest masses of the unflavored mesons which are of the order of 2300 MeV. The phenomenon to be explained there referred to the observed striking hydrogen-like degeneracies between states distinct through their spins but of same isospin (number of states with different electric charges), and same CP quantum numbers (the product of spatial (P), and charge conjugation (C) parities). The explanation suggested was based on two fundamental principles. We assumed validity of,

- deSitter dS_4 special relativity [6],
- interactions defined by Laplacians on geometric manifolds [7].

According to the first principle,

- the space-time in which the material bodies, in our case the hadrons, are propagating in real time, and that corresponds to the "external space-time", is described by means of all the events placed in the interiors of all Minkowski's light cones, constructed for anyone of the local observers on the four-dimensional dS_4 hyperboloid of one shell, $\mathbf{H}_1^4$, via its intersections by 4D planes passing parallel to the symmetry axis and containing the observer's location,
- the space-time in which the virtual bodies, in our case the quark and gluon constituents of the hadrons, can interact instantaneously, and that corresponds to the "internal space-time", is described by means of all the events placed outside of the aforementioned light cones.

According to the second principle, fundamental interactions are defined by Green functions of Laplace operators on manifolds, the latter being hypothesized by us to be suitably chosen dS_4 geodesics.

> The principle advantage of fixing the geometry of the internal space-time of hadrons to the dS_4 hyperboloid is that among its rich geodesic structure there exists in particular one geodesic space that inevitably and necessarily describes a confinement phenomenon akin to the color charge neutrality in QCD. This is the unique closed space-like hyper-spherical manifold, S^3, located at the "waist" of the hyperboloid.

Indeed, it can be shown that the Gauss theorem and the superposition principle allow only charge neutral configurations to exist on such a surface, the lowest configurations being charge dipoles. Moreover, free quantum motions of quark-anti-quark dipoles on S^3, when perturbed by the potential generated by a gluon-anti-gluon dipole, give rise to an instantaneous interaction capable of explaining a large amount of data on the reported meson excitations [5], [8]. To be specific, the aforementioned interaction presents itself as the color charge analogue to the "curved" Coulomb potential on S^3, given by [9]

$$V_C^{S3} = \frac{\ell(\ell+1)}{\sin^2 \frac{\widehat{r}}{R}} - \alpha Z \cot\frac{\widehat{r}}{R} \overset{R\to\infty}{\longrightarrow} V_C^{E_3} = R^2\frac{\ell(\ell+1)}{r^2} - R\frac{\alpha Z}{r}, \tag{1}$$

where R is the S^3 hyper-radius, $\widehat{r}$ denotes the arc of the geodesic (to be specified below) while $V_C^{E_3}$ is the standard Coulomb potential in flat three dimensional Euclidean space, E_3, the limiting $V_C^{S^3}$ case for $R \to \infty$. Then, the strong interaction under discussion is obtained from $V_C^{S^3}$ by replacing the product, αZ, of the fundamental constant α in QED, and the charge number, Z, by $\alpha_s N_c$, the product of the strong coupling α_s in QCD and the number of color charges, N_c. The strong interaction potential derived in this way allows one to put the description of mesons, the simplest (two-body) composite system in QCD, at comparable footing with that of the H Atom, the simplest (two-body) system in QED. In fitting by the above potential data on the observed degeneracy patterns of the unflavored meson spectra, predictions for α_s at different meson masses could be obtained in [8] that matched values reported by other sources. In [10] the method under discussion has been further successfully tested in the evaluation of the nucleon electromagnetic form-factors. In the present study we first briefly review the proposal of [5], [8] and then put it at work in the description of the previously not considered a_1 and f_1 meson spectra, and also in the explanation of the pion electric charge form factor, finding very satisfactory agreement with data.

The article is structured as follows. In the next section we highlight the idea of [6] on the extension of Einstein's special relativity to deSitter relativity. In section 3 we recall the elements of the algebraic description of the deSitter geometry, which allows us to consider there free quantum motions on the geodesics of the unique closed hyper-spherical space-like manifold, as well as on open hyperbolic time-like geodesics. In section 4 we transform the aforementioned free motions into one-dimensional quantum mechanical wave equations with the respective trigonometric Scarf well-, and the hyperbolic Pöschl-Teller barrier potentials. There, we present a scenario for a duality between the descriptions of hadrons as states bound within the Scarf potential, on the one side, and as resonances transmitted through the Pöschl-Teller barrier, on the other, two methods suited for handling particle physics phenomena in low- and high-energy regimes. In section 5 we explore consequences of the innate charge neutrality of the hyper-sphere on the internal structure of systems on this space and review the prediction of [5] on a confinement phenomenon. In section 6 we associate the predicted confinement phenomenon with the experimentally detected color confinement in QCD and analyze data on the f_1 and a_1 mesons, together with the pion electric charge form factor, before closing by a short Conclusion section.

2. The Geometric Ansatz. Immersing Minkowski- into deSitter Space-Time

All theoretical descriptions of processes in the micro-world have to obey the theory of Special Relativity which requires the values of the physical observables to be same in all

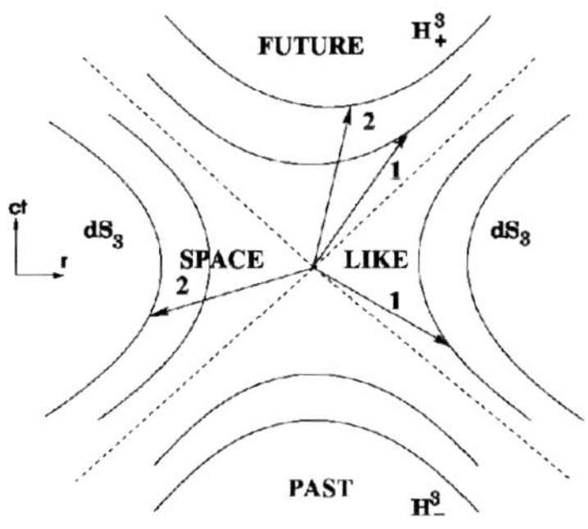

Figure 1. Schematic presentation of Minkowski's space time, reduced to a plane. Two examples of moved reference frames are given and their axes labeled by "1", and "2". The axes of a frame in motion do not remain mutually orthogonal, the angles between them being dependent on the velocity. A fixed interval, $s^2 > 0$, defines a three-dimensional hyperboloid of a vertical time-like symmetry axis and of two-sheets represented by the hyperbola of two branches on the figure, marked by $\mathbf{H}^3_-$ and $\mathbf{H}^3_+$, respectively, or, of one sheet, a so called deSitter dS_3 space-time, for $s^2 < 0$. The dashed lines mark the null-hyperboloid (light cone) corresponding to $s^2 = 0$.

reference frames related to each other by Lorentz transformations, known to conserve the "intervals", s^2, in a Minkowskian space-time, $\mathcal{M}_{1+3}$, of one time-like (ct) and three space-like $(\mathbf{r})$ dimensions according to,

$$\mathcal{M}_{1+3}: \; c^2(t_1 - t_2)^2 - (x_1 - x_2)^2 + (y_1 - y_2)^2 + (z_1 - z_2)^2 = s^2, \tag{2}$$
$$s^2 \gtreqless 0.$$

Here, x_i, y_i, and z_i (with $i = 1, 2$) are the Cartesian position coordinates of two bodies in the ordinary three-dimensional Euclidean space, E_3, while t_1 and t_2 are in turn their time coordinates, and c is the speed of light.

The intervals in (3) for $s^2 > 0$ can be viewed as locuses of points lying on three dimensional hyperboloids of two sheets, denoted by $\mathbf{H}^3_\pm$, encapsulated inside the "null hyperboloid", $c^2t^2 - \mathbf{r}^2 = 0$, a cone also termed to as the "Light Cone", as schematically illustrated in Fig. 1. The null-intervals remain invariant under five more transformations termed to as "conformal", the resulting symmetry being the "kinematic conformal symmetry" [11]. The conformal symmetry is associated with all the transformations which leave the interval, $ds^2 = g_{\mu\nu}(x)dx^\mu dx^\nu$, invariant, modulo a multiplicative "conformal factor" of $e^{\omega(x)}$.

In Special Relativity it is argued that the hyperboloids inside the upper part of the cone describe the causal "Future" while those of the bottom part describe the causal "Past". The meaning of "causal" is that events in these regions describe material bodies moving with subluminal velocities, because their time-orderings are preserved by all Lorentz transformations and thus allow one to define the arrow of time in this domain. In contrast, the surrounding hyperboloids of one sheet represent the acausal region, where no arrow of time can be defined because here Lorentz transformations can change the time-orderings of the events. Such events can be used only in the description of virtual degrees of freedom of the material bodies, such as their constituents, among them the nucleons in a nucleus, the quarks and gluons in hadrons etc.

Measurements in physics can be carried out only on material bodies, as described by means of events located in the causal region, the processes being termed to as "real", or "on mass-shell", in reference to the definition of the invariant mass, M, by means of the so called "on-mass shell" relation, $E^2 - \mathbf{p}^2c^2 = M^2c^4$, between the energy, E, and linear momentum, $\mathbf{p}$. Instead, the constituents of the material bodies and their instantaneous interactions are exclusively described by means of events located in the domain outside the light cone. Notice that the causal and acausal regions are disconnected and described by means of independent coordinates, a reason for which transformations and operators from the two regions are commuting with each other.

Thinking now geometrically, one can ask the question, posed in [6], on the type of physics one could encounter by imagining the light cone and the enclosed $\mathbf{H}^3_\pm$ hyperboloids immersed locally into a four-dimensional hyperboloid of one sheet as schematically represented in Fig. 2 . Stated differently, one could entertain the idea of extending the inaccessible to direct measurements space-like region of the Minkwski space-time by one extra space-like (infinite) dimension, x_4, and requiring outside the causal region invariance of the larger intervals,

$$dS_4: \quad c^2t^2 - \mathbf{r}^2 - x_4^2 = \mathcal{S}^2, \quad \mathcal{S}^2 = -R^2 < 0. \tag{3}$$

Here, R is a new real length parameter. A space-time defined in this way is known under the name of "four-dimensional deSitter space-time", abbreviated, dS_4. It represents a four dimensional hyperboloid of one sheet, denoted by $\mathbf{H}^4_1$, with the time line as a symmetry axis. Within this set up, causal hyperboloids in Einstein's special relativity correspond to so called causal patches on dS_4 obtained through intersections (slicings) of $\mathbf{H}^4_1$ by four-dimensional planes parallel to the time axis as depicted in Figure 2. The interval in (3) is conserved besides under Lorentz transformations also under transformations which ensure the transitivity of the dS_4 space-time and combine in a particular way translations with four of the five conformal transformations, the so called special conformal transformations, a reason for which dS_4 features conformal symmetry [6]. In the following we will explore consequences of this idea.

3. Building up the Algebraic Description of the dS_4 Geometry

In this section, which partially follows our previous work [5], we recall once again the usefulness of the dS_4 geometry for quantum physics. For this purpose we start building up the algebraic description of quantum motions there, beginning with free motions. Free quantum motions on any space-time are described by the eigenmodes of the kinetic energy operator there, given by the respective Laplacian. On dS_4, parameterized in global coordinates as

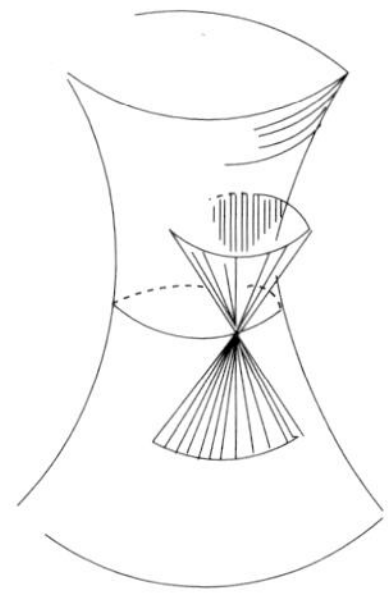

Figure 2. Schematic illustrative presentation of a causal Minkowski light cone relevant to a local observer placed at the "waste" of the dS_4 hyperboloid, for concreteness.

[12],

$$\begin{aligned}
x^0 &= R\sinh\rho, \quad \rho\in(-\infty,+\infty),\\
x^4 &= R\cosh\rho\sin\chi, \quad \chi\in\left[-\frac{\pi}{2},+\frac{\pi}{2}\right],\\
x^1 &= R\cosh\rho\cos\chi\sin\theta, \quad \theta\in\left[-\frac{\pi}{2},+\frac{\pi}{2}\right],\\
x^2 &= R\cosh\rho\cos\chi\cos\theta\sin\varphi, \quad \varphi\in[0,2\pi],\\
x^3 &= R\cosh\rho\cos\chi\cos\theta\cos\varphi,
\end{aligned} \tag{4}$$

one encounters the Laplacian, denoted by $\Delta_{dS_4}(\rho,\chi,\theta,\varphi)$ as,

$$\Delta_{dS_4}(\rho,\chi,\theta,\varphi) = \frac{1}{R^2\cosh^3\rho}\frac{\partial}{\partial\rho}\cosh^3\rho\frac{\partial}{\partial\rho}+\frac{\mathcal{K}^2(\chi,\theta,\varphi)}{R^2\cosh^2\rho}. \tag{5}$$

The $\mathcal{K}^2(\chi,\theta,\varphi)$ part of $\Delta_{dS_4}(\rho,\chi,\theta,\varphi)$ in (5) stands for the squared four dimensional ($4D$) angular momentum operator,

$$\mathcal{K}^2(\chi,\theta,\varphi) = -R^2\Delta_{S^3}(\chi,\theta,\varphi) = -\frac{1}{\cos^2\chi}\frac{\partial}{\partial\chi}\cos^2\chi\frac{\partial}{\partial\chi}+\frac{\mathbf{L}^2(\theta,\varphi)}{\cos^2\chi} \tag{6}$$

where $\Delta_{S^3}(\chi,\theta,\varphi)$ is the Laplace operator on a geodesic of the unique closed space-like manifold [13] of dS_4 which is a three dimensional hyper-sphere, S^3, of (hyper)radius R. The $\Delta_{S^3}(\chi,\theta,\varphi)$ eigenmodes are determined by the hyper-spherical harmonics, $Y_{K\ell m}(\chi,\theta,\varphi)$, as

$$\begin{aligned}
-\hbar^2c^2\Delta_{S^3}(\chi,\theta,\varphi)Y_{K\ell m}(\chi,\theta,\varphi) &= \frac{\hbar^2c^2}{R^2}K(K+2)Y_{K\ell m}(\chi,\theta,\varphi),\\
|m|\in[0,\ell],
\end{aligned} \tag{7}$$

explicitly given by,

$$Y_{K\ell m}(\chi,\theta,\varphi) = \mathcal{S}_{n\ell}(\chi)Y_\ell^m(\theta,\varphi), \tag{8}$$

$$\mathcal{S}_{n\ell}(\chi) = \cos^\ell\chi\,\mathcal{G}_n^{\ell+1}(\sin\chi), \quad n=K-\ell. \tag{9}$$

Here, $\mathcal{G}_n^{\ell+1}(\sin\chi)$, and $Y_\ell^m(\theta,\varphi)$ in turn denote the Gegenbauer polynomials, and the ordinary spherical harmonics. They describe rigid "dumbbells" (" rotators") in four Euclidean dimensions, freely hovering around their mass centers, with the ends tracing S^3 great circles.

The free waves of the dS_4 Laplacian in (5) are then described in terms of 5D pseudo-spherical harmonics, denoted by $Y_{\bar{K}K\ell m}(\rho,\chi,\theta,\varphi)$, the solutions to,

$$-\hbar^2c^2\Delta_{dS_4}(\rho,\chi,\theta,\varphi)Y_{\bar{K}K\ell m}(\rho,\chi,\theta,\varphi) = -\frac{\hbar^2c^2}{R^2}\bar{K}(\bar{K}+3)Y_{\bar{K}K\ell m}(\rho,\chi,\theta,\varphi), \tag{10}$$

and are defined as,

$$\begin{aligned} Y_{\bar{K}K\ell m}(\rho,\chi,\theta,\varphi) &= \phi_{\bar{K}}(\rho)Y_{K\ell m}(\chi,\theta,\varphi), \quad \bar{K},K=0,1,2...,\\ \ell &= 0,1,2,...,K, \end{aligned} \tag{11}$$

where $\bar{K}$ and K can take all non-negative integer values.

Here, the $\phi_{\bar{K}}(\rho)$ functions express in terms of Jacobi's polynomials, $P_{n_r}^{ab}(x)$, of equal real parameters, and a pure imaginary argument according to,

$$\phi_{\bar{K}}(\rho) = \cosh^{-\frac{a}{2}-\frac{3}{2}}\rho P_{n_r}^{-a-\frac{1}{2},-a-\frac{1}{2}}(i\sinh\rho), \tag{12}$$

$$a = \frac{1}{2}+(K+1), \tag{13}$$

$$P_{n_r}^{-a-\frac{1}{2},-a-\frac{1}{2}}(i\sinh\rho) = \frac{\left(-a+\frac{1}{2}\right)_n}{n_r!}{}_2F_1\left(-n_r,n_r-2a;-a+\frac{1}{2};\frac{1-i\sinh\rho}{2}\right) \tag{14}$$

Furthermore, ${}_2F_1$ is the hyper-geometric function, $(...)_t$ is the Pochhammer symbol, ρ is the arc of an open time like hyperbolic geodesic on dS_4, and n_r denotes the polynomial degree.

In summary, the equations (10) and (7) describe free quantum motions along the respective open time-like and closed space-like geodesics on dS_4 in the sense that the arguments of the wave functions $\phi_{\bar{K}}(\rho)$, and $\mathcal{S}_{\ell n}(\chi)$ in the respective equations (12) and (9), change along the arcs, $R\rho$, and $R\chi$, of great circles on such geodesics. The open time-like geodesics considered here are hyperbolic deSitter space-times of one less dimension, i.e. dS_3 space-times, while the closed space-like geodesic are great circles on the three dimensional (hyper)spherical surface S^3 located at the "waist" of the dS_4 hyperboloid.

4. From Free Quantum Motions on dS_4 to One-Dimensional Stationary Wave Equations with Centrifugal Potentials

The free quantum motions on the open time-like–, and the closed space-like dS_4 geodesics can be transformed in their turn into stationary quantum mechanical wave equations with the one-dimensional hyperbolic Pöschl-Teller potential, $V_{\mathrm{PT}}(\rho)$, and the trigonometric Scarf potential, $V_{\mathrm{Sc}}(\chi)$.

4.1. Quantum Motions on Open Time-Like Geodesics and the Hyperbolic Pöschl-Teller Barrier

To prove the aforementioned statement, the following variable change in (12) has to be performed,

$$F_{\bar{K}n_r}(\rho) = \cosh^{\frac{3}{2}}\rho\,\phi_{\bar{K}}(\rho), \quad \bar{K} = K - n_r, \tag{15}$$

in which case the equation (10) is similarity transformed toward,

$$\begin{aligned} -\hbar^2c^2 \left[\cosh^{\frac{3}{2}}\rho\,\Delta_{dS_4}(\rho,\chi,\theta,\varphi)\cosh^{-\frac{3}{2}}\rho\right]\left[\cosh^{\frac{3}{2}}\rho Y_{\bar{K}K\ell m}(\rho,\chi,\theta,\varphi)\right] \\ = -\frac{\hbar^2c^2}{R^2}\bar{K}(\bar{K}+3)\left[\cosh^{\frac{3}{2}}\rho Y_{\bar{K}K\ell m}(\rho,\chi,\theta,\varphi)\right]. \end{aligned} \tag{16}$$

Introducing now the notation, $\mathcal{H}_{\mathrm{PT}}(\rho,\chi,\theta,\varphi)$, for the Laplacian, $\left[-\hbar^2c^2\Delta_{dS_4}(\rho,\chi,\theta,\varphi)\right]$, upon its similarity transformation by the $\cos^{\frac{3}{2}}\chi$ function,

$$\mathcal{H}_{\mathrm{PT}}(\rho,\chi,\theta,\varphi) = \cosh^{\frac{3}{2}}\rho\left[-\hbar^2c^2\Delta_{dS_4}(\rho,\chi,\theta,\varphi)\right]\cosh^{-\frac{3}{2}}\rho, \tag{17}$$

and making use of (5) and (11), the equation (16) equivalently rewrites as,

$$\begin{aligned} &\mathcal{H}_{\mathrm{PT}}(\rho,\chi,\theta,\varphi)F_{\bar{K}n_r}(\rho)Y_{K\ell m}(\chi,\theta,\varphi) = \\ &\frac{\hbar^2c^2}{R^2}\left[-\frac{d^2}{d\rho^2}+V_{PT}(\rho)+\frac{3^2}{2^2}\right]\mathbf{I}(\chi,\theta,\varphi)\times F_{\bar{K}n_r}(\rho)Y_{K\ell m}(\chi,\theta,\varphi) \\ &= -\frac{\hbar^2c^2}{R^2}\bar{K}(\bar{K}+3)F_{\bar{K}n_r}(\rho)Y_{K\ell m}(\chi,\theta,\varphi). \end{aligned} \tag{18}$$

Here, $\mathbf{I}(\chi,\theta,\varphi)$ is the identity operator in the $S^3(\chi,\theta,\varphi)$ subspace, $V_{\mathrm{PT}}(\rho)$ stands for the Pöschl-Teller (PT) potential given by,

$$\mathcal{V}_{\mathrm{PT}}(\rho) = \left[(K+1)^2-\frac{1}{4}\right]\mathrm{sech}^2\rho, \quad \rho\in(-\infty,+\infty). \tag{19}$$

Notice that the similarity transformation conserves the $\left[-\hbar^2c^2\Delta_{dS_4}(\rho,\chi,\theta,\varphi)\right]$ eigenvalues in (10). A way of rephrasing the similarity transformation is to cast it in the form of an intertwinement by the $\cosh^{\frac{3}{2}}\rho$ function of $\mathcal{H}_{\mathrm{PT}}(\rho,\chi,\theta,\varphi)$ with the $\left[-\hbar^2c^2\Delta_{dS_4}(\rho,\chi,\theta,\varphi)\right]$ Laplacian according to,

$$-\hbar^2c^2\cosh^{\frac{3}{2}}\rho\,\Delta_{dS_4}(\rho,\chi,\theta,\varphi) = \mathcal{H}_{\mathrm{PT}}(\rho,\chi,\theta,\varphi)\cosh^{\frac{3}{2}}\rho\,, \tag{20}$$

and to recall that intertwined Hamiltonians are isospectral [14]. Also notice that the PT potential can be equivalently rewritten to,

$$\mathcal{V}_{\mathrm{PT}}(\rho) = \left[(K+1)^2-\frac{1}{4}\right]\frac{1}{\cosh^2\rho} = -\left[(K+1)^2-\frac{1}{4}\right]\tanh^2\rho+\frac{\hbar^2c^2}{R^2}, \tag{21}$$

where the potential on the r.h.s. is known under the name of the "Higgs oscillator" on dS_4 in reference to the fact that the leading term in the series expansion of $\tanh^2\rho$ goes as the square, ρ^2, of the argument meaning that to leading order the hyperbolic Pöschl-Teller potential behaves as an oscillator. In this sense, free motions on the open hyperbolic time-like geodesics on dS_4 can be viewed as "Higgs" oscillations. An interesting phenomenon that can be related to free quantum motion on the dS_4 space is predicted upon the complexification, $(K+1) \longrightarrow i(K+1)$, of the parameter defining the potential strength in (19), equivalently, upon the analytical continuation of the four-dimensional angular momentum to complex values. In this case $\mathcal{V}_{\mathrm{PT}}(\rho)$ is transformed into a barrier and one can consider the complex energies of the resonances transmitted through it. To do so, one has to calculate the transmission scattering matrix, $T(k)$. A scheme for such a calculation has been presented in detail in [15]. The expression obtained along the lines of [15] for the case of our interest reads,

$$T(k) = \frac{\sinh^2 \pi k}{\cosh^2 \pi k + \sinh^2 \pi (K+1)}. \tag{22}$$

Then the poles, which appear at $k = \left[(K+1) + i\left(n_r - \frac{1}{2}\right)\right]$, define the squared complex energies, $\left(\mathcal{E}^{(res)}\right)^2$, of the resonances transmitted through the barrier as [5],

$$\left(\mathcal{E}^{(res)}\right)^2 = \frac{\hbar^2 c^2}{R^2} k^2, \quad \mathcal{I}m\left(\mathcal{E}^{(res)}\right)^2 = 2\frac{\hbar^2 c^2}{R^2}(K+1)\left(n_r - \frac{1}{2}\right). \tag{23}$$

$$\begin{aligned} \mathcal{R}e\left(\mathcal{E}^{(res)}\right)^2 - c_0 &= \frac{\hbar^2 c^2}{R^2}(K+1)^2, \\ c_0 &= -\frac{\hbar^2 c^2}{R^2}\left(n_r - \frac{1}{2}\right)^2 + A^2, \end{aligned} \tag{24}$$

where A^2 is some additive ad hock constant which on long term could be helpful in adjustments of the potential parameters to data. Let us interpret the real parts of the squared complex energies, $\left[\mathcal{R}e\left(\mathcal{E}^{(res)}\right)^2 - c_0\right]$, at the poles of $T(k)$, as a squared invariant mass M^2, i.e. let us set,

$$\begin{aligned} M^2 &\equiv \mathcal{R}e\left(\mathcal{E}^{(res)}\right)^2 - c_o = \frac{\hbar^2 c^2}{R^2}(K+1)^2 = \frac{\hbar^2 c^2}{R^2}(n+\ell+1)^2, \\ K &= n+\ell. \end{aligned} \tag{25}$$

The expression gives rise to linear "trajectories" (linear dependencies of the angular momentum on the mass) of the art,

$$\ell = \alpha(R)M - n - 1, \quad \ell = 0,1,2,..., \quad \ell + n = K, \ K = 0,1,2,... \tag{26}$$

with n being the number of nodes in the Gegenbauer polynomials in (9).

4.2. Quantum Motion on the Closed Space-Like Hyper-Spherical Geodesics and the Trigonometric Scarf Potential

In a similar way, through the variable change,

$$\begin{aligned} \cos\chi Y_{Klm}(\chi,\theta,\varphi) &= \cos\chi \mathcal{S}_{n\ell}(\chi) Y_{\ell m}(\theta,\varphi) \\ &= \mathcal{U}_{n\ell}(\chi) Y_\ell^m(\theta,\varphi), \quad \mathcal{U}_{n\ell}(\chi) = \cos\chi \mathcal{S}_{n\ell}(\chi), \end{aligned} \tag{27}$$

with $\mathcal{S}_{n\ell}(\chi)$ from (9), the equation (7) is similarity transformed toward,

$$-\hbar^2c^2\left[\cos\chi\Delta_{S^3}(\chi,\theta,\varphi)\cos^{-1}\chi\right]\left[\cos\chi Y_{K\ell m}(\chi,\theta,\varphi)\right] = \frac{\hbar^2c^2}{R^2}K(K+2)\left[\cos\chi Y_{K\ell m}(\chi,\theta,\varphi)\right]. \tag{28}$$

In now introducing the notation, $\mathcal{H}_{\mathrm{Sc}}(\chi,\theta,\varphi)$, for the Laplacian, $\left[-\hbar^2c^2\Delta_{S^3}(\chi,\theta,\varphi)\right]$, upon its similarity transformation by the $\cos\chi$ function,

$$\mathcal{H}_{\mathrm{Sc}}(\chi,\theta,\varphi) = -\hbar^2c^2\left[\cos\chi\Delta_{S^3}(\chi,\theta,\varphi)\cos^{-1}\chi\right], \tag{29}$$

making use of (8), (9), and (6), and upon some algebraic manipulations, the equation (29) equivalently rewrites as,

$$\begin{aligned} &H_{\mathrm{Sc}}(\chi,\theta,\varphi)U_{\ell n}(\chi)Y_{\ell m}(\theta,\varphi) = \\ &\frac{\hbar^2c^2}{R^2}\left(-\frac{d^2}{d\chi^2}+V_{\mathrm{Sc}}(\chi)-1\right)\otimes\mathbf{I}(\theta,\varphi)U_{\ell n}(\chi)Y_{\ell m}(\theta,\varphi) = \\ &\frac{\hbar^2c^2}{R^2}K(K+2)U_{\ell n}(\chi)Y_{\ell m}(\theta,\varphi), \end{aligned} \tag{30}$$

where $\mathbf{I}(\theta,\varphi)$ is the identity operator on the $S^2(\theta,\varphi)$ subspace of $S^3(\chi,\theta,\varphi)$, and $V_{\mathrm{Sc}}(\chi)$ is the trigonometric Scarf potential

$$V_{\mathrm{Sc}}(\chi) = \ell(\ell+1)\sec^2\chi, \quad \chi\in\left[-\frac{\pi}{2},+\frac{\pi}{2}\right]. \tag{31}$$

Notice that the barrier potential in (19) can be obtained from the well potential in (31) through complexification of the argument, modulo multiplicative constant.

Same as in the case of the dS_4 Laplacian, also the similarity transformation of the S^3 Laplacian conserves its eigenvalues in (7) and because these eigenvalues are bound from below, they are referred to as the Laplacian's "spectrum" .

A way of rephrasing the similarity transformation in (29) is to cast it in the form of an intertwinement by the $\cos\chi$ function of $\mathcal{H}_{\mathrm{Sc}}(\chi,\theta,\varphi)$ and the $\left[-\hbar^2c^2\Delta_{S^3}(\chi,\theta,\varphi)\right]$ Laplacian according to

$$-\hbar^2c^2\cos\chi\,\Delta_{S^3}(\chi,\theta,\varphi) = \mathcal{H}_{\mathrm{Sc}}(\chi,\theta,\varphi)\cos\chi\,, \tag{32}$$

and recall isospectrality of intertwined operators known from the super-symmetric quantum mechanics [14]. The considerations from above show that in terms of one-dimensional wave equations, the free quantum motions along the open time-like, and the closed space-like dS_4 geodesics give in turn rise to the hyperbolic Pöschl-Teller, $V_{\mathrm{PT}}(\rho)$, equivalently, the Higgs oscillator on dS_4, and the trigonometric Scarf, $V_{\mathrm{Sc}}(\chi)$ potential, which are well known from the super-symmetric quantum mechanics [14] to be exactly solvable. More important, the Hamiltonians $\mathcal{H}_{\mathrm{PT}}(\rho,\chi,\theta,\varphi)$ and $\mathcal{H}_{\mathrm{Sc}}(\chi,\theta,\varphi)$, in being intertwined with the respective dS_4 and S^3 Laplacians, conserve the symmetry of the free geodesic motions which is the conformal symmetry, a reason for which the one dimensional potentials in (19)-(21), and (31) are said to be conformal. Such occurred because the latter potentials correspond to "centrifugal" terms on the respective hyperbolic and hyper-spherical manifolds. We now rename by $\left(\mathcal{E}^{(bound)}\right)^2$ the $H_{\rm Sc}(\chi,\theta,\varphi)$ eigenvalues in (30), i.e. we introduce the notation,

$$\left(\mathcal{E}^{(bound)}\right)^2=\frac{\hbar^2c^2}{R^2}(K+1)^2,\quad K=n+\ell. \tag{33}$$

Because the energy depends on the K quantum number alone, and due to the branching rules,

$$\ell\in[0,K],\quad |m|\in[0,\ell], \tag{34}$$

as explained in the equation (11), one immediately notices that the levels of the trigonometric Scarf potential are $\sum_{\ell=0}^{\ell=K}(2\ell+1)=(K+1)^2$-fold degenerate. This type of degeneracy is known from the Hydrogen Atom, where it is attributed to the conformal symmetry of the Maxwell equations, or, more general, to the conformal symmetry of the Relativistic Electrodynamics [16]. Therefore, the eigenmodes of the free quantum motion along the closed S^3 hyper-spherical geodesic on dS_4 fall into a conformal spectrum.

4.3. Equality between the Masses of the Resonances Transmitted through the Hyperbolic Pöschl-Teller Barrier and the Excitation Energies of the States Bound within the Trigonometric Scarf Potential

Comparison of the real parts of the complex masses of the resonances transmitted through the Pöschl-Teller barrier in (25) with the real excitation energies of the states bound within the trigonometric Scarf potential in (33) reveals their equality as,

$$M^2=\mathcal{R}e\left(\mathcal{E}^{(res)}\right)^2=\frac{\hbar^2c^2}{R^2}(K+1)^2. \tag{35}$$

Within the geometric context pursued through the text, the duality between the mass-formulas in (26) and (33), translates into duality between quantum motions on open hyperbolic time-like, and the closed hyper spherical space-like geodesics on dS_4.

> In conclusion, with the aid of the dS_4 structure of the domain located outside the causal Minkowski light cones, hypothesized by us to provide the relevant geometry of internal space-time, and employed in the definition of interactions, a pair of potentials could be encountered with the remarkable properties that the real energies of the states bound within the Scarf well potential equal the real parts of the complex energies of the resonances transmitted to the Pöschl-Teller barrier. Moreover, these energies carry same hydrogen-like degeneracies.

In this manner, by virtue of the hypothesis on validity of dS_4 special relativity [6], we were able to establish in [5] a duality between the free quantum motions on open hyperbolic time-like–, and closed hyper-spherical space-like geodesics of the dS_4 space-time. To the end of the article, the physical significance of such a duality will be revealed. Before, in the next section, a deeper insight into the physics on the closed hyper-spherical surface S^3 will be gained.

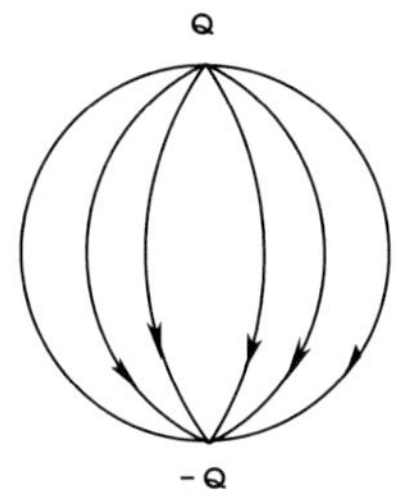

Figure 3. The inevitable charge neutrality of the sphere.

5. Behind the Curtains of Confinement. Charge Neutrality on the Hyper-Spherical dS_4 Geodesic and a Conformal Color-Dipole Confining Potential

The main point of the present section is to explore consequences of the fact that no single charge can be consistently defined on spheres [17]. Indeed, the lines pouring out of a charge located at a particular point on this space intersect at the antipodal point, creating there a fictitious opposite charge (see Fig. 3). In order to satisfy the Gauss theorem and the principle of superposition on S^3, fictitious charges have to be avoided and replaced by real charges, thus creating necessarily and inevitably the charge neutrality of the sphere. In order to see this, we recall that the Gauss theorem predicts a $\sec^2\chi$ functional form [18] for the electric field for $\chi \in [-\pi/2, +\pi/2]$. On the other side, the superposition principle on S^3 predicts same field to be the gradient of the sum of the pod- and anti-pod single charge potentials. In the following we shall show that the E field of a "charge-dipole" (for the time being without specification of the nature of the charge) predicted by the superposition principle on S^3 coincides with the corresponding predictions of the Gauss theorem. In this way we shall prove that a consistent "charge-static" on the hyper-sphere is guaranteed by its charge neutrality.

For this purpose, one needs to know the Laplacian on the complete space-time $R^1 \otimes S^3$, with R^1 standing for the time line, and then to consider it at constant times. This can be done within the framework of the so called Radial Quantization Technique which prescribes to first Wick-rotate via the complexification, $ct \rightarrow ict$, the Minkowski space time of $(1+3)$ dimensions to four Euclidean dimensions, switch to global hyper-spherical coordinates, and introduce "time" via the logarithm of the S^3 radius [19]. In so doing one finds the following Laplacian,

$$-\left[\frac{\partial^2}{\partial x_4^2} + \frac{\partial^2}{\partial x_1^2} + \frac{\partial^2}{\partial x_2^2} + \frac{\partial^2}{\partial x_3^2}\right] \longrightarrow -\left[\frac{1}{R^3}\frac{\partial}{\partial R}R^3\frac{\partial}{\partial R} - \frac{1}{R^2}\mathcal{K}^2(\chi,\theta,\varphi)\right] \quad (36)$$

$$\stackrel{R=e^\tau}{\longrightarrow} e^{2\tau}\left[-\frac{\partial^2}{\partial\tau^2} + \left(\mathcal{K}^2(\chi,\theta,\varphi) + 1\right)\right]. \quad (37)$$

The τ parameter is known as "conformal time". To build up charge-statics, one has to set τ=const and to calculate the Green function to the remaining piece, the so called conformal Laplacian, $\Delta^1_{S^3}(\chi,\theta,\varphi)$, on S^3, read off from (37) as

$$\Delta^1_{S^3}(\chi,\theta,\varphi) = \mathcal{K}^2(\chi,\theta,\varphi) + 1, \quad (38)$$

where we have chosen $\tau = 0$, equivalent to $R = 1$. In order to construct the potentials of a charge $\mathcal{Q}$ placed at, say, the West "pole", and of an anti-charge, $(-\mathcal{Q})$, at the East " pole", one needs to know the respective Green functions, in turn denoted by $\mathcal{G}_{-\frac{\pi}{2}}(\chi)$, and $\mathcal{G}_{+\frac{\pi}{2}}(\chi)$, which have been calculated, among others, in [20], however for $\chi \in [0, \pi]$, and need to be shifted to $\chi \in \left[-\frac{\pi}{2}, +\frac{\pi}{2}\right]$. After this small algebraic manipulation, the potentials, constructed from the Green's function in the standard way known from potential theory [7], emerge as,

$$V_{-\frac{\pi}{2}}(\chi) = \mathcal{Q}\mathcal{G}_{-\frac{\pi}{2}}(\chi) = -\frac{\mathcal{Q}}{4\pi^2}\left(\chi - \frac{3\pi}{2}\right)\tan\chi, \tag{39}$$

$$V_{+\frac{\pi}{2}}(\chi) = -\mathcal{Q}\mathcal{G}_{+\frac{\pi}{2}}(\chi) = -\frac{(-\mathcal{Q})}{4\pi^2}\left(\chi - \frac{\pi}{2}\right)\tan\chi,\ \chi \in \left[-\frac{\pi}{2}, +\frac{\pi}{2}\right] \tag{40}$$

In now assuming validity of the superposition principle, one encounters a Charge Dipole (CD) potential to emerge at a point χ on S^3 according to,

$$V_{CD}(\chi) = \mathcal{Q}\mathcal{G}_{-\frac{\pi}{2}}(\chi) - \mathcal{Q}\mathcal{G}_{+\frac{\pi}{2}}(\chi) = \frac{\mathcal{Q}}{4\pi}\tan\chi. \tag{41}$$

The electric field to this dipole is obtained in the standard way through differentiation as,

$$E(\chi) = -\frac{\partial}{\partial\chi}V_{CD}(\chi) = -\frac{\mathcal{Q}}{4\pi}\frac{1}{\cos^2\chi}. \tag{42}$$

On the one side, this is the precise expression prescribed by Gauss's theorem [18], and on the other, one recognizes in it the functional form of the Scarf well. As an important reminder, $V_{CD}(\chi)$ is conformal because the Green functions have the symmetry of the Laplacian, that on its side has the symmetry of the space-time on which it defines the kinetic energy operator.

One of the advantages of defining non-relativistic potentials in terms of Green functions is that elements of causality can be brought to the interactions through considering the retarded functions, a point that so far will not be attended here.

> As a result, the hypothesis on the validity of dS_4 special relativity within the interaction domain [6] amounted to the `prediction of the new phenomenon of conformally invariant charge neutral systems confined to closed hyper spherical spaces.`

The particular tangent form of the dipole potential is not universal but rather due to the choice of the parametrization for the second polar angle on S^3, namely, $\chi \in \left[-\frac{\pi}{2}, +\frac{\pi}{2}\right]$. This choice is arbitrary and could be changed to $\chi \in [0, \pi]$ in which case the tangent goes into a cotangent and the $\sec^2$ into $\csc^2$, as originally considered by [20]. From now onward we shall switch to this very parametrization which turns out to be more suited for physical interpretations, to be presented in the subsequent section.

Notice that $\mathcal{Q}$ stands for dimension-less charges. In terms of dimensional charges, q, related to Q via,

$$\mathcal{Q} = \frac{q}{\sqrt{\hbar c}}, \tag{43}$$

the potential perceived by another charge, $q/\sqrt{\hbar c}$, is

$$V(\chi) = \frac{q^2}{4\pi\hbar c\epsilon_0}\cot\chi, \quad \chi \in [0, \pi]. \tag{44}$$

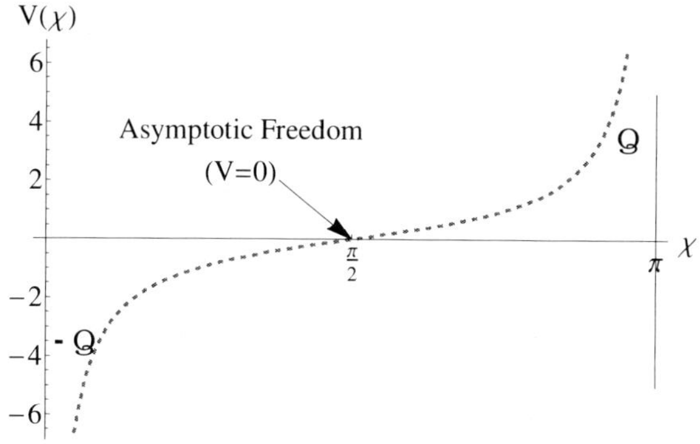

Figure 4. The conformal color dipole confining potential in the $\chi \in [0, \pi]$ parametrization.

For example, in the case of electrostatic, q is taken as the electron charge, e, in which case the special notation of,

$$\alpha = \frac{e^2}{4\pi\hbar c\epsilon_0} \tag{45}$$

known as the fundamental coupling constant of Electrodynamics, is introduced. In case the dipole is acting on a Ze charge, one finds the standard Coulomb interaction,

$$V_C^{(\alpha Z)}(\chi) = \alpha Z \cot\chi, \quad \chi \in [0, \pi]\,. \tag{46}$$

In Fig. 5 we display a dipole interaction, now in the $\chi \in [0, \pi]$ parametrization.

The potential in (44) can be applied as a perturbance of the free quantum motion on S^3 in (31), in which case the following wave equation emerges,

$$\mathcal{H}_{\text{Pert}}(\chi, \theta, \varphi)U_{\ell n}^{(\alpha Z)}(\chi)Y_{\ell m}(\theta, \varphi) = \tfrac{\hbar^2 c^2}{R^2}\left(\epsilon_{\ell n}^{(\alpha Z)}\right)^2 U_{\ell n}^{(\alpha Z)}(\chi)Y_{\ell m}(\theta, \varphi) \tag{47}$$

$$\mathcal{H}_{\text{Pert}}(\chi, \theta, \varphi) = \left(-\tfrac{\hbar^2 c^2}{R^2}\tfrac{\mathrm{d}^2}{\mathrm{d}\chi^2} + V_C^{(\alpha Z)}(\chi)\right)\mathbf{I}(\theta, \varphi), \tag{48}$$

$$V_C^{(\alpha Z)}(\chi) = V_{\text{Sc}}(\chi) + V(\chi) = \tfrac{\hbar^2 c^2}{R^2}\tfrac{\ell(\ell+1)}{\sin^2\chi} - \alpha Z\tfrac{\hbar^2 c^2}{R^2}\cot\chi, \tag{49}$$

where $\mathbf{I}(\theta, \varphi)$ is the identity operator in the (θ, φ) variables. Because the perturbance of the free quantum motion on S^3 takes place only in the χ variable, while leaving intact the motion in the θ and φ variables, the $\mathcal{H}_{\text{Pert}}$ operator depends non-trivially only on the χ variable and is in reality one-dimensional. The energies, $\frac{\hbar^2 c^2}{R^2}\left(\epsilon_{\ell n}^{(\alpha Z)}\right)^2$, of the states bound within this potential are known [14] and given in as,

$$\begin{aligned}\frac{\hbar^2 c^2}{R^2}\left(\epsilon_{\ell n}^{(\alpha Z)}\right)^2 &= \frac{\hbar^2 c^2}{R^2}\left[-\frac{\alpha^2 Z^2}{4\,(K+1)^2} - (K+1)^2\right],\\ K = \ell + n, \quad & K = 0, 1, 2, \ldots.\end{aligned} \tag{50}$$

The "spectral" mass formula resulting then from this potential reads,

$$M^2 = \frac{\hbar^2 c^2}{R^2}\left(\epsilon_{\ell n}^{(\alpha Z)}\right)^2 = \frac{\hbar^2 c^2}{R^2}(K+1)^2 - \frac{\hbar^2 c^2 \alpha^2 Z^2}{4R^2(K+1)^2} + c_0. \tag{51}$$

In eqs. (47)-(49) one recognizes the one-dimensional wave equation with a version of the trigonometric Rosen-Morse potential, known from the super-symmetric quantum mechanics to be exactly solvable [14].

Identifying now the "charge" with the color charge, α is replaced by the QCD strong coupling, α_s, while Z is replaced by the number of colors, N_c [5], [8]. In this way, the following potential has been found,

$$V^{(\alpha_s N_c)}(\chi) = \frac{\hbar^2 c^2}{R^2}\frac{\ell(\ell+1)}{\sin^2\chi} - \alpha_s N_c \frac{\hbar^2 c^2}{R^2}\cot\chi, \tag{52}$$

to be relevant in quark models. Correspondingly, the mass formula emerges as,

$$M^2 = \frac{\hbar^2 c^2}{R^2}\left(\epsilon_{\ell n}^{(\alpha_s N_c)}\right)^2 = \frac{\hbar^2 c^2}{R^2}(K+1)^2 - \frac{\hbar^2 c^2 \alpha_s^2 N_c^2}{4R^2(K+1)^2} + c_0. \tag{53}$$

To recapitulate, the result is that in hypothesizing validity of dS_4 special relativity [6], it became possible to predict the form of a potential generated by equal numbers of charges and anti-charges locked (confined) on a closed hyper-spherical space, much alike the phenomenon of color confinement in QCD.

6. The Experimental Verification

The phenomenon of charge confinement predicted by the dS_4 special relativity in the previous section is well known from the theory of strong interactions, the Quantum Chromodynamics (QCD) [21], in which the fundamental matter degrees of freedom, the quarks, are endowed with so called "color" charges. There are three color charges, and three color anti-charges, although the strongly interacting particles, the hadrons, are by themselves all color charge neutral. For example, the hadrons of integer spin, the mesons, are constituted by effective degrees of freedom, termed to as "constituent" quarks, and "anti-quarks", which are equipped by opposite color charges. Within this context it is legitimate to ask the question as to what extent the scenario developed in the preceding sections and especially of the previous section 5 could apply to the description of the experimentally reported meson masses. The question we are posing is as to what extent the dual "trajectory"- and "spectral" mass formulas in the respective eqs. (26), and (33) are suited for the description of the meson mass dependencies on their angular momenta. Such an analysis has been performed in [5] and we here limit ourselves to only briefly highlight the results.

In [5] we analyzed 71 reported unflavored mesons belonging to the four families of the f_0, a_0, π, and η mesons. Data convincingly confirmed the conformal hydrogen-like $(K+1)^2$-fold degeneracy of the levels in accord with (53). In addition, their splittings followed pretty much those of the states bound within the trigonometric Scarf potential in (33). Also the dual mass formula in (26) turned out well applicable. The only splittings which could not come out well from the formulas under discussions were the splittings between the ground states and the first excited states. They could be accounted for by the help of the extension in (52).

Usage of (44) as a perturbance was justified by the suggestion to consider mesons as made of two types of color-dipoles, one constituted by a quark and an anti-quark, and the other by a gluon and an anti-gluon. The meson Hamiltonian, $\mathcal{H}_{\text{Pert}}$, of such a system, given in (48), can then be approximately separated into a dominant free 4D rotational motion of say, the supposedly light quark-anti-quark pair, described by the Hamiltonian $\mathcal{H}_{\text{Sc}}$ in (29), perturbed by the "residual" color-dipole interaction, $V_{CD}(\chi)$ in (44), generated by a background gloun-antigluon dipole. Such a picture is consistent with the idea of the constituent quarks as effective degrees of freedom, i.e. as fundamental quarks "dressed" by gluons. In adopting $\mathcal{H}_{\text{Pert}}$ as a working hypothesis, one is of course neglecting the perturbance of the glueball by the color-dipole potential generated by the quark-anti-quark pair, and the tensor interaction between the two dipoles [22].

Notice that the interpretation of hadrons as states bound within a well potential is adequate near the rest frame, while their interpretation as resonances transmitted through a barrier is suited for high to ultra-relativistic energies. The rest frame is acceptable as an approximation at low transferred momenta in the so called infrared (IR) regime of QCD, while the adequate tool for hadron description at high-and ultra-high transferred momenta,i.e. close to the ultraviolet (UV) regime of QCD, is provided by the scattering matrix. Stated differently, within the QCD context, the quantum motions on the open time-like geodesics can be associated with processes approaching the ultraviolet regime, while those on the closed space-like geodesics can be associated with near rest frame processes, i.e. with the infrared regime of QCD. In this sense, the achievement of the model advocated in [5] is to have

- associated (to a good approximation) each one of the infrared and the ultraviolet regimes of QCD with free motions along the respective closed hyper-spherical space-like –, and the open hyperbolic time-like geodesics on a conformally symmetric dS_4 space-time,
- established with respect to meson description a duality between states bound within a well-(IR), and resonances transmitted through a barrier (UV) potentials,
- gained the new understanding about the inevitable color-neutrality of hadrons in the infrared as a consequence of the innate charge-neutrality of the closed hyper-spherical space-like dS_4 geodesics and predicted a confinement phenomenon known in QCD,
- revealed relevance of conformal symmetry in both regimes of QCD.

Notice that color-neutrality in the ultraviolet occurs by virtue of the dual meson description and is not inevitable there because on open space-times free charges and therefore a deconfined phase, are allowed to exist.

Finally, it needs to be emphasized that the infrared $(-\mathcal{Q})\,\mathcal{G}_{\frac{\pi}{2}}(\chi)$ single-color potential in (40) relates by an argument shift toward, $\chi=\chi'+\pi/2$, followed by the complexification, $\chi'\to i\chi'\to\rho$, to the expression $(-\mathcal{Q})\,(-i\chi')\cot i\chi'\to-(-\mathcal{Q})\,\rho\coth\rho$, giving a single-color potential generated in the ultraviolet by a soft-gluon emission of a fast quark, stopped at a cusp [23], and in support of the IR-UV duality advocated here and in [5]. The Wilson loop calculation of such a potential, in making use of radial quantization, is practically equivalent to our calculation in section 5.

6.1. The Spectra of the a_1 and f_1 Mesons

As already mentioned above, the fundamental matter degrees of freedom in the theory of strong interaction, the Quantum Chromodynamics, the quarks, can carry three different color charges. The color charge, denoted by $g(Q^2)$, depends on the square, q^2, of the transferred space-like four-momentum, via $Q^2 = -q^2$, and the related fundamental coupling in QCD, $\alpha_s(Q^2)$, defined in analogy to (45) as,

$$\alpha_s(Q^2) = \frac{g^2(Q^2)}{4\pi\hbar c}, \quad Q^2 = -q^2 \geq 0, \tag{54}$$

changes with Q^2, a reason for which it is is termed to as "running".

In the following we complement the analyzes of the mass distributions of the four meson families previously studied in [5] (71 masses) by two more meson families, those of the f_1, and a_1 mesons, which contribute 18 more masses. Afterwards we extract the corresponding α_s values. Compared to the f_0, η, π, a_0 mesons, analyzed in [5], less is known about the f_1 and a_1 mesons. The compiled data on the last two families are presented in Fig. 5. It needs to be said that only the $a_1(1260)$, $a_1(1640)$, and $f_1(1285)$ mesons are contained in the Summary Table of the Meson Particle Listings in [1] meaning that their existence is considered as fully reliable. All the other particles in Fig. 5 are either included in the Meson Particle Listings, or into the list of the less reliably established particles referred to as " Further States". We here favored the $f_1(1510)$ meson as the first excited state of $f_1(1285)$ over the $f_1(1420)$ state from the Summary Table. In this way we are suggesting that the internal $f_1(1510)$ structure may be closer to a quark–anti-quark color dipole, perturbed by the dipole potential due to gluon–anti-gluon dipole, than the $f_1(1420)$ internal structure, which may contain strange-anti-strange quarks. Our point is that our conformal symmetry classification scheme, in needing states omitted from the Summary Table, provides an additional legitimization to their observation and consolidates their status. Same as in [5], we here employ the mass formula in (53) to perform a least square fit to the meson mass distributions in Fig. 5. The results on the three parameters, the hyper-sphere radius, R, and the two dimensionless parameters b, and c_0 in (53), are listed in the columns second to fourth of Table 1. The fifth column contains the average deviation values of the corresponding fits, while in the last column we present the values of the strong coupling constant, α_s of QCD (normalized to π) which we extracted by virtue of the formula (53). The Table shows that the two α_s/π values, in being of the order of $\alpha_s/\pi \sim 0.80$, remain significantly less than 1, and in complete agreement with data reported in [24], whose principle result is that the strong coupling constant tends to approach a fixed value in the infrared, thus opening up a so called "conformal window" and thereby hints on relevance of conformal symmetry in the infrared.

6.2. The Electric Form Factor of the Pion

In our previous work [10] we formulated on S^3 the Dirac equation with the color-confining dipole potential in (44), found the spinor solutions in closed forms, and used them in the calculations of the proton's and neutron's electric-charge, $G^E_{p/n}(Q^2)$, and magnetic dipole, $G^M_{p/n}(Q^2)$ form factors, together with the respective Dirac- and Pauli form factors, $F_1^{p/n}(Q^2)$, and $F_2^{p/n}(Q^2)$.

Table 1. Parameters of the least square fit to the 18 masses of the a_1 and f_1 mesons, distributed over the two trajectories indicated in the first column, and calculated employing the mass formula in (53). The last column contains the value of the extracted strong coupling constant, α_s/π, identified by the equation (53), with $2b = \alpha_s N_c$. The explicit expression of σ^2 reads $\sigma^2 = \frac{(\sum \Delta M^2)^2}{N-1}$, where ΔM^2 stands for the deviation of the squared experimental from the squared theoretical masses, and N is the number of states in a family.

trajectory	R	b	c_0	σ^2 [GeV]^4	$\frac{\alpha_s}{\pi}$
f_1 for $K \geq 0$	0.58 fm	3.788	5.125	0.1652839 GeV^4	0.80
a_1 for $K \geq 0$	0.58 fm	3.726	5.071	0.09191531 GeV^4	0.79

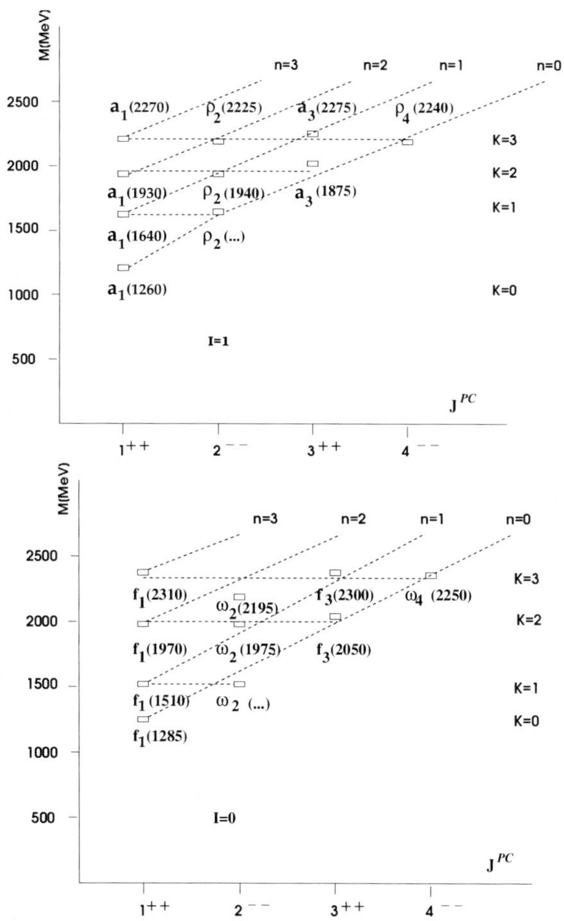

Figure 5. The excitations of the isovector $a_1(1260)$ meson (left) and the isoscalar $f_1(1285)$ meson (right). "Missing" states are denoted by (...). Here, J is the total angular momentum value of the quark-diquark system obtained from coupling the total spin $S = 1$, to the relative angular momentum, $\ell = 0, 1, 2, 3$, and CP give in their turn the C and P quantum numbers. Recall the definition of $K = n + \ell$ in (26).

The electric charge form factor of the proton is determined by the Fourier transform of the electric-charge density, $\rho^p(\vec{r})$, as

$$G_p^E(Q^2) = \int_{V(\vec{r})\to\infty} \rho^p(r) e^{i\vec{q}\cdot\vec{r}} d^3\vec{r}, \quad \rho^p(\vec{r}) = |\Psi_{gst}(\vec{r})|^2, \tag{55}$$

$$Q^2 = -q^2, \tag{56}$$

where Q^2 equals the negative squared of the transferred space-like momentum, $q^2 < 0$, $V(\vec{r})$ is the space integration volume, and $\Psi_{gst}(\vec{r})$ is the Dirac spinor of ground state. The proton's magnetic dipole form-factor is given by the Fourier transform of the magnetic dipole density, $\rho^p_{mag}(\vec{r})$ as,

$$G_p^M(Q^2) = \int_{V(\vec{r})\to\infty} \rho^p_{mag}(r) e^{i\vec{q}\cdot\vec{r}} d^3\vec{r}, \quad \rho^p_{mag}(\vec{r}) = \bar{\Psi}_{gst}(\vec{r})\gamma_5\Psi_{gst}(\vec{r}). \tag{57}$$

The relation between $G_p^E(Q^2)$, and $G_p^M(Q^2)$ to the Dirac and Pauli form factors, $F_1^p(Q^2)$, and $F_2^p(Q^2)$, is given by,

$$F_1^p(Q^2) = \frac{G_E^p(Q^2) + \tau_p G_M^p(Q^2)}{1+\tau_p}, \quad \tau_p = \frac{Q^2}{2M_p^2},$$

$$\kappa_p F_2^p(Q^2) = \frac{G_M^p(Q^2) - G_E^p(Q^2)}{1+\tau_p}, \quad \kappa_p = \mu_p - 1, \tag{58}$$

where μ_p is the proton's magnetic dipole moment. The neutron form factors are expressed by means of the proton's form factors as,

$$G_n^E(Q^2) = -\frac{\mu_n \tau_n}{1+B\tau_n} G_p^E(Q^2), \quad \tau_n = \frac{Q^2}{2M_n^2}$$

$$\frac{G_M^n(Q^2)}{\mu_n} = \frac{G_M^p(Q^2)}{\mu_p}, \tag{59}$$

where B is a parameter, while μ_n is the magnetic dipole moment of the neutron. In [10] expressions in closed form for all the above form-factors have been reported, which we opted not to repeat here for the sake of not overloading the presentation. On the other side, all the form factors can also be alternatively expressed in terms of the respective form factors of the constituent u and d quarks. Specifically for the Dirac and Pauli form factors the following expressions hold valid,

$$F_i^u(Q^2) = 2F_i^p(Q^2) + F_i^n(Q^2), \quad i = 1, 2, \tag{60}$$

$$F_i^d(Q^2) = F_i^p(Q^2) + 2F_i^n(Q^2). \tag{61}$$

These form factors, extracted from the $F_p^1(Q^2)$ and $F_n^2(Q^2)$ expressions reported in [10] are plotted in Fig. 6.

We observe that our predictions on the light quark form factors are of same quality as those reported in [25] and worked out within the framework of the related ADS_5/CFT_4

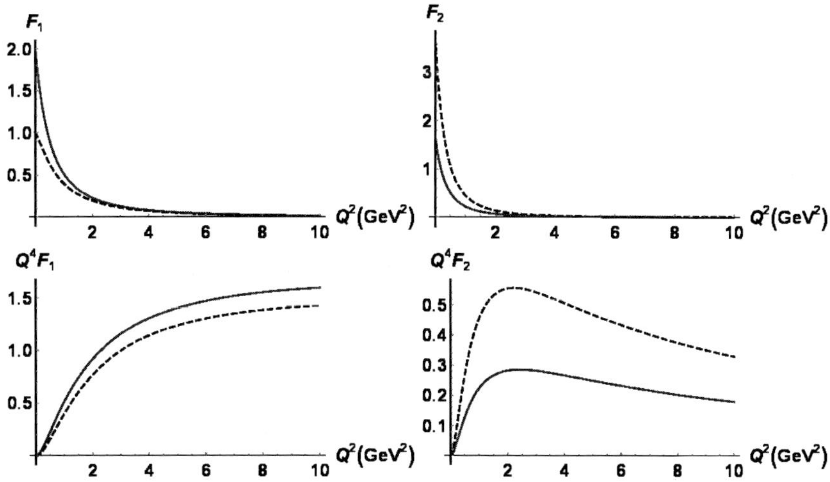

Figure 6. The Dirac and Pauli form factors of the u (solid lines) and d (dashed lines) quarks (left and right top figures), and their respective scalings (left and right bottom figures).

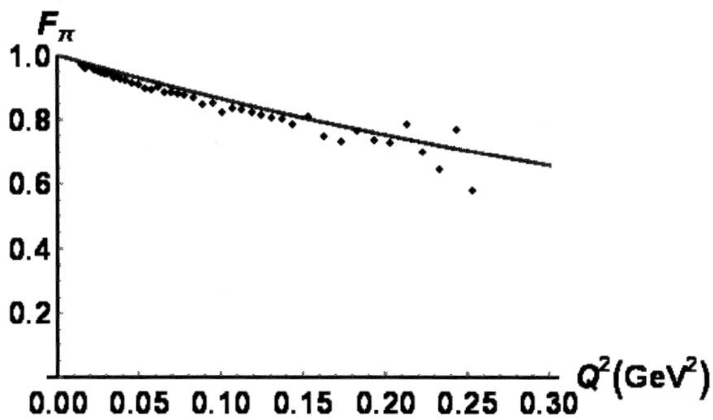

Figure 7. The pion electric form factor, $F_\pi(Q^2)$, calculated with the aid in (62) (solid line) in comparison to data taken form [26] (dots).

formalism. Knowing the quark form factors allows one to obtain the form factor $F_\pi(Q^2)$ of the positively charge pion, π^+, as

$$F_\pi(Q^2) = \frac{1}{2}\left[\frac{2}{3}F_1^u(Q^2)\right] - \left[-\frac{1}{3}F_1^d(Q^2)\right], \tag{62}$$

where we maintained same potential parameters as those fitted to the proton's and neutron's form factors.

Here, the last term on the r.h.s. in the latter equation stands for the $\bar{d}$ form factor. The satisfactory result, displayed in Fig. 7 together with its comparison with data taken from [26], provides a further support for the convenience of our suggested approach to hadron spectroscopy.

Conclusion

Our analyzes show that the experimentally reported conformal symmetry patterns of the light flavored meson mass distributions are well modelled by a conformal dS_4 internal space-time, considered as located outside of the Minkowskian light cones attached to local observes on this space, and hypothesized as the geometry relevant for the description of the virtual degrees of freedom inside hadrons. Along this line, the interaction in the infrared regime of QCD among the virtual hadron constituents has been defined by the Laplacian of the closed space-like hyper-spherical geodesics of S^3 as part of dS_4, where quark and gluon systems are inevitably and necessarily colorless, In effect, hadrons in the infrared have been viewed as color-anti-color dipoles in free conformal four-dimensional quantum rotation, described by the $\ell(\ell+1)\csc^2\chi$ potential, or, rotation slightly perturbed by the also conformal color-dipole potential, $(-\alpha_s N_c \cot\chi)$, attributed to a gluon-anti-gluon color dipole. The magnitude of the perturbance is proportional to the product of the number of colors, N_c by the strong coupling constant, α_s, by virtue of (52), a circumstance that allowed us to extract from data on meson masses reasonable α_s-values, and to conclude on the relevance of a dynamical conformal symmetry in both the extreme regimes of QCD.

Thus, the dynamical conformal symmetry, via dS_4 special relativity, backs up the internal color gauge group, $SU(3)_C/Z_3$, in QCD. However, on open spaces, be them the open time-like geodesics on dS_4, or its casual patches, on which no charge-neutrality is required, color excess could exist within this scenario and become detectable. Our findings favor a space-time, with one extra (infinite) dimension in the internal coordinate space. The scheme is Fourier transformable and can be re-formulated in momentum space. Models of this type are free from infrared divergences [23], a reason for which considering QCD on closed spaces is appealing (also see [4] for further reading). To the best of our knowledge, the model under discussion is so far the only one which allows to estimate the fundamental coupling constant of QCD from analyzes of global (integral) properties of hadron systems such as spectra. Usually, these values are extracted from data on the spin structure function [24]. Our major conclusion is that behind the curtains of conformal symmetry and color-confinement in QCD one may encounter the deSitter dS_4 special relativity as a string-puller. Finally, the quality reached in the description of the electromagnetic form factor of the pion indicates that the u quark in the pion is same as in the proton, while the $\bar{d}$ quark in the pion is related to the d quark in the proton by charge conjugation.

References

[1] K. A. Olive et al. (Particle Data Group), Review of Particle Physics, *Chinese Physics* C **38** 090001 (2014) and 2015 update.

[2] Fayyazuddin and Riazuddin, *A Modern Introduction to Particle Physics*, 2nd edition (World Scientific, Singapore, 2000)

[3] A. Chodos, R. L. Jaffe, K. Johnson, and C. B. Thron, *Phys. Rev.* D **10** (1974) 2599.

[4] D. Kharzeev, E. Levin, and K. Tuchin, *Phys. Rev.* D **70** (2004) 054005 .

[5] M. Kirchbach and C. B. Compean, *Eur. Phys. J.* A **52** (2016) 210.

[6] R. Aldrovandi, J. P. Beltrán Almeida, and J. G. Pereira, Class. Quant. Grav. **24** (2007) 1385.

[7] O. D. Kelogg, *Foundations of Potential Theory* (Dover, New York, 1953).

[8] M. Kirchbach and C. B. Compean, *Eur. Phys. J.* A **53** (2017) 65.

[9] A. O. Barut and R. Wilson, *Phys. Lett.* A **110** (1983) 351.

[10] M. Kirchbach and C. B. Compean, *Nucl. Phys.* A **980** (2018) 32.

[11] Slava Rychkov, *EPFL Lectures on Conformal Field Theory in* $D \geq 3$ *dimensions*, E-print arXiv:1601.05000[hep-th].

[12] Yoonbai Kim, Chae Young Oh, and Namil Park, J. *Korean Phys. Soc.* **42** (2003) 573 .

[13] Andrew Pressley, *Elementary Differential Geometry* (Springer, London Dordrecht Heidelberg New York, 2012).

[14] F. Cooper, A. Khare, and U. P. Sukhatme, *Supersymmetry in Quantum Mechanics* (World Scientific, Singapore, 2001).

[15] D. Cevik, M. Gadella, S. Kuru, and J. Negro, *Phys. Lett.* A **380** (2016) 1600.

[16] H. Bateman, *Proc. London Math. Soc.* (ser. 2) **7** (1909) 70.

[17] L. D. Landau and E. M. Lifschitz, *The Classical Theory of Fields*, Vol. 2 of A Course of Theoretical Physics, 3d edition (Pergamon Press 1971) p.335.

[18] Pouria Pedram, *Am. J. Phys.* **78** (2010) 403.

[19] S. Fubini, A. J. Hanson, and R. Jackiw, *Phys. Rev.* D **7** (1972) 1732.

[20] B. Alertz, Ann. *Inst. Henri Poincaré,* **53** (1990) 319.

[21] Kerson Huang, *Quarks, Leptons and Gauge Fields* (World Scientific, Singapore, 1982).

[22] J.-M. Gaillol and M. Trulsson, J. *Chem. Phys.* **141** (2014) 124111.

[23] A. V. Belitsky, A. S. Gorsky, and G. P. Korchglemsky, *Nucl. Phys.* B **67** (2003) 3.

[24] A. Deur, V. Burkert, J. P. Chen, and W. Korsch, *Phys. Lett.* **665** (2008) 349.

[25] Dipankar Chakrabart, Chandan Mondal, *Eur. Phys. J.* C **73** (2013) 2671.

[26] S. R. Amendolia et al., *Nucl. Phys.* B 277 (1986) 168.

In: Future Relativity, Gravitation, Cosmology
Editors: V. V. Dvoeglazov et al.
ISBN: 979-8-88697-455-3

Chapter 2

SOME GEOMETRICAL ASPECTS OF SPACE-TIME AND RELATIVITY

Rolf Dahm*
Beratung für Informationssysteme und Systemintegration,
Mainz, Germany

Abstract

In order to ask for future concepts of relativity, one has to build upon the original concepts instead of the nowadays common formalism only, and as such recall and reconsider some of its roots in geometry. So in order to discuss 3-space and dynamics, we recall briefly Minkowski's approach in 1910 implementing the nowadays commonly used 4-vector calculus and related tensorial representations as well as Klein's 1910 paper on the geometry of the Lorentz group. To include microscopic representations, we discuss few aspects of Wigner's and Weinberg's 'boost' approach to describe 'any spin' with respect to its reductive Lie algebra and coset theory, and we relate the physical identification to objects in P^5 based on the case $(1,0) \oplus (0,1)$ of the electromagnetic field. So instead of following this – in some aspects – special and misleading 'old' representation theory, based on 4-vector calculus and tensors, we provide and use an alternative representation based on line geometry which – besides comprising known representation theory – is capable of both describing (classical) projective geometry of 3-space as well as it yields spin matrices and the classical Lie transfer. In addition, this geometry is capable of providing a more general route to known Lie symmetries, especially of the su(2)$\oplus$i su(2) Lie algebra of special relativity, as well as it comprises gauge theories and affine geometry. Thus it serves as foundation for a future understanding of more general representation theory of relativity based, however, on roots known from classical projective geometry and P^5. As an application, we discuss Lorentz transformations in point space in terms of line and Complex geometry, where we can identify them as a subset of automorphisms of the Plücker-Klein quadric M_4^2 of P^5. In addition, this description provides an identification as a special, but singular parameterization of the tetrahedral Complex, too. As such, we propose to generalize and supersede the usual rep theory of relativity by an embedding into the general geometry of P^5, and the use of appropriate concepts of projective and algebraic geometry in Plücker's sense by switching geometrical base elements and using transfer principles.

*Corresponding Author's Email: dahm@bf-IS.de; homepage: http://www.researchgate.net/profile/Rolf_Dahm.

PACS 02.20.-a, 02.40.-k, 03.70.+k, 04.20.-q, 04.50.-h, 04.62.+v, 11.10.-z, 11.15.-q, 11.30.-j, 12.10.-g

Keywords: relativity, unification, quantum field theory, Dirac theory, Lie algebras, Lie groups, geometry, projective geometry, line geometry, line Complex, Complex geometry, congruences, null systems

SOMMERSPROSSEN SIND DOCH KEINE GESICHTSPUNKTE!

Ekkhard Verchau, Historisches Seminar, JGU Mainz, 1990

1. Introduction

Having been asked to contribute to this topical collection on the future of relativity on occasion of Valeriy's 60th birthday anniversary, it is a pleasure to appreciate the jubilarian and his long-term work and interest, and it is – not less – an honour to contribute some aspects which might be interesting with respect to geometrical (and possibly alternative) views on relativity which are lost nowadays in the community's prevailing memory. While on the one hand, content being capable of addressing an alternative viewpoint grew and grew throughout working with classical projective geometry (below briefly 'PG') during the last years, and even throughout writing this text, we've experienced on the other hand the standard arguments with respect to relativity over and over again during conferences, in papers, and especially in public networks to protect and shield often a naive textbook reasoning, sometimes even to avoid scientific[1] discussion by claiming that 'everything is well established and settled'. Now, it is NOT that we want to contradict on the subsequent pages the physical aspects of relativity, however, we feel it necessary AND worth to discuss several aspects of the various formalisms in use and their associated interpretations. We think that those aspects are different facets of a more general geometrical description which is known for more than two centuries now, and which – formally starting with Hamilton's formulations and his work on optics – reappeared from time to time in few aspects but is nowadays covered by a patchwork of formal approaches and additional – sometimes implicit – assumptions. As such, this background in geometry most often is not really visible on its own. So we want to provide at next an extended outline, melting some background with the organizational outline of the following pages towards a possible alternative view on relativity.

1.1. Outline

As there is no 'future of ... ' without founding on its past, we want to recall briefly in sec. 2. two of the central 'classical papers' on the formalism of relativity – Minkowski 1908/1910

[1] And in this context, we do NOT address or reference the exchange on some current 'social networks' where the signal-to-noise ratio with respect to discussions on relativity, at least according to our perception, is close to zero.

and Klein 1910 – as well as Weinberg's papers 1964/65 on 'quantum' reps in terms of 'any spin' in order to explain where we want to work the switches later on towards geometry. Minkowski's and Klein's papers focus mainly on invariant theory of the 'Lorentz invariant' quadric $x^2 + y^2 + z^2 - t^2$ in point space and related aspects, and they yield an interpretation of coordinates and their transformations in point space, and appropriate rep theory. If we neglect for a moment the coordinate-oriented approaches closer to experimental observations like pursued by Maxwell, Lorentz, Poincaré and Einstein, then the standard reference with respect to 'classical relativity' nowadays is Minkowski's paper[2] [1] (where he introduced the 4-vector calculus and some related tensorial representations[3]) which marks the center of interest. Not less important to pave the way for Minkowski's 4-dim rep theory were Klein's papers 1910 [7], and 1872 [8] – the older one creating obviously the 4-dim metric foundations on 4-dim space concepts, however, founding on a different but in the beginning of the 20th century apparently forgotten or misinterpreted geometrical background.

Weinberg's papers are necessary in this context because they attach Lie symmetries to a skew 6-dim 'tensor' $\omega_{\mu\nu}$ and provide 'quantum' rep theory which we are going to discuss in sec. 4.2. using a different interpretation of $\omega_{\mu\nu}$.

As such, after having sketched the standard trail very briefly[4] by discussing few geometrical aspects and identifications, we use sec. 3. to extract the main ideas which we are going to use in order to rearrange and generalize the discussion. Thus, we comment in sec. 3. also on Klein's review 1910 [7] – marked by both highlights and lowlights – which seems to have emphasized the historical mathematical discussion in favour of Minkowski's rep theory, and as such also as a foundation of Weyl's concepts [9]. There, we turn briefly back to Minkowski's paper [1] and recall our alternative identification [10] in order to approach physics and geometry much closer. In [10], it turned out that Minkowski's calculus can be seen as a special case of line and Complex geometry, i.e. based on 5-dim objects $a_\alpha p_\alpha = 0$ – so-called 'line Complexe', or 'Complexe' for short, which 'live' in P^5 – where $1 \leq \alpha \leq 6$, $a_\alpha \in$, and p_α denote six line coordinates of P^3. Very recently, we've indeed found additional work published much later by von Laue [11] which emphasizes this concept, too, and which discusses some consequences with respect to form and structure of the energy tensor. Moreover, our approach underpins Einstein's ongoing, lifelong quest for a common geometrical description after having committed to Minkowski's 4-dim reps not earlier than 1912.

Projective line and especially Complex geometry allow to rearrange and supersede var-

---

[2]Originally published in 1908 in the 'Göttinger Nachrichten', p. 1, Klein being the editor of 'Mathematische Annalen' decided to republish it there again in 1910 after Minkowski's sudden death. Due to better availability of 'Mathematische Annalen', we've decided to reference the latter publication, so the page references given below relate to [1].

[3]We suppress the discussion of differential geometry and Riemannian spaces here, as we have discussed already various of those aspects thoroughly [2], [3] by the coset approach summarized by Gilmore [4] and detailed by Helgason's books [5], [6]. So we consider this discussion as a subsidiary concept, and focus instead on the synthesis of geometry and physics in the spirit of Einstein's efforts towards an unified concept comprising both. As before, we use the shorthand notation 'rep' to denote both representations as well as realizations as long as the context is obvious.

[4]There is an enormous number of textbooks floating around and treating the formalism, typically based on Weyl's affine thoughts [9]. However, we are not aware of even one focusing on the *different characters* and *possible interpretations* of space-time coordinates beyond typical Euclidean/metric or affine – and thus intrinsically linear – concepts in the context of 'relativity'.

ious 'well-known' formalisms and symbolisms used throughout physics, and to put them at their places – as the analytic counterpart of the geometry of 3-space. We address few aspects throughout sec. 4.. However, the description we are going to use differs in the respective choices of representations by using six line coordinates $p_{\mu\nu}$ by means of ray coordinates, or using directly the six coordinates p_α of P^5 given above instead of the usual sets of four point and four plane coordinates. The six line coordinates, in order to work in P^3, have to obey the Plücker condition $p_{01}p_{23} + p_{02}p_{31} + p_{03}p_{12} = 0$ (or equivalently $p_1p_4 + p_2p_5 + p_3p_6 = 0$) as an additional quadratic constraint. Line coordinates grant immediate use of their comprised planar and spatial geometry in P^3 or to its subsets, or of line Complexe and Complex geometry in P^5, and as such they induce quadratic constraints in P^3 which originate from the Plücker-Klein quadric M_4^2 in P^5 and additionally reflect in line-line duality in P^3. This, however, is helpful to understand physical aspects as we'll discuss in sec. 4.4.. Needless to mention, that due to the different objects and dimensions, the respective transformation theories as well as the reps of their respective invariant theories differ, too! However, this aspect is nothing but one of the essential clues of Klein's 'Erlanger Programm', and shouldn't worry people.

The 'quantum' aspects, especially Bargmann's and Wigner's concepts and the various spin reps, can be related [12] to the Lie transfer of line geometry [13] (or even using more sophisticated transfer principles like e.g. Laguerre's geometry, or Fiedler's cyclography, and their related rep theories). Thus, it is straightforward to relate Plücker's Euclidean line rep to spin reps [12] and (projective) point reps to include absolute elements (and spinorial reps). While emphasizing Plücker's fifth line coordinate η – a determinant and formally quadratic in the four original line coordinates r, s, ρ, and σ – we have introduced Pauli matrices and quaternions [12], and constructed a parallel, matrix-based formalism to represent the original, projective geometry of 3-space in terms of commutators, and thus by means of a formal su(2) Lie algebra. However, the respective physical interpretations require sophisticated geometrical interpretations and appropriate care to proceed because we discuss objects and algebra of the transferred space [12]. Using this identification as building block, common 'quantum concepts' like the projective line rep in relation to a complex number (by means of an angle, or 'phase', i.e. by pencil coordinates like visualized in appendix A, figure 3), the construction of Clifford algebras by Pauli matrices, Heisenberg models, point identifications on the line and Hesse's transfer principle vs. binary forms, etc. may be recast in terms of (or at least related to) classical line geometry, and as such to projective geometry (or subsequently for short just 'PG') of 3-space.

Here, due to the celebration of Valeriy's anniversary and his great work on rep theories and equations of motion, we do not want to carry coals to Newcastle, so we skip most rep details here, but relate later (see sec. 4.2.) few of our aspects and identifications to Valeriy's overviews [14], [15], [16], [17], [18] on different reps and formalisms, and especially the 2(2S+1)-Formalism, or $(S,0) \oplus (0,S)$-reps, with $S = 1$.

Instead, in order to approach future aspects of relativity, we want to pave the way for an alternative geometrical rep. So by presenting some background and some examples, we propose in sec. 4. to focus on P^5 as well as on projective and algebraic geometry. After having recalled null systems in sec. 4.1., we have to discuss the Lie algebra approach as a subsidiary concept of line geometry, and of projective geometry when generating surfaces by lines, and when generating higher order objects in general by means of projective mechanisms[5]. So in sec. 4.3., we rewrite Lorentz transformations of point space by line (or Complex) transformations, and show how they fit into properties of tetrahedral Complexe and the Plücker-Klein quadric M_4^2 in P^5.

Especially Complexe and their geometry serve as unifying concepts as they are related linearly to null systems[6], i.e. to the very description of 6-dim forces ('Dynamen'), or their possible decomposition into 3-dim 'forces' and 'moments' like e.g. in the theory of the top [21], [22], as well as to the 'gauge description' $F_{\mu\nu}$ of the 'photon field' in terms of a special linear Complex[7] (see [24] II.A, [25], or [10]). More general, regular linear Complexe describe null systems and forces (see e.g. sec. 4.1.), so our reasoning with respect to two Complexe (or a 'Congruence' [26]) as outlined and discussed in [24] comprises von Laue's work, but one has to look much deeper into this kind of analytic reps of P^5, and – even more important – reconnect this rep theory to physics in P^3 again in order to gain a deeper understanding after all the formalism as well as the, most often, empty formal calculus of the past century since Einstein's work. And as three Complexe describe ruled surfaces (reguli, or 'Configurations' in Plücker's original notion [26]), we are thrown back onto the necessity to discuss surfaces and even more intricate objects using lines as building blocks in P^3, or Complexe in P^5. A typical example based on three Complexe yields the two generator sets of real lines of the one-sheeted hyperboloid, and a special linear Complex can be used in this scenario to describe its symmetry axis by identification of the axis/directrix[8]. So there is immediately a deep connection of linear Complexe to surfaces of 2nd order and surfaces up to 2nd class, as well as obvious relations of Complexe to surfaces and their projective construction, and to polar and focal theory.

However, in this context it is also immediately obvious, that complex numbers (as well as hypercomplex number systems in general) are nothing but an algebraical symbolism (and as such a mathematical, *not a physical* tool!) to represent various geometrical cases analytically and algebraically in a unified manner. So although we appreciate complex analysis and function theory, differential geometry and Riemannian spaces – as we want to focus here on some physical roots in order to propose an additional geometrical trail, we have to extend the usual discussion by additional facets. And although our summary so far is even far from presenting *new* knowledge, due to an exuberant technical focus in these days on mathematical concepts like complex numbers, point spaces, or differential geometry, people often use different understanding and emphases, especially when applying Grassmann numbers and associated point space concepts[9].

On the other hand, we know of the central rôle of the Plücker-Klein quadric M_4^2, i.e. a quadratic constraint in P^5 (the Plücker condition) which guarantees that points on

---

[5]There are even much deeper background and possibilities to relate projective geometry to the human observation ability, i.e. to neural, medical and psychological aspects and the adoption of the brain and its recognition patterns to real physical observations [19], but further details are off-topic here. Nevertheless, I'm very grateful to B. Schmeikal for pointing me to A. Trehub's [20] work on the 'Retinoid System' and for helpful private discussions.

[6]And as such they relate mathematically to correlations and involutions!

[7]The parameter spaces of the transformations exhibit even more and higher-dimensional symmetries, as described e.g. in [23] by 'Somen', 'Protosomen', 'Pseudosomen', etc.

[8]German: Treffgerade/Leitgerade

[9]Note, that this differs considerably from Plücker's, Lie's ([13], p. 151), and Study's [23] reasoning in geometry!

the quadric in P^5 map to lines in P^3. So whereas this yields a mechanism *to derive* line geometry of P^3 as a subset of geometry of P^5 and related (quadratic) constraints, this background emphasizes the existence and occurrence of quadratic (and as such right from the beginning nonlinear!) structures like known from involutions, or polar and focal theory related to 2nd order surfaces which reflects in quadratic algebras like known from Pauli or Clifford rep theory. Considering P^5 and the quadric M_4^2, this automatically rises more questions beyond just linear point reps of P^5 in that we have to consider at least quadratic Complexe and their rep theory in P^3, too[10]. As such, the tetrahedral Complex has played a prominent and important rôle throughout physics and geometry[11], and for our discussion here – besides being related to the quadratic structure of P^3 rep theory and yielding the Lorentz-invariant quadric $x^\mu x_\mu = 0$ in point space [24] – the tetrahedral Complex yields an invariant theory in that it classifies all lines in P^3 by the anharmonic ratio of their four intersection points with the planes of the fundamental coordinate tetrahedron. Grouping the lines according to this ratio yields one free parameter, the double ratio, and because we know that projective transformations preserve this ratio, we may immediately stress Klein's 'Erlanger Programm', and attach and apply Lie's theory of continuous groups and their algebras to these equivalence classes. The invariant theory, however, a priori is a geometrical one.

So on one hand, the tetrahedral Complex is deeply connected to the very definition of point and plane coordinates of P^3 (and to polar theory, if we circumscribe the fundamental coordinate tetrahedron by a sphere), on the other hand, it induces structures on the lines in P^3, or their different reps – either in terms of complex numbers when we interpret the line as a special case of a circle or when discussing point sets on such lines like in Hesse's transfer principle ([29], or [30], §51) or using Clebsch's binary forms [31].

Below we'll give some more arguments and discuss few more aspects. In general, we want to emphasize the idea that what we denote by 'relativity' in P^3 is part of Complex geometry in P^5, especially when represented in terms of lines in P^3 and treated by classical projective line geometry. As an example, we discuss briefly von Laue's identification, and relate those aspects to the tetrahedral Complex and 'field reps' in terms of point coordinates in that we construct field lines by an involution on a line. This is well-known from classical projective geometry but here we can relate such field configurations additionally to the tetrahedral Complex and appropriate classes of lines.

So thinking of 'relativity' in terms of transformations on point coordinates is much too short[12] and restricted to catch the background, especially when attributing affine geometry, and its transport and connection concepts, only. Accordingly 'miraculous effects' like Thomas precession, 'space-time mixture – but only in velocity direction' –, etc. appear which can be simply resolved by linear Complex geometry, and which thus can be understood as artefacts of P^5-geometry [10], [24] in P^3.

With respect to this background, it is important to recall the quadratic character of line geometry. As necessary background, we discuss in sec. 4.4. senary quadratic forms in line (or Complex) space P^5 (which are associated quaternary forms in point space [24]) and two of their applications: the six Klein coordinates $\vec{E} \pm i\vec{B}$, and Klein's $3 \oplus 3$ right- and

[10]The general (or pure) approach via P^5 will be addressed in sec. 4.6..

[11]See e.g. Lie's discussion in [27], or related discussions in [28].

[12]... because these symmetry properties are intrinsic properties of lines and Complexe, see below!

left-handed linear fundamental Complexe.

In the last section 4.5., we recall briefly some related aspects of Hudson's book [32] which besides discrete transformations based on point-plane incidence, the Heisenberg group and K3-surfaces, connects a wealth of additional aspects of projective and algebraic geometry.

In order to prepare a future discussion of the associated physics and of the 'future of relativity', we close this document in sec. 4.6. with an outlook by emphasizing the urgent necessity to 'reunify' geometry and geometrical objects again with the respective rep theories by means of Complexe, i.e. we argue to focus again on Klein's 'Erlanger Programm' and the lessons learned from advanced projective geometry and invariant theory[13] by choosing lines as base elements of geometry [26] instead of points only.

2. Known Aspects of Rep Theory

To discuss possible modifications, changes and reinterpretations later on, we summarize throughout this section briefly some of the major aspects of the standard approaches to relativity and some relevant aspects of rep theory. As such, this reflects – of course – our personal reasoning and concentrates on aspects which we want to emphasize with respect to the upcoming discussion of geometrical identifications and reps. Especially, we do not want to discuss formalisms by themselves like Riemannian geometry or Hilbert's axiomatization of Einstein's general relativity [34], [35], [36], [37] here in depth. With respect to physical modeling and reasoning, we think it is at first necessary to gain an impression on the physical objects and to find an adequate mathematical rep in order to model our observations sufficiently. Only afterwards, we can go and see what we can get from the respective formalisms. A nice example is Weinberg's statement ([38], p. B1319) where he refuses 'ab initio' to present field equations or Langrangians – even nowadays a standard question on conferences and during discussions – simply because they are not needed. He emphasizes the evident fact that as soon as one has found reps which fulfil covariance and irreducibility, everything is known, and that there is no more need to suppress superfluous components of the reps. In other words, just consider symmetry and find suitable reps of objects and transformation groups.

So in this section, we address aspects of the coordinate definition which can be seen as the fundamental rep for applications of the invariant and group theory of the Lorentz group, and as such, one has to address aspects of second order surfaces and especially the invariance of a quaternary quadric, too. As emphasized above, we discuss some aspects of the historical approach by means of coordinate and physical identifications, on the one hand in order to keep track of associated physics, on the other hand to avoid pure mathematical formalism and prevent the physical aspects from being buried by abstract generalizations or axiomatizations like Berlin- or Bourbaki-type formalizations, or Hilbert's or Weyl's axiomatic approaches, which sometimes loose connection to physics, or cover physical aspects by mathematical formalism (and sometimes formal artefacts). It is sufficient to

[13] In this context, Klein's footnote ∗∗∗ in [33] on page 419 naturally comments on an ongoing simplification by pure formalisms and by shrinking concepts to subsidiary problems for the sake of analytical presentability, only.

know that in case we have the need to calculate analytically, we can always find appropriate mechanisms from group theory, algebra or differential geometry to write things down. As such, we remember Minkowski's paper(s) on 4-vector formalism, and Klein's background, and with respect to discussions in 'quantum' theories, we want to mention Weinberg's formalism because we can use it later to attach geometry.

2.1. Standard Approach – Minkowski 1908

As with respect to Minkowski's paper [1], in his introductory remarks he claims to derive the basic equations of motion from the 'principle of relativity'[14] in a manner, determined uniquely by this principle. To proceed, in [1], §1, p. 475, he *defines* a coordinate system x, y, and z without explicitly claiming the *type* of the chosen coordinates, and in addition, he defines 'time' in terms of a fourth and a priori *independent* rectangular coordinate t. As we'll see later, although this sounds familiar today, in terms of Euclidean (or 'metric') coordinates, it is ambiguous. Implicit later use of the coordinates in his paper indicates that he uses an Euclidean coordinate interpretation[15].

After having defined the projections of the point 'vector' $\vec{r}$ onto a general vector $\vec{v}$, $|\vec{v}| = q < 1$, his eqns. (10) and (12) in §4,

$$\vec{r}_{||} = \frac{\vec{r}_{||} - qt}{\sqrt{1-q^2}}, \quad t' = \frac{-q\vec{r}_{||} + t}{\sqrt{1-q^2}}, \tag{1}$$

yield what he denotes as 'special Lorentz transformation' while the orthogonal components of $\vec{r}$ with respect to the velocity $\vec{v}$ remain invariant. q at that time has been defined in a complicated manner related to a transformation parameter ψ according to $q = -i\,\tan i\psi$ ([1], eq. (2)).

However, Minkowski's reasoning so far is based on some kind of 'cut and paste' transfer of electromagnetism which he derived after identifying (§2, p. 476, bottom) the six skew symmetric components $f_{\alpha\beta}$, $1 \leq \alpha, \beta \leq 4$, with electromagnetic field components by

$$\begin{aligned} & f_{23}, f_{31}, f_{12}, f_{14}, f_{24}, f_{34} \\ \cong\ & m_x, m_y, m_z, -ie_x, -ie_y, -ie_z \end{aligned} \tag{2}$$

without explaining the background of f. Then, by a lot of arguments, he derives the transformation equations

$$e'_{x'} = \frac{e_x - qm_y}{\sqrt{1-q^2}}, \ m'_{y'} = \frac{-qe_x + m_y}{\sqrt{1-q^2}}, \ e'_{z'} = e_z\,, \tag{3}$$

$$m'_{x'} = \frac{m_x + qe_y}{\sqrt{1-q^2}}, \ e'_{y'} = \frac{qm_x + e_y}{\sqrt{1-q^2}}, \ m'_{z'} = m_z\,, \tag{4}$$

of the fields (see [1], eqns. (6) and (7)). Although he remarks that the structure of these two equation sets may be superseded by vectorial representations[16] according to $\vec{e} + [\vec{v}\vec{m}]$, and $\vec{m} - [\vec{v}\vec{e}]$, even here he doesn't discuss the higher and distinct background structure of f.

---

[14]German: Prinzip der Relativität.

[15]See beginning of §3, p. 477, and beginning of §4, p. 480, where he argues with rotations of the three rectangular space axes.

[16]His symbol $[\cdot\cdot]$ denotes the 3-dim vector product, where $\vec{v} = q\hat{z}$.

Instead, Minkowski discusses the quadratic invariant $x^2+y^2+z^2-t^2 \longleftrightarrow x'^2+y'^2+z'^2-t'^2$ in point coordinates (x,y,z,t), some issues of rep theory, and after having introduced total differentials $\mathrm{d}x$, $\mathrm{d}y$, $\mathrm{d}z$, and $\mathrm{d}t$ of the point coordinates, he identifies velocities and some physics.

It is only in §5, p. 484, eq. (23),

$$\begin{aligned} & f_{23}(x_2u_3-x_3u_2)+f_{31}(x_3u_1-x_1u_3) \\ + \; & f_{12}(x_1u_2-x_2u_1) \\ + \; & f_{14}(x_1u_4-x_4u_1)+f_{24}(x_2u_4-x_4u_2) \\ + \; & f_{34}(x_3u_4-x_4u_3) \end{aligned} \tag{5}$$

that Minkowski begins to construct 'space-time vectors of 2nd kind'[17] f with six components out of two 'space-time vectors of 1st kind'[18] x and u which he requires to transform invariantly, i.e. eq. (5) has to transform into

$$\begin{aligned} & f'_{23}(x'_2u'_3-x'_3u'_2)+f'_{31}(x'_3u'_1-x'_1u'_3) \\ + \; & f'_{12}(x'_1u'_2-x'_2u'_1) \\ + \; & f'_{14}(x'_1u'_4-x'_4u'_1)+f'_{24}(x'_2u'_4-x'_4u'_2) \\ + \; & f'_{34}(x'_3u'_4-x'_4u'_3)\,. \end{aligned} \tag{6}$$

In addition to this observation, it is important for our later use that in his eqns. (25) and (26), he claims that the *two* quantities

$$\vec{m}^{\,2}-\vec{e}^{\,2} \;=\; f_{23}^2+f_{31}^2+f_{12}^2+f_{14}^2+f_{24}^2+f_{34}^2\,, \tag{7}$$

and

$$\vec{m}\vec{e} \;=\; i\;(f_{23}f_{14}+f_{31}f_{24}+f_{12}f_{34})\;, \tag{8}$$

derived from components which constitute the 'space-time vectors of 2nd kind' $(\vec{m},-i\vec{e})$, *are invariant under Lorentz transformation* ([1], p. 485).

Here, we do not follow his further and sometimes weird formal constructions and discussions as he obviously didn't notice[19] (or didn't want to notice[20]) that eq. (5) relates to

[17] German: Raum-Zeit-Vektoren II. Art

[18] German: Raum-Zeit-Vektoren I. Art

[19] We leave a thorough discussion of these strange historical circumstances to more qualified people like science historians. Indeed, it is *very strange* to see that Minkowski either didn't notice or even desperately avoided the geometrical background of Complexe (or '6-vectors') – after having studied in Königsberg under Weber and especially Voigt, having studied in Berlin under Kummer, having been advised by von Lindemann during his PhD thesis, having worked in Bonn – Plücker's long-term domain – for seven years, having been professor of *geometry* in Zurich during Einstein's studies there, and – last not least – having been in permanent local contact with Klein in Göttingen since his professorship started there in 1902 – Klein, who himself acted as editor and contributing author to publish Plücker's heritage on line geometry and Complexe in conjunction with Clebsch [26], and who published additional basic work on line geometry, Complexe and transformation groups with Lie in the early 1870s (see e.g. [8] or [39].) However, it's a fact that Minkowski didn't use the existing, very elegant line and Complex geometry established back since the 1860's and 1870's, which moreover had been emphasized much stronger by Study's exhaustive contemporary work [23] on Dynamen! This lack is underpinned by his statement [1], p. 499, #6, where he constructs line coordinates out of his two vectors w and s – whether point or plane coordinates – without mentioning the line geometrical background or null systems. However, he noticed that he had to introduce a dual $[w,s]^*$ being relevant for subsequent physical discussions. We have given a more detailed (but still introductory) treatment and partial transfer to line and Complex geometry in [10], however, based on formal and algebraic arguments only.

[20] see [10], footnote 10.

the very definition of a line Complex[21] according to Plücker (see [26], or eq. (30) here, or e.g. [40], or [41]). Moreover, according to his introduction of complex phases in eq. (2) above, he didn't use real Complex components, but switched implicitly to Klein coordinates instead of Plücker coordinates[22]. Minkowski's invariance requirement with respect to eq. (5) and the transformation into eq. (6) constrains the objects[23]. He identifies this Complex by its six components ('coordinates') $(\vec{m}, -i\vec{e})$, and if we rewrite eqns. (7) and (8) according to

$$\vec{m}^{\,2} + (-i\vec{e})^{\,2} \;=\; f_{23}^2 + f_{31}^2 + f_{12}^2 + f_{14}^2 + f_{24}^2 + f_{34}^2\,, \tag{9}$$

and

$$\vec{m}(-i\vec{e}) \;=\; f_{23}f_{14} + f_{31}f_{24} + f_{12}f_{34}\,, \tag{10}$$

we are led automatically into Complex geometry of P^5 and the related invariants[24] of a linear Complex $(\vec{m}, \vec{e})$. As such, the rhs of eq. (10) is the so-called 'parameter'[25] of the linear Complex which in the case of a special linear Complex $\vec{m}(-i\vec{e}) = 0$ yields the Plücker condition related to the line coordinates $f_{\alpha\beta}$ of an axis/a directrix[26]. In this case, the special linear Complex $f \sim (\vec{m}, \vec{e})$ is described by a line/axis hit by the lines of the Complex. Physically, we have used this picture to describe a linearly/uniformly moving point emitting light rays (see [24], sec. 2.2, and ibd., appendix A, and [28], sec. 3).

The rhs of eq. (9) has its deeper background directly in P^5 and possible coordinatizations which we discuss later in vicinity of Klein's remarks in [30], especially §§22 and 23. Here, by referring also to footnote 22, it is obvious that according to complexifications/phases of the underlying P^3 point/plane coordinates one can treat the usual quadratic invariant in 4 variables by considering the whole 'family' of symmetry groups SO(α,β), $0 \leq \alpha, \beta \leq 4$, $\alpha + \beta = 4$ in point as well as in plane (or 'momentum') space. However,

[21] The 3-vectors $\vec{E}$ and $\vec{B}$ may be formally arranged as linear Complex, or special 'Dyname' when interpreted as 'force'.

[22] We postpone the phase discussion and the reality considerations of the fields – due to Minkowski's artificial introduction of an imaginary 4-component of the point reps x and u – to a later stage, see sec. 4.4.. As a consequence, in eq. (5) the coefficients f_{14}, f_{24} and f_{34} have to be imaginary, too, to represent meaningful geometry which would convert the signature of the squares in eq. (7) to SO(3,3) instead of their formal SO(6) invariance. Alternatively, instead of real coefficients with Klein coordinates, one could choose imaginary coefficients with Plücker coordinates, or – as the best and most transparent approach – start from scratch with line and Complex geometry in terms of tetrahedral homogeneous coordinates which fixes the phases and the coordinate sets *uniquely*. Moreover, as within this context it is necessary to introduce conjugation of Complexe as well, like (at least formally) performed e.g. in von Laue 'reprise' [11], one thus approaches the background concepts of P^5, too. However, both authors – Minkowski and Klein – do not create the necessary relationships to an underlying Complex geometry, and thus miss this original background of P^5 and the rôle of the Plücker-Klein quadric.

[23] Being a priori a constraint on the form, in other words with respect to irreducibility of the chosen reps in eqns. (5) and (6), one can, of course, identify the expressions in parenthesis there as well as their accompanying coefficients f geometrically. If we identify later the coordinate expressions in parenthesis as reps of line coordinates in terms of point reps, i.e. as 'ray' reps, their invariance forces the linear Complex to remain invariant, too. We have discussed special cases already in [25], the general discussion is given in 4.3..

[24] This relation is discussed in sec. 4.4. in more detail.

[25] With respect to textbooks, both notations with x_4 and x_0 are in use, however, the rhs of eq. (10) differs by an overall minus sign when one switches to x_0 instead of x_4 according to the usual ordering of the coordinates and the antisymmetry of the line coordinates. This has consequences with respect to respective physical interpretation(s).

[26] German: Treffgerade

related to their associated six line coordinates, we have to discuss also the accompanying 'family' SO(n,m), $0 \leq n, m \leq 6$, $n + m = 6$, [24], [28], as well as quadratic mappings like the Plücker-Klein quadric, the Veronese mapping, birational maps, etc.

2.2. An Early 'Relativistic' Example

For now, we want to mention (as the first and as an early example) only the relation to generating lines of the one-sheeted hyperboloid[27]. Each generating family of lines of the hyperboloid can be derived from three Complexe, so by Lie transfer [12] or by the differential rep discussed in appendix E, we may relate as well the two operator sets L_i and K_j, each comprising three generators[28].

Thus, we may use the operator discussion in [42], ch. 1, sec. 11.3, and especially in [42], ch. 1, appendix III, as well as the rep theory discussed there in terms of su(2)⊕i su(2) by means of the 3-dim operator sets J_i and K_i, and the associated invariants F and G, where $F = \frac{1}{2}(\vec{J}^2 - \vec{K}^2) = \frac{1}{4}M_{\mu\nu}M^{\mu\nu} = \frac{1}{2}\left(N(N+2) + M^2\right)$, and $G = \vec{J} \cdot \vec{K} = \frac{1}{4}\epsilon_{\mu\nu\rho\sigma}M^{\mu\nu}M^{\rho\sigma} = iM\,(N+1)$. M and N denote 'quantum numbers' (see [42], ch. 1, appendix III, eq. (III.14), or see sec. 2.6. and compare to eqns. (9) and (10)), and the 'translation' of $\vec{K} = i\vec{B}$ (see [42], ch. 1, appendix III, eq. (III.5), or sec. 2.6.) yields the complexification of the original su(2) operators, and thus to line geometry when recalling the line generators of second order surfaces (see appendix E). Moreover, it connects to Minkowski's artificial complexification of the fourth space-time coordinate above, and as such of the second 3-dim generator set K_i (the 'boosts') which involves the zero- or four-coordinate. So the reference to Gegenbauer polynomials as 'basis functions' of such polynomial reps is evident, however, one should take care of the projective character and an appropriate identification of the coordinates involved in such a rep theory. This reflects once more the relation of the quadrics in terms of four homogeneous point or plane coordinates, and in terms of their six homogeneous line coordinates, i.e. P^3 and P^5, or SO(4) and SO(6) (and their respective reps) when discussing transformation groups and invariants. The real (Plücker) case relates SO(3,1) and SO(3,3) (see e.g. [28], sec. 1.6, and the discussion there).

An analogous reasoning can be extracted from [43], sec. 2.1, where a skew 4×4 Lie algebra rep $M_{\mu\nu}$ is mapped (or decomposed by separating space coordinates 1,2,3 from a 'time' coordinate 0) to two operator triples $\vec{M} = (M_{23}, M_{31}, M_{12})$ and $\vec{N} = (M_{01}, M_{02}, M_{03})$. Both are usually treated in terms of su(2) Lie algebras and attached to operator reps of the inhomogeneous Lorentz group on a Hilbert space[29].

---

[27] This case of this second order surface yields the most practical access to discuss real lines within a surface, and by departing from this special surface – which is analytically easy to handle and geometrically easy to understand for being a ruled surface in 3-space – we can relate further discussions of general 2nd order or class surfaces by complexifying individual point/plane or line coordinates.

[28] Due to possible alternative identifications and ambiguities, we have summarized some details in appendix E

[29] Not only by this approach, it is evident that people wanted to construct Hilbert space reps in order to handle the skew 6-dim object $M_{\mu\nu}$, however, the formal treatment so far ended in su(2)⊕su(2) or su(2)⊕i su(2) discussions while neglecting the geometrical background of the 6-dim Complex rep. As this is the 'classical' decomposition of the line rep resp. the organization of the line coordinates by point coordinates, for us the only open issue is to discuss the occurrence of the skew 4×4 rep and the differential reps in appendix E. In a naive approach, in order to switch from classical geometry to differential reps, one can replace the line coordinates $p_{\mu\nu}$ by operators $M_{\mu\nu}$ (and vice versa) and check the consequences with respect to linear reps and second order surfaces.

The important issue in this context for now is Minkowski's claim that for $\vec{m}^2 - \vec{e}^2 = 0$ and $\vec{m}\vec{e} = 0$, i.e. according to our 'new' notation due to a singular Complex, these two properties remain invariant under every Lorentz transformation ([1], p. 485). So here, with respect to our Complex approach [10], we've recovered the two Lorentz invariants related formally to su(2) algebras, which e.g. in the su(2)$\oplus$i su(2) Lie algebra rep approach [42] are related to the invariant 'quantum numbers' of the reps. We discuss in appendix E briefly the necessary operator representation derived by classical polarity and line geometry, whereas the 'quantum' notion can be introduced according to [12], [44] based on Lie transfer of lines to spheres and their reps in terms of the Pauli algebra. However, already here, it is immediately evident that the linearization and the rep by infinitesimal su(2) generators have to be traced back to their origins and background in PG in order to understand more details of relativity, symmetry and 'quantum' reps. In other words, the example rises the question on how to find linear reps, given a quadric, which can – of course – be answered immediately from the viewpoint of PG (at least for elements of low grades). Later, we can enhance this discussion on generating higher order (or class) elements.

2.3. Further Aspects

As is nowadays common textbook knowledge, one can develop both theories of relativity – special and general[30] – based on Minkowski's space-time (or point-geometrical) approach and related differential geometry, although in cosmology there are various models with symmetry groups SO(n,m) under discussion where $n + m \leq 6$. Whereas the Poincaré group – which is also often mixed up with transformation groups acting on homogeneous coordinates – can be obtained *only after contraction* from SO(3,2) or SO(4,1)[31], its rep theory accordingly requires Euclidean/metric coordinates and semidirect products.

Beginning with Bateman's and Cunningham's work in 1909 on Maxwell's equations, the discussions of conformal symmetries related to SO(4,2) – and also to invariant 6-dim quadrics with different signatures – showed up and entered the field and the physical discussions. In our understanding, this topic is completely resolved and superseded by line geometry and Complexe as soon as one accepts the P^5 background represented in terms of *six homogeneous* coordinates, their related symmetry groups and subgroups, and – last not least – various transfer principles and projections techniques to lower dimensional spaces (see e.g. [30] 'Zweiter Hauptteil'). In other words: just strictly apply Klein's 'Erlanger Programm' from scratch!

Classical relativity has been elaborated further by Einstein using Minkowski's formalism, emphasized by Hilbert's (re-)formulation of Einstein's general theory in a short series of papers [34], [35], [37] in exchange with Klein [36], and also by Weyl with his axiomatization and the affine approach [9]. The standard textbook procedure treating and handling electromagnetism in terms of 4-vectors and -tensors starts here, too. Lots of textbooks in

[30] see e.g. Einstein's papers (or the notes in appendix D, or Hilbert's papers cited above.

[31] For a detailed discussion of contractions, their geometrical background and the relevant references, see e.g. [4], especially ch. 10. Gilmore's detailed approach by coset spaces as well as Helgason's marvelous work [5], [6] can be seen as a mechanism and framework to comprise, generalize and supersede Wigner's approach by boosts [45] which has been used by Weinberg's formalism ([38], [46], [47], and see also [14], [15], [16], [17], and [18]). We have summarized some aspects in 2.6. later in order to identify the operators and the reductive algebra structure.

mathematics and physics so far followed this trail, always accepting and using those coordinate and differential definitions without deeper investigations. Now, we agree, of course, with the fact that this formalism works in point space while using 4-vectors (or 'space-time vectors of 1st kind') only, and by specifying circumstances as well as obeying obvious and usual restrictions. However, answers with respect to transformation properties of extended physical objects, even with respect to basic mathematical objects like lines or planes, most often are not treated at all, but the focus is set to point transformations only[32].

Moreover, the general background of Complex geometry and its relation to physics, and especially to relativity and electromagnetism, has been forgotten and has been lost in literature throughout the years. So for us, the milestone of the Minkowski paper [1], and especially von Laue's paper [11] much later, show how to attach the differential representation of relativity and electromagnetism to the geometry of (singular) linear Complexe and their invariant theory [10]. However, from the viewpoint of Complex geometry (or PG in general), this is a singular and very special case, so the main switch we are going to work later is to generalize special/singular Complexe to general Complexe – regular/general linear Complexe as well as higher order Complexe and their respective geometrical properties, instead of generalizing the dimensions of the irreps or 'quantum numbers' of the SU(2) subgroups as usual.

The 'twofold antisymmetric tensor' $F^{\mu\nu}$, of course, occurs in various formalized contexts in physics like classical field theory, affine connection concepts, covariant derivatives and differential geometry, even in quantum field theory when formulating gauge theories. However, its foundation in null systems and the analytic description of forces which yields the physical background and – according to our opinion – the correct trail to unify representation theory by geometry, has been forgotten and suppressed by applying tensor calculus, formal algebra and differential geometry, although the framework of (skew) Lie algebra reps and Lie symmetries in general is a common tool-set in these days.

Also in Einstein's papers, we've found[33] only few statements and contexts mentioning the 'six-vector formalism', and only one of them [49] addressing the electromagnetic field rep in conjunction with the gravitational field in detail. However, all of his approaches in appendix D are based on Minkowski's paper [1] by using only one six-vector and its dual, however, with major use of 4-vector and various derived tensorial reps. So we see von Laue's paper [11], by now introducing *two* 'six-vectors' to describe field *and* matter, as the next relevant step towards Complex geometry. The associations of the 3-vector fields $-i\vec{E}$, $\vec{B}$ to a 'six-vector' $\mathfrak{M}$, and $-i\vec{D}$, $\vec{H}$ to a 'six-vector' $\mathfrak{B}$ is used [11] to derive Maxwell's equations, and matter properties are introduced by coupling the two 'six-vectors' by

$$\mathfrak{B}_{\alpha\beta}Y_\beta = \epsilon\,\mathfrak{M}_{\alpha\beta}Y_\beta\,, \quad \mathfrak{M}^*_{\alpha\beta}Y_\beta = \mu\,\mathfrak{B}^*_{\alpha\beta}Y_\beta\,, \tag{11}$$

[32]Even in this context, people focus on point coordinates instead of recalling that e.g. length contraction or time dilation intrinsically argue with *differences* of points (or their coordinate reps). So it is not sufficient to include an 'origin' of a coordinate rep by the null *vector* which is neglected but one has to recall that the very coordinate definition of such 'vectors' already includes geometrical assumptions e.g. on linearity as in the case of Euclidean or affine coordinates, and on the metric, i.e. in these cases on the polar system in the absolute plane. So one should consider naive generalizations from 3 to 4 coordinates very carefully, like e.g. in the case of gauge theories.

[33]So far, we've checked mainly publications in 'Annalen der Physik' and the published/printed versions of Einstein's academy contributions in Berlin, (re-)printed in 'Sitzungsberichte der Preußischen Akademie der Wissenschaften', 1914-1932 [48]. We've given a brief summary of what we've found in appendix D.

where von Laue used the '4-velocity' Y_β, and conjugation of the 'six-vectors' denoted by $\cdot^*$. Details can be found in [50].

Moreover, in all of these references, we've found no remark or discussion of the original background of null systems and forces. In contrary, all authors base their reasoning and their calculations upon the rep incidence of antisymmetric twofold tensor reps with the ray rep of singular linear Complexe in terms of 4×4-matrices, however, they subsume this rep as part of a tensorial calculus based on 4-vector reps. This resembles using the 4-vector as a fundamental (ir-) rep[34] and constructing higher order reps of a Lie algebra or group, and they proceed performing formal algebra.

Thus, a lot of physical background has been overlooked in favor of technocracy and stiff algebra applications, and was lost throughout the last 100 years for general discussion(s) of the underlying physics and geometry.

2.4. Standard Approach – Klein 1910

Although we have already discussed the major switch – Minkowski's 1910 paper [1] – which we want to work below in that we want to use general line and Complex geometry to comprise and supersede this paper and its concept, it is necessary to consider another important paper in 1910 which emphasized Minkowski's coordinate identifications above: Klein's paper [7] on the Lorentz group! To forestall one of the major results – his arguments have not only emphasized Minkowski's 4-vector formalism, but in addition they have forced the coordinate discussion to generalize this formalism to '5-dim' point reps by expressing the four coordinates of space and time in terms of five homogeneous point coordinates. This has had dramatic impact on the formulation of various 5-dim theories (Kaluza-Klein, or see e.g. Pauli [51], [52]) as well as on approaches in projective (differential) geometry like Veblen's efforts (see e.g. [53], [54], or [55], and the associated literature and discussion). In order to gain more control on the context, we want to recall and discuss briefly some aspects of Klein's paper [7] on the geometry of the Lorentz group.

So whereas in the first part Klein summarizes the historical development so far, beginning with Cayley's 6th memoir on quantics in 1859, and Klein's personal geometrical approach towards his 'Erlanger Programm' in 1872, he focuses there mainly on collineations and invariant theory. However, in the first part, he keeps track on the correct identification of the *different types* of coordinates and their respective transformation groups which we see as a real highlight of the paper! There, he distinguishes Euclidean ('metric') point coordinates explicitly from (ordinary) homogeneous coordinates[35], he discusses some aspects of order vs. class from classical projective geometry of 3-space, he addresses some issues related to conics as an application to discuss the Cayley-Klein approach to metrics, and – last not least – he focuses on affine transformations.

However, on [7], p. 293, and this – in the light of his achievements – is from our viewpoint not one of the highlights of [7], he begins 'to generalize' point coordinates, and it is here, where we'll have to switch later on to another reasoning.

Following Klein's arguments, he interprets the four space and time coordinates x, y, z, and t as Euclidean ('metric') coordinates which can be seen in his next paragraph where he

[34] Which, of course, works only severely limited in this context!

[35] Klein doesn't mention tetrahedral homogeneous coordinates in this paper at all!

introduces five homogeneous coordinates x_1, x_2, x_3, x_4, and x_5. He relates both sets by

$$x = \frac{x_1}{x_5},\ y = \frac{x_2}{x_5},\ z = \frac{x_3}{x_5},\ t = \frac{x_4}{x_5}, \tag{12}$$

in the 'classical manner' to define ordinary homogeneous coordinates, and as such naively recalls and transfers the affine approach of 3-space, or P^3, to be applied to some kind of affine P^4, represented inhomogeneously by the coordinates x, y, z, and t, and by the homogeneous coordinates x_1, x_2, x_3, x_4, and x_5. So, in addition, he gains direct contact to his old 4-dim metric representation of line geometry in the early 1870s, however, the need for a satisfactory interpretation of the four 'metric' coordinates appears immediately and should have been addressed there already.

We do not want to discuss his reasoning with respect to his 4-dim 'metric' space-time spanned by x, y, z, and t further because we have given arguments already (see e.g. [24], [28], [10], or appendix A here) that one should consider 'time' in a different manner[36] than just as an additional 'coordinate' mixed up with point space. Moreover, by assuming the simple case of uniform motion, we've shown (see e.g. [24], sec. 3.5, or [28], or [25] sec. 2) that it is projective geometry which – represented in point space – allows to understand the Lorentz factor just by relations of velocities[37] when identifying ct with the absolute element. In appendix A, we discuss some more examples which show the relation of point space and metric versus the identification of 'time'. From within this context, the important message is that point reps and velocities (considering uniform motion and an origin on the line of the motion, or the trajectory, to simplify the vector description) are proportional due to choosing the usual linear time rep, so time represents a line parameter as soon as one introduces additionally individual points on lines.

In the derivation of the tetrahedral Complex (see [28], sec. 3, especially sec. 3.3), and in appendix A, it is easy to construct cases where the 'time' is derived from other coordinatization, e.g. only using line geometry and line intersections. It is just the geometrical picture which changes the interpretation of the parameter, and especially of 'a time' associated to points and their motion. So in sec. 4., we'll try to adjust Klein's problematic conception of the five homogeneous coordinates founding on the point interpretation and a linear absolute element above, and instead we argue in favour of Complex geometry of P^5 in terms of *six* homogeneous coordinates. The projection to usual space-time can be realized by the Plücker condition, i.e. by a quadratic constraint, so whereas in 3-space it is a generalization to switch from point/plane to line representations, the physical background discussion should reconsider Complex geometry in P^5 as is known from mechanics (null

---

[36] This is especially important from the physical point of view because one has to keep track of physical dimensions. It is NOT straightforward to set a 'velocity' $c = 1$ without caring about the consequences. Already physical dimension considerations require to treat *all* the coordinates x, y, z, and t either as space coordinates, however, while taking care of their *identification and interpretation* (see below), or to change the interpretation and introduce *a time* conceptually into the picture which is typically done by restriction to a Euclidean point setup or to a setup compatible with special relativity, i.e. requiring uniform and as such linear motion. This, however, introduces linear velocities and their relations, appropriate transformations, as well as 'an upper bound' in order to handle absolute elements. So implicitly this corresponds to the axioms of standard projective geometry of 3-space. It is then evident that 'time' has to connect the point and the velocity picture [24], [28], [25], and we find projective relations e.g. between the velocities to justify β or the Lorentz factor γ.

[37] If expressed by anharmonic ratios, or 'Würfe', the assumption of a common 'time' (or 'time' difference, or projection) allows to relate points and velocities, and to transfer projective relations.

systems/forces), optics (rays), or the electromagnetic description (see e.g. von Laue [11]). Note already here, that this framework generalizes the idea of forces by 3-dim line sets distributed over the whole 3-dim space, however, respecting a certain geometrical distribution [26]. So whereas on the one hand this is an alternative picture of 'fields' and potential theory by identifying the individual members of such line sets in 3-space which are passing through the individual point coordinates[38], on the other hand, one has to switch notion from points (without extension) and 'point-based' interactions into a picture which includes spatial extension (or non-locality) right from the beginning. This framework allows for a plethora of different reps, only one being points and their accompanying planes in 3-space in order to complete dualism as transfer principle. Accordingly, we've found a wealth of 'transfer principles' between different reps and different interpretations, and one should be careful to attach physics naively to only one special rep or picture, especially one – the 'point' – which is the perfect candidate to prescind. So a 'wave-particle dualism' is far from being mystical, but it reflects only one out of several transfer possibilities between lines and spheres (like e.g. in Lie transfer), and each of the patterns holds as long as the underlying transfer principle holds mathematically (i.e. the rep 'exists' faithfully). Coined in Plücker's, early Klein's, Lie's, and Study's words, we may change the base elements of the respective geometry appropriately.

An immediate identification within Lie's description of line elements ([27], Abschnitt II), however, is evident where according to Lie we may use ('metric') point coordinates and the two ratios p and q of their differentials $\mathrm{d}x : \mathrm{d}y : \mathrm{d}z$, so Lie has presented a 5-dim rep (x, y, z, p, q) which he uses to develop his rep theory of line and area elements (and which, of course, appears in related scenarios like dynamics and differential geometry having his contact geometry as background).

2.5. Standard Approach – Klein before 1910

In order to gain some more insight into possible origins of a 4-dim 'metric' concept of special relativity, its embedding into PG and of the discussion of the 'well-known' invariant quadric $x^2 + y^2 + z^2 \pm t^2 = 0$, it is worth to mention some earlier aspects of Lie's and Klein's work on transfer principles as well as Klein's and Lie's work of sustaining and summarizing Plücker's ideas on line and Complex geometry, both a vista of years before 1908/1910.

As such, we can start our discussion with [8], footnote $**$ on page 257, in 1872. There, Klein claims in the text his goal to generalize Lie's transfer principle [13] – which maps two 3-dim spaces r and R of lines and line Complexe to spheres and sphere Complexe under special metric restrictions – to a 4-dim *metric* space because line geometry itself is 4-dim[39].

[38] In detail, given a point x, one can ask which lines of a Complex, a Congruence, a Configuration (or ruled surface), or other sets of lines, pass this point. This is answered immediately in the case of a Complex by an associated planar line pencil, and in the case of a Congruence (or a 'ray system (German: 'Strahlensystem erster Ordnung und erster Classe') by a 2-dim set of lines in space where one space-point is passed by exactly one line of the set. In this case, there is only one line of the set within each spatial plane. An example is the set of lines being incident with two skew lines. Moreover, one can consider cones or general second order surfaces – having Monge's equation, Lie's sphere geometry, or second order partial differential equations in mind – which are related to the point x in the respective geometry.

[39] It is restricted by the additional Plücker condition, i.e. the Plücker-Klein quadric in P^5, to 3-dim, and it is

In the accompanying footnote $**$, he introduces four rectangular point coordinates x, y, z, and t, which according to his statement define an absolute quadric $x^2 + y^2 + z^2 + t^2 = 0$ and thus provide the basis of the metric in this 4-dim *metric* space. So, vice versa, the only possible, valid coordinate interpretation in this scenario is to understand the four rectangular point coordinates x, y, z, and t in his text as *ordinary homogeneous* point coordinates within a set of *five* coordinates where the fifth coordinate has already been set to 0 in order to realize the absolute quadric in the absolute 'plane', and accordingly the 'metric' by means of a direct generalization of the 3-dim case. So although Klein's explanation – even in german – is difficult to read and understand, he obviously[40] discusses a P^4 where in the philosophy of the Cayley-Klein mechanism the metric properties have to be derived by projective relations of objects with respect to the absolute quadric $x^2 + y^2 + z^2 + t^2 = 0$. In other words, the conceptual focus is set on *linear* elements when defining coordinates and attached to a special identification of absolute elements[41].

By the way, having discussed this construction scheme with respect to coordinate definitions and the construction of the metric via a quadric now for a second time, we want to rephrase this context with respect to rep theory, and especially to linear reps. As such, it is interesting to ask how 'to resolve' a quadric by identifying two appropriate linear reps (like the rep and its 'adjoint' or 'dual', or the roots of a square as a special case of relating linear reps and their adjoints). Especially the way around to construct quadrics from linear reps, or in general the generation of elements of higher grade (order or class) can be performed using PG. In the case of linear elements, prominent examples are the generation of conics by line pencils in the plane (see e.g. [56]) or the generating line families of second order surfaces (see e.g. [30]), but this can be extended, of course, to schemes using higher order elements right from the beginning and classifying e.g. the intersection results, or unions.

Klein's coordinate identification as discussed above and the related reasoning can be verified also in another context – although published by Klein more than 50 years later – if we refer once more to Klein's book on advanced geometry [30]. There in §§37–39, Klein develops the four historical stages of the foundations of projective geometry. Especially what he calls 'the third stage' in §39, it is where he comments on the historic approach to derive metric properties from projective geometry in terms of homogeneous coordinates. There, he uses appropriate analytic examples to justify the coordinate identification/interpretation we have discussed already above.

His planar example uses analytic point coordinates x, y, and t, whereas his spatial example uses point coordinates x, y, z, and t. In the first case, he derives the two absolute points by starting from the quadric $x^2 + y^2 + t^2 = 0$, so the absolute line $t = 0$ yields $x^2 + y^2 = 0$ which results in two conjugate imaginary point solutions.

Then, he has related the 'metric' 3-dim case to $x^2 + y^2 + z^2 + t^2 = 0$ in terms of

thus suited to describe 3-space geometrically besides using points and planes only.

[40] See also the discussion in the next paragraph based on his later publication [30]!

[41] In [24], we have 'derived' the hyperbolic quaternary quadric in point space from its association to the senary quadric in line geometry. In other words, the 'light cone' may also be associated with a certain quadratic Complex, or an element of the Plücker-Klein quadric in P^5. This, at the same time has some influence on the coordinate definition by self-polar fundamental tetrahedra, however, that's beyond scope here. Another well-known relation of hyperbolic geometry is known by considering three linear Complexe which constitute a Configuration, i.e. a ruled surface and as such an hyperboloid [26]. In elementary geometry, this is reflected e.g. by generating the hyperboloid by lines, and discussing the two families of generating lines.

four ordinary homogeneous coordinates, and he thus obtains the absolute conic section $x^2 + y^2 + z^2 = 0$ in the absolute plane $t = 0$. His reasoning is attached to the projective arguments that if the analytic expression $(x^2 + y^2 + z^2) + 2\alpha x + 2\beta y + 2\gamma z + \delta = 0$ of a sphere is given in terms of Euclidean coordinates x, y, and z, and four parameters $\alpha, \beta, \gamma, \delta$, the related expression in *homogenous coordinates*[42] reads as $(x^2 + y^2 + z^2) + 2\alpha xt + 2\beta yt + 2\gamma zt + \delta t^2 = 0$. Its intersection with the absolute plane $t = 0$ thus yields the conic section $x^2 + y^2 + z^2 = 0$ which is common to *all* spheres because the set of equations $x^2 + y^2 + z^2 = 0, t = 0$, evidently doesn't depend on the parameters $\alpha, \beta, \gamma, \delta$ comprising the sphere description. So this conic section is suitable to classify spheres as those second order surfaces which leave the absolute conic invariant. Note, that throughout all those discussions, t is a homogeneous coordinate and NOT 'time'. Further contemporary examples of this notation of the fourth homogeneous variable can be found e.g. in [32], or [57], so it is necessary to distinguish standard reasoning in PG from a physical 'time' identification of 't' as done by Einstein[43].

As such, the case of four 'metric' coordinates expressed by means of five ordinary homogeneous point coordinates x, y, z, t, and s thus would give rise to the related set of corresponding equations in P^4, i.e. $x^2 + y^2 + z^2 + t^2 = 0$ where $s = 0$. Note however, that in order to obtain 'metric' coordinates in an 'affine picture', i.e. fixing a *linear* 'plane' in P^4, one has to divide the four coordinates by s linearly, and thus absorb the singularity of the absolute element already in the 'metric' (or Euclidean) coordinate definition[44].

Throughout all this reasoning, the related concepts are driven by an invariant *linear* absolute element so that e.g. in the case of spatial affine transformations the absolute plane $t = 0$ remains invariant under linear transformations[45]. This requirement, however, is a strong restriction which e.g. in case of non-linear transformations in general can be fulfilled only in infinitesimal limits or by special considerations. And this aspect yields exactly an approach towards future discussions of space-time and relativity in that e.g. line geometry is a priori quadratic. So in order to discuss finite transformation and geometry, we necessarily have to consider non-linear transformations, and as well invariants of higher order or class. Although invariant theory and form systems underwent mathematical and physical research and extensions over the decades since Clebsch, Gordan, Study, Hilbert, and Weitzenböck, the necessary quaternary or even senary formalism is tedious and error-prone. Various algebraic approaches introduce formal problems, or physical and mathematical ambiguities, or even lack physical evidence completely. Typical examples are large Lie groups, or associated algebras of various kinds, where final SU(2) reductions (e.g. by using cosets or coset

[42] It is important to note that Klein thus implicitly switched the *character* and *interpretation* of the originally Euclidean point coordinates x, y, and z to the set (x, y, z, t) of ordinary homogeneous coordinates without changing notation appropriately!

[43] Klein attributes this identification of 't' and time to Einstein ([36], p. 474) when writing letters to Hilbert on occasion of Hilbert's papers on 'Die Grundlagen der Physik' [34], [35], [37]. Extracts taken from the letters exchanged between Hilbert and Klein were published in Klein's paper [36].

[44] We have started this discussion in [24] as we see this as a possible mechanism as long as one considers linear transformations and related invariants, however, we have given there an alternative (and to our opinion more general) definition which has to be extended if we discuss non-linear transformations and spaces but it includes the linear case (see [24], sec. 3.5). So it can serve as basis to generalize the coordinate description and related rep theory.

[45] This, recalling duality vice versa as a correlative mapping, involves the origin of the coordinate system, too.

chains) and the necessary physical identifications are ambiguous[46], supersymmetry which so far seems to have its place in nuclear physics according to Iachello's work and research, however, NOT in the description of elementary particles [58], or string (or 'superstring') models which remind of some aspects of Plücker's reasoning when interpreting geometry and their base elements, however, in different clothing. Last not least, so far we are not aware of significant evidence in physics.

To summarize, Klein's reasoning introduces the concept of a P^5 (or four 'metric' coordinates) which has been reused[47] by Klein in 1910 and in [30]. However, it is important to emphasize Klein's identification of the two 4-dim spaces, i.e. whereas one can of course start in the first space by using ray/point or axis/plane coordinates to introduce antisymmetric line coordinates there, he identifies the second space again with a metric space which is relevant to physical observations, too. As we don't want to discuss related aspects and problems in detail in this section, but only want to report the approaches, we close here and address in the next subsection 2.6. one more common rep which yields some insight into the origin of the appearance of the Lie algebras su(2)$\oplus$i su(2) and su(2)$\oplus$su(2). Please note right here, that the derivation is valid also from the classical viewpoint, and that – in contrary to common belief – there is no need at all to discuss this rep solely in terms of 'quantum' theory or a 'quantum' interpretation although this framework has been developed there due to the common use of the theory of Lie groups and Lie algebras[48].

2.6. Feynman Rules and Spin

A lot of work has been done on 'quantum' reps and concepts to treat relativity on the quantum level. Especially Valeriy has published a lot of great work on various aspects of the different rep theories, so we feel free to skip most of the discussion here. Instead, we focus in this subsection on the overlapping conceptual aspects which are relevant for our later discussion here, and as such, feeling guided by Valeriy's papers [14], [15], [16], and [17], we want to refer here only to Weinberg's presentations [38], [46], and [47] on spin reps. Why?

On the one hand, they give a strong hint with respect to the identification of the two su(2) algebras[49] related to his 'infinitesimal Lorentz transformation',

$$\Lambda^{\mu}{}_{\nu} = \delta^{\mu}{}_{\nu} + \omega^{\mu}{}_{\nu}\,, \qquad \omega_{\mu\nu} = -\omega_{\nu\mu} \tag{13}$$

as given by [38], eqns. (2.18) and (2.19), and 'quantum' reps in this Hamiltonian approach[50]. His 'proper homogeneous orthochronous Lorentz transformations' were de-

---

[46]The various possibilities can be 'counted' phenomenologically e.g. in the Dynkin diagram.

[47]As of today, we cannot verify whether Minkowski used the same underlying geometrical concept while discussing the quadric $x^2 + y^2 + z^2 - t^2 = 0$.

[48]Or as hostile researchers named it: 'Gruppenpest'.

[49]More precise: with respect to the association of the skew 4×4 matrix rep of real transformation parameters $\omega_{\mu\nu}$ and the differential reps of the operators by means of two su(2) algebra reps.

[50]Moreover, it seems to be the major source of [42], ch. 1, appendix III, cited in our first example above.

fined[51] by [38], eq. (2.1),

$$x^{\mu} = \Lambda^{\mu}{}_{\nu} x^{\nu} \,, \quad g_{\mu\nu} \Lambda^{\mu}{}_{\lambda} \Lambda^{\nu}{}_{\rho} = g_{\lambda\rho}$$
$$\det \Lambda = 1 \,, \quad \Lambda^{0}{}_{0} > 0 \,, \tag{14}$$

which can be understood as well as a covariance requirement of the 'metric tensor', $\Lambda^{\mu}{}_{\lambda} g_{\mu\nu} \Lambda^{\nu}{}_{\rho} = g_{\lambda\rho}$.

On the other hand, Weinberg identifies physics in terms of the electromagnetic field reps within his $2(2j+1)$-dim reps, here with $(1,0) \oplus (0,1)$ while $j = 1$. Because this physics is a direct and a really well-established and accepted showcase by identifying and relating the different rep concepts, we can use it later 'vice versa' to break the geometrical aspects down to 'known' rep theory[52].

Now, addressing the first aspects of 'infinitesimal Lorentz transformations' in eq. (13), Weinberg decomposes the transformation into a symmetric and antisymmetric part, $\delta^{\mu}{}_{\nu}$ and $\omega^{\mu}{}_{\nu}$. Whereas within the notion and terminology used nowadays in Lie theory, people usually attribute this to the exponential and 'infinitesimal' displacements, implicitly this comprises already the background of Lie's logarithmic mapping and the polar system, i.e. the metric properties of the coordinates. The feature addressed by Weinberg, however, is the antisymmetric part $\omega^{\mu\nu}$ which he denotes[53] as 'an infinitesimal 'six-vector'', and which he interprets by associating *unitary* operator reps ([38], eqns. (2.20) and (2.21)) according to

$$U[1+\omega] = 1 + \frac{i}{2} J_{\mu\nu} \omega^{\mu\nu} \,. \tag{15}$$

There, due to having introduced the additional complexification i, he extracts the skew and hermitean operators $J_{\mu\nu}$ which he decomposes into two operator triples ('Hermitean three-vectors')[54],

$$J_i = \frac{1}{2} \epsilon_{ijk} J_{jk} \,, \quad K_i = J_{i0} = -J_{0i} \,, \tag{16}$$

i.e. in the spirit of requiring *unitary* reps, he sticks to *Hermitean* generator triples J_i and K_j. Please note, that at this point this rep theory maps Hermitean generators by using the mapping $\exp i(\cdot)$, i.e. by using an additional i to map Hermitean operators to unitary reps of the group, contrary to standard use in mathematics where it is easier to map anti-Hermitean operators by exp, thus avoiding excessive phases i without specific meaning[55].

---

[51] Compared to our picture, this Ansatz describes a subset of collineations whereas correlations are handled only implicitly here in the context of the adjoint rep, or conjugation, raising and lowering indices, or in some implicit aspects of invariant theory.

[52] ... although, in consequence, we'll see how the need arises to modify some of those reps and interpretations.

[53] Throughout his paper series cited above, the notation occurs twice without any comment on the background. So it is not clear whether this 'six-vector' is named according to counting just the six degrees of freedom of the skew 4×4 transformation rep $\omega_{\mu\nu}$ or as a reference to the old notion 'six-vector' as used e.g. by Minkowski and other people.

[54] Whereas Wigner [45] argues with the Poincaré group and relativistic angular momentum $M_{\mu\nu}$, Joos decomposes $M_{\mu\nu}$ into two operator triples $\vec{M}$ and $\vec{N}$ [43]. With respect to a quantum-mechanical interpretation, Joos associates $\vec{M}$ with operator reps of angular momentum, and the triple $\vec{N} = (M_{01}, M_{02}, M_{03})$ with a 'barycenter of energy' [43], sec. 2.1, p. 69.

[55] Therefore, we've included some remarks in sec. 4.2. to achieve a more transparent approach to reps of the algebras involved by avoiding superfluous phases, and moreover to attach directly to Helgason's work [5], [6], and benefit from its enormous power.

It is especially the requirements of *unitary* reps due to comparison within eq. (14) of the finite Lorentz transformations which induces signs and phases like e.g. in [38], eq. (2.26) of the 'boost' commutator.

Once more in other words: We are acting with unitary (or 'compact') reps on complex rep spaces, and it is the 4×4 skew matrix rep Λ in eq. (14) acting as a collineation on space-time points – this time $\omega_{\mu\nu}$ if we associate the parameters with physics – which is the origin of a certain model or rep theory. In Weinberg's case, he has chosen an su(2) based approach with complex coefficients to reflect the skewness and dim 6 of the generator algebra. Whereas this 'decomposition' of so(4)=so(3)$\oplus$so(3) fulfils the algebraic and formal requirements, and yields a formal 4×4 skew matrix rep, throughout his papers [38], [46], [47], there is only one strict and physically accepted identification when treating the source-free electromagnetic field, i.e. in the 6-dim case where the adjoint Lie algebra rep(s) map directly to a special linear Complex using Klein coordinates[56]. As we'll see below, this causes a lot of trouble with the correct identification of phases due to a lot of additional is appearing throughout rep (and as such perturbation) theory.

He then discusses some operator algebra of J_i and K_j, and commutators similar to reductive algebra commutators, and he decouples them by linear combinations $\vec{A} = \frac{1}{2}(\vec{J} \pm i\vec{K})$ into two commuting su(2) (or A_1) algebras,

$$\begin{aligned} [A_i, A_j] &= i\epsilon_{ijk} A_k \,, \qquad [B_i, B_j] = i\epsilon_{ijk} B_k \,, \\ [A_i, B_j] &= 0 \,. \end{aligned} \tag{17}$$

Decades ago, we've covered that formalism by our effective SU(4) approach[57], because right from the Dynkin it can be seen that A_3 comprises $A_1 \otimes A_1$ [59]. Further reasoning lead us to consider and discuss linear rep theory and the inclusion of axial charges [60], [61], [62], [63], to find suitable irreps. This led us later to the non-compact group SU$*$(4) (which is isomorphic to SL(2,) and describes projective transformations of quaternions) as well as to some associated real forms of SO(6,) [64], [2], [3], but details are beyond scope here.

For now, in order to stay in touch with Weinberg's notation and the reps within perturbation theory, it is important to remember the first 'overall' i which has been introduced into the exponential mapping (15) in order to have unitary operator reps U expressed by means of the skew, but hermitean 'operators' $J_{\mu\nu}$, despite of representing a non-compact group. Then, we have to remember the second, relative i between the 3-dim operator sets $\vec{J}$ and $\vec{K}$ in order to obtain commuting sets $\vec{A}$ and $\vec{J}$ in eq. (17). Whereas formally this is almost straightforward, we have postponed the discussion of problems to a later stage when we have to identify the operators physically and when we have to take care of compact vs. non-compact group transformations despite the unitary character of the reps. We'll discuss a typical example later in terms of the field components $\vec{E} \pm i\vec{B}$ by Plücker versus Klein coordinates, and their relation to the 3$\otimes$3 $\pm$-handed fundamental Complexe introduced by Klein which have been 'reused' and reinterpreted in both rela-

[56] For our purposes and the following reasoning, this is sufficient, especially since the skew transformation reps of the Complex and the generator set of the Lie algebra coincide.

[57] However, covering at first compact SU(2)$\otimes$SU(2) from Chiral Symmetry, or Chiral Dynamics, by compact SU(4) [59].

tivistic theories as well as chiral SU(2)×SU(2) symmetry discussions during the late 1960s.

The second useful aspect is Weinberg's identification of the electromagnetic field within the $2(2j+1)$-dim reps $(j,0)\oplus(0,j)$. Choosing $j=1$, he associates the field components within the 6-dim rep $(1,0)\oplus(0,1)$, and discusses various aspects throughout his paper series (see [38], sec. VII, for general properties and e.g. [46] (especially the end of sec. IV and of sec. XI), and [47] sec. II ff.). Finally, he associates certain field combinations $\vec{E}\pm i\vec{B}$ with the rep $(1,0)\oplus(0,1)$. However, as we've pointed out already a couple of times, we are discussing the invariance condition of the quaternary quadric and its related symmetry group SO(3,1) here in terms of *unitary* reps, so Weinberg's additional requirement of *unitary* reps Λ by eq. (14) and the additional phases i introduce additional complications. We postpone this phase discussion to sec. 4.2..

For our considerations later, the second aspect of Weinberg's 'idea' has already been represented in Valeriy's overviews appropriately[58] ([14], [15], [16], and [17]) by rearranging the field components $\vec{E}$ and $\vec{B}$ into a 'six-spinor' Ψ according to

$$\Psi = \begin{pmatrix} \chi \\ \phi \end{pmatrix} \tag{18}$$

(see [15], eqns. (1) and (2), or [17], eqns. (2.1) and (2.2)), and by acting with transformations $\exp \pm\theta\,\hat{\vec{p}}\cdot\hat{\vec{J}}$ on the two 3-dim subcomponents of Ψ. For $S=1$, the operators $\hat{\vec{J}}$, of course, have to be associated in this approach to the adjoint rep of so(3), or su(2), and the subcomponents take the form $\chi=\vec{E}+i\vec{H}$, $\phi=-\vec{E}+i\vec{H}=-(\vec{E}-i\vec{H})$. Some more aspects discussed in [15] are helpful to guide and compare the different reps:

- Valeriy mentions the fact that, according to Weinberg, [15], eq. (3), Ψ can be transformed to the equations for left- and right-circularly polarized radiation in the case $S=1$ of massless fields[59]. See also his discussion in [17], eqns. (2.24) and (2.25), where $\vec{E}$ and $\vec{H}$ are Pauli vectors, and we may stress Lie transfer [12] to locate the six components in line space.

- The 4-vector Lagrangean density is constructed by means of the field tensor and its dual ([15], eq. (6)), and by using a 'current' j^μ which has to be discussed separately.

- The energy-momentum tensor is given by the vector Lagrangean and the 6-dim rep Ψ, only, and with respect to the current tensor, there exist Sudbery's 'duality rotations'

[58]Although – based on the idea of two fundamental SU(2)s and 'spinors', obviously induced by 'quantum' notion – Weinberg has considered 'any spin'. For our purposes, especially because the free electromagnetic field is the only case of this spin approach without additional assumptions when relating physics and spin/spinor reps, we prefer to fix $S=1$ in order to compare. We've discussed the fundamental spinor rep and its origin in PG sufficiently in [12], [44] whereas we postpone the discussion of 'higher spins' to PG, e.g. in the case of twisted cubics and third order reps.

[59]Whereas in the 2(2S+1)-formalism this relates to the identification of χ and ϕ with electromagnetic field components as given above, in our picture when considering the geometrical base objects this relates to Klein's fundamental Complexe of P^5 [10] with 3⊕3 handedness [10], or [30], p. 99, with respect to involutions of linear Complexe.

which mix the field strength tensor and its dual ([15], eq. (15))[60]. In the source-free case, the action is invariant under this 'duality transformation' ([65], eqns. (36) and (37)), and by Noether's theorem one obtains the conservation laws for a symmetric energy-momentum tensor ([65], p. 7).

- One can find new 'gauge' transformations $F_{\mu\nu} \longrightarrow F_{\mu\nu} + \partial_\nu \Lambda_\mu - \partial_\mu \Lambda_\nu$ ([15], eq. (31)), i.e. by adding an antisymmetric tensor $A_{[\mu\nu]} = \partial_\nu \Lambda_\mu - \partial_\mu \Lambda_\nu$. Later, this addition of a second Complex rep $A_{[\mu\nu]}$ yields[61] the discussion of pencils of linear Complexe, finding lines within the pencil or line sets of quadratic Complexe, or analyzing Congruences.

- In [16], eq. (3), the decomposition of Ψ is associated with vector and pseudo-vector parts. Here, we want to mention only the geometrical limit used in sec. 3.1., or discussed in depth in [30], §16, which decomposes the antisymmetric line coordinates *only after* having performed the affine or Euclidean limit. So it is the transition from homogeneous coordinates to affine or Euclidean coordinates, which introduces the notion of vector and pseudo-vector, or of parity, which is subsidiary in homogeneous and projective theories because projective transformations in general mix the coordinates. There is no 'absolute element' yet, or a distinguished coordinate x_0 or its dual plane to introduce parity, or a linear orientation. In other words, as long as we perform PG, we benefit from pretty 'symmetric' line coordinates (because twofold in point reps) with respect to the order of composition, i.e. using *two* homogeneous point coordinates and their antisymmetry when exchanging the points. It is only *after* having fixed an absolute plane (or even after having defined an absolute circle within this plane) that we have to take care of polar and axial vectors with their 'new' asymmetry of being linear or quadratic in point coordinates because now the absolute element is fixed and absorbed in the coordinate definition (i.e. by formally setting $x_0 = 1$ in the affine case, or by switching to Euclidean coordinates which 'absorb' the linear singularity $x_0 = 0$ in the fraction by their very definition).

- [17], eq. (3.31), defines a 'new magnetic momentum vector' which is traced back to another skew tensor $\Sigma_{(\mu\nu)}$, and the parameter triples $\vec{\Theta}$, $\vec{\Xi}$ of $\hat{H}$ are derived from the three 0- or 4-components on the one hand, $\vec{\Theta} = (\Sigma_{(41)}, \Sigma_{(42)}, \Sigma_{(43)})$, and from $\vec{\Xi} = (\Sigma_{(23)}, \Sigma_{(31)}, \Sigma_{(12)})$ on the other hand. This time, this skew structure is constructed by means of the Pauli-Lubanski-vector $W_\mu(\vec{p})$, and $\Sigma_{(\mu\nu)}$ reads as $\Sigma_{(\mu\nu)}(\vec{p}) = \frac{1}{2}\left(W_\mu(\vec{p})W_\nu(\vec{p}) - W_\nu(\vec{p})W_\mu(\vec{p})\right)$

We'll address both major aspects from above – the Lie algebra and the 6-dim rep of electromagnetic field – later after having established more geometrical and analytical background in sec. 4.2.. For now, by summarizing the standard reps, it is important to note that a generalization in these days is usually achieved by generalizing the 'quantum numbers' of the reps associated to the spin algebras. However, we'll see later that identifying the

[60] In other words, we have to discuss the position of two dual lines given by a special linear Complex and represented by $F_{\mu\nu}$ and its conjugate, and we thus enter the geometrical discussion of Congruences/ray systems and of conjugation with respect to polar and null systems, in P^3 as well as in P^5.

[61] See also the related aspects discussed in sec. 4.4..

6-dim structure Ψ geometrically leads back to PG and yields generalizations in multi-linear algebra and algebraic geometry, and it is thus more interesting for us with respect to physics to follow the geometrical trail, parts of which we've addressed here and in the cited publications, than generalizing the rep dimensions of su(2) reps, or switching to irreps with higher multiplicities and 'quantum numbers'. We have given examples in the physics of the pion-nucleon-delta system [59] which due to use of SU(4) can be traced back to spatial geometry [28], [50], and there are lots of further examples where current formalisms may be derived easily from classical geometry of P^3 and P^5.

3. Standards Revisited

Throughout sec. 2., we've presented the major foundations of the usual and known formalism to treat relativity (without claiming completeness of the presentation!). And so far, we have introduced only few comments with respect to this description of physics by means of the point picture during this overview on 'standards'. However, before revisiting some central aspects of this rep theory (expressed in point coordinates) and working the switches as we've announced in the beginning, in order to be not misunderstood or misinterpreted, we want to emphasize that we do *NOT* want to argue or even reject the wealth of experimental evidence so far on relativity confirming Einstein's physical picture! In contrary, we argue that switching to a description and treatment in terms of PG and Complex geometry yields the observed effects more naturally, and that we may realize common textbook notation just by switching to antisymmetric point or plane reps of lines (or Complexe) in 3-space at any time. Our following arguments relate to the nowadays usual rep theory in terms of point reps, the 4-vector calculus and differential geometry which cover only limited aspects of a full-fledged physical description.

From the viewpoint of projective geometry, it is well-known that a point description without plane reps is far from complete to describe even 3-space and its physics because it lacks the oldest known transfer principle(s): duality, or reciprocity. Moreover, in order to introduce the metric or an absolute quadric, one usually needs additional assumptions e.g. on absolute elements or on the polar system, which are typically put in by additional requirements, and they do not result a priori from geometry like in the case of projective geometry when using the Cayley-Klein mechanism. However, here it is not our line of reasoning that switching to line and Complex geometry cures *all* the open questions but first of all, we see this geometry as a very efficient ordering and rep scheme to collect well-known point descriptions, and which allows to simplify, complete and generalize then geometrically. Moreover, a description of the observation process is automatically included if we think of optical lines ('rays') connecting observer and observation in order to represent physical observations [24], [28]. In linear scenarios, this concept comprises or even requires classical projective geometry; in non-linear scenarios, one has to include objects of higher orders[62] and classes as well as non-linear transformations like in advanced geometry or in algebraic geometry.

[62] An example we've mentioned already a couple of times, is the coupling of the mesons to the nucleon-delta systems where it is the 20-dim, 3rd order symmetric rep which serves as a basis for vector *and* axial currents which we have shown to be necessary in order to calculate the axial coupling constant. Other examples like skew cubics, Plücker's [26] or Kummer's (see e.g. [66] or [32]), or K3 surfaces are well-known as well.

To approach an alternative – and more general – description without getting lost in the enormous world of projective, or algebraic geometry, we use the preceding section as a certain guideline to transfer the aspects here in sec. 3. step by step and subsume them into the more general framework. In sec. 4. – after having switched to the more general framework – we'll discuss some more aspects, however, without claiming completeness of the outline.

3.1. Minkowski Revisited

As a first example, we may rewrite the linear Complex $\sum a_\alpha p_\alpha$, $1 \leq \alpha \leq 6$, from above in terms of homogeneous point coordinates with x_μ and y_μ, $1 \leq \mu \leq 4$, and we associate the P^5 coordinates p_α to Plücker's line coordinates in P^3 by[63]

$$\begin{array}{l} p_1 \longleftrightarrow p_{23},\ p_2 \longleftrightarrow p_{31},\ p_3 \longleftrightarrow p_{12}, \\ p_4 \longleftrightarrow p_{14},\ p_5 \longleftrightarrow p_{24},\ p_6 \longleftrightarrow p_{34} \end{array} \tag{19}$$

A general linear Complex $\mathfrak{C}$ in the ray representation may then be written according to $\mathfrak{C} = \sum a_\alpha p_\alpha \longrightarrow \sum f_{\mu\nu} p_{\mu\nu}$ with line coordinates $p_{\mu\nu} = x_\mu y_\nu - x_\nu y_\mu$, $1 \leq \mu, \nu \leq 4$, and $a_\alpha \in$, i.e. by

$$\begin{aligned} \mathfrak{C} = &\ f_{23}(x_2 y_3 - x_3 y_2) + f_{31}(x_3 y_1 - x_1 y_3) \\ + &\ f_{12}(x_1 y_2 - x_2 y_1) \\ + &\ f_{14}(x_1 y_4 - x_4 y_1) + f_{24}(x_2 y_4 - x_4 y_2) \\ + &\ f_{34}(x_3 y_4 - x_4 y_3) \end{aligned} \tag{20}$$

which we can compare to eqns. (5) and (6). It is evident that eq. (5) above (up to the additional phase i of the fourth component which has been introduced by Minkowski artificially into point space to change the signature of the quadric) yields exactly the very definition of a line Complex using ray coordinates[64] if we replace $y \longleftrightarrow u$. Therefore, Minkowski's claim of form invariance when rewriting the result *after* Lorentz transformations in terms of primed components like in eq. (6) reflects nothing but the requirement of the invariance of the respective linear Complex $\mathfrak{C}$. So according to the components (or the coordinates) $f_{\alpha\beta}$

---

[63]In order to compare directly to Minkowski's notation, eq. (19) shows the mapping due to Staude [67], §59.1, p. 322, which differs from our usual notation in e.g. [24] or later, where we typically use the point coordinate x_0 instead of x_4. This change introduces a sign in three of the line components p_4, p_5, and p_6 by reordering the indices of the antisymmetric line rep in ray coordinates. Whereas the Plücker condition itself is not altered, one has to keep track when working with Complex parameters, signatures and polar conditions in P^5, etc., and related linear coordinate sets. Note also, that Klein sometimes varies his notation at all in that he uses also other coordinates than x_0 or x_4 as linear absolute elements. This notation in terms of P^5 coordinates p_α also introduces the notation of two coordinate triples, usually denoted by $k = 1, 2, 3$ and $\overline{k} = 4, 5, 6$ which comprise 3-dim forces and moments in the force picture, or relate to axial and polar 'vectors' in the point picture. Please note also the connection to the decomposition of $M_{\mu\nu}$ into $\vec{M}$ and $\vec{N}$ as discussed in [43], or [42], see also sec. 2.1..

[64]The linear Complex expressed in Plücker coordinates with *real* coefficients/coordinates corresponds to an SO(3,3) invariant quadric and to an associated SO(3,1) invariant quadric in quaternary ('4-dim') point coordinates [24], sec. 1.6. So with respect to introducing additional phases $'i'$ one should be careful about interpretation and use of coordinates! Moreover, we are obviously involved once more in a phase discussion like in the case of Weinberg's 'unitary reps' of the Lorentz group which we've discussed in sec. 2.6.. This emphasizes the need 'to straighten' and unify this rep theory as we'll try to do in sec. 4.2..

of the linear Complex $\mathfrak{C}$, we have to consider transformations which leave this Complex invariant. This can be accomplished by transformations leaving $\mathfrak{C}$ invariant as a whole e.g. by interchanging the line coordinates and thus constraining the components $f_{\alpha\beta}$, or by leaving line coordinates individually invariant, etc. Minkowski's requirement thus has to be associated with the second, restricted case which results in the stricter constraints $f_{\alpha\beta} \longrightarrow f'_{\alpha\beta}$, and $p_{\alpha\beta} \longrightarrow p'_{\alpha\beta}$.

In all cases, we may use Klein's 'Erlanger Programm' as the appropriate framework (or tool-set) for continuous as well as for discrete symmetries. One example is $r + \sigma = 0$, r and σ denoting inhomogeneous Plücker coordinates [12] [13], which yields Lie transfer when stabilizing the linear Complex $r + \sigma = 0$. In terms of homogeneous coordinates ([27] p. 283), $r + \sigma = 0$ can be represented by $r + \sigma = \frac{p_{41}}{p_{43}} + \frac{p_{23}}{p_{43}} = \frac{p_{01}}{p_{03}} + \frac{p_{23}}{p_{03}} = 0$ which – besides the special rôle of $p_{01} + p_{23}$ – also emphasizes the special rôle of the individual line coordinate $p_{43} = p_{03}$ with respect to the definition of inhomogeneous line coordinates $(r, s, \rho, \sigma, \eta)$. This can be seen using Lie's rep of inhomogeneous versus homogeneous line coordinates ([27], p. 283),

$$\begin{array}{rlcrl} r = \frac{p_{41}}{p_{43}} = \frac{p_{01}}{p_{03}} & , & s = \frac{p_{42}}{p_{43}} = \frac{p_{02}}{p_{03}} \, , \\ \rho = -\frac{p_{31}}{p_{43}} = -\frac{p_{31}}{p_{03}} & , & \sigma = \frac{p_{23}}{p_{43}} = \frac{p_{23}}{p_{03}} \, , \\ \eta = \frac{p_{12}}{p_{43}} = \frac{p_{12}}{p_{03}} & , & \end{array} \tag{21}$$

where we've expressed the coordinates in addition by using the 0-component-notation instead of 4-components. It is noteworthy, that the six coordinates $p_{\alpha\beta}$ transform homogeneously whereas the five coordinates $(r, \rho, s, \sigma, \eta)$ undergo projective transformations if space itself is transformed by projective transformations (see [27], p. 285, Satz 5). Moreover, eq. (21) visualizes perfectly that the line coordinates involving 0-components of point reps determine the slopes r and s of the lines in terms of inhomogeneous coordinates, i.e. we may associate $p_{0\alpha}$ (or r and s) with the direction (information) of the line.

Lie gives the corresponding definitions in terms of point coordinates by

$$\begin{array}{rcl} p_{12} = x'y'' - y'x'' & \longrightarrow & p_{12} = x_1y_2 - x_2y_1 \\ p_{23} = y'z'' - z'y'' & \longrightarrow & p_{23} = x_2y_3 - x_3y_2 \\ p_{31} = z'x'' - x'z'' & \longrightarrow & p_{31} = x_3y_1 - x_1y_3 \\ p_{41} = x'' - x' & \longrightarrow & p_{01} = y_1 - x_1 = x_0y_1 - x_1y_0 \\ p_{42} = y'' - y' & \longrightarrow & p_{02} = y_2 - x_2 = x_0y_2 - x_2y_0 \\ p_{43} = z'' - z' & \longrightarrow & p_{03} = y_3 - x_3 = x_0y_3 - x_3y_0 \, . \end{array}$$

Whereas on the lhs we've used Lie original notation, we've re-expressed the rhs using $()' \longrightarrow x$ and $()'' \longrightarrow y$ with quaternary homogeneous coordinates of the points x and y, and we've used 0-components with $x_0 = y_0 = 1$ like in the affine picture. So on the one hand, it is evident that line coordinates $p_{\alpha\beta}$ comprise Minkowski's transformation theory. On the other hand, triples $\sim p_{0j}$ seem to be related to 'polar vectors' whereas triples $\sim (p_{12}, p_{23}, p_{31})$ seem related to 'axial vectors', or moments. This decomposition $3 \oplus 3$ of the 6-dim homogeneous line rep, however, occurs only AFTER having performed the transition from homogeneous to affine, or Euclidean, coordinates. So with respect to discussions of 'parity' symmetry, they have to be applied (or located) in rep theories only AFTER having performed this transition, and they are misleading in homogeneous rep theory. This can be seen easily if we exchange opposite lines (or edges) of the fundamental tetrahedron by simple coordinate exchange without altering the geometry. Moreover, linear combinations of the edges like $(p_{01} \pm p_{23})$, $(p_{02} \pm p_{31})$, or $(p_{03} \pm p_{12})$ which we've used with respect to Congruences or Onsager theory (see e.g. [44], sec. 4.3), remain invariant by themselves, and thus yield additional symmetries.

Earlier, we've re-expressed Lie's transfer principle already in terms of Pauli matrices and thus attached the usual quantum notion to line geometry of P^3 [12] [44]. Moreover, we've thus founded the constructing scheme of Clifford algebras $\sigma^{(1)} \otimes \sigma^{(2)} \otimes \ldots \sigma^{(n)}$ to projective constructions based on lines (and Complexe) instead of point notion and Euclidean concepts as usual.

In terms of transformation or group theory, we are thus at the very origin (or at least close to the heart) of symplectic transformations and of quadratic constraints in P^3 which originate e.g. from the Plücker-Klein quadric of P^5.

So the major guideline of our reasoning within the next sections is the idea to switch back to line and Complex geometry in order to catch the complete background instead of discussing only certain aspects of point reps and their differential geometry. This automatically reestablishes the almost lost connection of formal and technical algebra to physics in terms of null systems and forces, and thus interconnects experiments of the real world and well-known geometry. As such, recalling the two *different* approaches by Grassmann and Plücker to higher dimensional reps of geometry, we want to underpin Lie's summary [27], p. 274/275, and emphasize Plücker's achievements, i.e. to express and understand higher-dimensional 'spaces' and their respective coordinate reps by appropriate n-dim base elements of the respective geometry. Practically, this is established by transfer principles connecting the different reps while changing the coordinate *interpretations* and usually the dimension of the reps, too. A simple example can be taken from Lie transfer above, or already from Plücker's observation [68] when connecting line reps with three types of associated second order surfaces which naturally explains otherwise mystified rep relations like the wave-particle dualism. We think, this identification and reasoning as well as the use of transfer principles is much closer to physical reasoning than using n-dim *point* sets in abstract spaces as exercised nowadays e.g. in string or gauge theories.

3.2. Line and Complex Geometry Revisited

To depart from this introductory example in sec. 3.1., featuring coordinate properties and the assembly of lines and linear Complexe, we now want to collect at least few of the technical debris floating around and relate them briefly to known geometrical concepts, whereas in the next section – based on projective and especially on line and Complex geometry – we want to catch up with the guiding topic of this book, with the future of relativity, however, based on old physical principles and geometrical roots.

So after having identified Minkowski's 'six-vector' (or 'vector of second kind') geometrically in sec. 3.1. with a special line set in 3-dim space (a special linear Complex, or – equivalently – with a special physical system, a null system), it is obvious that people have introduced a special representation theory by means of 4-vector calculus (and related tensor calculus) to treat a small (sub-)aspect of line geometry by an independent, different and a priori disconnected formalism. Geometry and the ray representation of lines in terms of homogeneous fourfold point coordinates have been replaced by 4-vector calculus, tensor algebra and differential geometry, and we are faced with 'special relativity' comprising implicit geometrical assumptions and singularities which cannot be resolved from within the 4-vector rep theory without additional and non-trivial assumptions. We've stressed in sec. 2.6. already Weinberg's rep approach and emphasized the case $j = 1$ because we are going to compare this with line geometry and the case of a special linear Complex. However, as we have discussed already above, it is mandatory to take care with respect to *the interpretation* of the point coordinates, and in the same manner, we have discussed Minkowski's paper and central aspects of our reasoning in more detail elsewhere [10].

In another paper [25], sec. 3.2, we have given analytic proofs of the invariance properties even of individual line coordinates $p_{\mu\nu}$ which comprise the 'Lorentz transformations' in point space known nowadays from textbooks. As such, by using homogeneous point coordinates, the usual Lorentz transformation 'leaving x and y invariant, and mixing z and t' corresponds to the two individual invariance properties of the line coordinates p_{12} and p_{03} (or p_{34}) which according to their definition above by $p_{\mu\nu} = x_\mu y_\nu - x_\nu y_\mu$ may be represented as 2×2-determinants, too[65]. However, the 2×2-determinants – besides exposing further invariance properties – do not change under identity transformations or hyperbolic mixture within the 2×2 blocks [25]. And because geometrically the two line coordinates p_{12} and p_{03} build two opposite, non-intersecting sides of the fundamental coordinate tetrahedron, each invariance corresponds to an invariant coordinate in line geometry, or to the invariance of a certain Congruence[66] in Complex geometry. As such, a switch to rep theory in terms of line geometry yields the additional benefit to avoid artificial invariance requirements in point space and yields an easier and more evident treatment [44], sec. 4.C. It is much easier to discuss invariance properties such as $(p_{12}, p_{03}) \longrightarrow (p'_{12}, p'_{03})$ in terms of line coordinates than figuring out the antisymmetric (and invariant) rep parts out of an arbitrary point rep tensor (or event point space dependent field reps) from a given point space object. Last not least, according to their very construction, the line reps are the reps most close to *linear* rep theory, and we may use Hesse transfer [29] and binary forms [31] to treat various point or spinor reps, and various related aspects of PG. Although Poincaré's/Minkowski's 4-vector calculus yields a symbolism and tensorial rules on how to proceed and calculate, it is a symbolism attached to the special identifications above, and although it is capable to represent special geometrical aspects by the 4-dim rep, it is not capable to cover the geometrical aspects of null systems and Complex geometry of P^5 completely. It is stuck in the affine picture using point-based algebra reps and an appropriately related formalism (and symbolism). As such we think, moving back to line and Complex geometry yields the profound background especially when generalizing later to higher orders or classes, and to transfer principles and non-linear behaviour, i.e. to the general aspects of projective and algebraic geometry. So the usual discussion of 'Lorentz transformations' or special relativity can be subsumed without loss of generality under the invariance requirement of linear Complexe, and as such, it is part of Complex geometry and projective geometry which provides

[65] See eq. (36) in sec. 4.3. on transformations later on. In this section, we provide the full discussion because up to this section, we'll have developed the context to discuss first aspects of Complex geometry beyond the old formalism and phenomenology.

[66] More precise: a 'ray system' ('Strahlensystem erster Ordnung und erster Classe').

a much larger mathematical framework. So a necessary generalization – or the discussion of the future of relativity – needs this more profound basis to overcome and supersede the concepts of affine and differential geometry.

3.3. Affine Picture Revisited

It is also evident from the physical viewpoint, that one shouldn't naively put $c = 1$ (and neglect the physical dimension of this velocity!) in order to obtain a 'time' and some 'space-with-time' mixture, or nowadays often for short 'space-time'. c has physical dimensionality, and if we want to interpret c phenomenologically as an upper bound on velocities (which paves the path to hyperbolic models), the only consequence we see is that if we interpret[67] the three space coordinates x_i in the same manner by $x_i = v_i t$, one can establish *for the same time interval* t a comparison of the velocities v_i with their upper bound c. However, in order to compare, one has to compare quantities of the same physical dimension! We have discussed some of those aspects in [24], sec. 3.5, and derived the Lorentz factor.

Moreover, we've pointed out [3], [50], that we thus catch up with Smorodinskij's beautiful treatment of velocities and the geometry of velocity space [70]. What is open today for discussion is a certain restorative return on the occurrence and interpretation of second order surfaces which Study has summarized [69] in relation with normal congruences and focal theory although they are deeply related to the underlying concept of Hamilton's formalism. This can be addressed here only very briefly but has to be worked out in detail later on due to its importance with respect to its universal applications throughout most fields of physics and Lie theory formalizing invariant theory and symmetry transformations. Nevertheless, we've discussed the 'corrected' interpretation of 4-dim 'momentum' reps already in terms of plane coordinates of the Hesse form [50], i.e. by '4-dim' normals, and it is the picture of normals of planes that suggests to put focus on geometry and the related tetrahedral Complex of surfaces. Last not least, considering ct as '0-component' with respect to Lorentz transformations, this coordinate is *mixed* with one of the spatial coordinates related to the axis or direction of motion, usually $z = x_3$. So we do *not* encounter an affine transformation (which should keep x_0 invariant), but we need projective transformations and PG to access the picture[68] of an invariant normal plane x, y and mixture of x_0 and $z = x_3$ with respect to the direction of motion. There is even the possibility to introduce new coordinates with quadratic dependence [24] and switch to a consistent formalism.

3.4. Coordinate Notation and Terminology

This forwards our discussion to Klein's paper [7] featuring the remarkable break in terms of the transition from 'ordinary projective geometry' of 3-space – according to Klein's use of ordinary homogeneous coordinates mostly from the viewpoint of an affine geometry – to the identification of four 'space-time coordinates' in the second part of [7] and by introducing five homogeneous coordinates x_β, $1 \leq \beta \leq 5$. Now, of course, it is formally

[67] With respect to the use of the Hamilton formalism [69], we feel free to do so at least with respect to equal time considerations and the usual physical reasoning.

[68] See sec. 4.1., and there especially figure 1!

and algebraically/analytically possible to start from four rectangular coordinates and by generalizing the *linear* affine approach like discussed in sec. 2.4., i.e. by thus introducing a *linear* absolute element $x_5 = 0$ into this space. However, this opens a lot of room to figure out the additional *implicit intrinsic* assumptions of these definitions and to discuss their relation to physics, the more as they impose further constraints and restrict objects and transformations.

Although we have already discarded to pursue this trail of using P^4 further, we nevertheless want to point to some historical but contemporary aspects of the coordinate notation t in conjunction with the three space coordinates x, y, and z in literature[69]. So to a certain extent, it is known from letters exchanged between Lorentz and Poincaré [71], that both discussed an additional quantity t' introduced into physical equations and descriptions, however, originally not connected to a physical 'time' identification. In addition, at the beginning of the 20th century, there existed a notation already in mathematical textbooks (see e.g. [32], [67], §47, or [57]) to describe the fourth component of *ordinary* homogeneous coordinates by the letter t, following x, y, and z. This seems to be the common notational convention of that time when using x, y, or z, instead of index notation, because on the other hand, the tetrahedral homogeneous coordinates were usually denoted by x_β, $1 \leq \beta \leq 4$, or $0 \leq \beta \leq 3$, dependent on the author. In Hudson's case [32], there is some more evidence of such a common notion because whereas in the first sections (see e.g. §2 or §10) he uses 'homogeneous coordinates' x, y, z, and t to denote individual point coordinates, beginning with §11 he explicitly argues to switch to tetrahedral coordinates which he also denotes by x, y, z, and t, and he requires to reinterpret this coordinate notation. Staude [67] treats the notation much stricter, compare e.g. §47 on ordinary homogeneous coordinates versus §§57 and 59 on tetrahedral homogeneous coordinates. Note with respect to the interpretation of 'coordinates' and 'masses' especially, that he recalls implicitly Möbius's barycentric ideas of weights and masses of such coordinate definitions!

So according to our current understanding, it is at least worth to keep track of the parallel but *different* notations in physics and mathematics, and take care of the coordinate interpretation and ambiguities. It is definitely *NOT* sufficient, to naively put $c = 1$, neglect dimensional considerations and hope for unique formal treatments in mathematics. So it is part of our ongoing work to separate the various, sometimes overlapping coordinate interpretations in detail, and we see projective geometry on the one hand as an appropriate tool to handle the different coordinate interpretations versus the related Euclidean and non-Euclidean geometries. On the other hand, it allows to consistently complete point descriptions of 3-space by plane descriptions and duality, or if we switch instead to the equivalent line representation (by lines and their conjugated lines) we may describe physics and comprise established formalism like the examples we'll discuss in the next section, or like in the case of gauge formulations [72].

---

[69] We are not enough qualified to discuss this aspect exhaustively, and moreover, here, we think it's not the place to discuss all the historical and especially the important sociological aspects of that time and Minkowski's community profoundly. Nevertheless, we want to mention these aspects, and we hope that some science historians or sociologists have interest to look deeper into these interesting relations and the historical development. As for us, today we can state only the occurrence of this notation in the three given book references before 1908/1910, but we have had no time so far to follow the occurrence of the coordinate notation t for the fourth ordinary homogeneous point coordinate back to its origin(s) in mathematical literature.

4. A Reinterpretation – Back to the Roots!

Throughout the preceding section, we've argued with respect to selected aspects and by various selected examples, only. However, as there exists a well-defined and well-established general context in terms of Lie's contact geometry [27], which – at least in its original form and its deep connection to physical descriptions – seems to be forgotten, we want to recall briefly this framework in the next subsection. Afterwards, we want to switch to this framework and invert the first two sections in that we want to arrange and integrate the aspects mentioned above into their appropriate places[70].

4.1. Historical Context and Null Systems

Lie, however, has pointed out (for details, see [27], ch. 6, §3) in the context of null systems that Pfaff's equations have ever played an important rôle. Using Euclidean coordinates, he summarizes the historical development and derives – using metric conditions in 3-space – that the most general infinitesimal motion in 3-space when expressed in terms of Euclidean coordinates x, y, and z is given by

$$\begin{aligned} \delta x = &\ (Bz - Cy + D)\ \delta t \\ \delta y = &\ (Cx - Az + E)\ \delta t \\ \delta z = &\ (Ay - Bx + G)\ \delta t\,. \end{aligned} \tag{22}$$

This reflects the theorem attributed to Euler (1775) and D'Alembert (1780) that every motion in 3-space can be replaced by appropriate infinitesimal rotations by the coordinate axes and appropriate translations along the axes. Later, it has been shown that such a description of motion can be replaced by a single screw (see e.g. Plücker [73], [74], Ball [75], Lie [27], Klein [33], §2, or Klein and Sommerfeld [21], [22]), and others.

In the next step, Lie argues that for the 1-dim set of planes

$$Ax + By + Cz = \mathrm{const} \tag{23}$$

due to the displacements given by eq. (22), the planes given by (23) undergo a constant increment on the lhs of their respective equations as can be seen immediately by $A\delta x + B\delta y + C\delta z = (AD + BE + CG)\delta t$. Thus, they are displaced in parallel[71] by the infinitesimal motions given by eq. (22).

For points moving *perpendicularly* to the planes (23), their displacements δx, δy, and δz, thus should be proportional to A, B, and C, respectively, and Lie thus obtains the equations

$$\begin{aligned} \rho A = &\ Bz - Cy + D \\ \rho B = &\ Cx - Az + E \\ \rho C = &\ Ay - Bx + G\,, \end{aligned} \tag{24}$$

[70] We've applied this background already to approach and discuss the treatment of gauge theories in terms of Complexe and P^5 geometry [72].

[71] This is why we may use *the normal* by means of Hesse's plane rep to describe translations, fluxes, currents by a 4-dim rep, or the '4-momentum', respectively, as long as we choose a common point 'to attach' the normal to the plane, i.e. a point being member of a common line. Note that besides the usual tangent constructions – thus founding on polar constraints and reciprocity as transfer principle intrinsically – we may also use null points and their associated linear Complexe e.g. in terms of skew matrix reps or Lie algebras.

in order to derive the important equation

$$AD + BE + CG = \rho\left(A^2 + B^2 + C^2\right) . \tag{25}$$

Whereas Lie uses this equation to determine the 'Factor ρ' and a certain 'reduced' geometry, we want to emphasize already here its connection to the Complex parameter[72]. Lie then emphasizes the reduction of this equation set from three to two equations by an appropriate choice of ρ, which in turn reduces the geometrical setup to a straight line being orthogonal to the plane(s) given by eq. (23). This straight line, as given by (24), is invariant and transformed (or translated) into itself under the transformations (22) as long as $A^2 + B^2 + C^2 \neq 0$.

In a third and last step to obtain the rep of a screw (or the related Complex), Lie rotates the coordinate system in a manner that the new $\hat{z}$-axis is identical with the invariant line from above. He obtains the simple analytical expression for the infinitesimal motion,

$$\delta x = -y\delta t, \delta y = x\delta t, \delta z = k\delta t\,, \tag{26}$$

however, it is important the understand that he had *to redefine* his 'time' coordinate δt in that he had to introduce a new δt by $C\delta t$, and a new parameter k by the original identification $k\delta t = G\delta t$. In other words, due to the transformation to let the invariant line of the motion coincide with the new $\hat{z}$-axis in order to obtain the screw rep and infinitesimally an orthogonal decomposition, he had 'to merge' the original constants C and G into the new 'time', and alter the constant in the z-coordinate from G to k. So according to our understanding, it is this picture with all its geometrical assumptions which yields the formal picture today of Lorentz transformations when leaving x and y invariant whereas z and t 'mix'. As such we propose to supersede it right from the beginning by Complex geometry, and to consider P^5 directly.

However, to enhance the historical remarks related to null systems, Lie has related the appropriate Pfaff equation by considering line elements orthogonal to the axis of motion. As per point of 3-space (x, y, z), there exist ∞^2 5-dim line elements[73] $(x, y, z, \mathrm{d}x : \mathrm{d}y : \mathrm{d}z)$, one obtains the equation

$$\delta x \cdot \mathrm{d}x + \delta y \cdot \mathrm{d}y + \delta z \cdot \mathrm{d}z = 0\,. \tag{27}$$

Using eq. (26), this results in the Pfaff equation

$$x\mathrm{d}y - y\mathrm{d}x + k\mathrm{d}z = 0\,, \tag{28}$$

[72] There exists also an 'infinitesimal version' by means of partial derivatives [27]. Moreover, it is evident that the lhs of eq. (25) corresponds to the Plücker condition, and that we have to treat the case $A^2 + B^2 + C^2 = 0$ by involving complex numbers, i.e. by invoking absolute elements and 'spinors' [12]. Last not least, rewriting (22) in terms of four (ordinary) homogeneous coordinates, the *common* δt on the rhs of the three equations reflects the same coordinate in the absolute plane. Replacing this related *space* coordinate by ct, and the respective coordinate displacements δx, δy, and δz by their appropriate counterparts in terms of velocities $v_i t$ yields an alternative approach to our derivation of the Lorentz factor [24], [25], and emphasizes our proposal to understand Lorentz transformations as a line (or Complex) counterpart when working with point coordinates in 3-space.

[73] We have used the old notation in order to preserve the dependence of the ratios, i.e. the three coordinate differentials $\mathrm{d}x$, $\mathrm{d}y$, and $\mathrm{d}z$ are *not* independent, and the expression $\mathrm{d}x : \mathrm{d}y : \mathrm{d}z$ symbolizes that we have to treat only *two* independent quantities, usually denoted as p and q, instead of three!

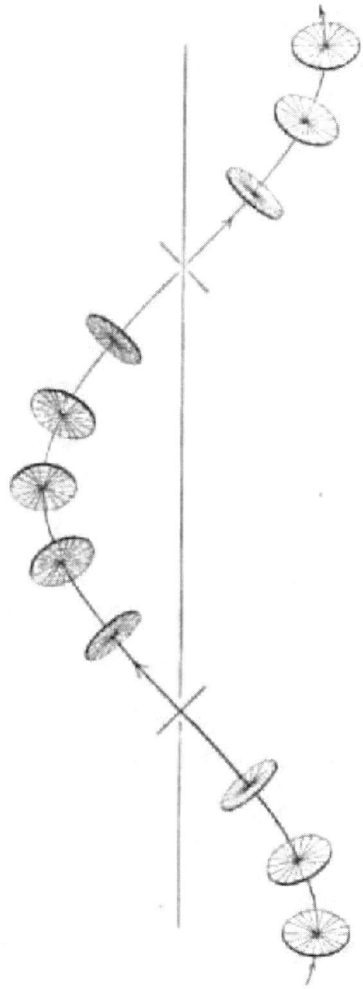

Figure 1. Figure of the screw motion, taken from [27], p. 210.

however, expressed in Lie's *NEW* coordinates which already are transformed with respect to the original coordinate system x, y, z, and thus implicitly respect the mixture emphasized above in the coordinates and parameters.

The last step to connect completely to linear Complexe can use Lie's proof (see [27], p. 210/211) that the integral curves of eq. (28) contain ∞^3 straight lines, i.e. integral lines as linear solutions, which results in the theorem ([27], p. 211, Satz 6) that related to every infinitesimal screw there exist ∞^3 straight lines whose points move *perpendicular* to the axis of motion. Per point of space (x, y, z), this results in a ∞^1 set of lines, i.e. a planar line pencil being orthogonal to the axis of motion. The corresponding Figure 1 is taken from [27], p. 210, figure 46.

Last not least, Lie presents the general Pfaff equation

$$\begin{aligned} &A(y\mathrm{d}z - z\mathrm{d}y) \\ &+B(z\mathrm{d}x - x\mathrm{d}z) \\ &+C(x\mathrm{d}y - y\mathrm{d}z) \\ &+D\mathrm{d}x + E\mathrm{d}y + G\mathrm{d}z \quad = 0\,, \end{aligned} \tag{29}$$

to describe the general infinitesimal motion in terms of perpendicular coordinates. So there exists always a set of ∞^3 straight lines as integral curves of this equation which can be seen by using Plücker's rep of the straight line given by a running coordinate z in terms of its (linear) projections $x = rz + \rho$ and $y = sz + \sigma$ which we've used [12], [44] to attach the spinor theory and 'quantum' notion in terms of the Pauli algebra, too. The line projections and intersections r, s, ρ, and σ, also known as Plücker coordinates of the straight line when completed by the fifth line coordinate $\eta = r\sigma - s\rho$, of the integral straight lines thus have to fulfil

$$A\sigma - B\rho + C(s\rho - r\sigma) + Dr + Es + G = 0\,, \tag{30}$$

which – in other words – constitute a line Complex. The Pfaff equation selects the planar pencil of perpendicular lines out of the lines through the point (x, y, z) of 3-space when using 4-dim line geometry as rep space. Note also, that we may understand such a pencil as a flat planar (or projected) 'version' of a cone, obtained e.g. by just shifting its vertex to the plane!

For later use in conjunction with the tetrahedral Complex, and for reference to [12] and [10], we want to emphasize the special selection $\eta = 0$. In this case, the definition yields $\eta = r\sigma - s\rho = 0$, so $r\sigma = s\rho$, or $\frac{r\sigma}{s\rho} = 1$. If we rescale e.g. s and ρ both by a parameter $\sqrt{\kappa}$, the ratio $\frac{r\sigma}{s\rho}$ reads as $\frac{r\sigma}{s\rho} = \kappa$. Now this ratio gains its interpretation and importance as anharmonic ratio later with respect to the parameter classifying tetrahedral Complexe.

Although Lie extends his discussion of null systems in [27] much further, as with respect to our discussion of special relativity here, we think that we've given enough aspects of the background as well as of the geometrical context of the usual rep of Lorentz transformations. So after having mentioned few of Lie's aspects of transformation properties in sec. 4.3. and discussed some more 'well-known' relativity axioms in sec. 4.4., we'll use sections 4. and 4.6. to assemble a guiding principle in terms of Complexe throughout topics in mechanics, optics, and relativity, and subsume typical formalism and 'axioms' into Complex geometry.

4.2. Feynman Rules and Spin Revisited

Before focussing on a further application of Complex geometry, we want to discuss two more aspects briefly – the rep theory of the 'quantum' picture mentioned in sec. 2.6. and the link to Onsager theory as discussed in [44] 4.3.

As we've extracted from Weinberg's Hamiltonian formalism and his modeling in sec. 2.6., one major issue is the consistent (and 'overall' unique) treatment of phases. Whereas from the practical viewpoint, we want to reduce phases and ambiguities within rep theory, from the mathematical viewpoint it would be nice to attach and use the enormous and extremely powerful machinery developed by Helgason [5], [6], and moreover, to work with consistent phases throughout rep theory especially when treating group chains, cosets, and algebras, and when identifying physics and measurements. So according to the notation in Helgason's books, we want to rewrite some operators in sec. 2.6. to use the exponential without additional 'i's, and extract a related operator rep. Meanwhile, we try to keep focus on physics and geometry.

As such, to recast Weinberg's operator structure in order to compare directly to Helgason, the first aspect is the exponential. Weinberg uses mappings $U \sim \exp \pm i\mathcal{O}$ of operators $\mathcal{O}$, $\mathcal{O}^+ = \mathcal{O}$, between Lie algebra and group. Formally, differentiation or series expansions of the exponential thus introduce 'i's into calculations. If, instead, we prefer to work directly with the operator action, there is no need to extract the i by hand, but we can use the exponential directly to map generators to group elements; the property we have to change (and what we have to keep in mind throughout calculations) is the Hermitean character of the generator which in conjunction with the additional i has to be joined into an anti-Hermitean operator[74] $\mathcal{O}' = i\mathcal{O}$, $\mathcal{O}'^+ = -\mathcal{O}'$, and $U \sim \exp \pm\mathcal{O}'$. For us, this rep is closer to the logarithmic mapping [27], ch. 8, §5, as well as closer to usual metric aspects by means of anharmonic ratios (or 'Würfe'), because we can now use the exponential directly to transfer products to sums (metric), and vice versa. Moreover, we think that we gain more control on symmetries because Hermitean conjugation consists of two operations, complex conjugation and transposition. Both have independent relevance, so we prefer the second picture of mapping anti-Hermitean operators to unitary reps. Last not least, we can thus avoid superfluous and mostly meaningless 'i's like in perturbation theory which are sometimes introduced by 'magical' (or even mythical) rules or recipes.

With respect to Weinberg's Hermitean operator reps J_i and K_j, we can rewrite the unitary rep U ([38], eq. (2.20), or eq. (15)) as

$$\begin{aligned} U[1+\omega] &= 1 + \tfrac{1}{2} \sum_{\mu\neq\nu} \omega^{\mu\nu}\, iJ_{\mu\nu} \\ &\sim 1 + \tfrac{1}{2} \sum_i (\omega_i^s\, iJ_i + \omega_i^0\, iK_i) \end{aligned}$$

where we've decomposed $\omega^{\mu\nu}$ into two triples of real parameters, ω^0 and ω^s. The notation reflects the occurrence of $0-$ and purely spatial coordinates, and using ω^0 and ω^s helps to remember this background throughout calculations especially when having to identify contributions from rotations and 'boosts' at a later stage within calculations. The operators J_i and K_j denote Weinberg's hermitean triples, $J_i^+ = J_i$ and $K_j^+ = K_j$. So the exponential maps iJ_i and iK_j to unitary U. The commutators derived by Weinberg via comparison with [38], eq. (2.3) by use of (2.1), yields

$$\begin{aligned} [J_i, J_j] &= i\epsilon_{ijk} J_k\,, \qquad [J_i, K_j] = i\epsilon_{ijk} K_k\,, \\ [K_i, K_j] &= -i\epsilon_{ijk} J_k\,. \end{aligned} \tag{31}$$

Due to the additional 'i's in the commutation relations, however, we define the 'new' anti-hermitean sets $\tilde{J}_i := -iJ_i$ and $\tilde{K}_i := -iK_j$, so that $\tilde{J}_i^+ = -\tilde{J}_i$ and $\tilde{K}_i^+ = -\tilde{K}_i$. Now, the

[74] For people using spectral theory, it is always possible to introduce hermiticity by multiplying a commuting $-1 = i^2$ to work with Hermitean operators relative to the remaining phase i, however, this can be done at any time.

commutators of the anti-hermitean sets $\tilde{J}_i$ and $\tilde{K}_i$ read as

$$\begin{aligned} \left[\tilde{J}_i, \tilde{J}_j\right] &= \epsilon_{ijk}\tilde{J}_k\,, & \left[\tilde{J}_i, \tilde{K}_j\right] &= \epsilon_{ijk}\tilde{K}_k\,, \\ \left[\tilde{K}_i, \tilde{K}_j\right] &= -\epsilon_{ijk}\tilde{J}_k\,. \end{aligned} \tag{32}$$

The sets $\tilde{J}_i$ and $\tilde{K}_i$ with *real* coefficients map to U, and the commutators do not introduce additional 'i's when decreasing the operator grade by 1. In a last step, we modify the third commutator only and want to get rid of the sign, so $\overline{K}_i := i\tilde{K}_i$ is a suitable re-definition (i.e. we revert to the original definition K_i), and we obtain (if we set in addition $\overline{J}_i := \tilde{J}_i$) the commutation relations

$$\begin{aligned} \left[\overline{J}_i, \overline{J}_j\right] &= \epsilon_{ijk}\overline{J}_k\,, & \left[\overline{J}_i, \overline{K}_j\right] &= \epsilon_{ijk}\overline{K}_k\,, \\ \left[\overline{K}_i, \overline{K}_j\right] &= \epsilon_{ijk}\overline{J}_k\,. \end{aligned} \tag{33}$$

Now we've reached the stage where we have a real, reductive algebra [5], and where only the product table of the algebra determines via exponential mapping the respective special functions while preserving a priori real coefficients. So it is purely the algebra itself which determines the reducibility of the result of the mapping and the reality conditions.

Nevertheless, we have the choice to calculate either with Hermitean J_i and K_i and the 'physical exponential' $\exp \pm i\,\cdot$, or with anti-Hermitean operators $\tilde{J}_i$ and $\tilde{K}_i$ where we have to remember the sign in the third commutator within the exponential expansion or when summarizing the series by means of special functions.

On the other hand, using $\overline{J}_i$ and $\overline{K}_i$ we may just apply and use the coset theory of the reductive 6-dim Lie algebra above [5], [6], and its decoupling into two commuting su(2) algebras. The relative, intrinsic i between the operator sets $\overline{J}_i = \tilde{J}_i$ and $\overline{K}_i = i\tilde{K}_i$ symbolizes the difference in the special functions between trigonometric and hyperbolic functions, or in physical language the rotations and the boosts. In other words, in both pictures it is visible that we are working with a non-compact group and its Lie algebra. So far, the operator relations can be summarized by

$$\overline{J}_i = \tilde{J}_i = -iJ_i\,, \quad \overline{K}_i = i\tilde{K}_i = K_i\,, \tag{34}$$

so the additional 'i's are evident. Whereas the first set $\overline{J}_i$ and $\overline{K}_i$ is mapped by $\exp\,\cdot$, the other sets are mapped by $\exp \pm i\,\cdot$ to obtain unitary reps of the group.

As side effects of Weinberg's publications on the Lorentz group, it is noteworthy that in 1968, he applied almost the same approach to chiral symmetry of the pion-nucleon system, however, this time without the additional i between the two su(2) algebras. So this time, instead of an su(2)$\oplus$i su(2) algebra and SO(3,1) invariance, he could use the same reasoning with respect to the p-wave coupling of the pion on the nucleon according to the rules of almost the same Lie algebra. This time the algebra su(2)$\oplus$su(2) corresponded to an SO(4) invariance, i.e. the boosts of the Lorentz group corresponded to the p-wave coupling character of the pion field according to the appropriate rep identifications [59]. The linear rep theories emerged as SO(4) invariant quadrics, the so-called σ-model, which comprised the power series of the Lie algebra parameters like above in terms of special functions. Due to the exponential mapping above, one could obtain also 'non-linear' reps, however, at the price to identify already the Lie algebra parameters as physical fields with observable properties, and not the group reps. Practically, however, Lagrangean expressions beyond quadratic terms were not really fruitful, inclusion of other particles and resonances became ambiguous even with respect to electromagnetic coupling, and in general one had to introduce more and more assumptions and parameters into such effective Langrangeans to describe physics and experiments phenomenologically.

Now, before we switch over to aspects of line coordinates and line geometry, we want to summarize few central aspects of Weinberg's description:

- He uses the Hamiltonian instead of a Lagrangean formalism because thus he doesn't need to eliminate superfluous field components, and because he knows the covariance and invariance properties which he applies to the field reps.

- He has modeled the skew 4×4 parameter set $\omega^{\mu\nu}$ by associating (necessarily skew) operators $J_{\mu\nu}$, which he re-expresses by two operator triples with su(2) commutation rules[75] and complex parameters instead of identifying and working with the 6-dim rep $\omega^{\mu\nu}$ and its associated object(s).

- He orientates himself with his modeling versus Wigner's boost reps and unitary rep theory, and the 6-dim rep is a special case of representing the electromagnetic field by a special parameter choice $j = 1$, induced by the two su(2) algebras and their individual rep theory, however, here combined into $(j, 0) \oplus (0, j)$.

For our further reasoning, we can use the case $j = 1$ of the adjoint reps in correspondence to the skew 6-dim rep. And as we know a geometrical identification of the 6-dim rep which was emphasized much earlier already by the six-vector calculus of the electromagnetic field rep, the real challenge is to keep track of the phases versus physical identifications, and the correct association of the respective transformation groups. As such, one has to consider very carefully the association of the components $\vec{E} \pm i\vec{B}$ to the rep $(1, 0) \oplus (0, 1)$ of the Lorentz group [38], [46], [47]. Although such an identification is reasonable from the physical point of view, and although we know of circularly polarized light, the components $\vec{E} \pm i\vec{B}$ are known to correspond to Klein coordinates of the special Complex (see sec. 4.4.), and moreover, this is related to the 'most compact' senary quadric (see e.g. eq. (7) with invariance group SO(6)) as well as to the oval quadric in point space with SO(4) symmetry. So in both cases, we have formally to work with compact symmetry groups. However, the real case using the six Plücker coordinates parametrized by $(\vec{E}, \vec{B})$ yields SO(3,3) symmetry, and its associated point space symmetry group is SO(3,1). Note, that we are talking of *real* parameters here! However, above we have seen, that one can easily redefine compact to non-compact generators, and vice versa, by complexifying the parameters by a phase i. So in order not to get lost and to track the correct phases between the generators, we have to start from scratch with line coordinates, so that at any time we can control whether there is need to complexify real parameters by multiplying with 'i's, and we can thus keep strict control on generators and coordinates.

[75] Essentially, choosing the formalism of Lie symmetries instead of null systems, in this 4×4-case we can just use the so(4) Lie algebra which can be rewritten in terms of so(3)$\oplus$so(3) (or su(2)$\oplus$su(2)), or even by reps of quaternions; all matrix reps, of course, have to respect the skew rep property to relate to their background in null systems.

As a phenomenological consequence (or result) of Weinberg's approach, we can conjecture already here a relation between Lie algebra generators and classical line coordinates, i.e. we understand the differential rep of a su(2) generator as a certain rep of a line coordinate $p_{\mu\nu}$ on function spaces. This can be read from eq. (15) if we understand the parametrization of the transformation as a linear Complex with six real parameters $\omega^{\mu\nu}$. So necessarily the corresponding 'operators' $iJ_{\mu\nu}$ have to represent the line coordinates in 'quantum' reps, or Hilbert space reps. In the appendix, we'll give a more precise derivation, and we discuss the action of such generators on plane waves and especially on quadrics, however, for now this phenomenological identification yields a certain interpretation and guideline.

4.3. Remarks on Transformations

So now, after having introduced null systems according to Lie's guideline [27] in sec. 4.1., and having connected and discussed some differential operator reps in sec. 4.2., it is time to have a deeper look on Lorentz transformations (here for short 'LT') and their origin. As such, we'll concentrate not only as usually on point reps and an invariant quadric in point space, but we derive the background of LT from P^5 and line geometry.

As a side effect, we want to mention here only briefly Lie's research on the 10-dim symmetry group of Complexe (more precise, of regular null systems, or dim 11 when treating special null systems). He has discussed this group as a subgroup of projective transformations in [27], ch. 6, exhaustively, and he has also related Pfaff's differential equations and discussed a lot of aspects. So later (if there is more time and space to write things down) one can attach symplectic considerations, modern differential geometry and dynamics/kinematics here. Monge's equations enter in his presentation as superset of Pfaff's equations, and one can think (later on) to understand the cones by Complex geometry, and relate them to quadratic Complexe. Such Complex cones automatically yield intersections with the absolute plane, so if we shift e.g. the center/vertex of the cone to the absolute plane, we can discuss Zindler's[76] rep at the beginning of vol. I [76], and introduce there geometrically $\hbar = \frac{h}{2\pi}$, h describing the height of a full turn of the curve (or screw) with respect to the cylinder symmetry. Otherwise without absolute vertex/center, we can investigate and use the conic intersection of the second order cone with the absolute plane throughout 'particle' motion, and thus we have immediate access to metric properties by projective relations to the absolute conic, i.e. by the Caley-Klein picture.

Last not least, the ongoing discussion of 'entanglement' in optics and 'quantum' physics is usually attributed to the intersection of second order ('light') cones, which – using Monge's equations and Complex geometry – results automatically when relating line and differential geometry by considering Complex curves ([27], ch. 7, §5), and the Complex cones of two infinitesimal close curve points (see [27], p. 304, and figure 63).

However, we leave all those associated and interesting aspects and details, and stop here, because due to the topic and our focus on relativity, we want to address the origin of Lorentz transformations which we've mentioned briefly in [25] by presenting there few details only.

[76] Zindler summarized at the beginning of the 20th century some aspects of line geometry in two volumes [76], [77].

The usual setup is nowadays the view on point space and point coordinate transformations, and the discussion of transformations according to[77]

$$\begin{array}{lcl} x_0 & \longrightarrow & x'_0 = \gamma x_0 - \gamma\beta x_1 \,, \\ x_1 & \longrightarrow & x'_1 = \gamma x_1 - \gamma\beta x_0 \,, \\ x_2 & \longrightarrow & x'_2 \,, \\ x_3 & \longrightarrow & x'_3 \,. \end{array} \tag{35}$$

Here, we've set ct, or ct' as x_0, or x'_0, and we need the constraint $\gamma^2(1-\beta^2) = \gamma^2 - \gamma^2\beta^2 = 1$ as known from LT. The usual physical association is to define $\beta = \frac{v}{c}$, a ratio of velocities[78], and $\gamma = \sqrt{1-\beta^2}^{-1} = \sqrt{1-(\frac{v}{c})^2}^{-1}$. A typical parametrization in terms of special functions is known to be $\beta = \tanh\eta = \frac{v}{c}$. Of course, this rep is sufficient and works, but as we'll see below, this *explicit* definition is not necessary as long as $\gamma^2(1-\beta^2) = \gamma^2 - \gamma^2\beta^2 = 1$.

Now before we start more detailed discussions, it is important to recall that we write the usual 'Lorentz invariant' in 4-vector notation and point coordinates according to $x_\mu x^\mu \stackrel{\text{LT}}{=} x'_\mu x'^\mu$. In other words, we require that a quaternary point rep remains on the invariant quadric $x_\mu x^\mu$. To gain some geometrical impression, we want to associate the picture of a sphere with a certain radius (i.e. a metric property) in order to use this picture later on to visualize some of the geometrical aspects.

Now, usually people do not take care of spatial extension but use models where x represents a point only. But if we associate the sphere picture (or represent an extension by a second point y), we have to think about representing and treating extension. So we need to know the action of the Lorentz group on other points, located e.g. close as well as far from the point x. The simplest approach is a second point y on the sphere (or more general on the quadric), because we know (or require) that the quadric is invariant. So one can discuss the transformation properties of this second point y as well by means of eq. (35). If we assume global transformations or if, with respect to the two points x and y we may assume them lying 'close together', then we can transform both points by the same transformation rules (35).

At next, if we define second order objects according to $p_{\alpha\beta} := x_\alpha y_\beta - x_\beta y_\alpha$ by using the coordinate enumeration of eq. (35), we may ask for their transformation properties under LT as well [25], sec. 3. Of course, we can identify the origin and background of $p_{\alpha\beta}$ in line geometry, but we postpone synthetic discussions for a while and follow algebra only. The easiest formal approach (besides just calculating the transformation properties brute force in individual coordinates) is to define 2×2-determinants,

$$p_{\alpha\beta} = \begin{vmatrix} x_\alpha & y_\alpha \\ x_\beta & y_\beta \end{vmatrix} , \tag{36}$$

and use the standard rules on determinants. We have discussed in [25] the invariance properties of two of the line coordinates. So it is evident, that the coordinate comprising the two point coordinates which are not affected by the LT do not change, i.e. in this case $p'_{23} = p_{23}$.

[77] For convenience, we've written transformations of the 0- and 1-coordinate, however, it doesn't matter to change the coordinates to other values of $0, 1, 2, 3$ as long as we leave a plane/two of the point coordinates invariant, e.g. 1 and 3 when transforming 0 and 2 (compare also to figure 1 in sec. 4.1.).

[78] See our related discussions in [24], [28], [25] with respect to PG.

To understand why the line coordinate comprised of the two transformed point coordinates 0 and 1 doesn't change, however, is not that obvious. In order to find a general expression, we use the determinant rep, too. So with the transformation set $x'_\alpha = \gamma x_\alpha - \gamma\beta x_\beta$, $x'_\beta = \gamma x_\beta - \gamma\beta x_\alpha$ for fixed $\alpha = 0, \beta = 1$, we have

$$p'_{\alpha\beta} = \begin{vmatrix} x'_\alpha & y'_\alpha \\ x'_\beta & y'_\beta \end{vmatrix} \tag{37}$$

$$= \begin{vmatrix} \gamma x_\alpha - \gamma\beta x_\beta & \gamma y_\alpha - \gamma\beta y_\beta \\ \gamma x_\beta - \gamma\beta x_\alpha & \gamma y_\beta - \gamma\beta y_\alpha \end{vmatrix} \tag{38}$$

Decomposing the last determinant by linearity and switching rows, we obtain

$$p'_{\alpha\beta} = (\gamma^2 - \gamma^2\beta^2) \begin{vmatrix} x_\alpha & y_\alpha \\ x_\beta & y_\beta \end{vmatrix} + (\gamma^2\beta - \gamma^2\beta) \begin{vmatrix} x_\alpha & y_\beta \\ x_\beta & y_\alpha \end{vmatrix} \tag{39}$$

Now, independent of values of β and γ, the second term on the rhs will always vanish. The remaining first term, however, can be considered also for transformations with other parameter β and γ, so that in general there is a priori no need for the LT constraint, $\gamma^2 - \gamma^2\beta^2 = 1$. If, however, we require $\gamma^2 - \gamma^2\beta^2 = 1$, this yields $p'_{01} = p_{01}$, and we've identified a second line coordinate which remains invariant although we are applying the nontrivial (and 'non-local') transformations (35) (see also [25]).

At this stage, we've identified already two out of six line coordinates which behave not only irreducibly, but each of them is individually invariant under LT.

Time to check the remaining four line coordinates $p'_{02}, p'_{12}, p'_{03}, p'_{13}$, and their respective transformation behaviour. With respect to the transformations (35), it is obvious that while one can leave the lower row in the determinant without changes, we have to treat the two cases p'_{02}, p'_{03}, and p'_{12}, p'_{13}, separately according to the two different transformation rules of x_0 and x_1 in eq. (35).

The first case $p'_{0\delta}$, $\delta = 2, 3$, reads as

$$p'_{0\delta} = \begin{vmatrix} \gamma x_0 - \gamma\beta x_1 & \gamma y_0 - \gamma\beta y_1 \\ x_\delta & y_\delta \end{vmatrix} = \gamma p_{0\delta} - \gamma\beta p_{1\delta}\,, \tag{40}$$

whereas the second case $p'_{1\delta}$ reads as

$$p'_{1\delta} = \begin{vmatrix} \gamma x_1 - \gamma\beta x_0 & \gamma y_1 - \gamma\beta y_0 \\ x_\delta & y_\delta \end{vmatrix} = \gamma p_{1\delta} - \gamma\beta p_{0\delta}\,. \tag{41}$$

As we see, the result can be still expressed in line coordinates (which features once more the irreducibility of the 6-dim set of line coordinates as an appropriate basis), however, this time, we have obtained additional terms, or a mixture of two coordinates by the LT. But because each δ-point coordinate remains invariant for $\delta = 2, 3$, the transformation induces a linear combination only in terms of the point coordinates affected by the LT. To summarize what we know so far with respect to LT acting on 0-/1-point coordinates, we can list:

$$\begin{aligned} p'_{01} &= p_{01} &&,\quad p'_{23} = p_{23}\,, \\ p'_{02} &= \gamma\, p_{02} - \gamma\beta\, p_{12} &&,\quad p'_{12} = \gamma\, p_{12} - \gamma\beta\, p_{02}\,, \\ p'_{03} &= \gamma\, p_{03} - \gamma\beta\, p_{13} &&,\quad p'_{13} = \gamma\, p_{13} - \gamma\beta\, p_{03}\,. \end{aligned} \tag{42}$$

So although the two individual coordinates p_{01} and p_{23} remain invariant (and of course, their combinations $p_{01} \pm p_{23}$, too), there is no similar *linear* mechanism if we consider individual coordinates $p'_{02}, p'_{03}, p'_{12}, p'_{13}$, or the combinations $p'_{02} \pm p'_{31}$, or $p'_{03} \pm p'_{12}$, which both result in sums and an additional mixture of the linear combinations[79].

Now to understand the background, we have to recall that the line coordinates fulfil the Plücker condition. In other words, we may ask what happens to this constraint on the line coordinates after having performed the LT on point coordinates. As such, if we consider at first the Plücker condition in transformed coordinates,

$$P' = p'_{01}p'_{23} + p'_{02}p'_{31} + p'_{03}p'_{12}\,, \tag{43}$$

the first summand on the rhs remains trivially invariant by (42), and we have to calculate the last two summands. The result is

$$P' = p_{01}p_{23} + (\gamma^2 - \gamma^2\beta^2)\,(p_{02}p_{31} + p_{03}p_{12})\,, \tag{44}$$

so that LT with $\gamma^2 - \gamma^2\beta^2 = 1$ guarantee the invariance of the Plücker condition throughout Lorentz transformations! In other words: lines remain lines under LT. However, this explains an additional aspect because we know about the background of the Plücker condition in P^5. There, this condition was used to map points of the Plücker-Klein quadric M_4^2 of P^5 to lines in P^3, and the Plücker condition has to be fulfilled by *each* line in P^3. Preserving this conditions under LT, i.e. $P' \overset{\text{LT}}{=} P$, therefore can be re-expressed by asking for the automorphism group of the Plücker-Klein quadric M_4^2 in P^5 (which is known to be a twofold 15-dim transformation group [30], §69). Vice versa, LT can be seen as the analogue of such automorphisms in P^3.

As another interesting aspect for later use, it is noteworthy that eq. (44) allows for an alternative identification. Therefore, we have to mention the general form of a tetrahedral Complex in line coordinates,

$$a\,p_{01}p_{23} + b\,p_{02}p_{31} + c\,p_{03}p_{12} = 0 \tag{45}$$

(see e.g. [30], p. 172, or p. 288, or [27], p. 319). The anharmonic ratio of the lines intersecting the planes of the coordinate tetrahedron corresponds to $\kappa = \frac{a-c}{b-c}$, where the ∞^3 lines[80] $x = rz + \rho$, $y = sz + \sigma$ fulfil $\kappa = \frac{s\rho}{r\sigma}$. The Complex cones are elementary cones of the differential equation

$$(b-c)x\mathrm{d}y\mathrm{d}z + (c-a)y\mathrm{d}z\mathrm{d}x + (a-b)z\mathrm{d}x\mathrm{d}y = 0 \tag{46}$$

of Monge-type [27], p. 319.

So we can trace and understand the action of Lorentz transformations also as modification of the tetrahedral complex (corresponding to the modification of the coordinate tetrahedron) by $a = 1$, and $b = c = \gamma^2 - \gamma^2\beta^2$. In the same manner, also the Plücker condition can be seen as a special case of a tetrahedral Complex with $a = b = c = 1$, i.e. related to a special choice of the coordinate tetrahedron in point space. Note, however, that in both

[79] Although this gives rise to further interesting symmetries, we postpone the discussion here. Parts have already given in [44], and [50].

[80] Remember that Lie uses another orientation of the coordinate axes.

cases due to $b = c$ the denominator of κ vanishes, i.e. $\kappa \longrightarrow \infty$, so that we can append our discussion given in [12] to derive the spinorial picture. There, we have discussed and used the fifth inhomogeneous Plücker coordinate $\eta = r\sigma - s\rho$ with the constraint $\eta = 0$ (see the discussion with respect to κ before).

So far, we've established a direct trail with respect to linear transformation behaviour of the line coordinates under LT (see eq. (42)) and the invariance of the Plücker condition. This invariance guarantees that we can rely on our line picture throughout calculations and LT in P^3. However, if we take a closer look on the transformation laws in eq. (35) and in eq. (42), we notice that these laws in the first two rows of (35) are similar to those in the last two rows of (42) if we replace $x \longleftrightarrow p$. Therefore, recalling Minkowski's 'two invariants' given in eqns. (7) and (8), we may calculate[81] the coordinate squares, too. Straightforward algebra yields:

$$p_{02}'^{\,2} - p_{12}'^{\,2} \stackrel{\mathrm{LT}}{=} \left(\gamma^2 - \gamma^2\beta^2\right)\left(p_{02}^2 - p_{12}^2\right) , \tag{47}$$

$$p_{03}'^{\,2} - p_{13}'^{\,2} \stackrel{\mathrm{LT}}{=} \left(\gamma^2 - \gamma^2\beta^2\right)\left(p_{03}^2 - p_{13}^2\right) , \tag{48}$$

so that the senary quadric yields under LT[82]

$$\begin{array}{l} \pm p_{01}'^{\,2} \pm p_{23}'^{\,2} \\ \pm(p_{02}'^{\,2} - p_{12}'^{\,2}) \pm (p_{03}'^{\,2} - p_{13}'^{\,2}) \\ \stackrel{(42)}{\longrightarrow} \\ \pm p_{01}^2 \pm p_{23}^2 \\ \pm\left(\gamma^2 - \gamma^2\beta^2\right)\left(p_{02}^2 - p_{12}^2\right) \\ \pm\left(\gamma^2 - \gamma^2\beta^2\right)\left(p_{03}^2 - p_{13}^2\right) . \end{array} \tag{49}$$

For LT, i.e. $\gamma^2 - \gamma^2\beta^2 = 1$, we have

$$\begin{array}{rl} & \pm p'^2_{01} \pm p'^2_{23} \pm (p'^2_{02} - p'^2_{12}) \pm (p'^2_{03} - p'^2_{13}) \\ \stackrel{\mathrm{LT}}{\longrightarrow} & \pm p_{01}^2 \pm p_{23}^2 \pm \left(p_{02}^2 - p_{12}^2\right) \pm \left(p_{03}^2 - p_{13}^2\right) . \end{array}$$

So by LT alone, we still have the freedom to absorb the one or other sign in an additional phase in point space, or in a transferred rep, e.g. by projections or special mappings. We cannot overemphasize that with each mapping or transfer, the respective 'objects' change physically, and we have to adjust the physical picture and its relevance appropriately. So pure axiomatization, or Bourbaki-style technocracy, may help with formal rep theory and yield a consistent 'grammar' of the language, however, to stay in this picture, there is still the need to use the correct words and pictures, and to write the epic works and poems of physics.

---

[81] Background details will be given in sec. 4.4..

[82] We do not discuss the additional freedom to switch the signs of the squares which is due to the LT being given only by point transformations and, as such, being not more restrictive. LT are a special case only, and the assumptions on transformations of single points in 3-space are not sufficient to separate a unique invariant senary quadric. As such, we find a couple of symmetry groups SO(m,n), $n + m = 6$, discussed throughout literature.

Last not least, to begin writing the poems, it is worth spending some thoughts on the geometrical setup we've introduced above, and on the associated picture[83] of the sphere, or quadric. If we understand the invariant quadric in point space as a sphere, then the invariance requirement of the quadric allows to shift points on the sphere (or in general along the quadric). If we look closer to the fundamental tetrahedron, we can define coordinates (e.g. by shifting the unit point appropriately) where the vertices lie on the circumscribed unit sphere, just in order to realize the invariant quadric. In this picture, the line coordinates correspond to edges of the tetrahedron which connect the vertices appropriately.

Now by fixing the two line coordinates p_{01} and p_{23}, we have identified two opposite edges of the tetrahedron with their endpoints each representing two of the four vertices of the tetrahedron (which by construction lie on the invariant sphere). So in order to fix these two edges/line coordinates, we can think of two stiff (or solid) edges whereas their endpoints can't leave the sphere (or quadric), and as such the ('Lorentz') transformations[84] shift the endpoints of one edge/stick p_{01} along the quadric whereas the second one, p_{23}, remains fixed. So in essence, while the endpoints of one[85] of the stiff edges can be moved along the quadric, the four remaining edges according to their non-trivial transformations under LT (or even in the case of more general transformations) can be seen as realized by rubber band, however, the sphere (quadric) will remain invariant.

Closing this section, it is necessary to integrate our aspects above, and to mention some more background and an additional picture used by the old geometers. As such, we can identify the skew edges/line coordinates p_{01} and p_{23} with two skew lines l_1 and l_2. Now by the definition of a 'ray system'[86] which we've used already above, we have exactly one spatial line per general space point x which intersects both lines l_1 and l_2 (see [27], p. 187, or p. 291ff, for details and background). In order to repeat not essential parts of Lie's book, it is sufficient to reference to [27], ch. 6, which yields the context of Lie's line and area elements, of null systems and various related aspects of Monge's and Pfaff's differential equations. Moreover, having identified the 'ray systems' above, we can connect to Kummer's work on such ray systems [66], and their general relevance in optics as well as in geometry. The line geometrical approach can be equally well applied by means of Complexe or Congruences, and it relates to Plücker's work.

So for now, we think that we've attached special relativity sufficiently to classical line geometry, and it is up to generalize to regular linear as well as higher order Complexe.

4.4. Quadratic Relations

So far in this section, we have discussed some basic properties and examples of *linear* Complexe as well as their relation to Pfaff's and Monge's equation, to Lorentz transformations in subsection 4.1. as well as aspects of related transformation groups in subsection 4.3..

[83] Before discussions start, please note that using homogeneous coordinates, we can apply the formalism to general quaternary quadrics also without using this geometrical picture in depth. As a further example of a quadric different from the sphere, we can e.g. discuss the 'Schmiegtetraeder' of an hyperboloid, see also [27], p. 348ff, with respect to generating lines and the associated tetrahedral Complex. Recall also, that we may change easily from Plücker to Klein coordinates in a controlled manner.

[84] In the general case the transformations leaving the quadric invariant.

[85] By more general transformations, the endpoints of both edges can be moved.

[86] German: Strahlensystem erster Ordnung und erster Klasse.

Most of the discussion was related to special linear Complexe and lines[87]. However, with respect to P^3 and known physics and geometry, we have to consider also briefly *quadratic* Complexe. There are multiple reasons to do this:

- The Plücker condition $P = p_{01}p_{23} + p_{02}p_{31} + p_{03}p_{12} = 0$ (or equivalently $p_1p_4 + p_2p_5 + p_3p_6 = 0$) is quadratic in line coordinates, so it is inevitable to consider quadratic Complexe besides just using linear Complexe and P^5 like in the last section. Moreover, if we work practically with lines, the coordinates of a special linear complex $\sum a_{\alpha\beta}p_{\alpha\beta}$ have to fulfil the Plücker condition, too. Intersection of lines can be attributed to incidence of two lines with a (common) plane, so if we take four points (two points per line) the 4×4-determinant of the four points has to vanish. Rewriting this determinant in terms of line coordinates $p_{\alpha\beta}$ and $p'_{\alpha\beta}$ of the two lines yields
$$\begin{aligned} & p_{12}p'_{03} + p_{31}p'_{02} + p_{01}p'_{23} \\ + \quad & p_{23}p'_{01} + p_{02}p'_{31} + p_{03}p'_{12} = 0\,, \end{aligned} \tag{50}$$
or using 'polarity', we may introduce partial derivatives and write $\sum \frac{\partial P}{\partial p_{\alpha\beta}} p'_{\alpha\beta} = 0$ where the sum has to be taken over all six coordinates. As Cayley has mentioned, due to linearity of the linear Complex $\sum a_{\alpha\beta}p_{\alpha\beta}$ in $p_{\alpha\beta}$ (which is essentially a 3-dim line set in 3-space), this intersection condition with respect to a line $(p'_{\alpha\beta})$ can be fulfilled by all member lines of the Complex, if they fulfil the quadratic relation $P_a = a_{01}a_{23} + a_{02}a_{31} + a_{03}a_{12} = 0$. This describes the special linear Complex of the electromagnetic field $(\vec{E}, \vec{B})$, and we may associate the picture given in [24], sec. 2.2 or appendix A, and [28], sec. 3 by a moving point light (if we simplify light emission to lines). In other words, the axis of the special linear Complex (i.e. the trajectory of the moving point) serves to collect those lines of 3-space into a Complex which hit this axis. This, of course, can be understood in a second picture as observers located throughout arbitrary points of 3-space and watching a moving point (once more, if we simplify the 'observation', or information transfer, by lines between observer location and moving point [24], [28].

- Another major aspect is the need to introduce coordinate systems into this world as soon as we want to talk about physics, measures and real world observations. Although we can use line geometry of P^3 and the intersections and unions of lines, it is often necessary to have additional possibilities to use individual point or plane reps in P^3, and compare to observations. However, introducing the fundamental coordinate tetrahedron in P^3, i.e. 4 non-planar points or 4 appropriate planes as tetrahedron sides, yields immediately a classification scheme of lines in P^3 because each of the lines hits the tetrahedron planes in four distinct points. This scheme almost automatically yields the definition of a quadratic Complex, the so-called tetrahedral Complex (see e.g. [24], sec. 3.2, or [28], sec. 3.3, and references there), and the associated anharmonic ratio κ from above.

[87] We discuss aspects of regular linear Complexe in [50] in conjunction with (null) planes when representing 4-momentum and Dirac spinors, i.e. instead of second *order* surfaces one can address as well second *class* surfaces, or transformation properties of planes with respect to null systems, and especially under additional tangential constraints.

- We've addressed the usual interpretation of the 4-dim rep of the '4-momentum' already a couple of times. Especially in [50], we address the interpretation in terms of Hesse coordinates of plane reps, and to keep the 6-dim rep for forces and momentum like given in [39]. This is based on the regular linear Complex as transfer mapping, or a correlation, of points to planes and vice versa [50]. As such, we have to consider normals of planes in the Hesse rep, and a system of normals related[88] to second order surfaces. In this environment, however, we benefit from the 'natural occurrence' of the tetrahedral Complex when treating normals (see [27], ch. 7, §2).

- Working with Complexe as base elements, it is natural to ask for common lines or line sets e.g. in Congruences or Configurations, i.e. we have to handle families or sets of Complexe, and related assemblies thereof. This results analytically in typical structures like $(\mathfrak{A})_{\alpha\beta} \sim a_{\alpha\beta} = \lambda_1 a^1_{\alpha\beta} + \lambda_2 a^2_{\alpha\beta}$, or $(\mathfrak{B})_{\alpha\beta} \sim b_{\alpha\beta} = \lambda_1 b^1_{\alpha\beta} + \lambda_2 b^2_{\alpha\beta} + \lambda_3 b^3_{\alpha\beta}$ etc. Now the quest for lines common to the pencil or the triple set of Complexe may be re-expressed by means of the quadratic Plücker condition, because the lines l in P^3 have to fulfil the condition $P_l = a_{01}a_{23} + a_{02}a_{31} + a_{03}a_{12} = 0$ from above. As such, we have to resolve quadratic equations in the coefficients $a_{\alpha\beta}$ or $b_{\alpha\beta}$ of the Complexe $\mathfrak{A}$ or $\mathfrak{B}$ which results in quadratic equations to determine the parameters λ_i from above (or their ratios, respectively). In the case of the pencil[89] $\mathfrak{A}$, this yields for example $(\lambda_1 a^1_{01} + \lambda_2 a^2_{01})(\lambda_1 a^1_{23} + \lambda_2 a^2_{23}) + \ldots + \ldots = 0$. In the same manner, without restricting the products on the rhs to 0, one can calculate the invariant of the pencil and relate the invariants of the original Complexe. This also justifies investigations of the relative position of such Complexe in P^5, and it gives rise to the notion of 'involutions' of Complexe [8]. For now, this aspect is sufficient to justify a deeper look onto quadratic Complexe in general.

- Last not least, because some aspects above already mention relations to the tetrahedral Complex, and because there are further contexts where this Complex emerges, it is worth keeping focus on this object (or the family of such objects). The tetrahedral Complex is a quadratic Complex (see eq. (45)), but although it is deeply connected to projective geometry and especially, due to the fundamental tetrahedron, with the coordinate definition, we found only few of its aspects discussed analytically in literature. Most of the discussion took place in synthetic geometry, originated by von Staudt and in parts by Reye, and there is a historical survey by Lie in [27], ch. 7, §2, which addresses some of its historical aspects and contexts. However, because our discussion of Lorentz transformations and automorphisms of the Plücker-Klein quadric M_4^2 throughout the last subsection, sec. 4.3., lead us to the tetrahedral Complex, it is worth to start a discussion of quadratic aspects, too.

So as a first step to approach analytic calculations with Complexe as base elements, we can follow Klein [30], §23, and start right from the beginning with P^5. As such, if we recall the basic check for lines to fulfil the Plücker condition, we can ask as well for rules and properties to work with the coefficients of such elements. More general, we have to

[88] This is a consequence of the mechanism/formalism how we treat dynamics and describe points (see [69]).

[89] We have two binary parameters λ_1 and λ_2, i.e. the assembly is dependent from *one* parameter and constitutes a pencil.

construct an invariant theory and appropriate forms and rules. As Klein remarks in [30], §23, one has to address quadratic forms already when working with linear Complexe, and one can consider in a first stage the transformation behaviour of quadratic (senary) forms when substituting the coefficients of the linear elements.

We skip Klein's discussion of inertia indices ([30], p. 97) here and focus on special substitutions to rearrange the Plücker condition. Due to the quadratic character and our results from the last subsection, it is evident that we can use binomial expressions to transfer the products to plain squares of a quadric, i.e. if we rename the coefficients $a_{\alpha\beta}$ by a senary index α into x_α, the invariant of the linear Complex in P^5 reads as[90] $\Omega_x = x_1x_2 + x_3x_4 + x_5x_6$.

By *real* and *linear* substitutions

$$\begin{aligned} x_1 &= y_1 + y_2, \quad x_3 = y_3 + y_4, \quad x_5 = y_5 + y_6, \\ x_2 &= y_1 - y_2, \quad x_4 = y_3 - y_4, \quad x_6 = y_5 - y_6, \end{aligned} \tag{51}$$

it is evident, that the quadratic form Ω_x from above now reads as $\Omega_y = y_1^2 - y_2^2 + y_3^2 - y_4^2 + y_5^2 - y_6^2$, so with respect to *real* and *linear* substitutions, we have reached the final stage of a sum of perfect squares and can read off their 'inertial indices'. The senary quadric Ω_y now shows formally its SO(3,3) symmetry which we've mentioned before as resulting from line geometry. Now a naive interpretation of this quadric in terms of (abstract) *point* coordinates allows to treat this SO(3,3) symmetry with formal arguments, however, as is obvious, in such a case the interpretation against the background of line and Complex geometry is completely lost.

However, in Minkowski's and Klein's papers 1910, the authors have left real substitutions and transformed the 0- (or 4-)component of the point coordinates by multiplying it with a phase i in order to represent the quaternary quadric as sum of squared components, and *in addition*, Minkowski introduces an 'imaginary' angle $i\psi$ as argument of trigonometric functions, i. e. in essence hyperbolic transformations with ψ. To sort things, Minkowski associates the real, hyperbolic transformations to the set x, y, z, t whereas x_4 and t have the relative imaginary phase i by [1], eq. (3). So we have a mixture of non-compact SO(3,1) transformations and compact SO(4) transformations, acting on point reps with real coefficients or point reps with 3 real and one purely imaginary coefficient which can only implicitly resolved because in addition Minkowski attaches the notion of a space-time point to both reps x, y, z, t AND x_α [1], §1, p. 475, last line[91]. Due to the character of Plücker's line coordinates, we have to address this change by 'absorbing' the additional i of the fourth point coordinate in a redefinition of the related coefficients of three of the line coordinates. So to keep the old definition of the line coordinates, one can shift the i to the coefficient, which results in sign changes of the quadric.

[90]To keep track with Klein and his book [30], we adopt to his special naming convention [30], p. 96, of the subscripts. Take care, that this notation differs from senary indices used elsewhere, and here Klein's 'even' coefficients x_2, x_4, and x_6 are related to absolute elements (i.e. to the index 0, or 4) in point space. In order to recall the appropriate coordinate set of the respective rep of Ω, we have attached the related subscript to Ω.

[91]The symmetry of his eqns. (A) and (B) [1], p. 477, and his eq. (17) on p. 481 associate the groups SO(4) to x_α, and SO(3,1) to x, y, z, t.

Klein starts the other way around and rewrites Ω by introducing imaginary transformations[92] of the y, i.e.

$$\begin{aligned} z_1 &= y_1, & z_3 &= y_3, & z_5 &= y_5, \\ z_2 &= iy_2, & z_4 &= iy_4, & z_6 &= iy_6, \end{aligned} \tag{52}$$

so that we now have $\Omega_z = z_1^2 + z_2^2 + z_3^2 + z_4^2 + z_5^2 + z_6^2$ with formal SO(6) symmetry. This basically reflects Hermite's trick to represent the sum of two squares of real numbers by the product of two conjugate complex numbers[93]. Rewriting the set (51), we obtain

$$\begin{aligned} x_1 &= z_1 - iz_2, & x_3 &= z_3 - iz_4, & x_5 &= z_5 - iz_6, \\ x_2 &= z_1 + iz_2, & x_4 &= z_3 + iz_4, & x_6 &= z_5 + iz_6. \end{aligned} \tag{53}$$

So with respect to Minkowski's '*two* invariants' given in eqns. (9) and (10) here, it is now obvious that we are talking on Ω_x and Ω_z of the same linear Complex f, however, related by the linear substitutions above. Therefore, we consider it more appropriate to discuss the Complex invariant with respect to the P^5 background instead of 'two invariants', or 'quantum numbers', related to SO(4), SO(3,1), or glued to assemblies of two su(2) 'spin' algebras with or without relative complexification. As such (see also [24], sec. 2) the source-free electromagnetic case featuring $\vec{E} \cdot \vec{B} = 0$ can of course be attached to vanishing values of invariants, however, geometrically it is more fruitful and constructive to discuss linear Complexe in general, i.e. in P^5, and points on the Plücker-Klein quadric as special cases thereof. Besides relating 'quantum numbers', we expect this to be the framework to discuss further some known relations and aspects like summarized in [78], ch. 6, §22, or by various aspects of physical applications in the following ch. 7, or in [79] e.g. with respect to the W-algebra in ch. 4, and the 'Special Topics' in ch. 5, especially the Regge trajectories ('Topic 5'), and SU(1,1).

To proceed further in our current context of quadratic Complexe, the simultaneous invariant of the linear Complexe in the case Ω_y then reads as

$$y_1y_1' - y_2y_2' + y_3y_3' - y_4y_4' + y_5y_5' - y_6y_6' \tag{54}$$

which Klein uses to discuss the ± 1-handedness of the six fundamental Complexe and their grouping into $3 \oplus 3$. With respect to Ω_z, one has to discuss Complexe with general complex coefficients, too, however due to the geometric interpretation of i and hypercomplex numbers as an algebraic symbolism, right here, we do not feel the need to do so. Here, in order to close this section on quadratic relations, it is important to mention Klein's subsequent arguments ([30], p. 100) that working with quadratic Complexe one automatically has to consider linear substitutions of *two* quadratic forms to result in pure squares as summands, because in addition to the quadratic Complex one has to respect the Plücker condition Ω, usually as Ω_y or Ω_z. In essence, also with respect to quadratic Complexe one can use a

[92]On demand, one can also symmetrize the normalization by using $\sqrt{\frac{1}{2}}$ in both sets, however, this in turn can be reabsorbed by the homogeneous coordinates definition. Nevertheless, for calculations using coordinate transformations it is helpful to perform the 'bookkeeping' correctly.

[93]Moreover, it leads to reps by 3 complex numbers, i.e. as used by Study in various contexts which we'll discuss elsewhere.

system of six *linear* Complexe, either purely real or imaginary [30], p. 100, and introduce concepts like confocal Complexe in analogy to confocal cyclids, etc.

In physics, because quadratic Complexe are related to second order cones ('Complex cones' [26]), this framework generically yields access to the requirements discussed with respect to relativity by Ehlers et al. [80], and we find immediately applications in entanglement discussions usually performed by intersecting two second order cones[94].

4.5. Discrete Transformations

There are further beautiful relations to discrete transformations as well as to mathematics which is nowadays usually discussed in conjunction with 'quantum' physics, namely the Heisenberg group and Kummer surfaces, or K3 theory in general, which in our current context is off-topic. Nevertheless, it is worth to recall these relations briefly as an 'interlude', or 'stint', as they are attached to ongoing discussions and to P^5 geometry which we'll propose later as a suitable framework to generalize and unify some of the *physical* ideas discussed above.

Now, here it is sufficient to follow Hudson [32] and his discussion of permutation symmetries of points in the (projective) point/plane incidence equation[95] $x \cdot u = 0$ of 3-space. He then derives the 16_6 configuration, some dimensional identifications from line and Complex geometry which are interesting with respect to our earlier work[96], and he discusses apolarity of Complexe. In his preface to Hudson's book [32], Wolf Barth translates this apolarity condition into modern terminology by means of tensors S and T representing anti-selfadjoint maps $\;^4 \longrightarrow (\;^4)^*$ while the apolarity can be represented by $S^{-1}T = T^{-1}S$ with $S^{-1}T = (T^{-1}S)^{-1}$, thus being an involutory map of $\;^4$. Whereas the rich content of the book yields various aspects of algebraic geometry, here we can focus on chapters IV - VI. Ch. IV can be used with respect to linear Complexe, apolarity when considering the fundamental Complexe, and the parity decomposition. Ch. V discusses the simultaneous diagonalization of quadratic forms in P^5 and the hyperplane geometry there with respect to Kummer's and Plücker's work in optics. The main topic, however, is the Kummer surface for the quadratic Complex and its model as a complete intersection of three quadrics in P^5: the Plücker quadric, the quadratic Complex itself, and a third quadric attached to both ([32], preface, p. xvi). Plücker's surface is comprised as a special case, however, especially with respect to optics, Hamilton theory and QFT, the further chapters of Hudson's book (or their content in modern terminology, respectively) will be part of upcoming papers after we've discussed more of the physical foundations to be settled more deeply in P^5.

---

[94] We have already mentioned the background of Monge's cones discussed by Lie [27] with respect to dynamics and contact transformations, and we want to mention with respect to our older work discussing the relation of quadratic Complexe and the 'light cone' in point, that throughout the discussion of confocal quadratic Complexe one can also find one special/singular quadratic Complex.

[95] According to our notion, we denote the point and plane by x and u, respectively, thus changing Hudson's notation in [32], ch. I, §3.

[96] As mentioned before, we've used SU(4) and related non-compact groups to discuss a linear covering of chiral and Lorentz group reps, linear as well as nonlinear. The su(4) root diagram when rotated to 3-space yields geometrically a tetrahedron and is associated to the fundamental rep $\underline{4}$. As such, SU(4) can serve to treat spatial geometry whereas SU(3) and the reps can serve to treat planar geometry by means of three homogeneous coordinates.

4.6. Classical PG and Dirac

We want to address briefly[97] another aspects of null systems as correlation, i.e. realizing a special mapping (or a transfer) between points and planes in P^3. As such, the rep of such a system (because either acting on point or plane reps of P^3 which are both quaternary) is necessarily a 4×4 matrix rep M, and in order to avoid singular cases, the determinant of a regular null system is not zero. On the other hand, the null point x is member of the null plane u (or incident with the null plane), so $x \cdot u = 0$, and for regular null systems (representing a correlation by e.g. $u_\mu = M_{\mu\nu} x_\nu$, or $x_\mu = (M')_{\mu\nu} u_\nu$), we have a constraint $x \cdot u = xMx = 0$ (or $x \cdot u = uM''u = 0$) which in essence is fulfilled for skew M (or M'') because $x \cdot u = M_{\alpha\beta} x_\alpha x_\beta = 0$ (or $x \cdot u = M''_{\alpha\beta} u_\alpha u_\beta = 0$) by symmetry and commutation of the two quaternary reps of x (or u)[98].

However, misinterpreting Mx as collineation, i.e. using $x \longrightarrow x' = Mx$ and interpreting the skew matrix rep M as rep of appropriate Lie generators acting on space coordinates, gives rise to a transformation theory of coordinates, or 'infinitesimal transformations' of a '4-dim space'. So 'by accident'[99], because this also fulfils the basic formal procedures of Lie theory in this special 4×4 case, and may be interpreted in terms of the related different geometry, the doors are wide open to discuss SO(4) or SO(3,1) transformations as well, or subgroups and subalgebras, or even complex coverings like SU(2)×SU(2), or SU(2)×i SU(2), etc., see above[100].

Now, if we rewrite in this context the 'Lorentz invariant' in point space, $x_\mu x^\mu$, which we might also fix to the 'light cone' $x_\mu x^\mu = 0$, by means of a regular null system, we have several possibilities to 'insert' a six-dimensional skew matrix M with an appropriate normalization by $M^T M =$, or $M^+ M =$, or even $M^2 =$. So the related second *class* surface may be written as $u^T M^T M u$, and in the general case, we may also treat additional symmetric transformations instead of using only skew 'M's related to null systems if we consider duality and treat the singular tangential case of the plane carefully with respect to the surface.

We have discussed in [50] the interpretation of the Dirac 'spinors' u and v in terms of 4-dim Hesse forms of planes, which reflects the idea to consider the correlations as projective mappings in 3-space, either by null systems or by polar systems (duality/reciprocity). There we can address easily well-known properties of Dirac theory and the $g-2$ discussion, and we find a consistent interpretation because we have switched from second *order* surfaces to second *class* surfaces. Moreover, we are consistent with Study's discussion

[97] Details are discussed in [50] with respect to null systems and as well with respect to polar systems and surfaces of second order and of second class. This yields also details with respect to the non-local 'Hilbert' rep of Lie generators which we've mentioned in appendix E to find operator counter parts of line generators vs. second order surfaces. Last not least, we address there [50] Dirac's papers on wave equations [81], [82] in de Sitter and conformal space where – similarly to Cartan's theory of spinors (see [12]) – reference to geometrical background and to existing relevant references are *NOT* given by the author.

[98] Note that in case of symmetric M or M'', this yields reps of second *order* surfaces or second *class* surfaces.

[99] Essentially, this rises the quest of an adequate identification of Mx with the adjoint x^T, or x^+ which has to be answered.

[100] To provide a *phenomenological* association, one can map one (classical) line to one su(2) generator in mind, linear independence of the 3 generators maps to (pairwise) skew lines, so su(2)⊕su(2), or su(2)⊕i su(2) map to two families of lines, and we may even understand i by orthogonality, i.e. in case of the hyperboloid the two line families are orthogonal.

of normal systems and Hamiltonians [69], and the four parameters of the '4-momentum' gain immediately natural explanations if we consider e.g. tangent planes to a sphere in the Hesse form which have distance $\pm p_0$ from the center. So also the wave picture of spherical waves vs. plane waves can be related easily. The general correlation mapping can be decomposed into a 10-dim symmetric/polar part $g_{\mu\nu}$ (which can be further decomposed into 6+4, or 6+(3+1) of the 4×4 symmetric matrix rep) and an antisymmetric 6-dim part $\omega_{\mu\nu}$ of the null system. This addresses also some of the questions and problems raised by Sexl and Urbantke in [83], ch. 4, with respect to a generalization of classical mechanics to '4-momentum vector'-calculus and related approximations. We discuss few more aspects on the '4-vector' and the rep by partial differential operators with respect to linear rep theory on Hilbert spaces in appendix E. All cases, however, show that by means of a '4-momentum' we discuss a plane rep and no 'vector' as part of the usual line rep (an orientation) of 3-space, and in order to use geometrical lines, we have to invoke the 6-dim homogeneous rep.

Conclusion and Outlook

By putting simple line geometry on top of our reasoning, we have derived an even more general description of 'special relativity' in point space where the restriction $\gamma^2(1-\beta^2) = \gamma^2 - \gamma^2\beta^2 = 1$ discussed in sec. 4.3. yields the usual Lorentz transformations in point space x_μ. On the one hand, this reflects the Plücker-condition (the M_4^2-constraint) in P^5 to realize lines in P^3, on the other hand the Lorentz symmetry is caused by fixing the line coordinates $p'_{23} = p_{23}$ and $p'_{01} = p_{01}$ of two skew lines individually, and it is immediately related to the coordinate tetrahedron, to Congruences (or ray systems), and as such to well-known electromagnetic theory and optics. Moreover, by fixing an absolute plane and applying classical Caley-Klein geometry, the usual spinor description is embedded, too, and we've identified the 6-dim transformation group in point space *NOT* as usual in terms of an so(4) (or su(2)$\oplus$su(2)) Lie algebra pointing to Weinberg's spin discussions but in terms of null systems and their linear Complex rep by 3 left- and 3 right-handed Complexe. This, however, sketches the scenario as well as the direction to develop further.

It is evident that we have discussed only few aspects of the general framework here, however, we think that the major idea to use different, but equivalently suited geometrical reps instead of the point picture has enough relevance to describe physics in 3-space. As such, we want to emphasize as a first stage what we've called in [24] a 'programmatic approach'. So in analogy to the figure given by Ehlers et al. [80], featuring certain requirements and grouped relations of 'relativity', we propose to start even earlier by pursuing two stages, and within the stages even two parallel tracks each of which should be compared with respect to each aspect within one track versus the other track: On the major level in stage I, to describe 3-space by PG, we should have one track using points and planes (to initially respect *linear* reciprocity/duality) in the context of usual physical/mathematical descriptions. Within a *parallel* track in stage I, one should address the same problems while focusing on reps in terms of line and Complex geometry. In both tracks, one can perform the steps from projective geometry and appropriate reps 'down to' affine and Euclidean space which allows on one hand to trace the results and their origin, on the other hand, we can compare both descriptions and relate them analytically to compare them to physical descriptions and reps. Both tracks can be confined to Klein's 'Erlanger Programm' so that we can identify objects and transformations, i.e. we can relate to known concepts and identifications, and we have strong transfer principles at hand to relate 'objects' like e.g. lines and spheres [13], §7, or properties, like e.g. special tangents and curvatures [13], §12.

In a second stage II, we can go beyond, and that's what we understand when talking of the 'future of relativity'. Having located e.g. Lorentz transformations in Complex geometry as above, we have the possibility to use other viewpoints and generalize aspects of the first stage with respect to e.g. rep theory of a more general object or a more general superset of transformations. So it is this stage II, where we see advanced concepts of projective and algebraic geometry, however, for our part we follow Plücker and Lie with respect to higher dimensions in that we interpret them as base elements of a different 'new' geometry, respectively, and not in the nowadays usual fashion to generalize formalism to arbitrary n, mostly following Grassmann and Riemann. In the example of Lorentz transformations, by starting the original invariance discussion of the quadric $x^2 + y^2 + z^2 - t^2$ in point space, one can – of course – discuss the standard form of LT.

However, even with point space reps, one has the relation to Ampère's, or to Monge's equation like discussed in ([27], see also sec. 4.1. and figure 1) which yields more background, or one can also think of exchanging quadratic subquadrics by mappings, or discussing the invariance e.g. of $x^2 + y^2$ with respect to conics, or even to absolute points by $x^2 + y^2 = 0$, etc. More interesting considerations appear, however, if we enter the P^5 discussion. Having started in line geometry from an originally 4-dim theory, Plücker has introduced his fifth inhomogeneous line coordinate η to preserve the grade of the coordinates with respect to *linear* transformations. As such, one has further introduced senary homogeneous line coordinates e.g. according to eq. (21) to work with a *linear* transformation theory. One possibility to check for alternatives accordingly is to work 'in the other direction' by loosening the conditions, especially with respect to non-linear mappings and applications. Another possibility using P^5 and the Plücker-Klein quadric is a more geometrical one: If we restrict general P^5 to the Plücker-Klein quadric by a quadratic constraint, we can select three special points on the quadric to represent a family of generating lines, and construct a line-based (i.e. *linear*) rep of a quadric in 3-space with rep counterpart in the point picture. Now whereas we can understand and represent the transformations of the respective quadric in point space (e.g. LT with respect to an 'hyperboloid') by transforming with respect to the generating linear elements ('lines'), we can in parallel remember the triangle of points on the Plücker-Klein quadric associated to the generating lines, and discuss transformations in P^5, also with respect to the different 'duality' situation. Last not least, we have thus reached the regime of general transformations acting on P^5 and of general non-linear mappings, which opens up for further research with respect to physical applications.

We want to close for now with two physical pictures which we've discussed this year [72] and which yield some insight into geometry vs. rep theory, however, here we just want to present briefly the synthetic part of both pictures.

The first picture relates to the construction of the one-sheeted hyperboloid, so we have a real case which can be visualized. We know that the hyperboloid can be generated by three lines of a family, or we can use the P^5 picture above, and select three Complexe, or a Configuration [26], to extract such lines by a linear constraint. Now, if we understand the lines as axes of three individual special linear Complexe, we may, of course, use the appropriate reps as well, i.e. we now have three 6-dim objects $F^a_{\mu\nu}$, $a = 1, 2, 3$, but completely equivalent to the electromagnetic case which we've discussed above. Now we know also, that the hyperboloid constructed in this way, has a symmetry axis with respect to a rotation around this line, i.e. an SO(2), or U(1) dependent on the rep. So we can introduce a fourth special linear Complex $F^0_{\mu\nu}$, accompanied by some parameters to qualify either a single hyperboloid (or we can admin even a family of hyperboloids with this symmetry axis). Last not least, we can even include an observer moving linearly, i.e. we have an additional line, and to make things easy and preserve the rotational symmetry of the hyperboloid[101], we assume that the line of the observer movement intersects the symmetry axis of the hyperboloid $\sim F^0_{\mu\nu}$. So we can construct our Langrangean in terms of line coordinates by the standard scheme of intersecting lines (see [39], or [24], or [25]), and thus end up with what is nowadays called a 'gauge theory'.

The second aspect relates to planar coordinates (or as well as differential operators) with respect to null systems. Although being used to discuss transformations or movements and dynamics in terms of points and (Euclidean) displacement, we can also take to points of the trajectory e.g. of a curve or from a tangential plane. But in this plane, we can associate a null point with the first point, and with point and plane fixed, we'll find a null system relating both. Now for the same Complex, we'll find no second null point in the same plane[102]. As we've required to connect the points *in the same plane*, we have necessarily to introduce a second Complex in order to associate its null point to the second point. This introduces immediately a second, skew matrix to describe the same plane, but as the null plane of the second point. So a point translation in the same plane (or its rep by 'derivatives') alters the rep in terms of classes because we have to consider the two skew reps for one and the same plane u. To preserve the skewness of the underlying null system(s), this can not be realized by simple matrix multiplication, but we need the commutator, so the whole description once more seems to require Lie algebras and groups like known from covariant derivatives. In addition, as with each null point, we may associate a line pencil, we may apply elementary PG (see e.g. Doehlemann [56]) which yields projective relations and a planar conic generated by mapping the two pencils. But moreover, we may as well switch to P^5 and associate Congruences and the related physical reps, or treat it completely with P^5 transformation theory.

A Coordinate Systems and Velocities

Now, in order to discuss some aspects of line geometry versus the point picture especially with respect to velocities and applications of Lorentz transformations, we do not want to perform all the – almost trivial and straightforward – algebra, but just recall few geometrical aspects and pictures.

[101] This is no restriction, because otherwise (like in the theory of the massive top) we can use Staude's reasoning and select the axis of the Staude rotation [84].

[102] Using the same Complex and shifting the point will result in a rotated plane at the other point. Although (due to the skew rep of the Complex), we can interpret this in terms of 'skew generator reps' and Lie theory, or in special 'derivatives' or some kind of precession, we are in the mind setting of (singular) tangential transformations and polar theory, i.e. using second order surfaces and symmetric reps.

As such, it is noteworthy, that the original analytical approach to line reps used a 'running coordinate' z in Euclidean 3-space, and described the two remaining space coordinates[103] x and y by $x = rz + \rho$ and $y = sz + \sigma$ [26] (see figure 2). Of course, this is

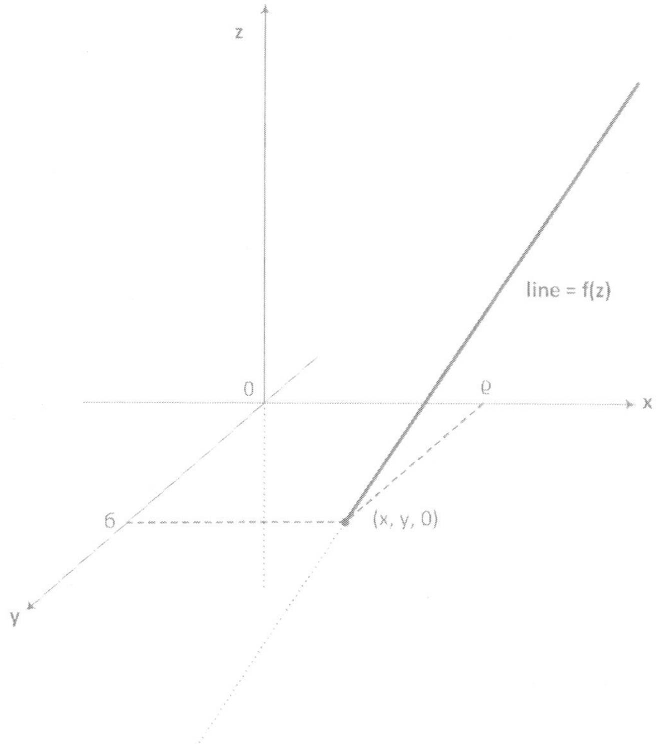

Figure 2. Notation used by Plücker, or Lie.

sufficient to describe the intersection point ρ, σ at $z = 0$, and due to the slopes r and s (i.e. the slopes of the planar projections of the line onto the respective coordinate planes) the line is determined throughout space and $\forall z$. One can also generalize this description versus analysis by setting $x = f_1(z)$ and $y = f_2(z)$ where $f_1(z) = rz + \rho$ and $f_2 = sz + \sigma$ describe the simplest, linear case, and one can think about other functional dependencies $f_i(z)$, also about switching to $z \in$ and performing analysis, relating geometrical pictures like the stereographic projection, etc. We have modified this approach using homogeneous coordinates to discuss spin and spinors, as well as aspects of special relativity with respect to the exceptional value $\eta = 0$ ([12] and [10], or sec. 4.1.).

As we want to keep our focus on lines and different reps only, we do not follow these aspects. Here, it is more interesting with respect to relativity – which can be understood also to make claims on velocities with respect to our surrounding space-time – to introduce a point on the line and discuss linear motion. This can be seen from two sides:

- Given a point, acting with a force on the point will put the point into motion. Instead, we can assume also existing linear uniform motion and ask for the trajectory rep

[103] Also at that time, people often used – like in Plücker's and Lie's case – a different orientation of the coordinate system.

throughout *time*, which of course should yield a line rep in 3-space. So today, the usual rep in 3-space will consist of a 3-vector pointing from the coordinate origin to a point on the line, and a second 3-vector pointing in the direction of the force or the movement[104]. In order to describe the point on the line, we thus have a 3-vector at 'time' t_0 when we've 'started' our process, and we can track the point using a parameter $t - t_0$, or Δt. So for all 'times' $t \geq t_0$, we 'know' the position of the point, however, we also know the integrated line as trajectory. So this picture 'lives' from the fact that a line is a point set, and that we can tag the points on the line individually by an appropriate (linear) parametrization if in addition we accept 'velocities' of the point on the line.

- An alternative geometrical view can be given by understanding a tagging of points on the line e.g. by its intersection with other known lines of a planar pencil, or even by known planes of a spatial pencil. So whereas on the one hand, we can construct lines as point sets (with at least two points), on the other hand the intersection of two lines, or a line by a plane, is by geometrical definition a point. In this picture, the movement of a point on a given line can be equally well parametrized by intersecting the line with a pencil of lines or planes, and individual points on the line are tagged by tagging the respective intersecting lines or planes, and by ordering the pencil elements appropriately. Instead of a 'coordinate system' on the original line e.g. by a 3-vector and 'time', we may use as well the coordinate system of the pencil and an additional rule (i.e. a mapping) related to the metric (if necessary). In the planar case, the line pencil (which we can e.g. take from a null system/Complex with the null point as center) and the original line should be coplanar, and the center of the pencil (in our example the null point) should not be incident with the original line (or analytically: $x_i l_i \neq 0$, where x_i are ternary homogeneous point coordinates and l_i are planar line coordinates). In other words, in this scenario the line l shouldn't be a null or Complex line. In case of the pencil of planes in 3-space, the axis of the pencil (i.e. the intersection of the planes) should be skew to the line of movement, so we can apply Congruences and null systems, too, or when introducing 'conjugation', besides using null systems, we can introduce second order surfaces/quadrics, and apply polar theory and duality/reciprocity (i.e. symmetric matrix reps and 'anti-commutators'). Now, the points on the lines are annotated by an 'angle' within the respective pencil related to fundamental rays or planes of the pencil, which we can describe by an anharmonic ratio (or better with von Staudt: as 'Wurf'). PG then yields several mechanisms to adopt this view to analytical and algebraical reps, typically by 'Würfe' or anharmonic ratios, and by deriving metrical properties thereof, see e.g. Doehlemann [56], III. Abschnitt.

Last not least, we may introduce and discuss 'velocities' to denote the motion of the point on the line by different 'coordinate systems', or simply by different reps. Whereas the original static picture $x = f_1(z)$ and $y = f_2(z)$ fails without further assumptions on a parameter dependence of z and enhanced notation, we are used to think of a linearly

[104] This picture shows impressively that the notion of a 'translation' in time is attached to this vectorial picture only. In essence, we consider different possibilities to define a parameter on the line, and there are other parameters in different geometrical setups suited as well.

moving point (i.e. a tagged point on a line!) in vector notation using point coordinates. So

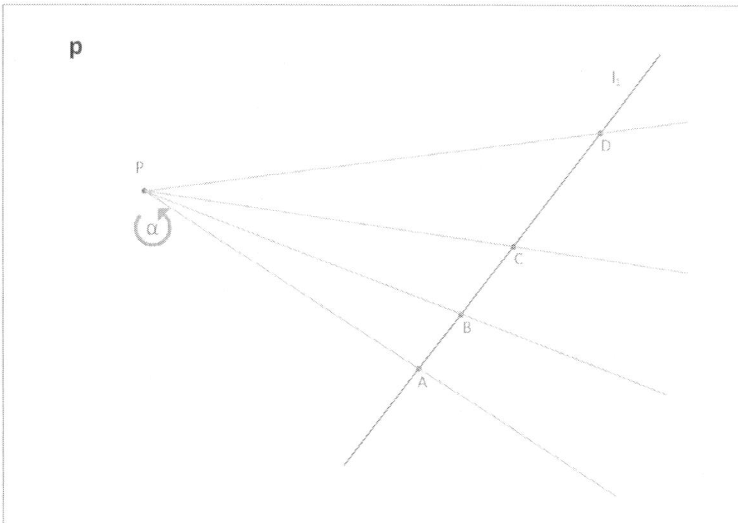

Figure 3. Points on a line l_1 versus a pencil coordinate α.

the adequate picture (see e.g. figure 3) – as soon as we *interpret* the direction of the line by associating it with a velocity vector[105]. In terms of the line pencil, however, switching to the different parametrization, we have *to interpret* the coordinate of the planar line pencil as an angular velocity related to a rotation so that 'over time' we measure different angles of α. If we interpret the original motion as continuous, we have introduced, of course, once more a dependence of a 'time' t, i.e. $\alpha(t)$. The same picture holds for null systems or Congruences if we use the angle between the planes[106].

Not that this switch is a problem, however, notation and the physical pictures change although they are, of course, related, and the relation between α and the position measured on the line – although nonlinear – can be given immediately because we know the respective functional dependencies from planar (or spatial) Euclidean geometry nowadays by heart, and we have even introduced names or naming of their groups like 'trigonometric functions', 'hyperbolic functions', 'HuppelDuppel' functions or polynomials, etc.

Using this simple setup, we can already discuss *various* possibilities describing point velocities on the line l_1 versus pencil parameter α:

- $\delta\alpha$ between pencil lines is constant per 'time' interval, i.e. $\omega = \alpha t$. Switching to metric properties on l and assuming common 'time' t for line and pencil yields that related distances on l between intersections in the line metric are a priori NOT equal. If we introduce 'physics' on the line (i.e. a velocity v on the line l with respect to

[105] It is important to understand that this is an intrinsic identification of one of the two vectors of the vector description above with a physical picture. In the linear case, this is well expressed and justified by the linear dependence of the parameter on the line which we interpret, relative to the velocity, as 'time'. There are, however, alternatives, e.g. by discussing points in the absolute plane.

[106] With respect to covariant derivatives and skew transformation terms, or effects like Thomas precession, see [50].

the metric and synchronize clocks at the intersection points), we find an accelerated motion $\dot{v} \neq 0$.

- We can change the usual metric to a 'non-Euclidean metric' on the line in that we require the distances between the intersection points on the line to be equidistant. Then in the linear picture, we can recover linear space and velocity, however, we have to pay the price with a strange (and for our common picture unusual) 'time definition'. Our understanding can be restored if we recall that a projective line can be understood as a special circle closing through 'infinity' (i.e. by enhancing the Euclidean picture by an absolute element). Then, the line behaves as a circle around the center of the planar pencil. However, we have to switch to projective instead of Euclidean geometry only[107].

- As an alternative, we can absorb this 'strange' behaviour by using different 'times' on l and on the rotating line of the pencil, so dependent on the metric on l whether we require the distance or the velocity to be constant, this can be absorbed as well by different non-constant rotations, or even by non-linear 'times'.

So in general the notion of velocities is generated by identifying and tagging a special (sub-)element of a lower stage – the point on the line or the line or plane of the respective pencil – and recalling projective construction schemes. The 'velocity' is introduced by the observation that the tagged elements evolve 'in time' through the allowed (or a priori existent) positions of the superior element, i.e. the mapping itself can be understood to change 'in time'. In essence, we thus obtain different mappings between the line and the pencil by different identifications and transfer rules.

In order to gain some more insight into such constructions, we may assume a second line l_2, and if – by chance – it happens (see figure 4, or the discussions in [24], [28]) that the motion of a point on l_2 can be tagged in the same manner like the point motion on l_1, we can apply some fundamental theorems of planar projective geometry. The points are determined by linear sections of the same pencil which relates them, i.e. we can write a well-defined mapping down by means of the anharmonic ratios of point sets and lines although the individual 'velocitities' on l_1 and l_2 are different. It is obvious, that this situation changes, if the points at the respective angles α_i do not coincide, or if – like in figure 5 – there is a second 'observer' at P_2. Nevertheless, we can discuss this planar setup by projective, or perspective mappings [56], and use anharmonic ratios to describe and relate the setup parameters.

Last not least, leaving the planar setup and switching to lines in space, we have even two scenarios at hand:

The general setup we have discussed so far, are two skew lines where points are moving (see [24], or [28]). In this scenario, to connect (or relate) points, we cannot in general use only *one* planar pencil, but we have to discuss Congruences, or 'ray systems', i.e. we need to consider at least two linear Complexe and their common lines. This illustrates why we can use the description of line geometry given above to treat Lorentz transformations and relativity.

[107] We do not want to discuss the Cayley-Klein mechanism and associating different polar systems here in general.

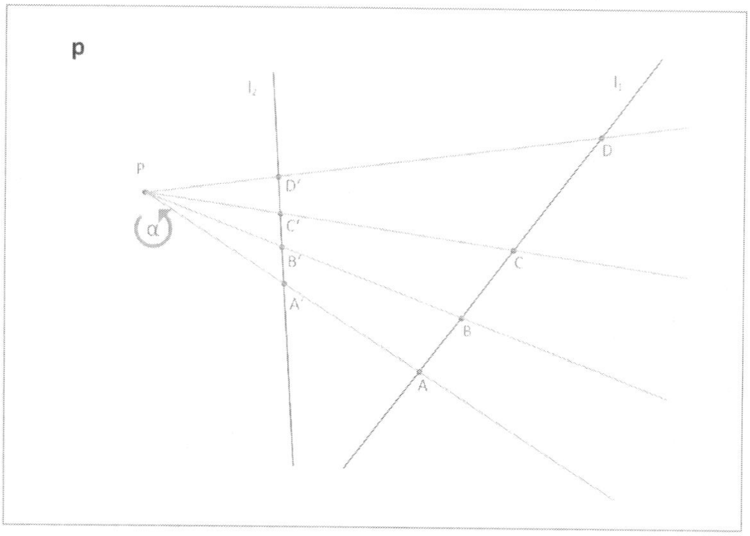

Figure 4. Points on two lines l_1 and l_2 with different velocities versus the same pencil coordinate α.

The second (and related) scenario can be developed right from figure 5. With respect to 'the second observer' at P_2, we may switch the lines l_1 and l_2 'out of plane' into different spatial positions. But now we can apply a construction scheme which Plücker has published [85] as an alternative to Monge's generation of a one-sheeted hyperboloid. Understanding P_2 as center/vertex of the original (planar) line pencil, the plane transforms into a one-sheeted hyperboloid while moving the lines out of the plane p. The two lines l_1 and l_2 (where their intersection points with the pencil remain fixed in the sense of a mechanical model/construction e.g. by filaments, or fibres), now positioned in space, belong to one of the family of generators whereas the intersecting lines belong to the second family. Plücker in [85] discussed explicitly this filament/fibre model, and he categorizes his *second class* construction versus Monge's construction as well as the rôle of duality. Especially with respect to cones, he emphasizes their singular/special character and comments on the relevance of duality.

As such, the notion of 'relativity' in general can be associated to construction schemes in PG, and as such to the respective transfer principles. Moreover, it can be subsumed under Klein's 'Erlanger Programm'. And it is obvious, that the parameter (or 'coordinate') 'time' has a different meaning by its tagging function than point reps in terms of pure spatial coordinates.

B Projective Generation and Fields

In addition, we want to recall a construction scheme of field reps in terms of circle pencils which is automatically introduced by the tetrahedral Complex and some transfer princi-

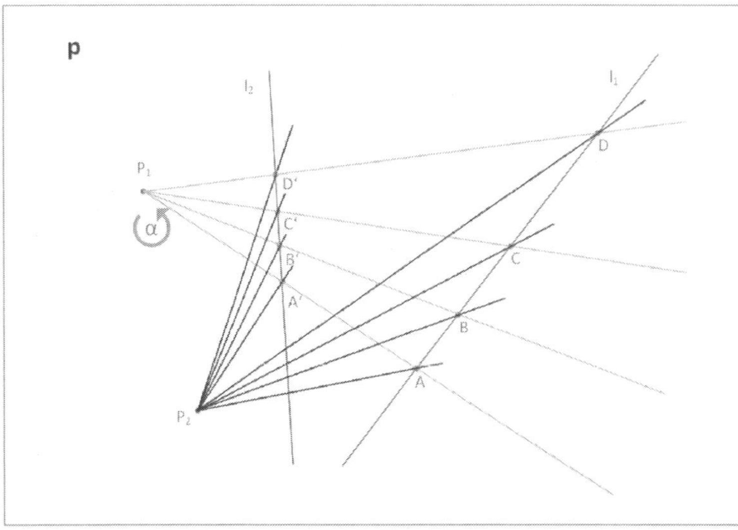

Figure 5. The same setup with two 'observers'.

ples[108]. We have mentioned above the relation of the tetrahedral Complex and the fundamental coordinate tetrahedron. It is also evident that lines in 3-space hit the planes of the coordinate tetrahedron in four points (the four planes opposite of the vertices of the tetrahedron, or within a self-polar fundamental coordinate tetrahedron the planes even dual to the vertices). Now by recalling that the tetrahedral Complex, a quadratic Complex, consists of all lines with the same anharmonic ratio κ of the four intersection points with respect to the tetrahedron, we find an associated set of ∞^1 such ratios depending on different intersections of lines with the fundamental coordinate tetrahedron. In other words, given a fundamental tetrahedron of coordinates and an additional parameter, interpreted as the anharmonic ratio of the intersection points, one tetrahedral Complex is selected. Note that the anharmonic ratio is invariant under projective transformations of projection and intersection!

Now, if in the next step we take a line with its four intersection points, and if – in addition – we intersect the line with a plane, i.e. the line with its four points are all elements of the plane (i.e. incident with the plane), we can apply an interesting geometrical construction

[108] We have omitted here the relation to W-curves and W-surfaces, because this is an interesting topic in itself with respect to Würfe and metric considerations. If of interest, please see the introductory remarks in [30] §41. We have postponed also Hesse transfer and rational curves, although by means of the tetrahedron, we should discuss cubics in general as well as twisted cubics. We've made first use with respect to reps of the quadric and spinors throughout [12], but it is evident that with respect to our old work on SU(4) and the tetrahedral symmetry of the symmetric reps **4**, **10**, and **20**, we may investigate appropriate cubics passing through the vertices and their relations. We've used **20** to treat the $N\Delta$-reps ('3 quarks') in order to complete vector and axial charge commutators in pion interactions, and PG seems to yield the appropriate background if we associate **4** with the fundamental tetrahedron of point reps, **4*** $\sim (1,1,1)$ as conjugate with plane coordinates, **15** with the adjoint rep, and construct reps as usual. Moreover, the transformations of points on the line using binary reps has a couple of interesting applications and identifications by means of Clebsch [31] and Hesse transfer [29], besides exchanging vertices of the tetrahedron of the tetrahedron as points of a C_3 when using the 3-dim transformation group [30] §51.

scheme [56] in order to attach pencils of circles and their orthogonal pencils[109]. Because Doehlemann's original purpose to investigate point involutions on the line by identifying the two carriers of the points is comprised, identifying point sets on the remaining common carrier, the line, yields an additional projective mapping, and a birational mapping of the individual 'parameters' [56] §§17ff. Moreover, we thus can apply immediately the original framework of Hesse's transfer principle to the point sets, use binary form reps and the associated differential equations [31] which occur in various contexts in physics and especially 'quantum' theories.

Here, we focus on the construction of circles and pencils of circles by using the line from the tetrahedral Complex with its four intersection points A, B, A', and B', with the fundamental tetrahedron[110]. In addition, we associate a point P off-line in the plane p to avoid analytical problems with degeneracies for the moment (see figure 6). Physically, the point P may be associated with a probe or an observer, and it is evident that with respect to the fundamental tetrahedron, we can either associate quaternary homogeneous space-coordinates of P^3, or switch to affine or Euclidean coordinates, i.e. we will find linear (or of order 1) reps of this point.

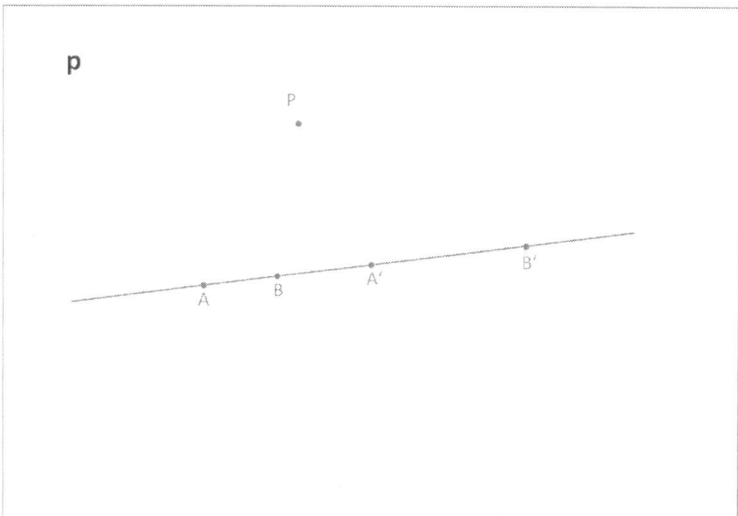

Figure 6. Step 1: Define point P and plane p with respect to the line, the four points A, A', B, B', and their anharmonic ratio κ.

However, if in the next step we intersect the point sets (P, A, A') and (P, B, B') by circles like in figure 7, we can denote the intersection point of the circles by Q (see figure 8). For the upcoming steps, the line with the points A, B, A', and B' is no longer of direct use, but we draw all circles through the points P and Q like in figure 9.

[109]From the physical viewpoint, we enter the discussion of potential lines and fields. Here, we leave the discussion open on whether the intersection points of the conics should be identified with physical charges or masses.

[110]To emphasize the point involution mentioned above, one can map $A \longrightarrow A'$, and $B \longrightarrow B'$, and follow Doehlemann's outline.

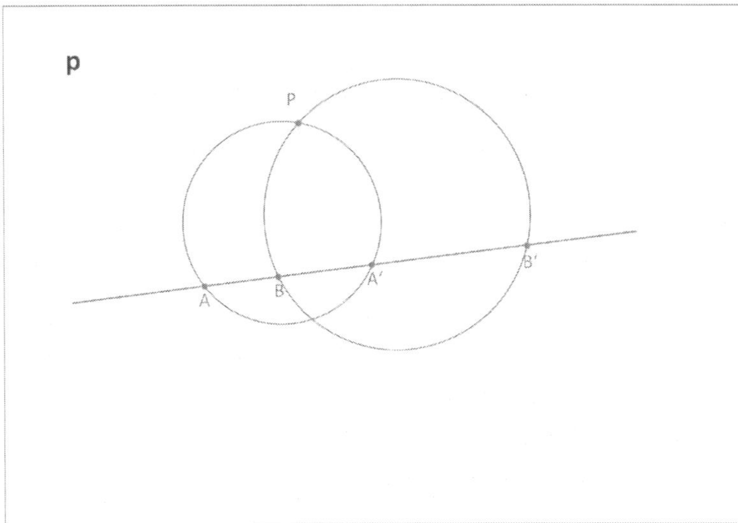

Figure 7. Step 2: Connect P with the two point sets (A, A') and (B, B') by circles.

In this figure, we have drawn only few lines to recall the principle but it is evident that one can find planar circles connecting each planar point with P and Q, i.e. we may attach (at least) one quadric rep to each planar point[111]. So in Plücker's sense, we may as well switch to a geometry using the circle as base element of this geometry[112].

In a last step to get closer to usual descriptions[113], it is helpful to change the planar coordinate system like in figure 10 where the first axis is a line incident with P and Q, and we may choose the origin by an orthogonal axis half between P and Q, i.e. in addition we can fix the orientation by the anharmonic ratio (Q,0,P,∞)=-1. As a comparison, we want to refer to the picture in figure 11 as given in [30] when presenting pencils of circles and of orthogonal circles[114].

Now if we recall P representing originally a probe or an observer, we have introduced a description where the 'forces' (or field line orientations') point towards Q, but we can introduce in addition orthogonal sets of circles or conics to introduce the notion of 'potentials'. As most of this is classical analytical geometry, we leave it to the reader to play with conics and partial differential operators using incidence relations (due to tangential conditions) in the linear and the quadric case. Moreover, it is interesting to play with the 'two infinities' of the rectangular axes and relate them to the scenario where the plane is the projection plane of a stereographic projection and both axes meet in a second point. Vice versa, one can think to replace P and Q by $\pm i$, and perform some algebra also with their anharmonic ratio versus 0, and ∞.

[111]Note the relation to second order partial differential equations.

[112]With respect to spheres in 3-space, we may use symmetry arguments for now, however, the analytic rep has to base on Lie transfer.

[113]In order to remove the circular restriction, one can even switch to general conics and 'deform' the description which complicates the setup considerably.

[114]We have omitted the orthogonal circles here, but like in the case of line generators of quadrics, one can of course consider the orthogonal system, too, which then usually appears as 'rotated' by 90 degrees, or i.

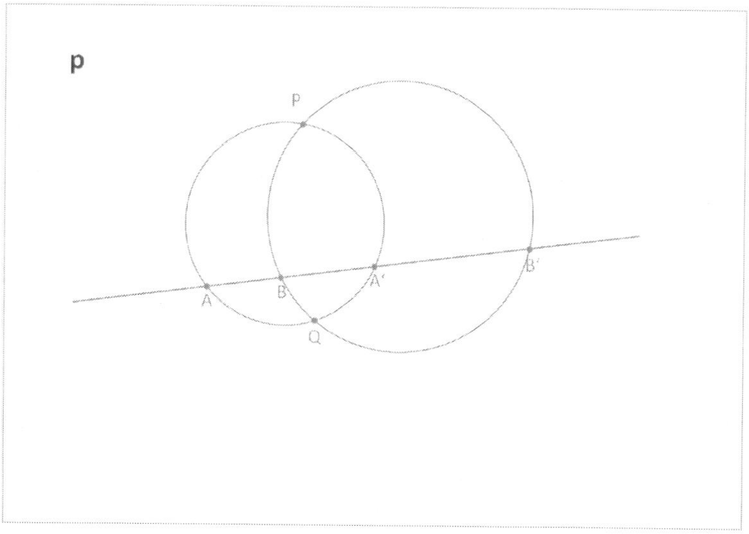

Figure 8. Step 3: Identify Q as second intersection of the two circles.

C Complex Numbers and Geometry

Although it sounds silly to discuss nowadays complex numbers because everybody is used to work with them from scratch, it is worth to briefly recall their use by means of a simple geometrical example[115].

As example, one can write down the intersection of a circle and a line in planar coordinates where it doesn't matter whether we use two Euclidean coordinates or three homogeneous planar coordinates. We obtain – of course – three cases to be distinguished:

a) the line hits the circle in two real points ('secant'),

b) the line is a tangent to the circle (one real double-point), or

c) the line misses the circle, i.e. there is no intersection in the Euclidean plane.

The analytic calculation to find the two intersection points by finding the two roots is well-known.

Now it is noteworthy, that exactly the physically not so interesting case where there is no intersection in the Euclidean plane requires to complex roots to treat the analytical rep in terms of coordinates. In other words, the complex numbers serve as an algebraic or analytic unification tool by introducing the symbol i and appropriate rules to work with this symbol. In the planar case, this extension serves to treat two 'conjugate' points on the absolute line, whereas in 3-space, we have to introduce additional symbols to treat a line or a plane vs. a second order surface. Based on Lie transfer and Study's work in line geometry, we have discussed the construction of spinors in [12] by means of homogeneous coordinates while respecting the absolute elements and treating 'null' elements. So it is justified to understand complex numbers (and quaternions, and in general hypercomplex

[115]Which – by the way – enforced the introduction of spinors and their special properties...

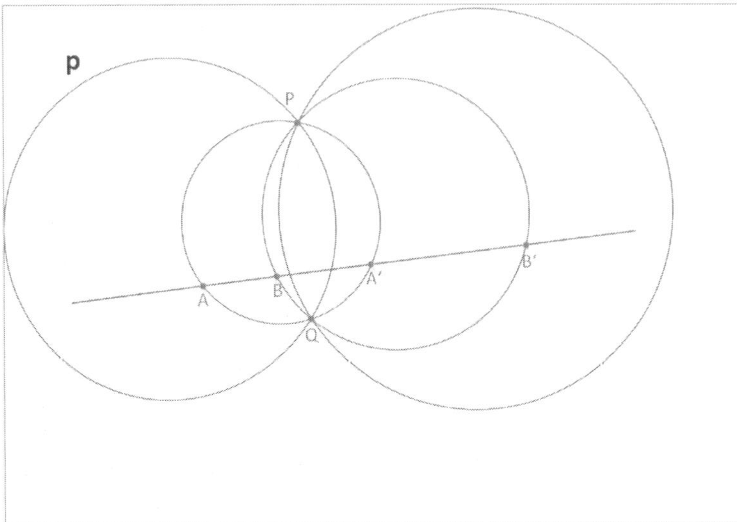

Figure 9. Step 4: Visualize the pencils of circles through P and Q.

numbers as well) as symbols to serve for algebraic/analytic completion when representing geometry and associating absolute elements in homogeneous coordinates to the respective geometries.

However, being physicist, when being asked to vote for the interesting case(s), I'd vote for a) and b) where the line meets the circle at least in one point and where the probability that 'something interesting happens' seems to be much higher than in the non-intersecting case c). It seems justified to regard and understand the case c) as being necessary to complete the algebraic/analytic description formally, and to see whether the mathematical tool sets work appropriately or not – the more, as we can apply a lot of more unphysical ideas to this picture like shrinking the circle or even the line to a point, etc. which introduces ambiguities or even singularities.

D Einstein and 'Six-Vectors'

As brief and quick access to Einstein's publications in order to guarantee a certain overview of the original concepts of special and general relativity, we've used publications of his talks in the transactions of the Prussian Academy of Science [48] during his time in Berlin as well as his publications in 'Annalen der Physik' which we'll discuss briefly with focus on the use of the 'six-vector' in order to attach Complex geometry to special and general relativity as discussed nowadays by 4-vector calculus and differential geometry.

D1. Prussian Academy of Science

Here, we've found three publications which we want to mention for later geometrical use.

In [86] Einstein mentions in a very small paragraph briefly 'six-vectors' as an alternative notation in the context of 2nd-rank antisymmetric tensors (p. 1037/1038) without physical

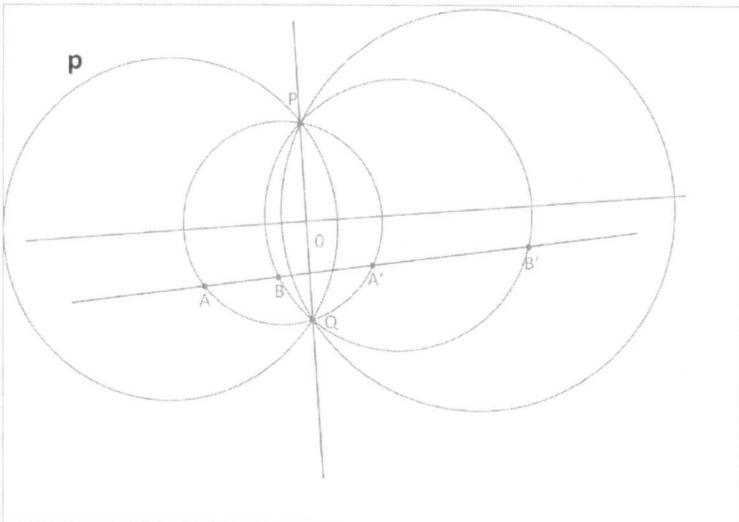

Figure 10. Step 5: Draw line through P and Q, and an orthogonal line at 0.

context. Only after having introduced the 'antisymmetric fundamental tensor of Ricci and Levi-Civita' yielding[116]

$$\begin{aligned} G_{iklm} &= \sqrt{g}\,\delta_{iklm}, \\ G^{iklm} &= \frac{1}{\sqrt{g}}\,\delta_{iklm}, \quad \text{and} \\ G_{ik}^{lm} &= \sum\nolimits_{\alpha\beta}\sqrt{g}\,\delta_{ik\alpha\beta}\,g^{\alpha l}g^{\beta m} \\ &= \sum\nolimits_{\alpha\beta}\frac{1}{\sqrt{g}}\,\delta_{lm\alpha\beta}\,g_{\alpha i}g_{\beta k} \end{aligned} \tag{55}$$

in eqns. (19), (21a), and (22), Einstein discusses the case of dual six-vectors in eq. (24),

$$(F^{\mu\nu})^* = \frac{1}{2}\sum_{\alpha\beta} G^{\mu\nu}_{\alpha\beta}F^{\alpha\beta} \tag{56}$$

with respect to the electromagnetic field rep. However, the whole discussion seems to focus mainly on tensor algebra. The major part of discussing 2nd-rank tensors throughout this contribution is related to symmetric tensors which yield 10 components and which in §9, p. 1055, is related as $T_{\sigma\nu}$ to the energy tensor. Because we do not want to discuss details here due to the focus on six-vectors (and Complexe or null systems), we just want to emphasize the connection of the symmetric part to polar systems and second order surfaces for later use.

In [49], Einstein starts from Minkowski's identification of the six electromagnetic field components as six-vector and discusses Maxwell's equations. Einstein defines a 'new' six-vector $\mathcal{F}^{\mu\nu} = \sqrt{-g}\sum_{\alpha\beta} g^{\mu\alpha}g^{\nu\beta}F_{\alpha\beta}$ which in case of special relativity reduces to the usual field components $\vec{h}$ and $-\vec{e}$ in order to incorporate electromagnetism into general relativity

[116]Here, Einstein uses δ instead of the nowadays common notation ϵ for the totally antisymmetric tensor with 4 indices.

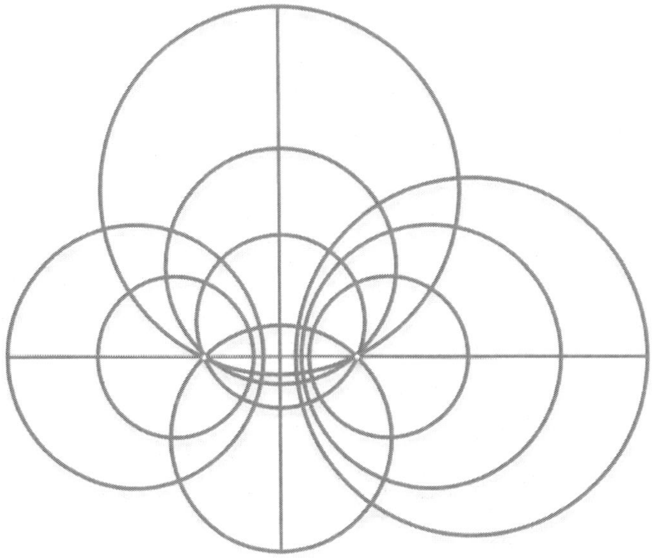

Figure 11. Comparison to FIG. 10: Pencils of original and orthogonal circles given by Klein [30], p. 42.

by means of the tensor $\mathcal{T}_\sigma^\nu$,

$$\mathcal{T}_\sigma^\nu = \sum_{\alpha\beta} \left(-\mathcal{F}^{\nu\alpha} F_{\sigma\alpha} + \frac{1}{4} \mathcal{F}^{\alpha\beta} F_{\alpha\beta} \delta_\sigma^\nu . \right) \tag{57}$$

From our point of view, it is important to note that throughout his calculations, he uses only the field tensor $F^{\alpha\beta}$, even in his generalized form $\mathcal{F}^{\alpha\beta}$ and in derived equations, i.e. *one* six-vector (or Complex).

In [87], Einstein and Mayer decompose in §2 the antisymmetric 4×4-tensor rep of R^4. By means of 'the totally antisymmetric tensor η^{iklm}' they exercise an algebra which basically represents working with pairs of four-indices, and which in line geometry is related to switching ray versus axis coordinates. η can be understood as a rep of the 4×4-determinant when working with point or plane coordinates. As such, one obtains a rep of line coordinates and in consequence naturally Lorentz transformations (see the discussion above or [10]).

D2. Annals of Physics

Within his publications throughout this journal, we've found only two places which seem to be related to Complexe, or to six-vectors. However, in both cases this seems to be a naming convention, and Einstein has made no real use of this geometry or related formalisms.

To be complete, we've found in his paper of 1905, titled 'Zur Elektrodynamik bewegter Körper', §8, p. 913f, five times the expression 'Lichtkomplex', i.e. a complex of light, however, without explanation or definition. So there is no evidence that Einstein referenced a line Complex, or whether he used some contemporary notion of describing light in a certain manner.

The other occurence, we've found in his paper on general relativity of 1916, titled 'Die Grundlagen der allgemeinen Relativitätstheorie'. There in §20, he discusses briefly the Maxwell equations of the electromagnetic field based on Minkowski's rep theory, as he mentions.

E Remarks on Representation Theory and Lie Symmetries

E1. Remarks

Throughout the text and the other appendices, we've switched 'on demand' between classical and 'quantum' reps of objects and operators. We have prepared in [50] a more detailed account, however, as we've used various aspects so far, we want to provide a brief overview(or summary) at least on the aspects in 3-space. From the viewpoint of classical geometry, the central aspects we've used are points, lines and planes, and we've discussed mappings where (besides a lot of other 'transfer principles') duality is a central aspect. As such, within classical geometry of 3-space, we have to quaternary reps x and u with respect to points and planes, and besides mapping points to points, and planes to planes, we know correlations mapping points to planes, and vice versa. Lines map under duality to lines, we may use senary reps as above (or nowadays 'Grassmannians' Gr(2,4)), so on the one hand, we have only a single line formalism, on the other hand, we automatically have to take care and work with quadratic forms a priori. As a sufficient framework, we can invoke classical projective and advanced geometry like discussed e.g. in [30] or literature cited there.

The classical approach, however, doesn't provide a rep for what people started to discuss as 'quantum' theories in terms of differential operators on Hilbert spaces, i.e. using function reps. First of all, we want to attach linear rep theory (or 'plane waves'), and in order to associate physical pictures, we need to find appropriate reps on function spaces, too. With respect to relativity, a suitable starting point in classical geometry of 3-space is the plane rep either in general form $Ax + By + Cz - D = 0$ or in terms of Hesse's implicit representation, i.e. if we define the plane as usual by all $\vec{x}$ which are normal to a given *normal* vector $\vec{n}$ in Euclidean 3-space, the plane $\mathfrak{p}$ reads as $\mathfrak{p} = \vec{n}\vec{x} - d$, where d denotes an *oriented distance* of plane and origin. Now, in this picture we can replace the Euclidean coordinates, of course, by the known fractions of homogeneous coordinates above, so this plane can be rewritten either as $\mathfrak{p} = Ax_1 + Bx_2 + Cx_3 - Dx_0 = 0$ for the general form, or by $\mathfrak{p} = n_1x_1 + n_2x_2 + n_3x_3 - dx_0$ in Hesse form, which both formally resemble a 'Lorentz invariant' product $\mathfrak{p} = p_\mu x^\mu$. Now the '4-momentum' p_μ comprises the components $(p_0 = d, n_1, n_2, n_3)$, however, it is a quaternary plane rep which in classical geometry we can treat also as u_α, and as such the 0-component p_0 representing the oriented distance changes sign appropriately. As an example, in the case of a sphere with radius r, we have two parallel tangential planes at distance $\pm r$ at antipodal points of the sphere, and we have to take care only whether we discuss spherical or elliptical geometry, but we do not have to take care of 'anti-particles' or 'time-reversal'. Note, however, that we have attached a special interpretation to the coordinates, and especially the 0-coordinate denotes a *metrical* property. So the calculus itself has to be attached to normals and normal congruences which relates this rep theory on the one hand to Study's summary paper [69], on the other hand we are back again on the original ground of maps of normal Complexe (see [27] and references therein).

If we specialize our coordinatization further with respect to a planar point in 3-space, in order to define the metric we have mentioned the usual procedure to use logarithms of anharmonic ratios of points. So using the above plane rep as argument of an exponential, and adding a phase i to compensate the phase in the metric, yields reps according to $\exp \pm ip \cdot x$. As check, if we take the tangential point as origin, the vector $\vec{x}$ of planar points in 3-space is always perpendicular to the normal $\vec{p}$, so we the logarithm of the metric mapping yields the remaining 0-component p_0 as metric distance. Acting with ∂_α on this rep yields p_α, and we may use standards rules like known from QFT to multiply exponentials. As such, we can see $i\partial_\alpha$ as a phenomenological (however, restricted) replacement of quaternary plane reps on function spaces, especially since the plane exponential remains but produces a linear quaternary rep $i\partial_\alpha \exp(-ip \cdot x) \sim p_\alpha \exp(-ip \cdot x)$, and moreover, it yields according to its later processing e.g. in QFT the necessary information on 'momentum conservation' within this rep. Formally, we can remember the geometrical relation of order and class, so we understand this 'plane wave' rep as class view, however, one has to be careful with the physical interpretation of the components. So the axiomatic approach by first quantization can be understood as well by transferring classical geometry to function spaces, and integer numbers or multiples of 2π emerge automatically for certain geometric conditions of Complexe (or 'screws') (see e.g. [76], ch. 1) or cones (second class surfaces), especially if the vertex is a point in the absolute plane, and we obtain a cylinder. The interpretation of 'currents' or 'lines' by means of such '4-momenta', however, is coincident with the senary line definition only if we respect Lie's picture discussed in sec. 4.1., and visualized by figure 1, i.e. if we understand the normal as the direction of the senary line rep, or in terms of line elements to follow Lie's presentation of differential geometry in 3-space.

Here, we've discussed this phenomenology with respect to identifications and linear reps used in various physical models and discussions. So far, we have reps for points and planes, however, with respect to Lie's and Study's refs cited above with respect to the normal systems of surfaces, we want to find also appropriate reps of lines and second order surfaces to gain control and to reproduce the construction schemes of projective geometry. So besides representing the line family sets of second order surfaces, it is interesting to have possibilities to investigate or even gain control on non-local behaviour. Wigner's 'relativistic angular momentum operator $M_{\mu\nu}$' in [45] shows dependence only with respect to one point coordinate rep, and Gilmore's generator definitions (see e.g. [4], p. 450/451, eqns. (1.28s) or (1.28h)), given for the case SO(n,1) according to $X_{ij} = x^i\partial_j - x^j\partial_i$, $X_{i,n+1} = x^i\partial_{n+1} + x^{n+1}\partial_i$, depend (like elsewhere when performing e.g. angular momentum algebra) only from one point rep x. However, right from the definitions above we see that the action of ∂_{n+1} in this case produces an additional sign, and if we recall that we act on second order surfaces, we obviously perform the algebra of a special or even singular rep[117].

[117] It is not that we want to abolish standard quantum mechanics exercises or standard reps of Lie generators on function spaces, because due to what is known from force systems or e.g. from Staude with respect to the Staude rotation of the massive top [84] which can be represented by one special axis, we can find effective reps attached to one point of space.

As such, we've defined for our use with respect to second order surfaces and (senary) line reps an a priori 'non-local' operator

$$L_{\alpha\beta} = \frac{1}{2}\left(x_\alpha \frac{\partial}{\partial y_\beta} - x_\beta \frac{\partial}{\partial y_\alpha}\right) \tag{58}$$

which we can justify generally for two points x and y of a quadric $S(x) = a_{\alpha\beta}x_\alpha x_\beta$ because acting with this operator (on a canonical/'diagonal' form of S) yields $L_{\alpha\beta}S = x_\alpha y_\beta - x_\beta y_\alpha$ which can be identified as $p_{\alpha\beta}$. So we have gained operator reps which by acting on second *order* surfaces behave as line coordinate reps $p_{\alpha\beta}$.

The typical differential discussion of taking two 'close' points x and y on the quadric S has to be modified due to the quaternary *homogeneous* coordinates used above, however, we can write at least symbolically $M_{\alpha\beta} = \lim_{y \longrightarrow x} L_{\alpha\beta}$ and relate to Wigner's 'relativistic angular momentum' and to usual angular momentum algebra. As such, we discuss details in [50] but right now, one can try to understand Schrödinger equations or Klein-Gordon equations by the class view and planar reps in 3-space.

If – as an example – we calculate the commutator $[L_{\alpha\beta}, L_{\gamma\delta}]$ of the 'line operators', we obtain

$$\begin{aligned} 4\left[L_{\alpha\beta}, L_{\gamma\delta}\right] = \quad & \delta_{\beta\delta}\left(x_\gamma \frac{\partial}{\partial x_\alpha} - x_\alpha \frac{\partial}{\partial x_\gamma}\right) \\ & + \delta_{\alpha\gamma}\left(y_\delta \frac{\partial}{\partial y_\beta} - y_\beta \frac{\partial}{\partial y_\delta}\right) . \end{aligned} \tag{59}$$

For $y \longrightarrow x$, this yields

$$\begin{aligned} 4\left[L_{\alpha\beta}, L_{\gamma\delta}\right] &= (\delta_{\beta\delta}\,\delta_{\gamma\rho}\delta_{\alpha\sigma} + \delta_{\alpha\gamma}\,\delta_{\delta\rho}\delta_{\beta\sigma}) \cdot \\ & \quad \cdot \left(x_\rho \frac{\partial}{\partial x_\sigma} - x_\sigma \frac{\partial}{\partial x_\rho}\right) \\ &= 2\left(\delta_{\beta\delta}\,\delta_{\gamma\rho}\delta_{\alpha\sigma} + \delta_{\alpha\gamma}\,\delta_{\delta\rho}\delta_{\beta\sigma}\right) L_{\rho\sigma} , \end{aligned} \tag{60}$$

and finally

$$\left[L_{\alpha\beta}, L_{\gamma\delta}\right] = \frac{1}{2}\left(\delta_{\beta\delta}\,\delta_{\gamma\rho}\delta_{\alpha\sigma} + \delta_{\alpha\gamma}\,\delta_{\delta\rho}\delta_{\beta\sigma}\right) L_{\rho\sigma} . \tag{61}$$

The remaining operators on the rhs show the form of standard Lie generators of compact groups (see also Gilmore's operator definitions cited above), and due to the quaternary indices, we are concerned with so(4). However, the intricate factor in front of the operators depends on Kronecker symbols of indices and restricts the algebra, and we have to resolve the result and its background by line geometry. This, however, is ongoing, and so far we do not have appropriate results from 'classical' line geometry to compare to.

References

[1] Minkowski H. 1910. Die Grundgleichungen für die elektromagnetischen Vorgänge in bewegten Körpern [The basic equations for the electromagnetic processes in moving bodies.] *Math. Ann.* **68**, 472.

[2] Dahm R. 2010.A Symmetry Reduction Scheme of the Dirac Algebra without Dimensional Defects. *Yad. Fis.* **73** 297; *Phys. Atom. Nuclei* **73** 276.

[3] Dahm R. 2012. On A Microscopic Representation of Space-Time. *Yad. Fis.* **75** 1244; *Phys. Atom. Nuclei* **75** 1173.

[4] Gilmore R. 1974. *Lie Groups, Lie Algebras and Some of Their Applications* (New York: John Wiley & Sons).

[5] Helgason S. 1978 *Differential Geometry, Lie Groups, and Symmetric Spaces* (San Diego: Academic Press).

[6] Helgason S. 1984 *Groups and Geometric Analysis.* (San Diego: Academic Press).

[7] Klein F. 1910. Über die geometrischen Grundlagen der Lorentzgruppe.[On the geometric foundations the Lorentz group.] *Jahresberichte Deutsche Math. Vereinigung* **19** 281.

[8] Klein F. 1872. Ueber Liniengeometrie und metrische Geometrie [About line geometry and metric geometry]. *Math. Ann.* **5**, 257.

[9] Weyl H. 1918. *Raum-Zeit-Materie.* [*Space-time matter.*] (6th ed. Berlin Heidelberg New York: Springer, 1970 (1918)).

[10] Dahm R. 2018 On A Microscopic Representation of Space-Time VIII – On Relativity. *Phys. Atom. Nuclei* **81** 819.

[11] von Laue M. 1950. Zur Minkowskischen Elektrodynamik der bewegten Körper [On Minkowski's electrodynamics the moving body]. *Z. Physik* **128**, 387 (1950), doi:10.1007/BF01339439.

[12] Dahm R. 2018. On A Microscopic Representation of Space-Time VII – On Spin. *J. Phys.: Conf. Ser.* **965** 012012.

[13] Lie S. 1872. Ueber Complexe, insbesondere Linien- und Kugel-Complexe, mit Anwendung auf die Theorie partieller Differential-Gleichungen [About complexes, especially line and Spherical complexes, with application to the theory of partial differential equations]. *Math. Ann.* **5** 145; Ueber einige partielle Differential-Gleichungen zweiter Ordnung [Over some partial differential equations second order]. *Math. Ann.* **5** 209.

[14] Dvoeglazov V V. 1993. Lagrangean Formulation of the Joos-Weinberg's 2(2S+1)-Theory and Its Connections with the Skew-Symmetric Tensor Description. Preprint:*IFUNAM FT93-016*, May 1993.

[15] Dvoeglazov V V. 1993. The 2(2S+1)-Formalism and Its Connection with Other Descriptions. *arXiv:hep-th/9305141v2*, 1993/2016.

[16] Dvoeglazov V V. 1993. Electrodynamics with Weinberg's Photons. *arXiv:hep-th/9306108v1*, 1993.

[17] Dvoeglazov V V. 1994. 2(2S+1)-Component Model and Its Connection with Other Field Theories. *arXiv:hep-th/9401043v3*; Preprint:*IFUNAM FT-94-36*, January 1994.

[18] Dvoeglazov V V. 2018. Methods for Derivation of Generalized Equations in the $(S, 0) \oplus (0, S)$ Representations of the Lorentz Group. *arXiv:1810.0458v1*, 2018.

[19] Schmeikal B. 2018. Private discussions.

[20] Trehub A. https://people.umass.edu/trehub/, as of 31.7.2019.

[21] Klein F., Sommerfeld A 1897. *Über die Theorie des Kreisels.* [*About the theory of the gyroscope.*] Heft I (Leipzig: B. G. Teubner).

[22] Klein F., Sommerfeld A 1898. *Über die Theorie des Kreisels.* [*About the theory of the gyroscope.*] Heft II (Leipzig: B. G. Teubner).

[23] Study E. 1903. *Geometrie der Dynamen.* [*Geometry of the dynamics.*] (Leipzig: Teubner).

[24] Dahm R. 2018 On a Microscopic Representation of Space-Time III. *Adv. Appl. Clifford Algebras* **29** 20; https://doi.org/10.1007/s00006-018-0936-x; see also http://arxiv.org/abs/1508.06872.

[25] Dahm R. 2016. On a Microscopic Representation of Space-Time V. *ISQS, Prague, Journal of Physics : Conference Series* **804** 012013 (2017).

[26] Plücker J. 1868/1869. *Neue Geometrie des Raumes.* [*New geometry of space.*] ed A Clebsch and F Klein (Leipzig: B. G. Teubner).

[27] Lie S. 1896. *Geometrie der Berührungstransformationen.* [*Geometry of touch transformations.*] (Leipzig: Teubner).

[28] Dahm R. 2015. On a Microscopic Representation of Space-Time IV. *Phys. Atom. Nuclei* **80**, 512 (2017) QTS 9, Yerevan, http://arxiv.org/abs/1512.07119.

[29] Hesse O. 1866. Ein Uebertragungsprincip [A transference principle]. *Journal für die reine und angewandte Mathematik* **66**, 15.

[30] Klein F. 1926. *Vorlesungen über höhere Geometrie.* [*Lectures on higher geometry.*] (Die Grundlehren der mathematischen Wissenschaft XXII [The basic teachings of mathematical science XXII]) (Berlin: Springer).

[31] Clebsch A. 1872. *Theorie der binären algebraischen Formen.* (Leipzig: B. G. Teubner).

[32] Hudson R W H T. 1905. *Kummer's Quartic Surface.* (Cambridge Mathematical Library) (Cambridge: Cambridge University Press, 1990).

[33] Klein F. 1906. Zur Schraubentheorie von Sir Robert Ball [Sir Robert Ball's screw theory]. *Math. Ann.* **62**, 419.

[34] Hilbert D. 1915. Die Grundlagen der Physik. (Erste Mitteilung.) [The basics of physics. (First message.)] *Göttinger Nachrichten* **1915** 395.

[35] Hilbert D. 1916. Die Grundlagen der Physik. (Zweite Mitteilung.) [The basics of physics. (Second notice.] *Göttinger Nachrichten* **1916** 53.

[36] Klein F. 1917. *Zu Hilberts erster Note über die Grundlagen der Physik [Hilbert's first note on the basics of physics]. Göttinger Nachrichten* **1917** 469.

[37] Hilbert D. 1924. Die Grundlagen der Physik [The basics of physics]. *Math. Ann.* **92** 1.

[38] Weinberg St. 1964. Feynman Rules for Any Spin. *Phys. Rev.* **133**, B1318 (1964).

[39] Klein F. 1871. Notiz betreffend den Zusammenhang der Linien-Geometrie mit der Mechanik starrer Körper [Note regarding the context of Linc geometry with the mechanics of rigid bodies]. *Math. Ann.* **4** 403.

[40] Lüroth O. 1867. Zur Theorie der windschiefen Flächen [On the theory of warped surfaces]. *Journal für die reine und angewandte Mathematik* **67**, 130 (1867).

[41] Clebsch A. 1870. Ueber die Plückerschen Complexe [On Plucker's complexes]. *Math. Ann.* **2** 1.

[42] De Alfaro V., Fubini S., Furlan G., and Rossetti C. 1973 *Currents in Hadron Physics.* (Amsterdam, London: North Holland Publishing Company).

[43] Joos H. 1962. Zur Darstellungstheorie der inhomogenen Lorentzgruppe als Grundlage quantenmechanischer Kinematik [On the representation theory of the inhomogeneous Lorentz group as the basis of quantum mechanical kinematics]. *Fortschr. Phys.* **10**, 65.

[44] Dahm R. 2019 On A Microscopic Representation of Space-Time IX – On Spin II. *J. Phys.: Conf. Ser.* **1194** 012023.

[45] Wigner E P. 1962. *Invariant Quantum Mechanical Equations of Motion.* In: Theoretical Physics, Lectures presented at the Seminar on Theoretical Physics, Trieste (Vienna, International Atomic Energy Agency 1963).

[46] Weinberg St. 1964.Feynman Rules for Any Spin. II. Massless particles. *Phys. Rev.* **134**, B882 (1964).

[47] Weinberg St. 1965. Photons and Gravitons in Perturbation Theory: Derivation of Maxwell's and Einstein's Equation. *Phys. Rev.* **138**, B988 (1965).

[48] Einstein A. 1914/1932. *Albert Einstein: Akademie-Vorträge.* [*Albert Einstein: Academy Lectures.*] ed D Simon (Weinheim: WILEY-VCH Verlag GmbH & Co. KGaA).

[49] Einstein A. 1916. Eine neue formale Deutung der Maxwellschen Feldgleichungen der Elektrodynamik. *Sitzungsberichte der königlich Preussischen Akademie der Wissenschaften, Berlin* [*Meeting reports of the royal Prussian Academy of Sciences, Berlin*], SB I, VII, 184.

[50] Dahm R. 2016. *On a Microscopic Representation of Space-Time VI – A Stopover.* to be published (2020).

[51] Pauli W. 1933 Über die Formulierung der Naturgesetze mit fünf homogenen Koordinaten. Teil I: Klassische Theorie [On the formulation of the laws of nature with five homogeneous coordinates. Part I: Classical Theory]. *Annalen der Physik* **18** 305.

[52] Pauli W. 1933 Über die Formulierung der Naturgesetze mit fünf homogenen Koordinaten. Teil II: Die Diracschen Gleichungen für die Materiewellen [with five homogeneous coordinates. Part II: Dirac's equations for the matter waves]. *Annalen der Physik* **18** 337.

[53] Veblen O., Hoffmann B. 1930. Projective Relativity. *Phys. Rev.* **36** 810.

[54] Veblen O. 1933. *Projektive Relativitätstheorie.* [*Projective Relativity Theory*] (Ergebnisse der Mathematik und ihrer Grenzgebiete [Results of mathematics and their border areas]) (Berlin: Springer).

[55] Veblen O. 1933. Spinors in Projective Relativity. *Proc. Nat. Acad. Sciences* **19** 979.

[56] Doehlemann K. 1905. *Projektive Geometrie in synthetischer Behandlung.*(Sammlung Göschen) [*Projective geometry in synthetic Treatment. (Collection Goschen)*] (3rd edition, Leipzig: G. J. Göschen'sche Verlagshandlung) (1898).

[57] Hilbert D. 1909. Ueber die Gestalt einer Fläche 4ter Ordnung [On the shape of a surface 4ter Order]. *Göttinger Nachrichten* **1909** 308.

[58] Catto S. 2019. *Closing plenary talk at ISQS 26*, Prague, July 2019, and private communication.

[59] Dahm R. 1996. *Spin-Flavour-Symmetrien und das $\pi N\Delta$-System* (Aachen: Shaker Verlag) ISBN 3-8265-1782-2.

[60] Dahm R., Kirchbach M., Riska D. O. 1992. *SU(4) Supermultiplet Description of the $\pi N\Delta$-System at Low Energies and Particle Interpretation in Effective Lagrangeans.* Proceedings, Dubna, 1993. ———————-

[61] Dahm R., Kirchbach M., Riska D. O. 1993. Wigner Supermultiplets in π^0–Photoproduction at Threshold. Invited Talk at the *International Workshop on 'Symmetry Methods in Physics', in memory of Professor Ya. A. Smorodinsky*, Dubna, Russia, July 6 – 10, 1993. Proceedings, Dubna, 1994.

[62] Dahm R., Kirchbach M. 1994. Linear Wave Equations and Effective Lagrangeans for Wigner Supermultiplets. *Int. J. Mod. Phys.* **A10**, 4225.

[63] Dahm R., Kirchbach M. 1995. *The Large N_c-Limit of QCD and Effective Lagrangeans for Spin-Flavour Supermultiplets.* ed. M. Kh. Khankhasayev, Zh. B. Kurmanov (Singapore: World Scientific) (1995).

[64] Dahm R. 1995 Relativistic SU(4) and Quaternions *Advances in Applied Clifford Algebra* **7(S)** 337.

[65] Tiwari S. C. 2012. On local duality invariance in electromagnetism. *arXiv:1110.5511v2 physics.gen-ph*, 2011.

[66] Kummer E. E. 1866. Über die algebraischen Strahlensysteme, in's Besondere über die der ersten und zweiten Ordnung [About the algebraic ray systems, in particular about those of the first and second order]. *Abh. Akad. Wiss. Berlin*, Math. Abhandlungen, 1.

[67] Staude O. 1905. Analytische Geometrie des Punktes, der geraden Linie und der Ebene. *(B. G. Teubner's Lehrbücher der Mathematischen Wissenschaften XVI)* [Analytical geometry of the point, the straight line line and the plane. *B. G. Teubner's Textbooks of Mathematical Sciences XVI).*] (Leipzig: B. G. Teubner, 1905).

[68] Plücker J. 1838 Discussion de la forme générale des ondes lumineuses [Discussion of the general form light waves]. *Journal für die reine und angewandte Mathematik* **19** 1; ibd. **19** 91.

[69] Study E. 1905. Über Hamiltons geometrische Optik und deren Beziehung zur Theorie der Berührungstransformationen [About Hamilton's geometrical optics and their Relation to the theory of touch transformations]. *Jahresberichte Deutsche Math. Vereinigung* **14** 424.

[70] Smorodinskij Ya A. 1965. Kinematik und Lobatschewski–Geometrie [Kinematics and Lobatschewski geometry.] *Fortschr. Phys.* **13**, 157 [*Atomnaja Energija* **14**, 110 (1963)].

[71] Poincaré H. http://henripoincarepapers.univ-lorraine.fr/, as of 31.7.2019.

[72] Dahm R. 2019. *Talk at ISQS 26*. Prague, July 2019.

[73] Plücker J. 1865. On a new Geometry of Space. *Phil. Trans. R. Soc. London* **155** 725.

[74] Plücker J. 1866. Fundamental Views regarding Mechanics. *Phil. Trans. R. Soc. London* **156** 361.

[75] Ball R. S. 1876. *The Theory of Screws: A Study in the Dynamics of a Rigid Body.* (Dublin: Hodges, Foster, and Co., Grafton-Street).

[76] Zindler K. 1902. *Liniengeometrie mit Anwendungen.* [*Line geometry with applications.*] (Sammlung Schubert XXXIV) (Leipzig: G. J. Göschensche Verlagshandlung).

[77] Zindler K. 1906. *Liniengeometrie mit Anwendungen.* [*Line geometry with applications.*] (Sammlung Schubert LI) (Leipzig: G. J. Göschensche Verlagshandlung).

[78] Biedenharn L. C., Louck J. D. 1981. Angular Momentum in Quantum Physics. Theory and Application. *(Encyclopedia of Mathematics and Its Applications)* ed. Gian-Carlo Rota. (vol. 8, Reading: Addison-Wesley Publishing Company) (1981).

[79] Biedenharn L. C., Louck J. D. 1981. The Racah-Wigner Algebra in Quantum Theory. *(Encyclopedia of Mathematics and Its Applications)* ed. Gian-Carlo Rota. (vol. 9, Reading: Addison-Wesley Publishing Company) (1981).

[80] Ehlers J., Pirani F. and Schild A. 1972. The Geometry of Free Fall and Light Propagation. In: *Studies in Relativity.* ed L O'Raifeartaigh (Oxford: Clarendon Press) pp 63–84.

[81] Dirac P. A. M. 1935. The Electron Wave Equation in De-Sitter Space. *Annals of Mathematics* **36** 657.

[82] Dirac P. A. M. 1936. Wave Equations in Conformal Space. *Annals of Mathematics* **37** 429.

[83] Sexl R. U., Urbantke H. K. 1992. *Relativität, Gruppen, Teilchen.* [*Relativity Groups, particle.*] (3rd edition, Wien New York: Springer).

[84] Staude O. 1894. Ueber permanente Rotationsaxen bei der Bewegung eines schweren Körpers um einen festen Punkt [On permanent axes of rotation in movement of a heavy body around a fixed point]. *Journal für die reine und angewandte Mathematik* **113**, 318.

[85] Plücker J. 1847 Über eine neue mechanische Erzeugung der Flächen zweiter Ordnung und Classe [About a new mechanical Generation of second-order and class surfaces]. *Journal für die reine und angewandte Mathematik* **34** 357.

[86] Einstein A. 1914. Die formale Grundlage der allgemeinen Relativitätstheorie [The formal basis of the general Theory of relativity]. *Sitzungsberichte der königlich Preussischen Akademie der Wissenschaften, Berlin*, SB II, XLI, 1030.

[87] Einstein A., Mayer W. 1932. Semi-Vektoren und Spinoren [Semi-Vectors and Spinors]. *Sonderausgabe aus den Sitzungsberichten der Preussischen Akademie der Wissenschaften, Physikalisch-Mathematische Klasse, Berlin*, SB I, XXXII, 522.

In: Future Relativity, Gravitation, Cosmology
Editor: V. V. Dvoeglazov et al.

ISBN: 979-8-88697-455-3

Chapter 3

Dyonic and Magnetic Black Holes with Rational Nonlinear Electrodynamics

S. I. Kruglov *
Department of Physics, University of Toronto,
Toronto, ON, Canada
Department of Chemical and Physical Sciences,
University of Toronto,
Mississauga, ON, Canada

Abstract

The principles of causality and unitarity are studied within rational nonlinear electrodynamics proposed earlier. We investigate dyonic and magnetized black holes and show that in the self-dual case, when the electric charge equals the magnetic charge, corrections to Coulomb's law and Reissner–Nordström solutions are absent. In the case of the magnetic black hole, the Hawking temperature, the heat capacity and the Helmholtz free energy are calculated. It is shown that there are second-order phase transitions and it was demonstrated that at some range of parameters the black holes are stable.

1. Introduction

The black holes (BHs) are real objects in the centers of many galactic and its physics is of great interest. Dyonic solutions in the string [1]-[4] and in the supergravity [5]-[8] theories for BHs with magnetic and electric charges were obtained. Such solutions are used in the theory of superconductivity and thermodynamics [9]-[11]. In this paper we obtain dyonic and magnetic BH solutions in the framework of rational nonlinear electrodynamics proposed in [12]. The attractive feature of this nonlinear electrodynamics (NED) is the absence of singularities in the center of charges and their finite self-energy. Similar properties of NED were firstly observed by Born and Infeld in another NED [13]. Quantum electrodynamics with loop corrections also leads to NED [14]. The singularity problems are absent

*Corresponding Author's Email: serguei.krouglov@utoronto.ca.

also in other NED models [15]-[19]. The general relativity (GR) and the thermodynamics of BH with NED was considered in [20]-[35]. The phase transitions in electrically and magnetically charged BHs were investigated in [36]-[40]. It worth noting that the universe acceleration also can be explained by NED coupled with GR [41]-[49].

The paper is organised as follows. In Sec. 2 we study the causality and unitarity principles. We obtain the dyonic solution in Sec. 3. In Sec. 4 we consider the magnetic BH. The metric function and their asymptotic as $r \to \infty$ are found. It was shown that the magnetic mass of BHs is finite and there are not singularities of the Ricci scalar as $r \to \infty$. The BH thermodynamics and the thermal stability of charged black holes are investigated in Sec. 5. We obtain the Hawking temperature, the heat capacity, the Helmholtz free energy and demonstrate that the phase transitions in BHs occur. Sec. 8 is devoted to a conclusion.

We use units with $c = 1$ and the metric signature diag$(-1, 1, 1, 1)$.

2. The Model and Principles of Causality and Unitarity

Here, we consider rational NED, proposed in [12], with the Lagrangian density

$$\mathcal{L} = -\frac{\mathcal{F}}{2\beta\mathcal{F}+1}, \tag{1}$$

where the parameter $\beta \geq 0$ possesses the dimension of (length)^4, $\mathcal{F} = (1/4)F_{\mu\nu}F^{\mu\nu} = (B^2 - E^2)/2$, $F_{\mu\nu} = \partial_\mu A_\nu - \partial_\nu A_\mu$ is the field tensor. The symmetrical energy-momentum tensor is given by [34]

$$T_{\mu\nu} = -\frac{F_\mu^{\ \alpha} F_{\nu\alpha}}{(1+2\beta\mathcal{F})^2} - g_{\mu\nu}\mathcal{L}. \tag{2}$$

From Eq. (2) we obtain the energy density

$$\rho = T_0^{\ 0} = \frac{\mathcal{F}}{1+2\beta\mathcal{F}} + \frac{E^2}{(1+2\beta\mathcal{F})^2}. \tag{3}$$

For healthy theory the general principles of causality and unitarity should hold. According to the causality principle the group velocity of excitations over the background has to be less than the light speed, and then tachyons are absent in the theory. The absence of ghosts is guaranteed by the unitarity principle. Both principles are satisfied for the case $\mathbf{E}\cdot\mathbf{B} = 0$ if the following inequalities hold [50]:

$$\mathcal{L}_\mathcal{F} \leq 0, \quad \mathcal{L}_{\mathcal{F}\mathcal{F}} \geq 0,$$

$$\mathcal{L}_\mathcal{F} + 2\mathcal{F}\mathcal{L}_{\mathcal{F}\mathcal{F}} \leq 0, \tag{4}$$

where $\mathcal{L}_\mathcal{F} \equiv \partial\mathcal{L}/\partial\mathcal{F}$. Making use of Eq. (1) we obtain

$$\mathcal{L}_\mathcal{F} = -\frac{1}{(1+2\beta\mathcal{F})^2},$$

$$\mathcal{L}_\mathcal{F} + 2\mathcal{F}\mathcal{L}_{\mathcal{F}\mathcal{F}} = \frac{6\beta\mathcal{F}-1}{(1+2\beta\mathcal{F})^3}, \quad \mathcal{L}_{\mathcal{F}\mathcal{F}} = \frac{4\beta}{(1+2\beta\mathcal{F})^3}. \tag{5}$$

With the help of Eqs. (4) and (5), the principles of causality and unitarity take place if $6\beta\mathcal{F} \leq 1$ ($\beta \geq 0$). When $\mathbf{E} = 0$, $\beta\mathbf{B}^2 \leq 1/3$.

3. The Dyonic Solution

The action of NED coupled with GR is given by

$$I = \int d^4x\sqrt{-g}\left(\frac{1}{16\pi G}R + \mathcal{L}\right), \tag{6}$$

where G is Newton's constant, $16\pi G \equiv M_{Pl}^{-2}$, and M_{Pl} is the reduced Planck mass. The Einstein equation is

$$R_{\mu\nu} - \frac{1}{2}g_{\mu\nu}R = -8\pi G T_{\mu\nu}. \tag{7}$$

Varying action (6) on electromagnetic potentials we obtain the fields equation for electromagnet fields

$$\partial_\mu\left(\sqrt{-g}F^{\mu\nu}\mathcal{L}_\mathcal{F}\right) = 0. \tag{8}$$

We consider the static and spherically symmetric metric with the line element

$$ds^2 = -A(r)dt^2 + \frac{1}{A(r)}dr^2 + r^2(d\vartheta^2 + \sin^2\vartheta d\phi^2), \tag{9}$$

where the metric function is given by

$$A(r) = 1 - \frac{2M(r)G}{r}, \tag{10}$$

and the mass function is

$$M(r) = m_0 + \int_0^r \rho(r)r^2dr = m_0 + m_{el} - \int_r^\infty \rho(r)r^2dr. \tag{11}$$

The total mass of the BH $m = m_0 + m_{el}$, where m_0 is the Schwarzschild mass and $m_{el} = \int_0^\infty \rho(r)r^2dr$ is the electromagnetic mass. The general solutions of field equations, found in [37], [38], are given by

$$B^2 = \frac{q_m^2}{r^4}, \quad E^2 = \frac{q_e^2}{\mathcal{L}_\mathcal{F}^2 r^4}, \tag{12}$$

where q_m and q_e are the magnetic and electric charges, respectively. With the help of Eqs. (1) and (12) one finds

$$E^2 = \frac{q_e^2(1+2\beta\mathcal{F})^4}{r^4}, \tag{13}$$

$$\beta\mathcal{F} = a - b(1+2\beta\mathcal{F})^4, \quad a = \frac{\beta q_m^2}{2r^4}, \quad b = \frac{\beta q_e^2}{2r^4}, \tag{14}$$

and we introduced the unitless variables a and b. Defining the unitless value $x \equiv \beta\mathcal{F}$, we obtain from Eq. (14) the equation as follows:

$$b(2x+1)^4 + x - a = 0. \tag{15}$$

Using unitless variables $t = r/\sqrt[4]{\beta q_m^2}$ and $n = q_m^2/q_e^2$, one finds from Eq. (15) the equation for $y = 2x + 1$:

$$y^4 + t^4y - n - t^4 = 0. \tag{16}$$

The real dyonic solution to Eq. (16) is

$$y=\sqrt{\frac{\sqrt[4]{3}t^4}{4\sqrt[4]{n+t^4}\sqrt{\sinh(\varphi/3)}}-\frac{\sqrt{n+t^4}\sinh(\varphi/3)}{\sqrt{3}}}-\frac{\sqrt{\sinh(\varphi/3)}\sqrt[4]{n+t^4}}{\sqrt[4]{3}},$$

$$\sinh(\varphi)=\frac{3^{3/2}t^8}{16(n+t^4)^{3/2}}. \tag{17}$$

Putting $n=0$ in Eq. (17) we come to the solution corresponding to the electrically charged BH [34]. We find the self-dual solution at $q_e=q_m$ ($a=b$) from Eq. (15). Then $x=0$ ($\mathcal{F}=0$, $E=B$), $E=q/r^2$ ($q\equiv q_e=q_m$) and with the help of Eqs. (3) and (11) we obtain the mass function

$$M(r)=m-\int_r^\infty\rho(r)r^2dr=m-\frac{q^2}{r}. \tag{18}$$

Making use of Eq. (10) one finds the metric function

$$A(r)=1-\frac{2mG}{r}+\frac{2q^2G}{r^2}. \tag{19}$$

The metric function (19) corresponds to the Reissner–Nordstöm (RN) solution with $2q^2=q_e^2+q_m^2$.

4. The Magnetic Black Hole

Let us consider the static magnetic BH [1]. Taking into account that $q_e=0$, $\mathcal{F}=q_m^2/(2r^4)$, we obtain from Eq. (3) the magnetic energy density

$$\rho_M=\frac{B^2}{2(\beta B^2+1)}=\frac{q_m^2}{2(r^4+\beta q_m^2)}. \tag{20}$$

With the help of Eqs. (11) and (20) one finds the mass function

$$M(x)=m_0+\frac{q_m^{3/2}}{8\sqrt{2}\beta^{1/4}}\left(\ln\frac{x^2-\sqrt{2}x+1}{x^2+\sqrt{2}x+1}\right.$$

$$\left.+2\arctan(\sqrt{2}x+1)-2\arctan(1-\sqrt{2}x)\right), \tag{21}$$

where $x=r/\sqrt[4]{\beta q_m^2}$. The BH magnetic mass is given by

$$m_M=\int_0^\infty\rho_M(r)r^2dr=\frac{\pi q_m^{3/2}}{4\sqrt{2}\beta^{1/4}}\approx 0.56\frac{q_m^{3/2}}{\beta^{1/4}}. \tag{22}$$

[1]In the paper M.-S. Ma, Ann. Phys. **362**, 529 (2015) the author also considered the static magnetic BH based on NED proposed in [12]. However, here we use unitless variables that are more convenient for the analyses of the BH thermodynamics. In addition, we analyse more general case when the BH besides the electromagnetic mass possesses the Schwarzschild mass (having non-electromagnetic nature).

The Schwarzschild mass m_0 is a free parameter and at $q_m = 0$ one has $m_M = 0$, and we arrive at the Schwarzschild BH. Making use of Eq. (10) we obtain the metric function

$$A(x) = 1 - \frac{2m_0 G}{\sqrt[4]{\beta q_m^2}\, x} - \frac{q_m G}{4\sqrt{2\beta}\, x}\left(\ln \frac{x^2 - \sqrt{2}x + 1}{x^2 + \sqrt{2}x + 1} \right.$$

$$\left. +2\arctan(\sqrt{2}x + 1) - 2\arctan(1 - \sqrt{2}x) \right), \tag{23}$$

As $r \to \infty$ the metric function (23) becomes

$$A(r) = 1 - \frac{2mG}{r} + \frac{q_m^2 G}{r^2} + \mathcal{O}(r^{-5}) \quad r \to \infty, \tag{24}$$

where $m = m_0 + m_M$. The correction to the RN solution, according to Eq. (24), is in the order of $\mathcal{O}(r^{-5})$. At $m_0 = 0$ and $r \to 0$, from Eq. (23) one finds the asymptotic with a de Sitter core

$$A(r) = 1 - \frac{Gr^2}{\beta} + \frac{Gr^6}{7\beta^2 q_m^2} - \frac{Gr^{10}}{11\beta^3 q_m^4} + \mathcal{O}(r^{12}) \quad r \to 0. \tag{25}$$

The solution (25) is regular because as $r \to 0$ we have $A(r) \to 1$. When $m_0 \neq 0$ the solution is singular and $A(r) \to \infty$. Let us introduce unitless constants $C = m_0 G/(\beta^{1/4}\sqrt{q_m})$, $B = q_m G/\sqrt{\beta}$. Then the horizon radii, that are the roots of the equation $A(r) = 0$ $(x_{+/-} = r_{+/-}/(\sqrt{q_m}\beta^{1/4}))$, are given in Tables 1 and 2.

Table 1. $B = 1$

C	0.6	0.7	0.8	0.9	1	2	3	4	5
x_+	1.75	2.02	2.27	2.52	2.75	4.91	6.97	9.00	11.02

Table 2. $m_0 = 0$

B	3.173	3.2	3.5	4	4.5	5	6	7	8
x_-	1.68	1.52	1.21	1.03	0.92	0.85	0.75	0.68	0.63
x_+	1.68	1.87	2.49	3.19	3.82	4.42	5.59	6.74	7.87

The plots of the metric function (23) are depicted in Figs. 1 and 2. According to Fig. 1 at $m_0 \neq 0$ ($B = 1$) there is only one horizon. For the bigger mass (the parameter C is greater) the horizon radius increases. Figure 2 shows that there are no horizons at $m_0 = 0$, $B < 3.17$), an extreme horizon occurs at $m_0 = 0$, $B \approx 3.173$, and two horizons hold at $m_0 = 0$ and $B > 3.173$. Making use of Eqs. (2) and (7) at $E = 0$, we obtain the Ricci scalar

$$R = 8\pi G T_\mu^{\ \mu} = \frac{16\pi G \beta q_m^4}{(r^4 + \beta q_m^2)^2}. \tag{26}$$

The Ricci scalar approaches to zero as $r \to \infty$ and spacetime becomes flat.

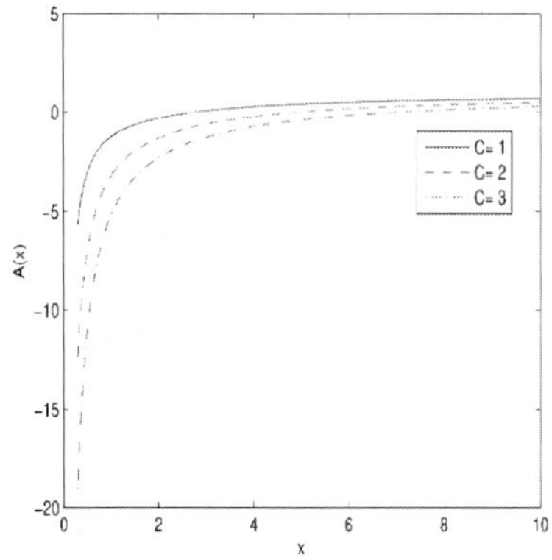

Figure 1. The plot of the function $A(x)$ for $B = 1$. The solid curve is for $C = 1$, the dashed curve corresponds to $C = 2$, and the dashed-doted curve corresponds to $C = 3$.

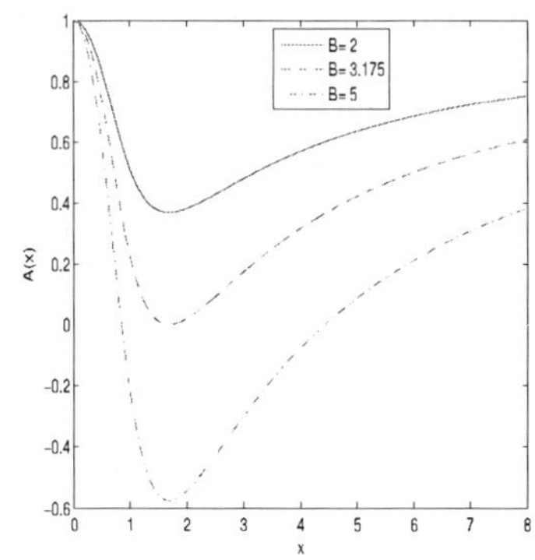

Figure 2. The plot of the function $A(x)$ for $m_0 = 0$. The solid curve is for $B = 2$, the dashed curve corresponds to $B = 3.175$, and the dashed-doted curve corresponds to $B = 5$.

5. The Black Hole Thermodynamics

To study the black holes thermodynamics and the thermal stability of magnetic BHs, we consider the Hawking temperature

$$T_H = \frac{\kappa}{2\pi} = \frac{A'(r_+)}{4\pi}, \tag{27}$$

where κ is the surface gravity and r_+ is the event horizon radius. With the help of Eqs. (10) and (11) one finds the relations

$$A'(r) = \frac{2GM(r)}{r^2} - \frac{2GM'(r)}{r}, \quad M'(r) = r^2\rho, \quad M(r_+) = \frac{r_+}{2G}. \tag{28}$$

Making use of Eqs. (3), (27) and (28) we obtain the Hawking temperature

$$T_H = \frac{1}{4\pi}\left(\frac{1}{r_+} - 2Gr_+\rho(r_+)\right) = \frac{1}{4\pi\beta^{1/4}\sqrt{q_m}}\left(\frac{1}{x_+} - \frac{Gq_m x_+}{\sqrt{\beta}(1+x_+^4)}\right). \tag{29}$$

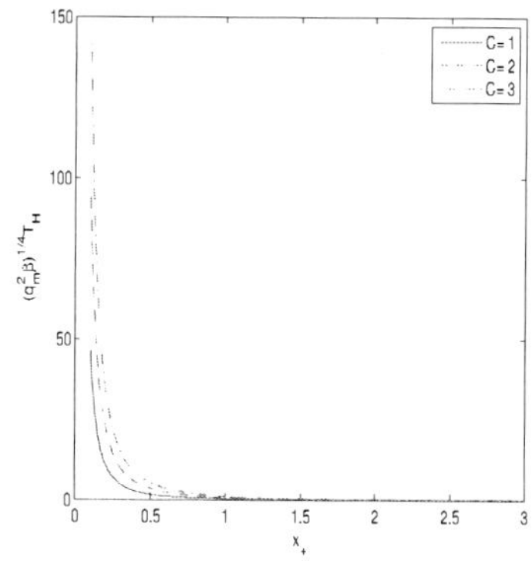

Figure 3. The plot of the function $T_H\sqrt{q_m}\beta^{1/4}$ vs x_+. The solid curve is for $C=1$, the dashed curve corresponds to $C=2$, and the dashed-doted curve corresponds to $C=4$.

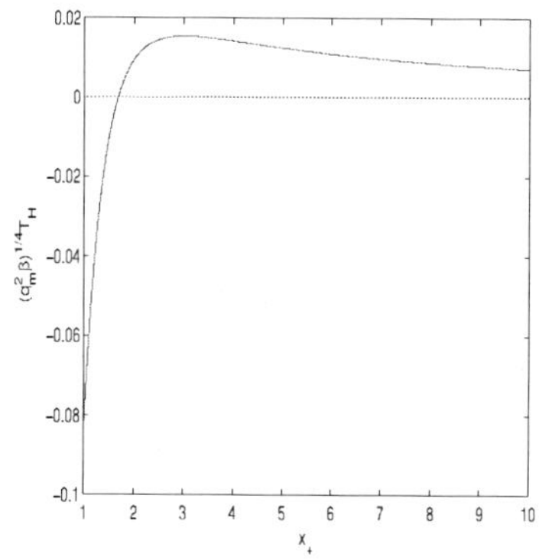

Figure 4. The plot of the function $T_H\sqrt{q_m}\beta^{1/4}$ vs x_+ for $C=0$ ($m_0=0$).

Using the equation $M(r_+)=r_+/(2G)$ and (21) we find

$$\frac{Gq_m}{\sqrt{\beta}}=\frac{4\sqrt{2}(x_+-2C)}{D},$$

$$D\equiv\ln\frac{x^2-\sqrt{2}x+1}{x^2+\sqrt{2}x+1}-2\arctan(1-\sqrt{2}x)+2\arctan(1+\sqrt{2}x). \qquad (30)$$

Replacing Eq. (30) into Eq. (29) one obtains the Hawking temperature as follows:

$$T_H==\frac{1}{4\pi\beta^{1/4}\sqrt{q_m}}\left(\frac{1}{x_+}-\frac{4\sqrt{2}(x_+-2C)x_+}{(1+x_+^4)D}\right). \qquad (31)$$

The plots of the functions $T_H(x_+)\sqrt{q_m}\beta^{1/4}$ are depicted in Figs. 3 and 4.

According to Fig. 3 the temperature is positive everywhere for the case $C\neq0$ ($m_0\neq0$). Figure 4 shows that the Hawking temperature for $C=0$ ($m_0=0$) is positive for $x_+>1.679$ and is zero at $x_+\approx1.679$. The BH is unstable when the temperature is negative. From Eq. (30) we obtain the value of $Gq_m/\sqrt{\beta}=3.173$ corresponding to

$x_+ = 1.679$. By studying the signs of the heat capacity and the Helmholtz free energy, we can observe the different stability phases of the BH [51]. Making use of the Hawking entropy of the BH $S = \text{Area}/(4G) = \pi r_+^2/G = \pi x_+^2 q_m\sqrt{\beta}/G$ we find the heat capacity

$$C_q = T_H\left(\frac{\partial S}{\partial T_H}\right)_q = \frac{T_H\partial S/\partial x_+}{\partial T_H/\partial x_+} = \frac{2\pi q_m\sqrt{\beta}x_+T_H}{G\partial T_H/\partial x_+}. \tag{32}$$

According to Eq. (32) the heat capacity possesses a singularity when the Hawking temperature has an extremum ($\partial T_H/\partial x_+ = 0$). The plots of the heat capacity versus the variable x_+ for different parameters C are depicted in Figs. 5. 6, and 7. Figure 5 shows the

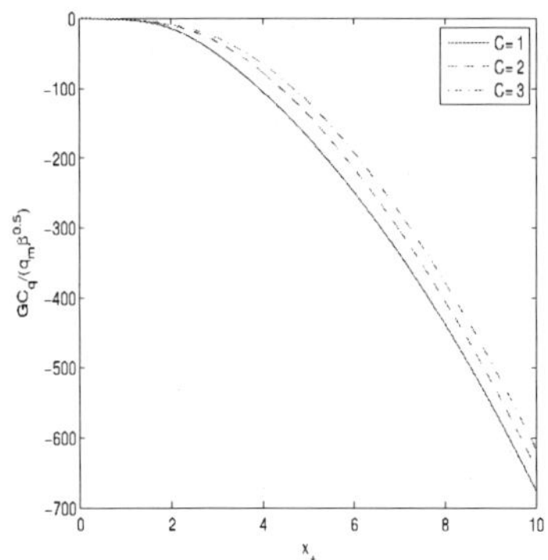

Figure 5. The plot of the function $GC_q/(q_m^2\beta)^{1/2}$ vs x_+. The solid curve is for $C = 1$, the dashed curve corresponds to $C = 2$, and the dashed-doted curve corresponds to $C = 3$.

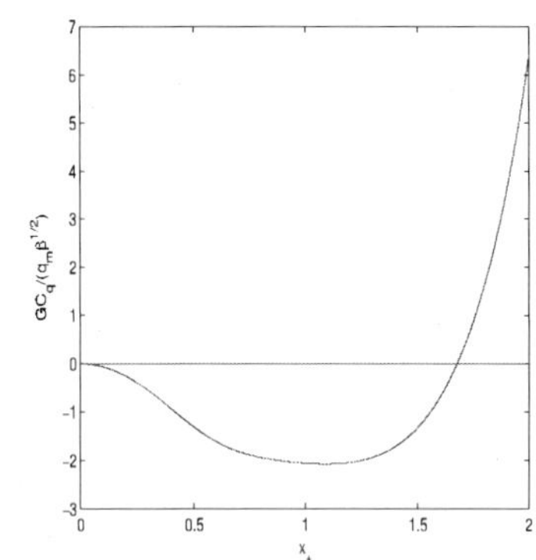

Figure 6. The plot of the function $GC_q/(q_m^2\beta)^{1/2}$ vs x_+ for $C = 0$.

Schwarzschild behaviour of the heat capacity for $C \neq 0$ ($m_0 \neq 0$), i.e. it is negative at $\partial T_H/\partial x_+ < 0$. As a result, the BHs are unstable for the case $m_0 \neq 0$. In accordance with Fig. 6 the BH is unstable at $1.679 > x > 0$ because the heat capacity is negative. Figure 7 shows a singularity in the heat capacity at the point $x \approx 3$ where the second-order phase transition occurs. When $m_0 = 0$ the heat capacity is positive at the range $3 > x > 1.679$ and the BH is stable.

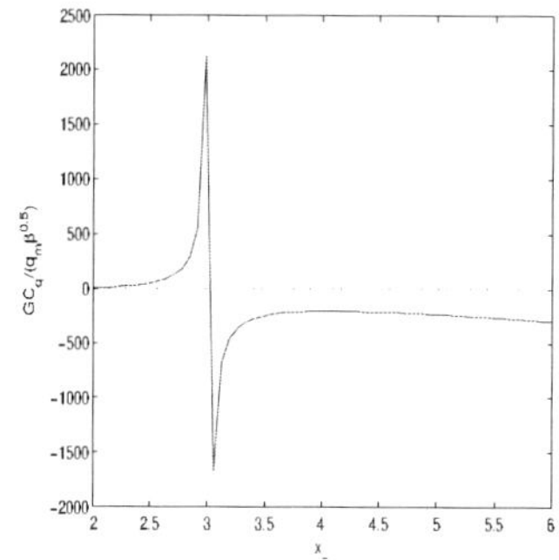

Figure 7. The plot of the function $GC_q/(q_m^2\beta)^{1/2}$ vs x_+ for $C = 0$.

To complete the analysis of phase transitions we consider the Helmholtz free energy which is given by

$$F = m - T_H S. \tag{33}$$

The mass of the BH m plays the role of the internal energy, and the Hawking entropy is $S = \pi r_+^2/G$. Making use of Eqs. (22). (31) and (33) we obtain

$$\frac{GF}{\sqrt{q_m}\beta^{1/4}} = B\left(C + \frac{\pi}{4\sqrt{2}}\right) - \frac{x_+}{4} + \frac{\sqrt{2}x_+^2(x_+ - 2C)}{(x_+^4 + 1)D}. \tag{34}$$

Substituting $B = q_m G/\sqrt{\beta}$ from Eq. (30) into (34) we find

$$\frac{GF}{\sqrt{q_m}\beta^{1/4}} = \left(C + \frac{\pi}{4\sqrt{2}}\right)\frac{4\sqrt{2}(x_+ - 2C)}{D} - \frac{x_+}{4} + \frac{\sqrt{2}x_+^2(x_+ - 2C)}{(x_+^4 + 1)D}\Bigg). \tag{35}$$

The plots of the unitless reduced free energy $GF/(\sqrt{q_m}\beta^{1/4})$ vs. x_+ are depicted in Figs. 8 and 9. The BHs with $F > 0$, $C_q < 0$ are unstable and the BHs with $F < 0$, $C_q > 0$ are

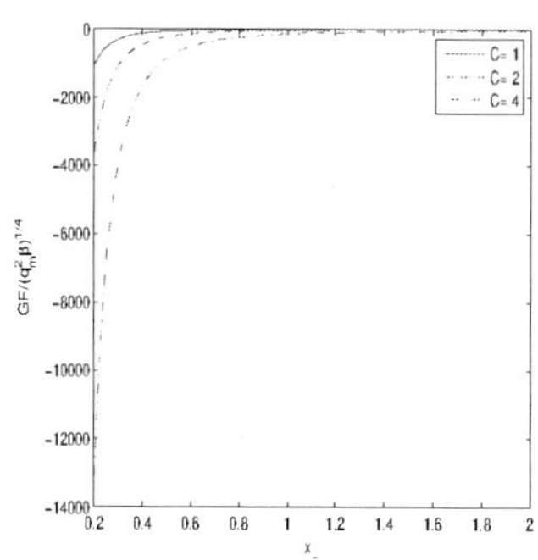

Figure 8. The plot of the function $GF/(\sqrt{q_m}\beta^{1/4})$ vs. x_+. The dashed curve corresponds to $C = 2$, the solid curve is for $C = 1$, and the dashed-doted curve corresponds to $C = 4$.

stable. In accordance with Figs. 5-9, there are other phases with $F > 0$, $C_q > 0$ and $F < 0$, $C_q < 0$. In the case $F < 0$, $C_q < 0$ the BHs are less energetic than the pure radiation and,

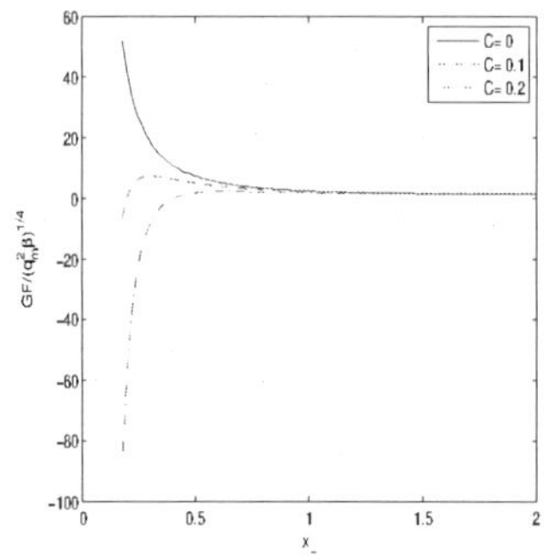

Figure 9. The plot of the function $GF/(\sqrt{q_m}\beta^{1/4}$ vs. $x_+)$. The dashed curve corresponds to $C = 0.1$, the solid curve is for $C = 0$, and the dashed-doted curve corresponds to $C = 0.2$.

as a result, BHs do not decay through tunneling. For large masses of BHs ($C > 1$) this phase holds. Because the heat capacities are negative, the BH temperature decreases when the mass of BH increases. Such phases are also realised in another model [52].

Conclusion

The correspondence principle of the NED model holds as for weak fields our model is converted into Maxwell's electrodynamics. It was demonstrated that at $\beta\mathbf{B}^2 \leq 1/3$ ($\mathbf{E} = 0$) the principles of causality and unitarity take place. In this model the singularity of the electric field at the center of charges is absent and the maximum electric field in the origin is $E(0) = 1/\sqrt{\beta}$. The dyonic and magnetic BHs in GR were studied. It was shown that in the self-dual case ($q_e = q_m$) the corrections to Coulomb's law and RN solutions are absent. The Ricci scalar does not have the singularity and as $r \to \infty$ space-time becomes flat.

The thermodynamics and the thermal stability of magnetized BHs were investigated. The Hawking temperature, the heat capacity and the Helmholtz free energy of BHs were calculated. It was demonstrated that the heat capacity diverges at some event radii r_+ (x_+) for the case when the total BH mass is the magnetic mass and the phase transitions of the second-order occurs. We shown that there is a new stability region of BH solutions when the heat capacity and the free energy are negative. In this case BHs are less energetic than the pure radiation and BHs do not decay via tunneling.

References

[1] A. D. Shapere, S. Trivedi, and F. Wilczek, *Mod. Phys. Lett.* A **6**, 2677 (1991).

[2] S. Mignemi, *Phys. Rev.* D **51**, 934 (1995).

[3] M. Cvetic and A. A. Tseytlin, *Phys. Rev.* D **53**, 5619 (1996); Erratum: *Phys. Rev.* D **55**, 3907 (1997).

[4] D. P. Jatkar, S. Mukherji, and S. Panda, *Nucl. Phys.* B **484**, 223 (1997).

[5] A. H. Chamseddine and W. A. Sabra, *Phys. Lett.* B **485**, 301 (2000).

[6] D. D. K. Chow and G. Compere, *Phys. Rev.* D **89**, 065003 (2014).

[7] P. Meessen, T. Ortin, and P. F. Ramirez, *JHEP* **1710**, 066 (2017).

[8] H. Lu, Y. Pang, and C. N. Pope, *JHEP* **1311**, 033 (2013).

[9] S. A. Hartnoll and P. Kovtun, *Phys. Rev.* D **76**, 066001 (2007).

[10] S. A. Hartnoll, P. K. Kovtun, M. Muller, and S. Sachdev, *Phys. Rev.* B **76**, 144502 (2007).

[11] S. Dutta, A. Jain, and R. Soni, *JHEP* **1312**, 060 (2013).

[12] S. I. Kruglov, *Ann. Phys.* **353**, 299 (2015).

[13] M. Born and L. Infeld, *Proc. R. Soc. Lond.* **144**, 425 (1934).

[14] W. Heisenberg and H. Euler, *Z. Phys.* **98**, 714 (1936).

[15] H. H. Soleng, *Phys. Rev.* D **52**, 6178 (1995).

[16] D. M. Gitman and A. E. Shabad, *Eur. Phys. J.* C **74**, 3186 (2014).

[17] C. V. Costa, D. M. Gitman, and A. E. Shabad, *Phys. Scripta* **90**, 074012 (2015).

[18] S. I. Kruglov, *Commun. Theor. Phys.* **66**, 59 (2016).

[19] S. I. Kruglov, *Mod. Phys. Lett.* A **32**, 1750201 (2017).

[20] R. Pellicer and R. J. Torrence, *J. Math. Phys.* **10**, 1718 (1969).

[21] H. P. de Oliveira, *Class. Quant. Grav.* **11**, 1469 (1994).

[22] E. Ayón-Beato and A. Garćia, *Phys. Rev. Lett.* **80**, 5056 (1998).

[23] K. A. Bronnikov, V. N. Melnikov, G. N. Shikin, and K. P. Staniukovich, *Ann. Phys.* **118**, 84 (1979).

[24] K. A. Bronnikov, *Phys. Rev.* D **63**, 044005 (2001).

[25] K. A. Bronnikov, *Phys. Rev. Lett.* **85**, 4641 (2000).

[26] K. A. Bronnikov, G. N. Shikin, and E. N. Sibileva, *Grav. Cosmol.* **9**, 169 (2003).

[27] A. Burinskii and S. R. Hildebrandt, *Phys. Rev.* D **65**, 104017 (2002).

[28] J. Diaz-Alonso and D. Rubiera-Garcia, *Phys. Rev.* D **81**, 064021 (2010).

[29] N. Breton, *Gen. Rel. Grav.* **37**, 643 (2005).

[30] M. Novello, S. E. Perez Bergliaffa, and J. M. Salim, *Class. Quant. Grav.* **17**, 3821 (2000).

[31] R. Garcia-Salcedo, T. Gonzalez, and I. Quiros, *Phys. Rev.* D **89**, 084047 (2014).

[32] J. P. S. Lemos and V. T. Zanchin, *Phys. Rev.* D **83**, 124005 (2011).

[33] L. Balart and E. C. Vagenas, *Phys. Rev.* D **90**, 124045 (2014).

[34] S. I. Kruglov, *Phys. Rev.* D **94**, 044026 (2016).

[35] S. I. Kruglov, *Ann. Phys. (Berlin)* **528**, 588 (2016).

[36] H. Yajima and T. Tamaki, *Phys. Rev.* D **63**, 064007 (2001).

[37] K. A. Bronnikov, *Grav. Cosmol.* **23**, 343 (2017).

[38] K. A. Bronnikov, *Int. J. Mod. Phys.* D **27**, 1841005 (2018).

[39] S. I. Kruglov, *Int. J. Mod. Phys.* A **33**, 1850023 (2018).

[40] S. I. Kruglov, *Ann. Phys.* **383**, 550 (2017).

[41] R. García-Salcedo and N. Breton, *Int. J. Mod. Phys.* A **15**, 4341 (2000).

[42] C. S. Camara, M. R. de Garcia Maia, J. C. Carvalho, and J. A. S. Lima, *Phys. Rev.* D **69**, 123504 (2004).

[43] E. Elizalde, J. E. Lidsey, S. Nojiri, and S. D. Odintsov, *Phys. Lett.* B **574**, 1 (2003).

[44] M. Novello, S. E. Perez Bergliaffa, and J. M. Salim, *Phys. Rev.* D **69**, 127301 (2004).

[45] M. Novello, E. Goulart, J. M. Salim, and S. E. Perez Bergliaffa, *Class. Quant. Grav.* **24**, 3021 (2007).

[46] D. N. Vollick, *Phys. Rev.* D **78**, 063524 (2008).

[47] S. I. Kruglov, *Phys. Rev.* D **92**, 123523 (2015).

[48] S. I. Kruglov, *Int. J. Mod. Phys.* A **32**, 1750071 (2017).

[49] S. I. Kruglov, *Int. J. Mod. Phys.* D **25**, 1640002 (2016).

[50] A. E. Shabad and V. V. Usov, *Phys. Rev.* D **83**, 105006 (2011) 5).

[51] S. W. Hawking and D. N. Page, *Commun. Math. Phys.* **87**, 577 (1983).

[52] J. A. R. Cembranos, A. Cruz-Dombriz, and J. Jarillo, *Universe*, **1**, 412 (2015).

In: Future Relativity, Gravitation, Cosmology
Editors: V. V. Dvoeglazov et al.
ISBN: 979-8-88697-455-3

Chapter 4

A Damped Oscillator Model of Relativistic Dynamics in Rapidity Space

R. M. Yamaleev *
Joint Institute for Nuclear Research,
LIT, Dubna, Russia

Abstract

The second order polynomial relationship between energy, mass and momentum, the mass-shell equation, is considered as the characteristic polynomial of the second order ordinary differential equation describing the damped oscillator model. It is shown, the quotient of two fundamental solutions of the differential equation describes an evolution of the energy and momentum of the relativistic particle with respect to hyperbolic and periodic angles. A geometrical interpretation is done within the framework of hyperbolic and elliptic geometries. Transitions between different kinds of the evolution parameters are interpreted as passes from one system of references onto the other. Evolution equations covariant with respect to these transitions are formulated within the framework of Jacobi elliptic functions.

1. Introduction

The polynomial of second order

$$P(X) = X^2 - 2p_0X + p^2 \tag{1.1}$$

is the other form of the presentation of the mass-shell equation

$$p^2 = (p_0 + mc)(p_0 - mc). \tag{1.2}$$

In fact, define the roots of $P(X)$ by

$$x_1 = p_0 - mc, \;\; x_2 = p_0 + mc, \;\; 2mc = x_1 - x_2, \tag{1.3}$$

*Corresponding Author's Email: yamaleev@jinr.ru.

then the *mass-shell equation* (1.2) gives arise as the Vieta's formula for the free coefficient of $P(X)$.

The polynomial $P(X)$ can be considered as the characteristic polynomial of the following differential equation

$$\frac{d^2\Phi}{d\phi^2} - 2p_0\frac{d\Phi}{d\phi} + p^2\Phi = 0. \tag{1.4}$$

The damped oscillator model is described by the second order differential equation (1.4) with periodic or aperiodic solutions in dependence of the discriminant of the characteristic polynomial $P(X)$.

The differential equation (1.4) describes an evolution of the function $\Phi(\phi)$ with respect to parameter ϕ, meanwhile, the coefficients remain to be constants.

The idea is to realize some transition from the dynamics of the solution of the differential equation to the coefficients of the characteristic polynomial. This transition is worked out via the Riccati equation associated with the very differential equation. In this way we elaborate an evolution law for the energy and momentum with respect to some natural parameter of evolution which geometrically can be interpreted as an distance in the velocity space of hyperbolic type.

The chapter consists of the following sections.

Section 2. Evolution with Respect to the Hyperbolical angle.
Section 3. Evolution with respect to the periodical angle.
Section 4. Geometrical interpretation.
Section 5. General representation by Jacobi elliptic functions.
Section 6. Conclusions.

2. Evolution with Respect to Hyperbolic Angle

Let $\Psi(\phi) = (g_0(\phi), g_1(\phi))^T$ with initial data $(1,0)^T$ be two fundamental solutions of the second order differential equation (1.4). For this pair of functions differential equation (1.4) can be transformed into the system of equations governed by companion matrix E of the characteristic polynomial $P(X)$ [1]. On making use the companion matrix

$$E = \begin{pmatrix} 0 & -p^2 \\ 1 & 2p_0 \end{pmatrix}, \tag{2.1}$$

the second order differential equation (1.4) is transformed to the system of equations

$$E\Psi = \frac{d}{d\phi}\Psi. \tag{2.2}$$

Simultaneously, each of the components of Ψ satisfies the second order differential equation

$$P(\frac{d}{d\phi})\Psi(\phi) = 0. \tag{2.3}$$

A matrix solution of the system (2.2) is presented by the matrix exponential

$$\Psi(\phi) = \exp(E\phi)\Psi(\phi_0). \tag{2.4}$$

Exponential matrix admits expansion (the Euler formula)

$$\exp(E\phi) = g_0(\phi) + E\, g_1(\phi), \tag{2.5}$$

which is decomposed into two equations

$$\exp(x_1\phi) = g_0(\phi) + x_1\, g_1(\phi), \quad \exp(x_2\phi) = g_0(\phi) + x_2\, g_1(\phi). \tag{2.6}$$

Inversion of this system defines explicit formulas for the trigonometric functions

$$g_1(\phi) = \frac{\exp(x_1\phi) - \exp(x_2\phi)}{x_1 - x_2},$$

$$g_0(\phi) = \frac{x_1\, \exp(x_2\phi) - x_2\, \exp(x_1\phi)}{x_1 - x_2}. \tag{2.7}$$

These functions belong to the class of trigonometric functions (in Ref. [2] named as "Trig"-functions). The class of "Trig" functions is the class of functions that preserves the elegance and simplicity of the trigonometric functions, they are sorted by the degree of the characteristic polynomial [3].

Let $\phi = \phi_0$ be a point where the equation $g_0(\phi_0) = 0$ is satisfied. Then, the formula (2.5) is reduced to

$$\exp(E\phi_0) = E\, g_1(\phi_0), \tag{2.8}$$

which at the eigenvalues of the matrix E is specified by the equations

$$x_1\, exp(x_2\phi_0) = x_2\, exp(x_1\phi_0), \tag{2.9}$$

and

$$\exp(x_1\phi_0)/x_1 = \exp(x_2\phi_0)/x_2 = g_1(\phi_0). \tag{2.10}$$

From these equations one eventually gets

$$\exp((x_1 - x_2)\phi_0) = \frac{x_1}{x_2}. \tag{2.11}$$

This is a basic relationship between exponential function and the fraction [4]. Furthermore, from this relationship we conclude that the roots x_1, x_2 can be viewed as functions of the parameter ϕ_0. This dependence is derived in such a way that remains the difference $(x_1 - x_2)$ independent of the parameter ϕ_0 [5]. Hence the mass

$$mc = \frac{1}{2}(x_1 - x_2)$$

remains invariant with respect to evolution, meanwhile coefficients p_0, p^2 have to be considered as functions of the parameter ϕ_0. By taking into account definitions (1.3) let us re-write formulas (2.7) as follows

$$g_1(\phi, \phi_0) = \exp(p_0(\phi_0)\phi)(\frac{\exp(mc\phi) - \exp(-mc\phi)}{2mc},$$

$$g_0(\phi, \phi_0) = \exp(p_0(\phi_0)\phi)\frac{x_1(\phi_0)\, \exp(-mc\phi) - x_2(\phi_0)\, \exp(mc\phi)}{2mc}. \tag{2.12}$$

By taking out the factor $\exp(p_0(\phi_0)\phi)$ we get the hyperbolic cosine-sine- like functions

$$S(\phi) = \frac{1}{mc}\sinh(mc\phi), \ \ C(\phi,\phi_0) = \cosh(mc\phi) - \frac{p_0(\phi_0)}{mc}\sinh(mc\phi). \tag{2.13}$$

Notice, the cosine-type function depends of two variables ϕ and ϕ_0, whereas the sine-type function depends of ϕ, only. Furthermore, at the point $\phi = \phi_0$ we have

$$C(\phi_0,\phi_0) = \cosh(mc\phi_0) - \frac{p_0}{mc}\sinh(mc\phi_0) = 0, \tag{2.14}$$

which yields

$$p_0(\phi_0) = mc\coth(mc\phi_0). \tag{2.15}$$

The basic formula (2.11) is a particular case of the general fraction which is derived from the system of equations (2.6). Form a quotient of two equations of (2.6) by

$$\exp(2mc\phi) = \frac{x_1(\phi_0) - U(\phi,\phi_0)}{x_2(\phi_0) - U(\phi,\phi_0)}, \tag{2.16}$$

where

$$U(\phi,\phi_0) = -\frac{g_0(\phi,\phi_0)}{g_1(\phi)}, \ \text{and} \ m = \frac{1}{2}(x_1 - x_2). \tag{2.17}$$

The function $U(\phi,\phi_0)$ obeys the Riccati equation [6]

$$\frac{d}{d\phi}U(\phi,\phi_0) = U^2 - 2p_0(\phi_0)U + p^2(\phi_0), \tag{2.18}$$

with initial data $U(\phi_0,\phi_0) = 0$.

Substitution

$$U(\phi,\phi_0) = p_0(\phi_0) - U_N(\phi), \tag{2.19}$$

transforms equation (2.18) to the normal form

$$\frac{d}{d\phi}U_N(\phi) = U_N^2(\phi) - m^2 = p^2(\phi). \tag{2.20}$$

Notice, the function $p_0(\phi_0)$ satisfies the same equation. Consequently,

$$U_N(\phi) = p_0(\phi),$$

and

$$U(\phi,\phi_0) = p_0(\phi_0) - p_0(\phi). \tag{2.21}$$

which completely clarifies a role of the Riccati equation.

Finally, the evolution of the energy p_0 and momentum p with respect to ϕ is written as

$$\frac{d}{d\phi}p(\phi) = -p_0(\phi)p(\phi), \tag{2.22}$$

$$\frac{d}{d\phi}p_0(\phi) = -p^2(\phi). \tag{2.23}$$

These equations can be recognized as the dynamical equations for coefficients characteristic polynomial of the second order differential equation (2.3).

3. Evolution with Respect to Periodic Angle

In that case we start from the relationship

$$p_0^2 = (p + imc)(p - imc) = y_1 y_2, \tag{3.1}$$

where y_1, y_2 are mutually conjugated complex roots $y_1 = p + imc,\ y_2 = p - imc$ of the polynomial

$$Y^2 - 2pY + p_0^2 = 0,\ \ p^2 < p_0^2. \tag{3.2}$$

It is seen, that in this approach the dynamical values p_0 and p are interchanged with their places. The differential equation corresponding to characteristic equation (3.2) is

$$\frac{d^2}{d\theta^2}\Psi(\theta) - 2p\frac{d\Psi(\theta)}{d\theta} + p_0^2\Phi(\theta) = 0. \tag{3.3}$$

This equation describes the oscillator accomplishing periodical motion with frequency ω. Since $p \sim \kappa,\ \ p_0 \sim \omega$, this model looks like as the L.Broglie's identifications

$$p = \hbar\kappa,\ \ p_0 = \hbar\omega. \tag{3.4}$$

Here the proper mass is identified with the proper frequency $\omega_0^2 = \omega^2 - \kappa^2$. Following the scheme of the previous section, firstly, let us write the Euler identity

$$\exp(y_k\theta) = f_0(\theta) + y_k f_1(\theta),\ \ k = 1, 2, \tag{3.5}$$

with cosine-sine like functions

$$f_0 = \exp(p\theta)(\cos mc\theta - \frac{p}{mc}\sin mc\theta),\ \ f_1 = \exp(p\theta)\frac{\sin mc\theta}{mc}. \tag{3.6}$$

These functions represent two linear independent solutions of the differential equation (3.3) with initial data $f_0(0) = 1,\ f_1(0) = 0$.

Let θ_0 be a point where $f_0(\theta_0) = 0$. Then,

$$\exp(E\theta) = Ef_1(\theta). \tag{3.7}$$

In that context, let us recall analogous formula from complex analysis, namely,

$$\exp(i\theta_0) = i\sin(\theta_0),\ \theta_0 = \frac{\pi}{2}. \tag{3.8}$$

From (3.7) it follows the principal formula with respect to periodical angle

$$\exp((y_1 - y_2)\theta_0) = \frac{y_1(\theta_0)}{y_2(\theta_0)}. \tag{3.10}$$

Since $i2mc = y_1 - y_2$, then,

$$\exp(i2mc\theta_0) = \frac{p + imc}{p - imc}. \tag{3.11}$$

This formula is the basic one in the geometrical interpretation of the relativistic dynamics via elliptic geometry. Remembering that the momentum is a vector, formula (3.11) is extended as follows

$$\exp(i2mc(\sigma\cdot\theta)) = \frac{(\sigma\cdot p)+imc}{(\sigma\cdot p)-imc}, \tag{3.12}$$

where $\sigma_k, k=1,2,3$ are Pauli matrices.

The variables θ and ϕ functionally are connected, this interconnection implicitly is expressed via the formulae

$$p_0 = mc\coth(\phi) = \frac{mc}{\sin(\theta)},\quad p = mc\cot(\theta) = \frac{mc}{\sinh(\phi)}. \tag{3.13}$$

From these formulae the following trigonometric relationship is derived

$$\tanh(mc\phi/2) = \tan(mc\theta/2). \tag{3.14}$$

Differential form of this formula is written as

$$\frac{d\theta}{d\phi} = \frac{p}{p_0} = \frac{V}{c}. \tag{3.15}$$

Compare this equation with

$$\frac{dr}{cdt} = \frac{p}{p_0} = \frac{V}{c}, \tag{3.16}$$

and take into account the similarity $p=\hbar\kappa,\ \ p_0=\hbar\omega$. Then, it is natural to speculate a relationship of the type

$$\theta = r/\hbar,\ \ \phi = ct/\hbar. \tag{3.17}$$

In the relativistic kinematics usually is used the following representation

$$V/c=\tanh(\alpha),\ p_0 = mc\cosh(\alpha),\ p = mc\sinh(\alpha). \tag{3.18}$$

This representation is interconnected with previous one as follows.

$$p_0 \pm p = mc\exp(\pm\alpha) = mc\frac{\cosh(\phi)\pm 1}{\sinh(\phi)}, \tag{3.19}$$

and

$$\tan(mc\theta/2) = \tanh(mc\phi/2) = \exp(\alpha). \tag{3.20}$$

Two forms of the quadratic polynomials implies the unique condition for the velocity $v<c$ and the momentum $p_0^2>p^2$. However, we can use unique form of the characteristic polynomial with the coefficients varying within the conditions

$$(a)\ \ p_0^2>p^2,\ c>v,$$

$$(b)\ \ p_0^2<p^2\ \ v>c.$$

Evidently, in that case the dynamics needs another kind of physical interpretation beyond the traditional one.

4. Geometrical Interpretation

In 1910, Klein took his revenge, proving that the Lorentz group is isomorphic to the group of isometries of the hyperbolic space [7]. The equation of a sheet of the hyperboloid is

$$k^2u^2 - x^2 - y^2 - z^2 = k^2, \tag{4.1}$$

on which $u > 0$. Isometric transformations of the Lobachevskii space are represented by linear transformations of the coordinates x, y, z, u that preserve the quadratic form (4.1). Such transformations constitute the Lorentz group. By taking the constant equal to the speed of light c, and the distance ρ is the particle rapidity s, one arrives at the concept of the velocity space as the Lobachevskii's space [8].

The differential equation (1.4) in a natural way is adopted to the kinematics. To this aim let us notice that two solutions of equation (1.4), $e^{x_k\phi}$, $k = 1, 2$, are eigenvalues of the matrix

$$\exp(E\phi) = g_0 I + E g_1. \tag{4.2}$$

Consequently, they are roots of the quadratic polynomial

$$F(\lambda) = \lambda^2 - 2(g_0 + p_0 g_1)\lambda + g_0^2 + 2p_0 g_0 g_1 + p^2 g_1. \tag{4.3}$$

Dividing equation $F(\lambda) = 0$ on the free coefficient we get [9]

$$v^2 - 2\frac{c}{V}v + 1 = 0, \tag{4.4}$$

where V is the velocity of the point-particle defined with respect to coordinate time. The rots of this equation are given by the exponential functions

$$v_{\pm} = \exp(\pm mc\phi). \tag{4.5}$$

The velocity V is defined by the equation

$$V = \frac{2v_{\pm}}{1 + v_{\pm}^2/c^2}. \tag{4.6}$$

In terms of the dynamical variables square of the roots of the equation (4.4) are defined by the formulae

$$\frac{v_{\pm}^2}{c^2} = \frac{p_0 \pm mc}{p_0 \mp mc}. \tag{4.7}$$

From the equation (4.6) the following formulae for roots are obtained

$$\frac{v}{c}|_{\pm} = \frac{c}{V}(1 \pm \sqrt{1 - \frac{V^2}{c^2}}). \tag{4.8}$$

Formula (4.8) coincides with the formula for distance between two points on the hyperbolic plane (see, for instance, [10], p.142). In this geometry distance between two point is defined by the logarithmic formula

$$\rho = \frac{k}{2}\log\frac{1 + \sqrt{1 - J^2}}{1 - \sqrt{1 - J^2}}, \tag{4.9}$$

where J in coordinates Beltrami is defined by the formula

$$J^2 = \frac{(z_1^2 - x_1^2 - y_1^2)(z_2^2 - x_2^2 - y_2^2)}{(z_1 z_2 - x_1 x_2 - y_1 y_2)^2}. \tag{4.10}$$

Formula (4.9) written in the exponential form coincides with (4.8), hence $V^2/c^2 = J^2$, i.e.,

$$V^2/c^2 = \frac{\Omega_{11}\Omega_{22}}{\Omega_{12}^2}. \tag{4.11}$$

Here,

$$\Omega_{kj} = z_k z_j - x_k x_j - y_k y_j \tag{4.12}$$

then, the following identity holds true

$$\Omega_{11}\Omega_{22} - \Omega_{12}^2 = (x_1 y_2 - x_2 y_1)^2 - (y_1 z_2 - y_2 z_1)^2 - (x_1 z_2 - x_2 z_1)^2. \tag{4.13}$$

In this model, the velocity V is presented by J, the rapidity ϕ with the distance ρ and speed of light c with curvature k. According to identity (4.13), the energy p_0, momentum p and the mass mc can be identified as follows

$$p^2 = (z_1^2 - x_1^2 - y_1^2)(z_2^2 - x_2^2 - y_2^2), \quad p_0 = z_1 z_2 - x_1 x_2 - y_1 y_2,$$
$$m^2 c^2 = (x_1 y_2 - x_2 y_1)^2 - (y_1 z_2 - y_2 z_1)^2 - (x_1 z_2 - x_2 z_1)^2. \tag{4.14}$$

Cross-ratio representation.

On the basis of the principal formula

$$\frac{q_1}{q_2} = \exp((q_1 - q_2)\phi_0),$$

let us consider two translations

$$\frac{q_1 + z_1}{q_2 + z_1} = \exp((q_1 - q_2)(\phi_0 + \delta_1)),$$

and

$$\frac{q_2 + z_2}{q_1 + z_2} = \exp((q_2 - q_1)(\phi_0 + \delta_2)).$$

The distance between initial and final points are defined by the product of these consequent translations:

$$\frac{q_1 + z_1}{q_2 + z_1}\frac{q_2 + z_2}{q_1 + z_2} = \exp((q_1 - q_2)(\delta_1 - \delta_2)). \tag{4.15}$$

This is nothing else than the cross-ratio. Distance is the logarithmic function of product of two translations:

$$\log[\frac{q_1 + z_1}{q_2 + z_1}\frac{q_2 + z_2}{q_1 + z_2}] = (q_1 - q_2)(\delta_1 - \delta_2). \tag{4.16}$$

5. General Representation by Jacobi Elliptic Functions

5.1 Elliptic Oscillator Model of Relativistic Dynamics

The dynamic variables p_0, p, including the kinematical variables V, V_0, in virtue to mass-shell equation can be expressed by two parameters q_1, q_2 – roots of the characteristic polynomial. In terms of variables q_1, q_2 they are defined by the following fractions

$$V_0 = \frac{q^2 - p^2}{q^2 + p^2}, \quad V = \frac{2qp}{q^2 + p^2},$$

$$P_0 = \frac{q^2 + p^2}{q^2 - p^2}, \quad P = \frac{2qp}{q^2 - p^2}. \tag{5.1}$$

In this section we use the units

$$m = 1, \ c = 1.$$

We have obtained four representations of the relativistic dynamical evolution written with respect to different rapidity parameter. Since these evolution equations are referenced to the same object, these equations can be considered as an unique evolution process written in different system of references.

All these four equations can be joined into unique evolution equation for elliptic functions by introducing an additional parameter k ranged between unite and zero: $1 \geq k \geq 0$.

The Jacobian elliptic functions are a natural extension of cosine-sine functions. The most important property of the Jacobian elliptic functions is double periodicity, in contrast to single periodicity of the circular functions. The Jacobian elliptic functions $sn(u,k)$, $cn(u,k)$, $dn(u,k)$ obey the following differential equations [11]

$$\frac{d}{du} sn(u,k) = cn(u,k) dn(u,k),$$

$$\frac{d}{du} cn(u,k) = -sn(u,k) dn(u,k),$$

$$\frac{d}{du} dn(u,k) = -k \ cn(u,k) sn(u,k), \tag{5.2}$$

The algebraic relations between these functions are given by

$$cn^2(u,k) + sn^2(u,k) = 1, \quad dn^2(u,k) + k \ sn^2(u,k) = 1. \tag{5.3}$$

When $k \to 0$ Jacobian elliptic functions give arise sine and cosine functions:

$$sn(u,k) \to \sin(u), \ cn(u,k) \to \cos(u), \ dn(u,k) \to 1. \tag{5.4}$$

5.2 Glaisher's Notation for Quotients.

A short and convenient notation has been invented by Glaisher to express reciprocals and quotients of the Jacobian elliptic functions; the reciprocals are denoted by reversing the order of the letters which express the function, thus

$$ns(u) = \frac{1}{sn(u)}, \quad nc(u) = \frac{1}{cn(u)}, \quad nd(u) = \frac{1}{dn(u)},$$

$$sc(u)=\frac{sn(u)}{cn(u)},\quad sd(u)=\frac{sn(u)}{dn(u)},\quad cd(u)=\frac{cn(u)}{dn(u)},$$

$$cs(u)=\frac{cn(u)}{sn(u)},\quad ds(u)=\frac{dn(u)}{sn(u)},\quad dc(u)=\frac{dn(u)}{cn(u)}. \tag{5.5}$$

From these definitions we obtain the following equations for the set of functions $ns(u),\ cs(u),\ ds(u)$:

$$\frac{d}{du}ns(u,k)=-cs(u,k)ds(u,k),\ \frac{d}{du}cs(u,k)=-cs(u,k)ds(u,k), \tag{5.6}$$

$$\frac{d}{du}ds(u,k)=-ns(u,k)cs(u,k).$$

The algebraic relations for this set of functions are given by

$$ns^2(u,k)-cs^2(u,k)=1,\quad ds^2(u,k)-cs^2(u,k)=1-k.$$

The Jacobi's imaginary transformation is defined by the following equations

$$isc(u,k')=sn(iu,k),\ cn(iu,k)=nc(u,k'),\ dn(iu,k)=dc(u,k'),\ k'=1-k \tag{5.7}$$

This set of functions obey the following equations

$$\frac{d}{du}sc(u,k')=nc(u,k')dc(u,k'),\ \frac{d}{du}nc(u,k')=sc(u,k')dc(u,k'), \tag{5.8}$$

$$\frac{d}{du}dc(u,k')=k'nc(u,k')sc(u,k').$$

The algebraic relations between these functions are given by

$$nc^2(u,k')-sc^2(u,k')=1,\quad dc^2(u,k')-k'sc^2(u,k')=1.$$

When $k'=0,\ k=1$ the set of functions $sc,\ nc,\ dc$ are reduced to hyperbolic sine and cosine functions:

$$sc(u,0)=\sinh(u),\ nc(u,0)=\cosh(u),\ dc(u,0)=1. \tag{5.9}$$

When $k'=1,\ k=0$ we get the following correspondence:

$$sc(u,1)=\frac{\sin(u)}{\cos(u)},\ nc(u,1)=\frac{1}{\cos(u)},\ dc(u)=\frac{1}{cos(u)}. \tag{5.10}$$

5.3 The Form of the Jacobi Equations Corresponding to the Hyperbolic and the Periodic Angles

These equations are written for the functions $nc(u,k), sc(u,k), dc(u,k)$:

$$\frac{d}{du}nc(u,k)=sc(u,k)dc(u,k),$$

$$\frac{d}{du}sc(u,k)=nc(u,k)dc(u,k),$$

$$\frac{d}{du}dc(u,k) = k'\; nc(u,k)sc(u,k), \tag{5.11}$$

Then, by introducing the functions

$$p = sc(u) = \frac{sn(u)}{cn(u)},\;\; p_0 = nc(u) = \frac{1}{cn(u)},\;\; p_1 = dc(u) = \frac{dn(u)}{cn(u)}, \tag{5.12}$$

we come to the following equations

$$\frac{d}{du}p_0(u,k) = p(u,k)p_1(u,k),$$

$$\frac{d}{du}p(u,k) = p_0(u,k)p_1(u,k),$$

$$\frac{d}{du}p_1(u,k) = k'\; p_0(u,k)p(u,k), \tag{5.13}$$

Under contractions

$$k = 1,\;\; k' = 0,$$

equations are reduced to

$$\frac{d}{du}p_0 = p,$$

$$\frac{d}{du}p = p_0, \tag{5.14}$$

where

$$p_0 = \cosh(u),\;\; p = \sinh(u).$$

Another form of the Jacobi equations corresponding to periodic rapidity are constructed as follows.

In that case under the contractions

$$k = 0,\;\; k' = 1,$$

equations (5.13) are reduced to equations

$$\frac{d}{du}p_0 = pp_0,$$

$$\frac{d}{du}p = p_0^2, \tag{5.15}$$

where

$$p_0 = \frac{1}{\cos(u)},\;\; p = \frac{\sinh(u)}{\cos(u)}.$$

5.4. The Forms of Contractions of the Jacobi Functions Corresponding to Hyperbolic and the Periodic Angles

Now we use equations for the functions $nc(u,k)$, $sc(u,k)$, $dc(u,k)$.

$$\frac{d}{du}ns(u,k) = -cs(u,k)ds(u,k),$$

$$\frac{d}{du}cs(u,k) = -ns(u,k)ds(u,k),$$

$$\frac{d}{du}ds(u,k) = -ns(u,k)cs(u,k),$$

$$ns^2(u,k) - cs^2(u,k) = 1, \quad ds^2(u,k) - cs^2(u,k) = 1 - k. \tag{5.16}$$

By denoting

$$p_0 = ns(u,k), \;\; p = cs(u,k), \;\; p_1 = ds(u), \tag{5.17}$$

we come to the following evolution equations

$$\frac{d}{d\phi}p = -p_0 p_1,$$

$$\frac{d}{d\phi}p_0 = -p p_1,$$

$$\frac{d}{d\phi}p_1 = -p p_0 \, k^2. \tag{5.18}$$

If $k^2 = 1$, then $p_1 = p_0$, and the system is reduced to

$$\frac{d}{d\phi}p = -p_0 p_0,$$

$$\frac{d}{d\phi}p_0 = -p p_0. \tag{5.19}$$

Under the contractions

$$k = 1, \;\; k' = 0,$$

equations are reduced to

$$\frac{d}{d\theta}p = -p_0 p,$$

$$\frac{d}{d\theta}p_0 = -p^2, \tag{5.20}$$

with periodic parameter of evolution θ.

In the general case, $1 > k^2 > 0$, solutions of the system of equations are given by the Jacobi elliptic functions.

Conclusion

We have started with second order differential equation

$$\frac{d^2}{dt^2}\Psi(t) - 2p_0\frac{d}{dt}\Psi(t) + p^2\Psi(t) = 0 \tag{6.1}$$

the solutions of which $\Psi(t)$ in some way are related with the coordinates of the oscillator model. We have shown that the differential equation are involved into dynamical evolution process of the coefficients via ratio of the fundamental solutions of the very differential equation. The situation in some sense is analogous to the interrelation between wave equations of the quantum mechanics and the classical Hamilton- Jacobi equations.

The distance of the hyperbolic space is define as the logarithm of the modulus of the general complex number

$$\phi = \frac{1}{a_1}\log(Det(Z)), \tag{6.2}$$

where

$$Det(Z) = Det(g_0 I + E g_1) = (g_0 + x_1 g_1)(g_0 + x_2 g_1) = \exp(2p_0\phi). \tag{6.1}$$

As it has been shown in [1], the Hamiltonian of the multicomplex oscillator model

$$H = \exp(-a_1\phi)Det(Z), \tag{6.4}$$

does not depend of ϕ.

The energy of this oscillator can be quantized displaying a new feature of the relativistic mechanics. The evolution of the coefficients of the differential equation in a same way can be applied to the oscillator model. In that case the principal formulae (2.16), (3.11) will take the form

$$\exp(2i\omega_0 t) = \frac{\kappa_0 + i\omega_0 - U(t,t_0)}{\kappa_0 + i\omega_0 + U(t,t_0)}, \tag{6.5}$$

$$\kappa(t) = \kappa_0 + i\omega - U(t,t_0),$$

where $U(t,t_0)$ is the solution of the Riccati equation (2.18).

References

[1] Yamaleev, R. M. Complex algebras on N-order polynomials and generalizations of trigonometry, oscillator model and Hamilton dynamics. *Advances in Applied Clifford algebras*, **15(1)**, (2005), 123.

[2] Dattoli, G., Di Palma, E., Nguen, F., Sabia, E. Generalized trigonometric functions and elementary applications. *Int. J. Appl. Comp. Math.* (2017) **3(2)** 445-458.

[3] Yamaleev, R. M. Multicomplex algebras on polynomials and generalized Hamilton dynamics. *J. Math. Anal. Appl.* **322** (2006) 815-824.

[4] Yamaleev, R. M. Geometrical and physical interpretation of evolution governed by general complex algebra. *J. Math. Anal. Appl.* **340** (2008) 1046-1057.

[5] Yamaleev, R. M. Extended relativistic mechanics of charged particle. Nova Science Publishers, Inc. ISBN 1-59033-981-9. In: *Relativity, Gravitation, Cosmology*. Eds. V. Dvoeglazov, A. Espinoza Garrido. 2004, pp.1-17.

[6] Yamaleev, R. M. Representation of solution of n-order Riccati equation via generalized trigonometric functions. *J. Math. Anal. Appl.* (2014) **420(1)** 334-347.

[7] Klein, F. On the geometric foundations of the Lorentz group. In: *New Ideas in Mathematics* [in Russian], No. 5, St. Peterburg (1914), pp. 144174. [Supplied by translator: This is the Russian translation of: F. Klein, ber die geometrischen Grundlagen der Lorentzgruppe, Jahresber. Deutsch. Math.-Verein., Bd. 19 (1910). It can be found also in: R. Fricke and A. Ostrowski (eds.), Felix Klein: Gesammelte Mathematische Abhandlungen, Erster Band, Verlag von Julius Springer, Berlin (1921), pp. 533552; reprinted, Springer-Verlag, Berlin-Heidelberg-New York (1973).]

[8] Chernikov, N. A. Introduction of Lobachevskii geometry into the theory of gravitation. *Sov. J. Pert. Nucl.* **23** (5),pp. 507-520. 1992.

[9] Yamaleev, R. M., Rodriguez-Dominguez, A. R. As regard the speed in medium of the electromagnetic radiatio field. *J. Mod. Phys.* **7** (2016) 1320-1330. http://dx.doi.org/10.4236/jmp.2016.711118

[10] Kagan, V. F. *Historical Foundations of Geometry.* (in Russian) Moscow, 1949.

[11] Akhiezer, N. I. *Elements of the Theory of Elliptic Functions*, M. Nauka, 1970.

In: Future Relativity, Gravitation, Cosmology
Editors: V. V. Dvoeglazov et al.
ISBN: 979-8-88697-455-3

Chapter 5

CLASSICAL MECHANICS WITH EVOLVING LOCAL METRICS

Martin Land *
Department of Computer Science
Hadassah College, Jerusalem

Abstract

We extend the canonical approach of Stueckelberg, Horwitz, and Piron in classical relativistic mechanics and electrodynamics to general relativity. Classical spacetime events $x^\mu(\tau)$, evolving as τ grows monotonically, trace out particle worldlines dynamically in geodesic motion determined by a 5D local metric $g_{\alpha\beta}(x,\tau)$, for $\alpha, \beta = 0,1,2,3,5$. The 5D metric obeys extended Einstein equations. To obtain a reasonable equivalence principle, the formal 5D tensor symmetries must break to tensor and scalar representations of O(3,1) by a consistent prescription. We take as an example the field produced by a τ-dependent mass $M(\tau)$, first as a perturbation in the Newtonian approximation and then for a Schwarzschild-like metric. The extended Einstein equations imply a flow of energy into spacetime corresponding to the changing source mass. The Hamiltonian, driven by terms proportional to $dM/d\tau$, is not generally conserved, but relaxes to a generalized Schwarzschild solution with vanishing Einstein tensor in τ-equilibrium.

1. Introduction

Stueckelberg-Horwitz-Piron (SHP) theory [1,2,3,4,5,6] is a covariant canonical approach to relativistic classical and quantum mechanics in which interacting events are defined with respect to coordinates x^μ $(\mu = 0,1,2,3)$ and an external evolution parameter τ. Events trace out particle worldlines as functions $x^\mu(\tau)$ or $\psi(x,\tau)$ under the monotonic advance of τ. SHP thus recognizes two distinct aspects of time: the coordinate time $x^0 = ct$, which along with the three space coordinates x^i

*Corresponding Author's Email: martin@hac.ac.il.

assigns a location in spacetime to each event, and the evolution parameter τ that provides a chronological ordering of event occurrence [7].

Historically, SHP electrodynamics [4] was constructed by generalizing the O(3,1)-symmetric action for a particle in Maxwell theory

$$S_{\text{Maxwell}} = \int d\tau \, \frac{1}{2} M \dot{x}^\mu \dot{x}_\mu + \frac{e}{c} \dot{x}^\mu A_\mu \left(x^\lambda \right) \qquad \mu, \lambda = 0,1,2,3 \tag{1}$$

by adding τ-dependence to the vector potential, along with a new scalar potential, to obtain the action

$$S_{\text{Maxwell}} \longrightarrow S_{\text{SHP}} = \int d\tau \, \frac{1}{2} M \dot{x}^\mu \dot{x}_\mu + \frac{e}{c} \dot{x}^\mu a_\mu \left(x^\lambda, \tau \right) + \frac{e}{c} c_5 a_5 \left(x^\lambda, \tau \right) \tag{2}$$

$$= \int d\tau \, \frac{1}{2} M \dot{x}^\mu \dot{x}_\mu + \frac{e}{c} \dot{x}^\beta a_\beta \left(x^\lambda, \tau \right) \tag{3}$$

where $\alpha, \beta, \gamma = 0,1,2,3,5$, and in analogy to $x^0 = ct$, we write $x^5 = c_5 \tau$, with $0 < c_5/c \ll 1$. Variation with respect to x^μ leads to the Lorentz force [8]

$$M \ddot{x}_\mu = \frac{e}{c} \left(\dot{x}^\nu f_{\mu\nu} + c_5 f_{\mu 5} \right) = \frac{e}{c} \dot{x}^\beta f_{\mu\beta} \tag{4}$$

$$\frac{d}{d\tau} \left(-\frac{1}{2} M \dot{x}^\mu \dot{x}_\mu \right) = e \frac{c_5}{c} \dot{x}^\beta f_{5\beta} \tag{5}$$

where (5) permits the exchange of mass between particles and fields. In this notation, the $\beta = 5$ index is understood to be a formal convenience, indicating O(3,1) scalar quantities, and not elements of 5D tensors. In particular, $\dot{x}^5$ is the constant scalar c_5. Terms of the type $\dot{x}^\beta a_\beta (x, \tau)$ suggest that (4) and (5) exhibit symmetry breaking, and one might be tempted to approach SHP theory by simply extending the Maxwell action (1) to 5D and restricting the set of solutions to those that include $\ddot{x}^5 = 0$.

However, treating $x^5(\tau)$ as a dynamical variable leads to inequivalent dynamics. SHP electrodynamics requires *first* imposing $x^5 = c_5 \tau$ and restricting the dynamical degrees of freedom to the 4D manifold, before applying a dynamical principle.

Classical and quantum SHP particle mechanics in a spacetime with local metric $g_{\mu\nu}(x)$ has been studied extensively by Horwitz [9, 10] and will not be discussed at length here. Our goal is to find a consistent prescription for extending general relativity to accommodate a metric $g_{\alpha\beta}(x, \tau)$ satisfying τ-dependent Einstein equations. (For a 5D ADM approach, see [11, 12].)

2. Canonical Mechanics

2.1. 5D Symmetry $\longrightarrow$ 4D+1

Before breaking the symmetry, we consider the free particle Lagrangian in 5D

$$L = \frac{1}{2} M g_{\alpha\beta} \dot{x}^\alpha \dot{x}^\beta \tag{6}$$

with local metric $g_{\alpha\beta}(x^\gamma)$, for $\alpha, \beta, \gamma = 0, 1, 2, 3, 5$, and an independent parameter τ. The Euler-Lagrange equations

$$0 = \frac{d}{d\tau}\frac{\partial L}{\partial \dot{x}^\gamma} - \frac{\partial L}{\partial x^\gamma} = \frac{d}{d\tau}\left(g_{\alpha\gamma}\dot{x}^\alpha\right) - \frac{1}{2}\frac{\partial}{\partial x^\gamma} g_{\alpha\beta}\dot{x}^\alpha \dot{x}^\beta \tag{7}$$

lead to the geodesic equations of motion

$$0 = \frac{D\dot{x}^\gamma}{D\tau} = \ddot{x}^\gamma + \Gamma^\gamma_{\alpha\beta}\dot{x}^\alpha \dot{x}^\beta \tag{8}$$

where $D/D\tau$ is the absolute derivative (in the notation of Weinberg [13]) and

$$\Gamma^\gamma_{\alpha\beta} = g^{\gamma\delta}\Gamma_{\delta\alpha\beta} = \frac{1}{2}g^{\gamma\delta}\left(\partial_\alpha g_{\delta\beta} + \partial_\beta g_{\delta\alpha} - \partial_\delta g_{\beta\alpha}\right) \tag{9}$$

is the standard Christoffel symbol in 5D. Writing the canonical momentum

$$p_\alpha = \frac{\partial L}{\partial \dot{x}^\alpha} = M g_{\alpha\beta}\dot{x}^\beta \qquad \longrightarrow \qquad \dot{x}^\alpha = \frac{1}{M}g^{\alpha\beta}p_\beta \tag{10}$$

the Hamiltonian

$$K = \dot{x}^\alpha p_\alpha - L = \frac{1}{2M}g^{\alpha\beta}p_\alpha p_\beta = L \tag{11}$$

is conserved, as seen directly through

$$\frac{Dg_{\alpha\beta}}{D\tau} = 0 \quad \longrightarrow \quad \frac{d}{d\tau}\left(\frac{1}{2}Mg_{\alpha\beta}\dot{x}^\alpha \dot{x}^\beta\right) = Mg_{\alpha\beta}\dot{x}^\alpha \frac{D\dot{x}^\beta}{D\tau} = 0. \tag{12}$$

This result may also be seen from the canonical equations of motion

$$\dot{x}^\alpha = \frac{dx^\alpha}{d\tau} = \frac{\partial K}{\partial p_\alpha} \qquad \dot{p}_\alpha = \frac{dp_\alpha}{d\tau} = -\frac{\partial K}{\partial x^\alpha} \tag{13}$$

and the Poisson bracket

$$\{F, G\} = \frac{\partial F}{\partial x^\alpha}\frac{\partial G}{\partial p_\alpha} - \frac{\partial F}{\partial p_\alpha}\frac{\partial G}{\partial x^\alpha} \tag{14}$$

so that

$$\frac{d}{d\tau}\left(\frac{1}{2M}g^{\alpha\beta}p_\alpha p_\beta\right) = \frac{dK}{d\tau} = \{K, K\} + \frac{\partial K}{\partial \tau} = \frac{1}{2M}p_\alpha p_\beta \frac{\partial g^{\alpha\beta}}{\partial \tau} = 0 \tag{15}$$

using the τ-independence of the metric.

To implement the symmetry breaking we put $x^5(\tau) = c_5\tau$ in (6), now fully equivalent to the evolution parameter and not a dynamical degree of freedom. The Lagrangian takes the form

$$L = \frac{1}{2}Mg_{\alpha\beta}\dot{x}^\alpha \dot{x}^\beta = \frac{1}{2}Mg_{\mu\nu}(x,\tau)\dot{x}^\mu \dot{x}^\nu + \frac{1}{2}Mc_5^2 g_{55}(x,\tau) \tag{16}$$

the equations of motion are

$$0 = \frac{D\dot{x}^\mu}{D\tau} = \ddot{x}^\mu + \Gamma^\mu_{\alpha\beta}\dot{x}^\alpha\dot{x}^\beta = \ddot{x}^\mu + \Gamma^\mu_{\lambda\sigma}\dot{x}^\lambda\dot{x}^\sigma + 2c_5\Gamma^\mu_{5\sigma}\dot{x}^\sigma + c_5^2\Gamma^\mu_{55} \tag{17}$$

$$0 = \frac{D\dot{x}^5}{D\tau} = \ddot{x}^5 + \Gamma^5_{\alpha\beta}\dot{x}^\alpha\dot{x}^\beta = \frac{d\dot{x}^5}{d\tau} = \frac{dc_5}{d\tau} \tag{18}$$

and the prescription for the symmetry-broken connection is

$$\Gamma^\mu_{5\alpha} = \frac{1}{2}g^{\mu\nu}\left(\partial_5 g_{\nu\alpha} + \partial_\alpha g_{\nu 5} - \partial_\nu g_{\alpha 5}\right) \qquad\qquad \Gamma^5_{\alpha\beta} \equiv 0\,. \tag{19}$$

We notice in particular the violation of the 5D symmetry $\Gamma_{5\mu\nu} = -\Gamma_{\mu 5\nu} + \partial_\nu g_{\mu 5}$ that no longer applies because the label 5 indicates a scalar quantity, and is not a tensor index. Because x^5 is not a dynamical quantity, it has no conjugate momentum, so the Hamiltonian takes the form

$$K = p_\mu\dot{x}^\mu - L = \frac{1}{2M}g^{\mu\nu}p_\mu p_\nu - \frac{1}{2}Mc_5^2 g_{55} = \frac{1}{2}Mg_{\mu\nu}\dot{x}^\mu\dot{x}^\nu - \frac{1}{2}Mc_5^2 g_{55} \tag{20}$$

which is not conserved for a τ-dependent metric, as seen from

$$\frac{dK}{d\tau} = \frac{\partial K}{\partial\tau} = -\frac{1}{2}M\dot{x}^\mu\dot{x}^\nu\frac{\partial g_{\mu\nu}}{\partial\tau} - \frac{1}{2}Mc_5^2\frac{\partial g_{55}}{\partial\tau} \tag{21}$$

where we used $\partial g^{\mu\nu}/\partial\tau = -g^{\mu\rho}g^{\nu\sigma}\partial g_{\rho\sigma}/\partial\tau$. Expression (21) is also found by taking the total τ-derivative of (20) and inserting the equations of motion (17).

2.2. Mass-Energy-Momentum Tensor

Considering non-thermodynamic dust, we define $n(x,\tau)$ to be the number of events per spacetime volume, so that the total number of events is

$$N = \int d\Omega\, n(x,\tau) \tag{22}$$

where $d\Omega$ is a scalar spacetime volume element, and

$$j^\alpha(x,\tau) = \rho(x,\tau)\dot{x}^\alpha(\tau) = Mn(x,\tau)\dot{x}^\alpha(\tau) \tag{23}$$

is the 5-component event current. The flat space continuity equation

$$\partial_\alpha j^\alpha = \partial_\mu j^\mu + \partial_5 j^5 = \partial_\mu j^\mu + \frac{\partial\rho}{\partial\tau} = 0 \tag{24}$$

is generalized with local metric by writing the covariant derivative (in the notation of Wald [14])

$$\nabla_\alpha j^\beta = \frac{\partial j^\beta}{\partial x^\alpha} + j^\gamma\Gamma^\beta_{\gamma\alpha} = 0\,. \tag{25}$$

Since the number density is a scalar on physical grounds, we must have

$$\nabla_5 j^5 = \frac{\partial j^5}{\partial x^5} + j^\gamma \Gamma^5_{\gamma 5} = \frac{\partial \rho}{\partial \tau} \tag{26}$$

consistent with prescription (19). The continuity equation leads to total event number is conservation through

$$\frac{\partial \rho}{\partial \tau} + \nabla_\mu j^\mu = 0 \longrightarrow \frac{d}{d\tau}\int d\Omega\, n(x,\tau) + \frac{1}{M}\int d\Omega\, \nabla_\mu j^\mu = \frac{dN}{d\tau} = 0\,. \tag{27}$$

Generalizing the 4D stress-energy-momentum tensor to 5D, we define the mass-energy-momentum tensor [4,15] as

$$T^{\alpha\beta} = Mn\dot{x}^\alpha \dot{x}^\beta = \rho \dot{x}^\alpha \dot{x}^\beta \longrightarrow \begin{cases} T^{\mu\nu} = Mn\dot{x}^\mu \dot{x}^\nu = \rho \dot{x}^\mu \dot{x}^\nu \\ T^{5\beta} = \dot{x}^5 \dot{x}^\beta \rho = c_5 j^\beta \end{cases} \tag{28}$$

combining $T^{\mu\nu}$ with j^α. The conservation equation

$$0 = \nabla_\beta T^{\alpha\beta} = \dot{x}^\alpha \nabla_\beta j^\beta + \rho \dot{x}^\beta \nabla_\beta \dot{x}^\alpha \tag{29}$$

vanishes by virtue of the continuity and geodesic equations

$$\nabla_\alpha j^\alpha = 0 \qquad\qquad \dot{x}^\beta \nabla_\beta \dot{x}^\alpha = \frac{D\dot{x}^\alpha}{D\tau} = 0 \tag{30}$$

for equations of motion (17) and (18) evaluated under prescription (19).

2.3. Einstein Equations

The Einstein equations are similarly extended to

$$G_{\alpha\beta} = R_{\alpha\beta} - \frac{1}{2} R g_{\alpha\beta} = \frac{8\pi G}{c^4} T_{\alpha\beta} \tag{31}$$

where the Ricci tensor $R_{\alpha\beta}$ and scalar R are obtained by contracting indices of the 5D curvature tensor $R^\delta_{\gamma\alpha\beta}$ found from the Christoffel symbols (9). Since conservation of $T^{\alpha\beta}$ depends on prescription (19), we must similarly suppress $\Gamma^5_{\alpha\beta}$ when constructing the Ricci tensor to insure $\nabla_\beta G^{\alpha\beta} = 0$. Splitting

$$R_{\alpha\beta} = R^\gamma_{\alpha\beta\gamma} = R^\lambda_{\alpha\beta\lambda} + R^5_{\alpha\beta 5} \tag{32}$$

we find $R_{\mu\nu} = \left(R_{\mu\nu}\right)^{4D}$ from

$$\begin{aligned} R^\lambda_{\mu\nu\lambda} &= \frac{\partial}{\partial x^\nu}\Gamma^\lambda_{\mu\lambda} - \frac{\partial}{\partial x^\lambda}\Gamma^\lambda_{\mu\nu} + \Gamma^\lambda_{\sigma\nu}\Gamma^\sigma_{\mu\lambda} - \Gamma^\lambda_{\sigma\lambda}\Gamma^\sigma_{\mu\nu} + \Gamma^\lambda_{5\nu}\Gamma^5_{\mu\lambda} - \Gamma^\lambda_{5\lambda}\Gamma^5_{\mu\nu} \\ &= \frac{\partial}{\partial x^\nu}\Gamma^\lambda_{\mu\lambda} - \frac{\partial}{\partial x^\lambda}\Gamma^\lambda_{\mu\nu} + \Gamma^\lambda_{\sigma\nu}\Gamma^\sigma_{\mu\lambda} - \Gamma^\lambda_{\sigma\lambda}\Gamma^\sigma_{\mu\nu} \end{aligned} \tag{33}$$

and

$$R^5_{\mu\nu 5} = \frac{\partial}{\partial x^\nu}\Gamma^5_{\mu 5} - \frac{1}{c_5}\frac{\partial}{\partial \tau}\Gamma^5_{\mu\nu} + \Gamma^5_{\sigma\nu}\Gamma^\sigma_{\mu 5} - \Gamma^5_{\sigma 5}\Gamma^\sigma_{\mu\nu} + \Gamma^5_{5\nu}\Gamma^5_{\mu 5} - \Gamma^5_{55}\Gamma^5_{\mu\nu} = 0. \tag{34}$$

Similarly,

$$\begin{aligned} R_{\mu 5} &= \frac{1}{c_5}\partial_\tau\Gamma^\lambda_{\mu\lambda} - \partial_\lambda\Gamma^\lambda_{\mu 5} + \Gamma^\lambda_{\sigma 5}\Gamma^\sigma_{\mu\lambda} - \Gamma^\lambda_{\sigma\lambda}\Gamma^\sigma_{\mu 5} \\ R_{55} &= \frac{1}{c_5}\partial_\tau\Gamma^\lambda_{5\lambda} - \partial_\lambda\Gamma^\lambda_{55} + \Gamma^\lambda_{\sigma 5}\Gamma^\sigma_{5\lambda} - \Gamma^\lambda_{\sigma\lambda}\Gamma^\sigma_{55}\,. \end{aligned} \tag{35}$$

3. Newtonian Approximation

The weak field approximation [16] is generalized for the 5D metric by writing

$$g_{\mu\nu} = \eta_{\mu\nu} + h_{\mu\nu} \longrightarrow g_{\alpha\beta} = \eta_{\alpha\beta} + h_{\alpha\beta} \longrightarrow \partial_\gamma g_{\alpha\beta} = \partial_\gamma h_{\alpha\beta} \tag{36}$$

where $(h_{\alpha\beta})^2 \approx 0$, leading to

$$R_{\alpha\beta} \simeq \frac{1}{2}\left(\partial_\beta\partial_\gamma h^\gamma_\alpha + \partial_\alpha\partial_\gamma h^\gamma_\beta - \partial_\gamma\partial_\gamma h_{\alpha\beta} - \partial_\alpha\partial_\beta h\right) \tag{37}$$

and $R \simeq \eta^{\alpha\beta}R_{\alpha\beta}$ with $h = \eta^{\alpha\beta}h_{\alpha\beta}$. Defining $\bar{h}_{\alpha\beta} = h_{\alpha\beta} - \frac{1}{2}\eta_{\alpha\beta}h$, the Einstein equations become

$$\frac{16\pi G}{c^4}T_{\alpha\beta} = \partial_\beta\partial_\gamma\bar{h}^\gamma_\alpha + \partial_\alpha\partial_\gamma\bar{h}^\gamma_\beta - \partial_\gamma\partial_\gamma\bar{h}_{\alpha\beta} - \partial_\alpha\partial_\beta\bar{h} \tag{38}$$

which takes the form of a wave equation

$$\frac{16\pi G}{c^4}T_{\alpha\beta} = -\partial^\gamma\partial_\gamma\bar{h}_{\alpha\beta} = -\left(\partial^\mu\partial_\mu + \frac{\eta_{55}}{c_5^2}\partial_\tau^2\right)\bar{h}_{\alpha\beta} \tag{39}$$

after imposing the gauge condition $\partial_\lambda\bar{h}^{\alpha\lambda} = 0$. The principal part Green's function [17] for this wave equation is

$$G(x,\tau) = -\frac{1}{2\pi}\delta(x^2)\delta(\tau) - \frac{c_5}{2\pi^2}\frac{\partial}{\partial x^2}\theta(-\eta_{55}g_{\alpha\beta}x^\alpha x^\beta)\frac{1}{\sqrt{-\eta_{55}g_{\alpha\beta}x^\alpha x^\beta}} \tag{40}$$

in which the first term is dominant at long distance, leading to

$$\bar{h}_{\alpha\beta}(x,\tau) = \frac{4G}{c^4}\int d^3x'\frac{T_{\alpha\beta}\left(t - \frac{|\mathbf{x}-\mathbf{x}'|}{c}, \mathbf{x}', \tau\right)}{|\mathbf{x}-\mathbf{x}'|}. \tag{41}$$

For a point source in a co-moving frame with $X = (cT(\tau), \mathbf{0})$, (28) becomes

$$T^{00} = mc^2\dot{T}^2\delta^3(\mathbf{x})\,\varphi(t - T(\tau)) \qquad T^{\alpha i} = 0 \qquad T^{55} = \frac{c_5^2}{c^2}T^{00} \tag{42}$$

where $\varphi(\tau)$ is a slowly varying function satisfying $\int d\tau\, \varphi(\tau) = 1$, and

$$\bar{h}^{00}(x,\tau) = \frac{4GM}{c^2R}\dot{T}^2 \qquad \bar{h}^{\alpha i}(x,\tau) = 0 \qquad \bar{h}^{55}(x,\tau) = \frac{c_5^2}{c^2}\bar{h}^{00} \tag{43}$$

where $M(\tau) = m\,\varphi(t - T(\tau))$. Using $h_{\alpha\beta} = \bar{h}_{\alpha\beta} - \frac{1}{2}\eta_{\alpha\beta}\bar{h}$ and neglecting terms in c_5^2/c^2, we see that $h^{00} = \bar{h}^{00}$. Since $g^{\alpha\beta}h_{\beta\gamma} \simeq \eta^{\alpha\beta}h_{\beta\gamma}$ the non-zero Christoffel symbols can be written

$$\begin{aligned} \Gamma^{\mu}_{00} &= -\frac{1}{2}\eta^{\mu\nu}\partial_\nu h_{00} & \Gamma^{\mu}_{0i} &= \frac{1}{2}\eta^{\mu\nu}\partial_i h_{\nu 0} \\ \Gamma^{\mu}_{50} &= \frac{1}{2c_5}\eta^{\mu 0}\partial_\tau h_{00} & \Gamma^{\mu}_{55} &= -\frac{1}{2}\eta^{\mu\nu}\partial_\nu h_{55} \end{aligned} \tag{44}$$

so the geodesic equations (17) split as

$$\ddot{t} = (\partial_\tau h_{00})\,\dot{t} + \dot{\mathbf{x}}\cdot(\nabla h_{00})\,\dot{t}^2 \qquad \ddot{\mathbf{x}} = \frac{c^2}{2}(\nabla h_{00})\,\dot{t}^2. \tag{45}$$

In spherical coordinates, putting $\theta = \pi/2$, the angular and radial equations are

$$2\dot{R}\dot{\phi} + R\ddot{\phi} = 0 \longrightarrow \dot{\phi} = \frac{L}{MR^2} \longrightarrow \ddot{R} - \frac{L^2}{M^2R^3} = -\frac{GM}{R^2}\dot{t}^2\dot{T}^2 \tag{46}$$

and the t equation is

$$\ddot{t} = \frac{2G\partial_\tau M}{c^2R}\dot{t} + \frac{4GM}{c^2R}\dot{T}\ddot{T}\dot{t} - \frac{2GM}{R^2c^2}\dot{R}\dot{T}^2 \approx \frac{2GM}{c^2R}\left(1 + \frac{\alpha(\tau)}{2}\right)\dot{\alpha}(\tau)\,\dot{t} \tag{47}$$

where we neglect $\partial_\tau\varphi \approx 0$ and $\dot{R}/c \approx 0$, and define

$$\dot{T} = 1 + \frac{\alpha(\tau)}{2} \longrightarrow \dot{T}^2 \simeq 1 + \alpha(\tau) \longrightarrow \dot{T}\ddot{T} \simeq \left(1 + \frac{\alpha(\tau)}{2}\right)\frac{\dot{\alpha}(\tau)}{2}. \tag{48}$$

In the Newtonian case, $\alpha = 0 \longrightarrow \dot{t} = 1$, but this t equation has the solution

$$\dot{t} = \exp\left[\frac{2GM}{c^2R}\left(\alpha + \frac{1}{4}\alpha^2\right)\right] \longrightarrow \dot{t}^2\dot{T}^2 \simeq 1 + \frac{1}{2}\left(1 + \frac{2GM}{c^2R}\right)\alpha \tag{49}$$

which, since $2GM/c^2R \ll 1$, leads finally to the radial equation in the form

$$\frac{d}{d\tau}\left\{\frac{1}{2}\dot{R}^2 + \frac{1}{2}\frac{l^2}{M^2R^2} - \frac{GM}{R}\left(1 + \frac{1}{2}\alpha(\tau)\right)\right\} = \frac{dK}{d\tau} = -\frac{GM}{2R}\frac{d}{d\tau}\alpha(\tau). \tag{50}$$

We recognize K on the LHS as the Hamiltonian of the particle moving in this local metric. The mass fluctuation of the point source is seen to induce a fluctuation in the mass of a distant particle through the field $g_{\alpha\beta}(x,\tau)$, producing a small modification of Newtonian gravity.

4. Extended Schwarzschild Solution

The $4D$ Schwarzschild metric is found by writing the spherically symmetric interval with $\theta = \pi/2$

$$ds^2 = -c^2 B(r) dt^2 + A(r) dr^2 + d\phi^2 \tag{51}$$

and solving $R_{\mu\nu} = 0$ to obtain

$$B(r) = \frac{1}{A(r)} = \left(1 - \frac{2GM}{rc^2}\right) \tag{52}$$

with the Minkowski metric and Newtonian gravity as boundary conditions. We extend this system in the manner of (16) as

$$L = \frac{1}{2}\left[-c^2 B(r,\tau)\dot{t}^2 + A(r,\tau)\dot{r}^2 + r^2\dot{\phi}^2 + c_5^2 Q(r,\tau)\right] \tag{53}$$

where $g_{55} = Q$. From the new non-zero Christoffel symbols

$$\Gamma^0_{50} = \frac{1}{2c_5}\frac{\partial_\tau B}{B} \qquad \Gamma^1_{55} = -\frac{1}{2}\frac{\partial_r Q}{A} \qquad \Gamma^1_{15} = \frac{1}{2c_5}\frac{\partial_\tau A}{A} \tag{54}$$

we find the Ricci tensor components by inserting these into (35). Generalizing the Schwarzschild solution by taking $M = M(\tau)$ in (52) and setting

$$Q = \int_r^\infty dr \, \frac{B}{r^2} = \frac{1}{r}\left(1 - \frac{M(\tau)G}{c^2 r}\right) = \frac{1}{2r}(3 - B) \tag{55}$$

we have $R_{\mu\nu} = \left(R_{\mu\nu}\right)^{4D} = 0$ as in the Schwarzschild case, but also

$$R_{05} = \frac{2\dot{M}G}{c_5 c^2 r^2}\left(1 - \frac{2MG}{c^2 r}\right)^{-1} \qquad R_{15} = -\frac{\dot{M}G}{c_5 c^2 r^2}\left(1 - \frac{2MG}{c^2 r}\right)^{-1}$$
$$R_{55} = 2\left(\frac{\dot{M}G}{c_5 c^2 r}\right)^2\left(1 - \frac{2MG}{c^2 r}\right)^{-2} \tag{56}$$

and

$$R = g^{\mu\nu} R_{\mu\nu} + g^{55} R_{55} = \frac{2}{r}\left(\frac{\dot{M}G}{c_5 c^2}\right)^2\left(1 - \frac{2MG}{c^2 r}\right)^{-2}\left(1 - \frac{MG}{c^2 r}\right)^{-1} \tag{57}$$

where $\dot{M} = dM/d\tau$. The Einstein tensor has components

$$G_{\mu\nu} = -\frac{1}{r}\left(\frac{\dot{M}G}{c_5 c^2}\right)^2\left(1 - \frac{2MG}{c^2 r}\right)^{-2}\left(1 - \frac{MG}{c^2 r}\right)^{-1} g_{\mu\nu}$$
$$G_{55} = \left(\frac{\dot{M}G}{c_5 c^2 r}\right)^2\left(1 - \frac{2MG}{c^2 r}\right)^{-2}$$
$$G_{50} = \frac{2\dot{M}G}{c_5^2 c^2 r}\left(1 - \frac{2MG}{c^2 r}\right)^{-1}\left(1 - \frac{MG}{c^2 r}\right)^{-1} \tag{58}$$
$$G_{51} = -\frac{\dot{M}G}{c_5^2 c^2 r}\left(1 - \frac{2MG}{c^2 r}\right)^{-1}\left(1 - \frac{MG}{c^2 r}\right)^{-1}$$

all proportional to $\dot{M}(\tau)$, and through $G^{5\alpha} = (8\pi G/c^4)T^{5\alpha} = (8\pi G/c^4)c_5 j^\alpha$ the non-zero current components j^0 and j^1 indicate the flow of mass density j^5 into the t and r directions. Treatment of the geodesic equations for a particle in this geometry follows the standard Schwarzschild case [13] closely, leading to solutions $\dot{t} = 1/B$ and $\dot{\phi} = J/r^2$ along with the radial equation

$$\frac{dK}{d\tau} = \frac{d}{d\tau}\frac{1}{2}\left(-c^2\frac{1}{B} + A\dot{r}^2 + \frac{J^2}{r^2} - c_5^2 Q\right) = -\frac{1}{2}\dot{r}^2\partial_\tau A - \frac{1}{2}c_5^2\partial_\tau Q. \tag{59}$$

We recognize K as the Hamiltonian in this parameterization, representing the dynamical mass of the moving particle which, as we saw in (21), is not generally conserved. As in the weak field approximation, a source with varying mass induces mass variation in a particle moving through the τ-dependent geometry described by the local metric. In equilibrium, $\dot{M} = 0$ recovers a massless system (away from the origin) with $R_{\alpha\beta} = G_{\alpha\beta} = \dot{K} = 0$ that permits solution of the radial equation as a first integral.

Conclusion

Stueckelberg-Horwitz-Piron (SHP) theory is a framework for posing classical and quantum relativistic physics in canonical form with an external parameter of evolution τ. This canonical structure preserves the tensorial character of the position $x^\mu(\tau)$, momentum $p_\mu(\tau)$, and the scalar Hamiltonian, which represents mass rather than energy. Events trace out particle worldlines under the monotonic advance of τ, providing a temporal foliation that avoids any decomposition into space and time coordinates, and thus remains observer independent. Interactions in SHP electrodynamics form an integrable system in which event evolution generates an instantaneous current defined over spacetime at τ, and in turn, these currents induce τ-dependent fields that act on other events at τ. We expect that in a similar way, a fully-developed SHP formulation of general relativity will describe how the instantaneous distribution of mass at τ expressed through $T_{\alpha\beta}(x,\tau)$ induces the local metric $g_{\alpha\beta}(x,\tau)$, which in turn, determines geodesic equations of motion for any particular event at $x^\mu(\tau)$.

References

[1] Stueckelberg E. (1941) *Helv. Phys. Acta* **14** 321–322 (In French).

[2] Stueckelberg E. (1941) *Helv. Phys. Acta* **14** 588–594 (In French).

[3] Horwitz L and Piron C. (1973) *Helv. Phys. Acta* **48** 316–326.

[4] Saad D, Horwitz L and Arshansky R. (1989) *Found. Phys.* **19** 11251149.

[5] Horwitz L P. (2015) *Relativistic Quantum Mechanics* (Dordrecht, Netherlands: Springer).

[6] Land M. (2018) Classical Stueckelberg-Horwitz-Piron Electrodynamics *Relativity, Gravitation, Cosmology: Beyond Foundations* Contemporary Fundamental Physics ed Dvoeglazov V (Nova Science Publishers).

[7] Horwitz L, Arshansky R and Elitzur A. (1988) *Found. Phys.* **18** 1159.

[8] Land M and Horwitz L. (1991) *Found. Phys. Lett.* **4** 61.

[9] Horwitz L P. (2019) *Journal of Physics: Conference Series* **1239** 012014 URL `https://doi.org/10.1088%2F1742-6596%2F1239%2F1%2F012014`.

[10] Horwitz L P. (2019) *The European Physical Journal Plus* **134** 313 ISSN 2190-5444 URL `https://doi.org/10.1140/epjp/i2019-12689-7`.

[11] Pitts J B and Schieve W C 1998 *Foundations of Physics* **28** 1417–1424 ISSN 1572-9516 URL `https://doi.org/10.1023/A:1018801126703`.

[12] Pitts J B and Schieve W C. (2001) *Foundations of Physics* **31** 1083–1104 ISSN 1572-9516 URL `https://doi.org/10.1023/A:1017578424131`.

[13] Weinberg S. (1972) *Gravitation and Cosmology: Principles and Applications of the General Theory of Relativity* (New York, NY: Wiley) URL `https://cds.cern.ch/record/100595`.

[14] Wald R M. (1984) *General relativity* (Chicago, IL: Chicago Univ. Press) URL `https://cds.cern.ch/record/106274`.

[15] Land M. (2019) *Journal of Physics: Conference Series* **1239** 012005 URL `https://doi.org/10.1088%2F1742-6596%2F1239%2F1%2F012005`.

[16] Misner C W, Thorne K S and Wheeler J A. (1973) *Gravitation* (San Francisco: W.H. Freeman and Co.).

[17] Land M and Horwitz L. (1991) *Found. Phys.* **21** 299–310.

In: Future Relativity, Gravitation, Cosmology
Editors: V. V. Dvoeglazov et al.
ISBN: 979-8-88697-455-3

Chapter 6

NON-RIEMANNIAN GEOMETRY, BORN-INFELD MODELS AND TRACE FREE GRAVITATIONAL EQUATIONS: MATHEMATICAL AND PHYSICAL CONSEQUENCES

Diego Julio Cirilo-Lombardo*
University of Buenos Aires,
Faculty of Exact and Natural Sciences, Department of Physics,
CONICET - University of Buenos Aires,
National Institute of Plasma Physics (INFIP). Buenos Aires, Argentina.
and Federal Research Center, Institute of Applied Mathematics,
M. V. Keldysh Institute of the Russian Academy of Sciences,
Moscow, Russian Federation

Abstract

Non-Riemannian generalization of the standard Born-Infeld (BI) Lagrangian was recently introduced and analized from a theory of gravitation with dynamical torsion field [14]. The field equations derived from the simplest proposed action lead to a trace free gravitational equation (non-riemannian analog to the trace free equation (TFE) from [1]-[3]) and the field equations for the torsion respectively. In this theoretical context, the fundamental constants arise all from the same geometry through geometrical invariant quantities (as from the curvature R). Important results involving generation of primordial magnetic fields and the link with leptogenesis and baryogenesis that were presented are reviewed and explained. The physically admisible matter fields can be introduced in the model via the torsion vector h_{μ}. Such fields include some dark matter candidates such as axion, right neutrinos and Majorana and moreover, physical observables as vorticity can be included in the same way. From a new wormhole soluton in a cosmological spacetime with torsion we also show that the primordial cosmic magnetic fields can originate from h_{μ} with the axion field (that is contained in h_{μ}) the responsible to control the dynamics and stability of the cosmic magnetic field and the magnetogenesis itself. The analisys of Grand Unified Theories

*Corresponding Author's Email: diego777jcl@gmail.com.

(GUT) in the context of this model indicates that the group manifold candidates are based in $SO(10)$, $SU(5)$ or some exceptional groups as $E(6)$, $E(7)$, etc. High energy astrophysical effects, as the magnetosphere structure of compact objects is reviewed in the same new theoretical context [74]. To this end we show that the new spherically symmetric solution, obtained in this theoretical framework before, physically represents a compact object of pseudoscalar fields (for example, axion field). The axially symmetric version of the Grad-Shafranov equation (GSE) is also derived in this context, and used to describe the magnetosphere dynamics of the obtained "axion star". The interplay between high-energy processes and the seed magnetic field with respect to the global structure of the magnetosphere is briefly discussed.

1. Introduction

The idea to construct a complete geometrization of the physics is very old. The drawback of the Einstein GR (General Relativity) equations is the RHS of them: $R_{\alpha\beta} - \frac{g_{\alpha\beta}}{2} R = \kappa T_{\alpha\beta}$ with the symmetric tensor (non-geometrical) $\kappa T_{\alpha\beta}$ that introduces *heuristically* the energy-momentum distribution. Similar drawbacks are contained by the unimodular gravity. It is well known that the unimodular gravity is obtained from Einstein-Hilbert action in which the *unimodular condition:* $\sqrt{-\det g_{\mu\nu}} = 1$ is also imposed from the very beginning [1]-[3], [41]. The resulting field equations correspond to the *traceless* Einstein equations and can be shown that they are equivalent to the full Einstein equations with the cosmological constant term Λ, where Λ enters as an integration constant and the equivalence between unimodular gravity and general relativity is given by the arbitrary value of lamda On the other hand the idea that the cosmological term arises as an integration constant is one of the motivations for the study of the unimodular gravity, for recent study, see [5] and [4] in the context of supergravity. The fact that the determinant of the metric is fixed has clearly several consequences on the structure of given theory. First of all, it reduces the full group of diffeomorphisms to invariance under the group of unimodular general coordinate transformations which are transformations that leave the determinant of the metric unchanged.

Similar thing happens in the non-Riemannian case, as pointed out in [12]-[16], where the corresponding affine geometrical structure induces naturally the following constraint: $\frac{K}{g} = constant$. This natural constraint impose a condition (ratio) between both basic tensors through their determinants: the metric determinant g and the fundamental one K (in the sense of a nonsymmetric theory that contains the antisymmetric structures), independently of the precise functional form of K or g. In this work our starting point will be precisely the last one, where a *metric affine structure* in the space-time manifold (as described in Section II) will be considered. We will also show that trace free gravitational equations can be naturally obtained when the Lagrangian function (geometrical action) is taken as a measure involving a particular combination of the fundamental tensors of the geometry:

$$\sqrt{|\det f\left(g_{\mu\nu}, f_{\mu\nu}, R_{\mu\nu}\right)|}$$

with the (0,2) tensors $g_{\mu\nu}, f_{\mu\nu}, R_{\mu\nu}$: the symmetric metric, the antisymmetric one (that acts as potential of the torsion field) and the generalized Ricci tensor (proper of the non Riemannian geometry). The three tensors are related with a Clifford structure of the tangent space (for details see[17]) where the explicit choice for $f\left(g_{\mu\nu}, f_{\mu\nu}, R_{\mu\nu}\right)$ is given in Section III This type of Lagrangians, because are non-Riemannian generalizations of the well known Nambu-Goto and Born-Infeld (BI) ones, can be physically and geometrically analyzed. Due the pure geometrical structure of the theory, induced energy momentum tensors and fundamental constants (actually functions) *emerge* naturally. Consequently, this fact allows the physical realization of the Mach principle that is briefly treated in SectionVIII after that the (trace free) dynamic equations in Section IV are obtained.

In Section V the trace free gravitational equations and the meaning of the cosmological term as integration constant are discussed from the physical point of view, meanwhile in Section VI the constancy of G (Newton constant) is similarly discussed. The important role played by the dual of the torsion field as structure of a geometrical energy-momentum tensor is given in Section VII. Some physical consequences of the model, as the geometrical origin of the $\alpha\Omega-$dynamo effect in astrophysics, is presented in Section IX establishing the link between the mathematical structure of the model of the first part of the article and the physics of the early universe and the particle physics of the second half part of this work. In Section X the direct relation between the torsion with axion electrodynamics and Chern-Simmons (CS) theory is discussed considering the geometrical structure of the dual vector of the torsion field. In Section XI an explanation about the magnetogenesis in FRW (Fermi-Robertson-Walker) scenario, the structure of the GUT where the Standard Model (SM) is derived and the role of the axion in the dynamics of the cosmic magnetic field is presented. Finally some concluding remarks are given in Section XII. Section XIII is devoted to the application of the generalized Grad-Shafranov equation (calculated in detail in [61] and described in the Appendix) in the case of the solution obtained in Section XIV to determine the dynamics and structure of the magnetosphere. For this purpose in Section XIV the new spherically symmetric solution (considering the dual tensor torsion h_μ from the gravitational theory based on affine geometry given in [51], [52]) representing a spherical configuration of pseudoscalars (e.g., axion star) is obtained and the analysis of the metric and the fields involved, carried out. Finally in Section XVI, we will present a concise discussion on the problem of the physics of magnetospheres based on the expressions obtained, the topology of the magnetic fields and the current knowledge regarding the intervention of high energy processes in these scenarios.

2. Basis of the Metrical-Affine Geometry

The starting point is a hypercomplex construction of the (metric compatible) spacetime manifold [17], [19]

$$M,\ g_{\mu\nu} \equiv e_\mu \cdot e_\nu \tag{1}$$

where for each point $p \in M$ there exists a local affine space A. The connection over A, $\widetilde{\Gamma}$, define a generalized affine connection Γ on M, specified by (∇, K), where K is an invertible $(1,1)$ tensor over M. We will demand for the connection to be compatible and rectilinear, that is

$$\nabla K = KT, \qquad \nabla g = 0 \tag{2}$$

where T is the torsion, and g the space-time metric (used to raise and lower the indices and determining the geodesics), that is preserved under parallel transport. This generalized compatibility condition ensures that the generalized affine connection Γ maps autoparallels of Γ on M into straight lines over the affine space A (locally). The first equation above is equal to the condition determining the connection in terms of the fundamental field in the UFT non-symmetric. Hence, K can be identified with the fundamental tensor in the non-symmetric fundamental theory. This fact gives us the possibility to restrict the connection to a (anti-)Hermitian theory.

The covariant derivative of a vector with respect to the generalized affine connection is given by

$$A^{\mu}{}_{;\nu} \equiv A^{\mu}{}_{,\nu} + \Gamma^{\mu}{}_{\alpha\nu} A^{\alpha} \tag{3}$$
$$A_{\mu;\nu} \equiv A_{\mu}{}_{,\nu} - \Gamma^{\alpha}{}_{\mu\nu} A_{\alpha} \tag{4}$$

The generalized compatibility condition (2) determines the 64 components of the connection by the 64 equations

$$K^{\mu}_{\nu;\alpha} = K^{\mu}_{\rho} T^{\rho}{}_{\nu\alpha} \qquad \text{where} \qquad T^{\rho}{}_{\nu\alpha} \equiv 2\Gamma^{\rho}{}_{[\alpha\nu]} \tag{5}$$

Notice that by contracting indices ν and α in the first equation above, an additional condition over this hypothetical fundamental (nonsymmetric) tensor K is obtained

$$K_{\mu\alpha;}{}^{\alpha} = 0$$

that, geometrically speaking, reads

$$d^{*}K = 0.$$

This is a current-free condition over the tensor K. Notice that the metric is used here to down the indices (metric compatible space-time) and consequently we can work also with $K_{\alpha\nu} = g_{\alpha\beta} K^{\beta}_{\nu}$

The metric is uniquely determined by the metricity condition, which puts 40 restrictions on the derivatives of the metric

$$g_{\mu\nu,\rho} = 2\Gamma_{(\mu\nu)\rho} \tag{6}$$

The space-time curvature tensor, that is defined in the usual way, has two possible contractions: the Ricci tensor $R^{\lambda}_{\mu\lambda\nu} = R_{\mu\nu}$, and the second contraction $R^{\lambda}_{\lambda\mu\nu} = 2\Gamma^{\lambda}{}_{\lambda[\nu,\mu]}$, which is identically zero due to the metricity condition (2).

In order to find a symmetry of the torsion tensor, let us denote the inverse of K by $\widehat{K}$. Therefore, $\widehat{K}$ is uniquely specified by condition $\widehat{K}^{\alpha\rho} K_{\alpha\sigma} = K^{\alpha\rho}\widehat{K}_{\alpha\sigma} = \delta^{\rho}_{\sigma}$.

As it was pointed out in [12]-[16], inserting explicitly the torsion tensor as the antisymmetric part of the connection in (5), and multiplying by $\frac{1}{2}\widehat{K}^{\alpha\nu}$, results, after straighforward computations, in

$$\left(Ln\sqrt{-K}\right)_{,\mu} - \Gamma^{\nu}_{(\mu\nu)} = 0 \tag{7}$$

where $K = \det\left(K_{\mu\rho}\right)$. Notice that from expression (7) we arrive at the relation between the determinants K and g:

$$\frac{K}{g} = \text{constant}$$

(strictly a constant scalar function of the coordinates). Now we can write

$$\Gamma^{\nu}{}_{\alpha\nu,\beta} - \Gamma^{\nu}{}_{\beta\nu,\alpha} = \Gamma^{\nu}{}_{\nu\beta,\alpha} - \Gamma^{\nu}{}_{\nu\alpha,\beta}, \tag{8}$$

as the first term of (7) is the derivative of a scalar. Then, the torsion tensor has the symmetry

$$T^{\nu}{}_{\nu[\beta,\alpha]} = T^{\nu}{}_{\nu[\alpha,\beta]} = 0 \tag{9}$$

This implies that the trace of the torsion tensor, defined as $T^{\nu}_{\ \nu\alpha}$, is the gradient of a scalar field

$$T_{\alpha} = \nabla_{\alpha}\phi \tag{10}$$

In reference [19] an interesting geometrical analysis is presented of non-symmetric field structures. There, expressions precisely as (1) and (2) ensure that the basic non-symmetric field structures (*i.e.* K) take on a definite geometrical meaning when interpreted in terms of affine geometry. Notice that the tensor K carries the 2-form (bivector) that will be associated with the fundamental antisymmetric form. Such antisymmetric form is introduced from the tangent space via the generalization of the Ambrose-Singer theorem by exponentiation.

3. Geometrical Lagrangians: The Generalized Born-Infeld Action

Let us start with the geometrical Lagrangian introduced in [[12]-[16]]

$$\mathcal{L}_g = \sqrt{\det\left[\lambda\left(g_{\alpha\beta} + F_{\alpha\beta}\right) + R_{\alpha\beta}\right]} \tag{11}$$

it can be rewritten as

$$\mathcal{L}_g = \sqrt{\det\left(G_{\alpha\beta} + \mathcal{F}_{\alpha\beta}\right)} \tag{12}$$

with the following redefinitions

$$G_{\alpha\beta} = \lambda g_{\alpha\beta} + R_{(\alpha\beta)} \quad \text{and} \quad \mathcal{F}_{\alpha\beta} = \lambda F_{\alpha\beta} + R_{[\alpha\beta]} \tag{13}$$

where a totally antisymmetric torsion tensor $T^{\alpha}_{\gamma\beta} = \varepsilon^{\alpha}_{\gamma\beta\delta}h^{\delta}$ is assumed (h^{δ} its dual vector field). Notice that the antisymmetric tensor $F_{\alpha\beta}$,that takes the role of the electromagnetic field, is proportional to the dual of the potential for the (totally antisymmetric) torsion field [12]-[16]. A brief review on the origin of this type of Lagrangians in the context of unified theories in reductive geometries is in the Appendix I. Consequently the generalized Ricci tensor splits into a symmetric and antisymmetric part, namely:

$$R_{\mu\nu} = \overbrace{\overset{\circ}{R}_{\mu\nu} - T_{\mu\rho}{}^{\alpha}T_{\alpha\nu}{}^{\rho}}^{R_{(\mu\nu)}} + \overbrace{\overset{\circ}{\nabla}_{\alpha}T_{\mu\nu}{}^{\alpha}}^{R_{[\mu\nu]}}$$

where $\overset{\circ}{R}_{\mu\nu}$ is the general relativistic Ricci tensor constructed with the Christoffel connection. The expansion of the determinant leads to the Born-Infeld generalization in the usual form :

$$\mathcal{L}_g = \sqrt{|G|}\sqrt{1 + \frac{1}{2}\mathcal{F}_{\mu\nu}\mathcal{F}^{\mu\nu} - \frac{1}{16}\left(\mathcal{F}_{\mu\nu}\widetilde{\mathcal{F}}^{\mu\nu}\right)^2} \tag{14}$$

$$= \Lambda^2\sqrt{|g|}\sqrt{1 + \frac{1}{2}\Lambda_1^2 F_{\mu\nu}F^{\mu\nu} - \frac{1}{16b^4}\left(\Lambda_2^2 F_{\mu\nu}\widetilde{F}^{\mu\nu}\right)^2} \tag{15}$$

where

$$\Lambda = \lambda + \frac{g_{\alpha\beta} R^{(\alpha\beta)}}{4} \tag{16}$$

$$\Lambda_1^2 = \lambda^2 \left(1 + \frac{2}{\lambda} \frac{F_{\mu\nu} R^{[\mu\nu]}}{F_{\mu\nu} F^{\mu\nu}} + \frac{1}{\lambda^2} \frac{R_{[\mu\nu]} R^{[\mu\nu]}}{F_{\mu\nu} F^{\mu\nu}} \right) \tag{17}$$

$$\Lambda_2^2 = \lambda^2 \left(1 + \frac{2}{\lambda} \frac{F_{\mu\nu} \widetilde{R}^{[\mu\nu]}}{F_{\mu\nu} \widetilde{F}^{\mu\nu}} + \frac{1}{\lambda^2} \frac{R_{[\mu\nu]} \widetilde{R}^{[\mu\nu]}}{F_{\mu\nu} \widetilde{F}^{\mu\nu}} \right) \tag{18}$$

Although the action is exact and have the correct limit, the analysis can be simplest and substantially improved using the following action

$$\mathcal{L}_{gs} = \sqrt{\det \left[\lambda g_{\alpha\beta} \left(1 + \frac{R_s}{4\lambda} \right) + \lambda F_{\alpha\beta} \left(1 + \frac{R_A}{\lambda} \right) \right]} \tag{19}$$

$$R_s \equiv g^{\alpha\beta} R_{(\alpha\beta)}; \qquad R_A \equiv f^{\alpha\beta} R_{[\alpha\beta]} \tag{20}$$

$\left(\text{with } f^{\alpha\beta} \equiv \frac{\partial \ln(\det F_{\mu\nu})}{\partial F_{\alpha\beta}}, \det F_{\mu\nu} = 2F_{\mu\nu}\widetilde{F}^{\mu\nu}\right)$.

If the induced structure from the tangent space $T_p(M)$ (via Ambrose-Singer theorem) it is intrinsically related to a (super)manifold structure, we have seen that there exists a transformation [10], [17] $U_A^B(P) = \delta_A^B + \mathcal{R}_{A\mu\nu}^B dx^\mu \wedge dx^\nu \to \delta_A^B + \omega^k (\mathcal{T}_k)_A^B$ with A,B.... generally a multi-index, having the same form as the blocks inside of the square root proposed Lagrangian (19): e.g.,$\lambda g_{\alpha\beta}\left(1 + \frac{R_s}{4\lambda}\right)$ where the Poisson structure is evident as the dual of the Lie algebra of the group manifold, leading the identification between the group structure of the tangent space with the space-time curvature as $\mathcal{R}_{A\mu\nu}^B dx^\mu \wedge dx^\nu \equiv \omega^k (\mathcal{T}_k)_A^B$

.

4. Field Equations

The geometry of the space-time Manifold is to be determined by the Noether symmetries

$$\frac{\delta L_G}{\delta g^{\mu\nu}} = 0, \quad \frac{\delta L_G}{\delta f^{\mu\nu}} = 0 \tag{21}$$

where the functional (Hamiltonian) derivatives in the sense of Palatini (in this case with respect to the potentials), are understood. The choice "measure-like" form for the geometrical Lagrangian L_G (reminiscent of a nonlinear sigma model), satisfy the following principles:

i) the principle of the natural extension of the Lagrangian density as square root of the fundamental line element containing also $F_{\mu\nu}$.

ii) the symmetry principle between $g_{\mu\nu}$ and $F_{\mu\nu}$(e.g., $g_{\mu\nu}$ and $F_{\mu\nu}$should enter into L_G symmetrically)

iii)the principle that the spinor symmetry, namely

$$\nabla_\mu g_{\lambda\nu} = 0, \qquad \nabla_\mu \sigma_{\lambda\nu} = 0 \tag{22,23}$$

with

$$g_{\lambda\nu} = \gamma_\lambda \cdot \gamma_\nu, \qquad \sigma_{\lambda\nu} = \gamma_\lambda \wedge \gamma_\nu \sim *F_{\lambda\nu} \tag{24,25}$$

should be derivable from L_G(21)

The last principle is key because it states that the spinor invariance of the fundamental space-time structure should be derivable from the dynamic symmetries given by (21). The fact that the L_G satisfies the 3 principles shows also that it has the simpler form.

Notice that the action density proposed by Einstein in [28] in his nonsymmetric field theory satisfies i) and ii) but not iii).

Remark 1 *Due the totally antisymmetric character of the torsion field it is completely determined by the fundamental (structural 2-form) antisymmetric tensor, and consequently the variations must acquire the form given by expression (21):* ***metric and torsion have each one their respective potentials that are in coincidence with the fundamental structure of the geometry.***

4.1. $\delta_g L_G$

The starting point for the metrical variational procedure is in the same way as in the standard Born-Infeld theory: from the following factorization of the geometrical Lagrangian :

$$\begin{aligned}\mathcal{L} &= \sqrt{|g|}\sqrt{\det(\alpha\lambda)}\sqrt{1 + \frac{1}{2b^2}F_{\mu\nu}F^{\mu\nu} - \frac{1}{16b^4}\left(F_{\mu\nu}\widetilde{F}^{\mu\nu}\right)^2} \\ &\equiv \sqrt{|g|}\sqrt{\det(\alpha\lambda)}\mathbb{R}\end{aligned} \tag{26}$$

where

$$b = \frac{\alpha}{\beta} = \frac{1 + (R_S/4\lambda)}{1 + (R_A/4\lambda)}, \qquad R_S = g^{\alpha\beta}R_{\alpha\beta}, \qquad R_A = f^{\alpha\beta}R_{\alpha\beta}, \tag{27,28,29}$$

and λ an arbitrary constant we perform the variational metric procedure with the following result (details see Appendix II)

$$\begin{aligned}\delta_g\mathcal{L} = 0 \Rightarrow R_{(\alpha\beta)} - \frac{g_{\alpha\beta}}{4}R_s &= \frac{R_s}{2\mathbb{R}^2\alpha^2}\left[F_{\alpha\lambda}F_\beta^{\ \lambda} - F_{\mu\nu}F^{\mu\nu}\frac{R_{(\alpha\beta)}}{R_s}\right] + \\ &+\frac{R_s}{4\mathbb{R}^2\alpha^2b^2}\left[F_{\mu\nu}\widetilde{F}^{\mu\nu}\left(\frac{F_{\eta\rho}\widetilde{F}^{\eta\rho}}{8}g_{\alpha\beta} - F_{\alpha\lambda}\widetilde{F}_\beta^{\ \lambda}\right) + \frac{F_{\eta\rho}\widetilde{F}^{\eta\rho}}{2}\frac{R_{(\alpha\beta)}}{R_s}\right] + \\ &+2\lambda\left[g_{\alpha\beta} + \frac{1}{\mathbb{R}^2\alpha^2}\left(F_{\alpha\lambda}F_\beta^{\ \lambda} + \frac{F_{\mu\nu}\widetilde{F}^{\mu\nu}}{2b^2}\left(\frac{F_{\eta\rho}\widetilde{F}^{\eta\rho}}{8}g_{\alpha\beta} - F_{\alpha\lambda}\widetilde{F}_\beta^{\ \lambda}\right)\right)\right],\end{aligned} \tag{30, 31}$$

Remark 2 *Notice that:*

1) The eq. (31) is trace-free *type, consequently the trace of the third term of the above equation (that is the cosmological one) is equal to zero. This happens trivially if $\lambda = 0$ or $4\mathbb{R}^2\alpha^2 = -\left(F_{\alpha\lambda}F^{\ \alpha\lambda} - \frac{\left(F_{\mu\nu}\widetilde{F}^{\mu\nu}\right)^2}{4b^2}\right)$ In terms of the Maxwell Lagrangian we have*

$(\mathbb{R}\alpha)^2 = \left(L_{Maxwell} + \frac{\left(F_{\mu\nu}\widetilde{F}^{\mu\nu}\right)^2}{16b^2}\right) \equiv \mathcal{W}\left(I_S, I_P, b\right)$ *that allow us to simplify the eq. (31) once more as follows*

$$R_{(\alpha\beta)} - \frac{g_{\alpha\beta}}{4}R_s = \frac{R_s}{2\mathcal{W}}\left[F_{\alpha\lambda}F_{\beta}^{\ \lambda} - F_{\mu\nu}F^{\mu\nu}\frac{R_{(\alpha\beta)}}{R_s}\right] +$$
$$+\frac{R_s}{4\mathcal{W}b^2}\left[F_{\mu\nu}\widetilde{F}^{\mu\nu}\left(\frac{F_{\eta\rho}\widetilde{F}^{\eta\rho}}{8}g_{\alpha\beta} - F_{\alpha\lambda}\widetilde{F}_{\beta}^{\ \lambda}\right) + \frac{F_{\eta\rho}F^{\eta\rho}}{2}\frac{R_{(\alpha\beta)}}{R_s}\right] +$$
$$+2\lambda\left[g_{\alpha\beta} + \frac{1}{\mathcal{W}}\left(F_{\alpha\lambda}F_{\beta}^{\ \lambda} + \frac{F_{\mu\nu}\widetilde{F}^{\mu\nu}}{2b^2}\left(\frac{F_{\eta\rho}\widetilde{F}^{\eta\rho}}{8}g_{\alpha\beta} - F_{\alpha\lambda}\widetilde{F}_{\beta}^{\ \lambda}\right)\right)\right],$$

2) b takes the place of limiting parameter (maximum value) for the electromagnetic field strength.

3) b is not a constant in general, in sharp contrast with the Born-Infeld or string theory cases.

4) Because b is the ratio $\frac{\alpha}{\beta} = \frac{1+(R_S/4\lambda)}{1+(R_A/\lambda 4)}$ involving both curvature scalars from the contractions of the generalized Ricci tensor: it is preponderant when the symmetrical contraction of $R_{\alpha\beta}$ is greater than the skew one.

5) The fact pointed out in ii), namely that the curvature scalar plays the role as some limiting parameter of the field strength, was conjectured by Mansouri in [20] in the context of gravity theory over group manifold (generally with symmetry breaking). In such a case, this limit was established after the explicit integration of the internal group-valuated variables that is not our case here.

6) In similar form that the Eddington conjecture: $R_{(\alpha\beta)} \propto g_{\alpha\beta}$, we have a condition over the ratios as follows:

$$\frac{R_{(\alpha\beta)}}{R_s} \propto \frac{g_{\alpha\beta}}{D} \tag{32}$$

that seems to be universal.

7)The equations are the simplest ones when $b^{-2} = 0$ $(\beta = 0)$,taking the exact "quasi-linear" form

$$R_{(\alpha\beta)} - \frac{g_{\alpha\beta}}{4}R_s = \underbrace{\frac{R_s}{2\alpha^2}\left[F_{\alpha\lambda}F_{\beta}^{\ \lambda} - F_{\mu\nu}F^{\mu\nu}\frac{R_{(\alpha\beta)}}{R_s}\right]}_{Maxwell-like} + 2\lambda\underbrace{\left[g_{\alpha\beta} + \frac{1}{\mathcal{W}}F_{\alpha\lambda}F_{\beta}^{\ \lambda}\right]}_{\widetilde{g}_{eff}}, \tag{33}$$

this particular case (e.g., projective invariant) will be used through this work. Notice that when $b^{-2} = 0$ $(\beta = 0)$ all terms into the gravitational equation (31) involving the pseudoscalar invariant, namely $F_{\mu\nu}\widetilde{F}^{\mu\nu}$ or $F_{\alpha\lambda}\widetilde{F}_{\beta}^{\ \lambda}$,vanishes. Consequently we arrive to the simplest expression (33) that will be used in Section XI for example.

4.2. $\delta_f L_G$

Let us to take as starting point the geometrical Lagrangian (19)

$$\mathcal{L}_{gs} = \sqrt{\det\left[\lambda g_{\alpha\beta}\left(1+\frac{R_s}{4\lambda}\right)+\lambda F_{\alpha\beta}\left(1+\frac{R_A}{4\lambda}\right)\right]} \tag{34}$$

$$= \sqrt{|g|}\lambda^2\alpha^2\left(\sqrt{1+\frac{1}{2}\mathcal{F}_{\mu\nu}\mathcal{F}^{\mu\nu}-\frac{1}{16}\left(\mathcal{F}_{\mu\nu}\widetilde{\mathcal{F}}^{\mu\nu}\right)^2}\right) \tag{35}$$

then, having into account that : $R_A = f^{\mu\nu}R_{\mu\nu}$ and $\frac{\partial \ln(\det F_{\mu\nu})}{\partial F_{\alpha\beta}} = f^{\alpha\beta}$ (due that b that contains R_A must be also included in the variation) we obtain

$$\frac{\delta L_G}{\delta F_{\sigma\omega}} = 0 \to \left(\frac{\sqrt{|g|}\lambda\beta}{2\mathbb{R}b}\right)\left[\mathbb{F}^{\sigma\omega}\beta - \frac{\mathbb{F}}{4\lambda}R_{[\mu\nu]}\chi^{\mu\nu\sigma\omega}\right] = 0 \tag{36}$$

where: $\mathbb{F} \equiv \left[F_{\mu\nu}F^{\mu\nu} - \frac{1}{4}b^{-2}\left(F_{\mu\nu}\widetilde{F}^{\mu\nu}\right)^2\right]$, $\mathbb{F}^{\sigma\alpha} \equiv \left[F^{\sigma\alpha} - \frac{1}{4}b^{-2}\left(F_{\mu\nu}\widetilde{F}^{\mu\nu}\right)\widetilde{F}^{\sigma\alpha}\right]$ and $\chi^{\mu\nu\sigma\omega} \equiv f^{\mu\omega}f^{\sigma\nu} - f^{\mu\sigma}f^{\omega\nu}$. Notice that the quantity $b = \alpha/\beta$ (concretely β) was also varied in the above expression given the second term in (36).

Contracting (36) with $F_{\alpha\beta}$, a condition over the curvature and the electromagnetic field invariants is obtained as

$$\left(\frac{\sqrt{|g|}\lambda\beta}{\mathbb{R}b}\right)\mathbb{F}\left[\beta - \frac{R_A}{2\lambda}\right] = 0$$

This condition is satisfied for $R_A = -4\lambda$ is the exact projective invariant case (that correspond with $\beta = 0$), and for $R_A = 2\lambda$.

Remark 3 *the variational equation (36) is a dynamic equation for the torsion field in complete analogy with the eqs. (31) for the curvature.*

5. Emergent Trace Free Gravitational Equations: The Meaning of Λ

Starting from the trace free equation (31) *that is not assumed* but arises from the model, consequently [3] we rewrite it as

$$\underbrace{\overset{\circ}{R}_{\alpha\beta} - \frac{g_{\alpha\beta}}{2}\overset{\circ}{R}}_{\equiv G_{\alpha\beta}} = 6\underbrace{\left(-h_\alpha h_\beta + \frac{g_{\alpha\beta}}{2}h_\gamma h^\gamma\right)}_{\equiv T^h_{\alpha\beta}} + \frac{g_{\alpha\beta}}{2}R_s + T^F_{\alpha\beta} + 2\lambda\rho_{\alpha\beta} \tag{37}$$

where

$$\rho_{\alpha\beta} \equiv g_{\alpha\beta} + \frac{1}{\mathcal{W}}\left(F_{\alpha\lambda}F_{\beta}^{\ \lambda} + \frac{F_{\mu\nu}\widetilde{F}^{\mu\nu}}{2b^2}\left(\frac{F_{\eta\rho}\widetilde{F}^{\eta\rho}}{8}g_{\alpha\beta} - F_{\alpha\lambda}\widetilde{F}_{\beta}^{\ \lambda}\right)\right) \tag{38}$$

$$T^F_{\alpha\beta} \equiv \frac{R_s}{2\mathcal{W}}\left\{\left(F_{\alpha\lambda}F_{\beta}^{\ \lambda} - F_{\mu\nu}F^{\mu\nu}\frac{R_{(\alpha\beta)}}{R_s}\right) + \right. \tag{39}$$

$$\left. + \frac{1}{2b^2}\left[F_{\mu\nu}\widetilde{F}^{\mu\nu}\left(\frac{F_{\eta\rho}\widetilde{F}^{\eta\rho}}{8}g_{\alpha\beta} - F_{\alpha\lambda}\widetilde{F}_{\beta}^{\ \lambda}\right) + \frac{\left(F_{\eta\rho}\widetilde{F}^{\eta\rho}\right)^2}{2}\frac{R_{(\alpha\beta)}}{R_s}\right]\right\}$$

the LHS of (37) is the Einstein tensor. The "GR" divergence $\overset{\circ}{\nabla}{}^{\alpha}$ of $G_{\alpha\beta}$ is zero because is a geometrical geometrical identity and in an analog manner $\overset{\circ}{\nabla}{}^{\alpha}\left(T^h_{\alpha\beta} + T^F_{\alpha\beta}\right) = 0$ because both tensors have the same symmetry that the corresponding GR energy momentum tensors of a vector field and electromagnetic field respectively:

$$\overset{\circ}{\nabla}{}^{\alpha} G_{\alpha\beta} = \overset{\circ}{\nabla}{}^{\alpha}\left(T^h_{\alpha\beta} + T^F_{\alpha\beta}\right) = 0$$

consequently the remaining part must be a covariantly constant tensor that we *assume* proportional to $g_{\alpha\beta}$:

$$\nabla^{\alpha}\left(\frac{g_{\alpha\beta}}{2}R_s + 2\lambda\rho_{\alpha\beta}\right) = 0$$

$$\Rightarrow \left(\frac{g_{\alpha\beta}}{2}R_s + 2\lambda\rho_{\alpha\beta}\right) = \Lambda g_{\alpha\beta} \rightarrow R_s = 2\Lambda \tag{40}$$

Coming back to the original trace free expressions we have the expected formula

$$\underbrace{\overset{\circ}{R}_{\alpha\beta} - \frac{g_{\alpha\beta}}{2}\overset{\circ}{R}}_{\equiv G_{\alpha\beta}} = \underbrace{6\left(-h_{\alpha}h_{\beta} + \frac{g_{\alpha\beta}}{2}h_{\gamma}h^{\gamma}\right)}_{\equiv T^h_{\alpha\beta}} + T^F_{\alpha\beta} + \Lambda g_{\alpha\beta} \tag{41}$$

Remark 4 *Tracing the first expression in (40) we have $R_s = 2\Lambda = \overset{\circ}{R} + 6h_{\mu}h^{\mu}$ linking the value of the curvature and the norm of the torsion vector field. Consequently, if the dual of the torsion field have the role of the energy-matter carrier, the meaning of lambda as the vacuum energy is immediately established.*

Remark 5 *Notice that the LHS in expression (40) instead to be proportional to the metric tensor it can be proportional to the square of a Killing-Yano tensor.*

6. On the Constancy of G

At this level, no assertion can state with respect to G or even with respect to c. The link with the general relativistic case is given by the identification of electromagnetic energy-momentum tensor with the term analogous $T^F_{\alpha\beta}$ in our metric variational equations:

$$\frac{8\pi G}{c^4}\left(F_{\alpha\lambda}F_{\beta}^{\ \lambda} - F_{\mu\nu}F^{\mu\nu}\frac{g_{\alpha\beta}}{4}\right) \rightarrow \frac{R_s}{2\mathcal{W}}\left(F_{\alpha\lambda}F_{\beta}^{\ \lambda} - F_{\mu\nu}F^{\mu\nu}\frac{R_{(\alpha\beta)}}{R_s}\right)$$

Consequently we have:

$$\kappa = \frac{8\pi G}{c^4} \to \frac{R_s}{2\mathbb{R}^2\alpha^2}$$
$$and \;\; \frac{g_{\alpha\beta}}{4} = \frac{R_{(\alpha\beta)}}{R_s}$$

The above expression indicates that the ratio must remains constant due the Noether symmetries and conservation laws of the field equations. Notice that, as in the case of b parameter in the standard BI theory, there exist a limit for all the physical fields coming from the *geometrical invariants quantities.*

7. The Vector h_μ and the Energy-Matter Interpretation

One of the characteristics that more attract the attention in unified field theoretical models is the possibility to introduce the energy and matter through its geometrical structure. In our case the torsion field takes the role of RHS of the standard GR gravity equation by mean its dual, namely h_μ.

Consequently, in order to explain the physical role of h_μ, we know (due the Hodge-de Rham decomposition [Appendix III]) that it can be decomposed as:

$$h_\alpha = \nabla_\alpha \Omega + \varepsilon_\alpha^{\beta\gamma\delta} \nabla_\beta A_{\gamma\delta} + \gamma_1 \overbrace{\varepsilon_\alpha^{\beta\gamma\delta} M_{\beta\gamma\delta}}^{axial\ vector} + \gamma_2 \overbrace{P_\alpha}^{polar\ vector} \tag{42}$$

where $\gamma_1$and γ_2 can be phenomenologically related to physical constants (e.g., $\gamma_1 = \frac{8\pi}{c}\sqrt{G}$ is a physical constant related to the Blackett formula [21]). The arguments in favour of this type of theories and from the decomposition (42) can be resumed as follows:

i) the existence of an angular momentum Helmholtz theorem [23], [54]: the theorem in analysis is exactly as in E_3 but, in the four dimensional case M_4 there exists an additional *axial vector*;

iii) the concept of $chirality$ is achieved in the model by the existence of polar and axial vectors in expression (42).

iv) if Ω, $A_{\gamma\delta}$ are the wave tensors and $\varepsilon_\alpha^{\beta\gamma\delta} M_{\beta\gamma\delta}$, P_α the particle vectors (vector and axial part respectively), the concept of an inertial-wave vector is introduced in the equation (42).

Consequently, from the eq.motion for the torsion namely: $\nabla_\alpha T^{\alpha\beta\gamma} = -\lambda F^{\beta\gamma}$and coming back to (42) we obtain the following important equation

$$\overset{\circ}{\Box} A_{\gamma\delta} - \gamma \left[\nabla_\alpha M^\alpha{}_{\gamma\delta} + (\nabla_\gamma P_\delta - \nabla_\delta P_\gamma) \right] = -\lambda F_{\gamma\delta} \tag{43}$$

Let us consider, in particular, the case when $\lambda F_{\gamma\delta} \to 0$:

$$\overset{\circ}{\Box} A_{\gamma\delta} = \gamma \left[\nabla_\alpha M^\alpha{}_{\gamma\delta} + (\nabla_\gamma P_\delta - \nabla_\delta P_\gamma) \right] \tag{44}$$

We can immediately see that, if $M^{\alpha}_{\ \gamma\delta}$ is identified with the intrinsic spin angular momentum of the ponderable matter, P_δ is its lineal momentum vector and $A_{\gamma\delta}$ is the gravitational radiation tensor, then eq.(44) states that the sum of the intrinsic spin angular momentum and the orbital angular momentum of ponderable matter is conserved if the gravitational radiation is absent., if $M^{\alpha}_{\ \gamma\delta}$ is identified with the intrinsic spin angular momentum of the ponderable matter, P_δ is its lineal momentum vector and $A_{\gamma\delta}$ is the gravitational radiation tensor, then eq.(44) states that the sum of the intrinsic spin angular momentum and the orbital angular momentum of ponderable matter is conserved if the gravitational radiation is absent.

7.1. Killing-Yano Systems and the Vector h_μ

Without enter in many details (these will be treated somewhere) the antisymmetric tensor $A_{\gamma\delta}$ in the h_β composition is related with the Killing [33] and Killing-Yano [34] systems. Consequently we can introduce two types of couplings into the $A_{\gamma\delta}$ divergence as follows.

i) Defining

$$A_{\gamma\delta} \equiv A_{[\gamma;\delta]} \tag{45}$$

such that

$$\overset{\circ}{\nabla}_\rho A_{[\gamma;\delta]} = \frac{4\pi}{3}\left(j_{[\gamma}\, g_{\delta]\rho}\right) \tag{46}$$

then , in this case we can identify $A_{\gamma\delta} = 2F_{\gamma\delta}$ because $F^{\delta}_{\ \gamma;\delta} = 4j_\gamma$ and $A_{[\gamma\delta;\rho]} = F_{[\gamma\delta;\rho]} = 0$

In this case the contribution of $A_{\gamma\delta}$ to h_β is null.

ii) Let us consider now a fully antisymmetric coupling as

$$A_{[\gamma;\delta];\rho} = \frac{4\pi}{3} j_{[\gamma}\, F_{\delta]\rho} \tag{47}$$

having into account the vorticity vector also

$$\omega_\mu \equiv u^\lambda \varepsilon_{\lambda\mu\nu\rho} \nabla^\nu u^\rho \tag{48}$$

and considering a plasma with electrons, protons etc.

$$j^\gamma \sim A^\gamma + q_s n_s u_s^\gamma \tag{49}$$

where A_μ is the vector potential and q_s is the particle charge, $n_s$is the number density (in the rest frame) and the four-velocity of species s is u_s^γ. In this case $h_\alpha$takes the form

$$h_\alpha = \nabla_\alpha \Omega + \varepsilon_\alpha^{\beta\gamma\delta} \nabla_\beta A_{\gamma\delta} + \gamma_1 \varepsilon_\alpha^{\beta\gamma\delta} M_{\beta\gamma\delta} + \gamma_2 P_\alpha \rightarrow \tag{50}$$

$$h_\alpha = \nabla_\alpha \Omega + \varepsilon_\alpha^{\gamma\delta\rho} \frac{4\pi}{3} j_{[\gamma}\, F_{\delta]\rho} - \gamma_1 u^\lambda \varepsilon_{\lambda\alpha\nu\rho} \nabla^\nu u^\rho + \gamma_2 P_\alpha \tag{51}$$

$$h_\alpha = \nabla_\alpha \Omega + \varepsilon_\alpha^{\gamma\delta\rho} \frac{4\pi}{3} \left[A + q_s n_s u_s\right]_{[\gamma}\, F_{\delta]\rho} - \gamma_1 u^\lambda \varepsilon_{\lambda\alpha\nu\rho} \nabla^\nu u^\rho + \gamma_2 P_\alpha \tag{52}$$

Consequently in 3+1 decomposition we have (overbar correspond to spacial 3-dim. vectors)

$$h_0 = \nabla_0 \Omega + \frac{4\pi}{3} \overline{j} \cdot \overline{B} + \gamma_1 \overline{u} \cdot \left(\overline{\nabla} \times \overline{u}\right) + \gamma_2 P_0 \tag{53}$$

$$h_0 = \nabla_0 \Omega + \frac{4\pi}{3} \left[\overline{A} \cdot \left(\overline{\nabla} \times \overline{A}\right) + q_s n_s \overline{u}_s \cdot \overline{B}\right] + \gamma_1 \overline{u} \cdot \left(\overline{\nabla} \times \overline{u}\right) + \gamma_2 P_0 \tag{54}$$

and

$$h_i = \nabla_i \Omega + \frac{4\pi}{3}\left[-\left(\overline{j}\times\overline{E}\right)_i + j_0 \overline{B}_i\right] +$$
$$+\gamma_1\left[u_0\left(\overline{\nabla}\times\overline{u}\right) + \left(\overline{u}\times\overline{\nabla}u_0\right) + \left(\overline{u}\times\dot{\overline{u}}\right)\right]_i + \gamma_2 P_i \tag{55}$$

$$h_i = \nabla_i \Omega + \frac{4\pi}{3}\left[-\left(\left(\overline{A}+q_s n_s \overline{u}_s\right)\times\overline{E}\right)_i + \left(\Phi + q_s n_s u_{0s}\right)\overline{B}_i\right] + \tag{56}$$
$$+\gamma_1\left[u_0\left(\overline{\nabla}\times\overline{u}\right) + \left(\overline{u}\times\overline{\nabla}u_0\right) + \left(\overline{u}\times\dot{\overline{u}}\right)\right]_i + \gamma_2 P_i$$

Notice that in h_0 component we can recognize the magnetic and vortical helicities

$$h_0 = \nabla_0 \Omega + \frac{4\pi}{3}\left[h_M + q_s n_s \overline{u}_s \cdot \overline{B}\right] + \gamma_1 h_V + \gamma_2 P_0 \tag{57}$$

The above expression will be very important in the next sections, in particular to discuss magnetogenesis and particle generation. Notice the important fact that the symmetry of the vorticity can be associated to a 2-form bivector in the context of the notoph field [35] theory.

8. Physical Consequences

In this Section we will make contact between the mathematical formulation and the physical consequences of the model. The first step is to introduce the 3+1 splitting for axisymmetric spacetimes that it is useful to analyze the electrodynamic equations with high degree of nonlinearity, as is in our case. Secondly we take the 3+1 field equations where the induction equations (dynamo) are obtained, in the linear limit, showing explicitly the important role of the torsion field as the generator of a purely geometric α-term. Finally, we derive the geometrical analog of the Lorentz force and the elimination of the electric field from the induction equations and the origin of the seed magnetic field via the geometrical α-term generated by the torsion vector is worked out.

8.1. Electrodynamic Structure in 3+1

The starting point will be the line element in $3+1$ splitting [6], [7] (Appendix IV): the 4-dimensional space-time is split into 3-dimensional space and 1-dimensional time to form a foliation of 3-dimensional spacelike hypersurfaces. The metric of the space-time is consequently, given by $ds^2 = -\alpha^2 dt^2 + \gamma_{ij}\left(dx^i + \beta^i dt\right)\left(dx^j + \beta^j dt\right)$ where γ_{ij} is the metric of the 3-dimensional hypersurface,α is the lapse function, and β^i is the shift function (see Appendix IV for details). For any nonlinear Lagrangian, in sharp contrast with the Einstein-Maxwell case, the field equations $d{*}\mathbb{F} = {*}\mathbb{J}$ and the Bianchi-geometrical condition $dF = 0$ (where we have defined the Hodge dual $*$ and $\mathbb{F} = \frac{\partial \mathcal{L}}{\partial F}$) are expressed by the vector fields

$$E, B, \mathbb{E} = \frac{\partial \mathcal{L}}{\partial E}, \mathbb{B} = \frac{\partial \mathcal{L}}{\partial B} \tag{58}$$

that live into the space-time slice. In our case they coming from the geometrical Lagrangian $\mathcal{L}_g$, a

$$\nabla \cdot \mathbb{E} = -\overline{h} \cdot \mathbb{B} + 4\pi \rho_e \tag{59}$$

$$\nabla \cdot B = 0 \tag{60}$$

$$\begin{aligned} \nabla \times (\alpha E) &= -(\partial_t - \mathcal{L}_\beta) B \\ &= -\partial_0 B + (\beta \cdot \nabla) B - (B \cdot \nabla) \beta \end{aligned} \tag{61}$$

$$\begin{aligned} \nabla \times (\alpha \mathbb{B}) + h_0 \mathbb{B} - \overline{h} \times \mathbb{E} &= -(\partial_t - \mathcal{L}_\beta) \mathbb{E} + 4\pi \alpha j \\ &= \partial_0 \mathbb{E} - (\beta \cdot \nabla) \mathbb{E} + (\mathbb{E} \cdot \nabla) \beta + 4\pi \alpha j \end{aligned} \tag{62}$$

being $\mathcal{L}_\beta$ the Lie derivative and h^μ is the torsion vector. Notice that, here and the subsequent Sections, the overbar indicates 3-dimensional space vectors.

8.2. Dynamo Effect and Geometrical Origin of $\alpha\Omega$ Term

In the case of weak field approximation and $\left(F^{01} \to E^i,\ F^{jk} \to B^i\right)$the electromagnetic Maxwell-type equations in 3+1 take the form

$$\nabla_\nu F^{\nu\mu} = T^{\mu\nu\rho} F_{\nu\rho} = \varepsilon^{\mu\nu\rho}{}_\delta h^\delta F_{\nu\rho} \qquad (d^* F = {}^* J) \tag{63}$$

$$\overline{\nabla} \cdot \overline{E} = -\overline{h} \cdot \overline{B} \tag{64}$$

$$\partial_t \overline{E} - \overline{\nabla} \times \overline{B} = h^0 \overline{B} - \overline{h} \times \overline{E} \tag{65}$$

and

$$\nabla_\nu {}^* F^{\nu\mu} = 0 \qquad (dF = 0) \tag{66}$$

$$\overline{\nabla} \cdot \overline{B} = 0 \tag{67}$$

$$\partial_t \overline{B} = -\overline{\nabla} \times \overline{E} \tag{68}$$

Putting all together, the set of equations is

$$\overline{\nabla} \cdot \overline{E} + \overline{h} \cdot \overline{B} = \rho_{ext} \tag{69}$$

$$\partial_t \overline{E} - \overline{\nabla} \times \overline{B} = h^0 \overline{B} - \overline{h} \times \overline{E} - \sigma_{ext} \left[\overline{E} + \overline{v} \times \overline{B} \right] \tag{70}$$

$$\overline{\nabla} \cdot \overline{B} = 0 \tag{71}$$

$$\partial_t \overline{B} = -\overline{\nabla} \times \overline{E} \tag{72}$$

where we have introduced external charge density and current. Following the standard procedure we take the rotational to the second equation above obtaining straightforwardly the modified dynamo equation

$$\begin{aligned} &\overline{\nabla} \times \partial_t \overline{E} + \overline{\nabla}^2 \overline{B} = \overline{\nabla} \times \left(h^0 \overline{B}\right) + \\ &+ \left(\overline{h} \cdot \overline{B} - \rho_{ext}\right) \overline{h} + \left(\overline{\nabla} \cdot \overline{h}\right) \overline{E} - \sigma_{ext} \left[\partial_t \overline{B} + \left(\overline{\nabla} \cdot \overline{v}\right) \overline{B}\right] \end{aligned} \tag{73}$$

where the standard identities of the vector calculus plus the first, the third and the fourth equations above have been introduced. Notice that in the case of the standard approximation and without any external or additional ingredients, we have

$$\overline{\nabla}^2 \overline{B} = h^0 \left(\overline{\nabla} \times \overline{B}\right) + \left(\overline{h} \cdot \overline{B}\right) \overline{h} + \left(\overline{\nabla} \cdot \overline{h}\right) \overline{E} \tag{74}$$

Here we can see that there exist and α- term with a pure geometrical origin (and not only a turbulent one) that is given by h^0 (the zero component of the dual of the torsion tensor).

8.3. The Generalized Lorentz Force

An important point in any theory beyond relativity is the concept of force. As is known, general relativity has deficiencies at this point. Now we are going to show that it is possible to derive from our proposal the Lorentz force as follows. From expression (32) the geometrical induced current is recognized

$$\partial_t \overline{E} - \overline{\nabla} \times \overline{B} = h^0 \overline{B} - \overline{h} \times \overline{E} \equiv \overline{J} \tag{75}$$

$$\overline{J} \times \overline{B} = \left(h^0 \overline{B} - \overline{h} \times \overline{E} - \overline{j}_{ext}\right) \times \overline{B} \tag{76}$$

$$= -\left[\left(\overline{h} \cdot \overline{B}\right) \overline{E} - \left(\overline{E} \cdot \overline{B}\right) \overline{h}\right] - \overline{j}_{ext} \times \overline{B} \tag{77}$$

we assume j_{ext} proportional to the velocity and other contributions. Consequently, reordering terms from above, a geometrically induced Lorentz-like force arises

$$\left(\overline{J} + \overline{j}_{ext}\right) \times \overline{B} = -\left[\underbrace{(\overline{h} \cdot \overline{B})\overline{E}}_{\rho_{geom}} - \left(\overline{E} \cdot \overline{B}\right) \overline{h}\right] \rightarrow \tag{78}$$

$$\underbrace{(\overline{h} \cdot \overline{B})}_{\rho_{geom}} \overline{E} + \underbrace{\left(\overline{J} + \overline{j}_{ext}\right)}_{j_{gen}} \times \overline{B} = \left(\overline{E} \cdot \overline{B}\right) \overline{h} \rightarrow \text{Lorentz induced force} \tag{79}$$

being the responsible of the induced force, the torsion vector itself. Notice, from the above equation, the following issues:

1) The external currents are identified with $\overline{J}$
2) We can eliminate the electric field in standard form

$$\overline{E} = \frac{\left(\overline{E} \cdot \overline{B}\right) \overline{h} - \overline{j}_{gen} \times \overline{B}}{\left(\overline{h} \cdot \overline{B}\right)} \tag{80}$$

being the above expression very important in order to replace the electric field into the dynamo equation, introducing naturally the external current in the model.

8.4. Generalized Current and α-Term

In previous paragraph we have derived a geometrical induced Lorentz force where the link between the physical world and the proposed geometrical model is through a generalized current j_{gen}. An important fact of that expression is that it is possible to eliminate the electric field (and insert it into the equation of induction) as follows.

From the formula of the induction, namely

$$\overline{\nabla}^2 \overline{B} + \underbrace{\overline{\nabla} \times \left(-h^0 \overline{B} + \overline{h} \times \overline{E}\right)}_{\varepsilon_{Geom}} = 0 \tag{81}$$

and using the eq. (49) to eliminate the electric field as function of the torsion, the generalized current and the magnetic field respectively:

$$\overline{h}\times\overline{E}=\frac{-\overline{h}\times\left(\overline{j}_{gen}\times\overline{B}\right)}{\left(\overline{h}\cdot\overline{B}\right)}=-\frac{\left(\overline{h}\cdot\overline{B}\right)\overline{j}_{gen}-\left(\overline{h}\cdot\overline{j}_{gen}\right)\overline{B}}{\left(\overline{h}\cdot\overline{B}\right)} \tag{82}$$

$$\overline{h}\times\overline{E}=-\overline{j}_{gen}+\frac{\left(\overline{h}\cdot\overline{j}_{gen}\right)}{\left(\overline{h}\cdot\overline{B}\right)}\overline{B}=\left|\overline{j}_{gen}\right|\left(-n_{\overline{j}_{gen}}+\frac{\cos\alpha}{\cos\beta}n_{B}\right) \tag{83}$$

being α the angle between the vector torsion $\overline{h}$ and the generalized current $\overline{j}_{gen}$and β the angle between $\overline{h}$ and thc magnetic field $\overline{B}$.Above, n_B and $n_{\overline{j}_{gen}}$ are unitary vectors in the direction of $\overline{B}$ and $\overline{j}_{gen}$ respectively. Notice the important fact that the RHS of (68) is independent of the torsion and the magnetic field. Consequently we obtain

$$\underbrace{\overline{\nabla}^2\overline{B}+\overline{\nabla}\times\left[-\overline{j}_{gen}+\left(\frac{\left(\overline{h}\cdot\overline{j}_{gen}\right)}{\left(\overline{h}\cdot\overline{B}\right)}-h^0\right)\overline{B}\right]}_{\mathcal{E}_{Geom}}=0 \tag{84}$$

We introduce the explicitly the physical scenario via the generalized current $\overline{j}_{gen}$

$$-\overline{j}_{gen}\sim\sigma_{ext}\left[\overline{E}+\overline{v}\times\overline{B}\right]+\left(\frac{c}{e}\frac{\overline{\nabla}p}{n_e}\right) \tag{85}$$

then

$$\underbrace{\overline{\nabla}^2\overline{B}+\overline{\nabla}\times\left[\sigma_{ext}\left[\overline{E}+\overline{v}\times\overline{B}\right]+\left(\frac{c}{e}\frac{\overline{\nabla}p}{n_e}\right)+\left(\frac{\left(\overline{h}\cdot\overline{j}_{gen}\right)}{\left(\overline{h}\cdot\overline{B}\right)}-h^0\right)\overline{B}\right]}_{\mathcal{E}_{Geom}}=0 \tag{86}$$

$$\underbrace{\overline{\nabla}^2\overline{B}+\sigma_{ext}\left[\left(-\partial_t\overline{B}\right)+\overline{\nabla}\times\left(\overline{v}\times\overline{B}\right)\right]+\overline{\nabla}\times\left(\frac{c}{e}\frac{\overline{\nabla}p}{n_e}\right)+\overline{\nabla}\times\left(\frac{\left(\overline{h}\cdot\overline{j}_{gen}\right)}{\left(\overline{h}\cdot\overline{B}\right)}-h^0\right)\overline{B}}_{\mathcal{E}_{Geom}}=0 \tag{87}$$

finally the expected geometrically induced expression is obtained:

$$\partial_t\overline{B}=\eta\overline{\nabla}^2\overline{B}+\overline{\nabla}\times\left(\overline{v}\times\overline{B}\right)+\eta\overline{\nabla}\times\left[\left(\frac{c}{e}\frac{\overline{\nabla}p}{n_e}\right)+\alpha\overline{B}\right]=\partial_t\overline{B} \tag{88}$$

$$\rightarrow\eta\underbrace{\overline{\nabla}^2\overline{B}}_{diffusive}+\underbrace{\overline{\nabla}\times\left(\overline{v}\times\overline{B}\right)}_{advective}+\eta\underbrace{\overline{\nabla}\times\left(\alpha\overline{B}\right)}_{\alpha-term}+\underbrace{\frac{c}{e}\frac{\overline{\nabla}p\times\overline{\nabla}n_e}{n_e^2}}_{Biermann\ battery}=-\partial_t\overline{B} \tag{89}$$

where $\eta\equiv\frac{1}{\sigma_{ext}}$ as usual and the geometric α:

$$\begin{aligned}\alpha&\equiv\left(\frac{\left(\overline{h}\cdot\overline{j}_{gen}\right)}{\left(\overline{h}\cdot\overline{B}\right)}-h^0\right)\\&=\left(\frac{\cos\alpha\left|\overline{j}_{gen}\right|}{\cos\beta\left|\overline{B}\right|}-h^0\right)\end{aligned} \tag{90}$$

8.5. Seed Magnetic Field

Notice from the last expression that $\alpha\overline{B}$ is explicitly

$$\alpha\overline{B} = \frac{\cos\alpha\left|\overline{j}_{gen}\right|}{\cos\beta} n_B - h^0\overline{B} \tag{91}$$

or (via elimination of the unitary vector)

$$\alpha\left|\overline{B}\right| = \frac{\cos\alpha\left|\overline{j}_{gen}\right|}{\cos\beta} - h^0\left|\overline{B}\right| \tag{92}$$

we see *clearly* the first term in RHS independent of the intensity of the magnetic field. Considering only the terms of interest without the diffusive and advective term in the induction equation (only time-dependence for the magnetic field is preserved) namely

$$\underbrace{\eta\overline{\nabla}\times\left(\alpha\overline{B}\right)}_{\alpha-term} = -\partial_t\overline{B} \tag{93}$$

$$\eta\overline{\nabla}\left(\frac{\cos\alpha\left|\overline{j}_{gen}\right|}{\cos\beta}\right) = -\partial_t\left|\overline{B}\right| \tag{94}$$

we see that the currents given by the fields (related to the geometry via h_α) originate the magnetic field.

If we consider all the currents of the fields of theory (fermions, bosons, etc.) the seed would be precisely these field currents. The other missing point is to derive the fluid (hydrodynamic) equations (which as is known does not have a definite Lagrangian formulation) from the same unified formulation. Notice that there are, under special conditions, analogous formulas for vorticity ω than for the magnetic field B. This would mean that the 2-form of vorticity must also be included in the fundamental antisymmetric tensor, together with the electromagnetic field.

8.6. Comparison with the Mean Field Formalism

Now we compare the obtained equations with respect to the mean field formalism [24]. Starting from expressions (69-72) as before, we have:

$$\eta\overline{\nabla}^2\overline{B} + \overline{\nabla}\times\left(\overline{v}\times\overline{B}\right) - \partial_t\overline{B} + \underbrace{\eta\overline{\nabla}\times\left(-h^0\overline{B} + \overline{h}\times\overline{E}\right)}_{\mathcal{E}_{Geom}} = 0 \tag{95}$$

$\mathcal{E}_{Geom}$ takes the place of electromotive force due the torsion field with full analogy as $\mathcal{E} = \langle u\times b\rangle$ that is the mean electromotive force due to fluctuations. Also as in the mean field case where there are a splitting

$$\mathcal{E} = \mathcal{E}^{\langle 0\rangle} + \mathcal{E}^{\langle\overline{B}\rangle} \tag{96}$$

with $\mathcal{E}^{\langle 0\rangle}$ independent of $\langle \overline{B}\rangle$ and $\mathcal{E}^{\langle \overline{B}\rangle}$ linear and homogeneous in $\overline{B}$, we have in the torsion case the following correspondence

$$-h^0\overline{B} \longleftrightarrow \mathcal{E}^{\langle \overline{B}\rangle}$$
$$\overline{h}\times\overline{E} \longleftrightarrow \mathcal{E}^{\langle 0\rangle}$$
$$geometrical \longleftrightarrow turbulent$$

Consequently, the problems of mean-field dynamo theory that are concerned with the generation of a mean EMF by turbulence, have in this model a pure geometric counterpart. In the past years, attention has shifted from kinematic calculations, akin to those familiar from quasilinear theory for plasmas, to self-consistent theories which account for the effects of small scale magnetic fields (including their back-reaction on the dynamics) and for the constraints imposed by the topological conservation laws, such as that for magnetic helicity. Thus, the dynamo problem is seen in our model as one of generalized helicity transport. A key element in this approach is to understand the evolution of the torsion vector field besides of the turbulence, energy and the generalized helicity profiles in space-time.

9. Torsion, Axion Electrodynamics vs. Chern Simons Theory

Let us review briefly the electromagnetic sector of the theory QCD based in a gauge symmetry $SU(3)\times U(1)$

$$L_{QCD/QED} = +\sum\overline{\psi}_f\left[\gamma^\mu\left(\partial_\mu - ig_f t^\alpha A^\alpha_\mu - iq_f A_\mu\right) - m_f\right]\psi_f - \tag{97}$$
$$-\frac{G^\alpha_{\mu\nu}G^{\alpha\mu\nu}}{4} - \frac{F_{\mu\nu}F^{\mu\nu}}{4} - \frac{g^2\theta G^\alpha_{\mu\nu}\widetilde{G}^{\alpha\mu\nu}}{32\pi^2} - \frac{g^2\theta F_{\mu\nu}\widetilde{F}^{\mu\nu}}{32\pi^2},$$

As is well know, electromagnetic fields will couple to the electromagnetic currents, namely:$J_\mu = \sum\limits_f q_f\overline{\psi}_f\gamma_\mu\psi_f$ consequently, there appear a term that will induce through the quark loop the coupling of $F_{\mu\nu}\widetilde{F}^{\mu\nu}$(anomaly) to the QCD topological charge. The effective Lagrangian can be written as

$$L_{MCS} = -\frac{F_{\mu\nu}F^{\mu\nu}}{4} - A_\mu J^\mu - \frac{c}{4}\theta F_{\mu\nu}\widetilde{F}^{\mu\nu} \tag{98}$$

where a pseudo-scalar field $\theta = \theta(\overline{x},t)$ (playing the role of the axion field) is introduced and $c = \sum\limits_f \frac{(q_f e)^2}{2\pi^2}$. This is the Chern-Simmons Lagrangian where, if θ is a constant, the last term is a total divergence: $F_{\mu\nu}\widetilde{F}^{\mu\nu} = \partial_\mu J^\mu_{CS}$.The question appears if θ is not a constant e.g., $\theta F_{\mu\nu}\widetilde{F}^{\mu\nu} = \theta\partial_\mu J^\mu_{CS} = \partial_\mu\left(\theta J^\mu_{CS}\right) - J^\mu_{CS}\partial_\mu\theta$

If, from the general decomposition of the four dimensional dual of the torsion field via the Hodge de Rham theorem, we retain b_α as gradient of a pseudoscalar (e.g: axion) these equations coincide in form with the respective equation for Maxwell-Chern-Simons (MCS) theory. Precisely because under this condition $h_\alpha = \nabla_\alpha\theta$, in flat space (torsion$\neq 0$) the equations become the same as in [8] namely

$$\overline{\nabla}\cdot\overline{E}-c\overline{P}\cdot\overline{B}=\rho_{ext} \tag{99}$$

$$\partial_t\overline{E}-\overline{\nabla}\times\overline{B}=-c\dot{\theta}\overline{B}+c\overline{P}\times\overline{E}-\sigma_{ext}\left[\overline{E}+\overline{v}\times\overline{B}\right] \tag{100}$$

$$\overline{\nabla}\cdot\overline{B}=0 \tag{101}$$

$$\partial_t\overline{B}=-\overline{\nabla}\times\overline{E} \tag{102}$$

provided:

$$h^0\rightarrow-c\dot{\theta} \tag{103}$$

$$\overline{h}\rightarrow-c\overline{P} \tag{104}$$

where from QCD the constant c is determined as $c=\frac{e^2}{2\pi}$ and the $\partial_\mu\theta=\left(\dot{\theta},\overline{P}\right)$ in the [8] notation. The main difference is that while in the case of photons in axion ED was given by [9] the Lagrangian where that above equations are derived is

$$L_{MCS}=-\frac{F_{\mu\nu}F^{\mu\nu}}{4}-A_\mu J^\mu+\frac{c}{4}P_\mu J^\mu_{CS},\qquad J^\mu_{CS}\equiv\varepsilon^{\mu\sigma\rho\nu}A_\sigma F_{\rho\nu} \tag{105}$$

in our case is the dual of the torsion field (that we take as the gradient of a pseudoscalar) responsible of the particular structure of the set of equations.

10. Magnetic Helicity Generation and Cosmic Torsion Field

Here we consider the projective invariant case: $\beta=0$ $(R_A=-4\lambda)$ where the gravitational and field equations are considerably simplified because $\mathbb{R}=1$ and $b^{-1}=0$. Scalar curvature R and the torsion 2-form field $T^a_{\mu\nu}$ with a $SU(2)$ −Yang-Mills structure are defined in terms of the affine connection $\Gamma^\lambda_{\mu\nu}$ and the $SU(2)$ valuated (structural torsion potential) $f^a_{\ \mu}$ by

$$R=g^{\mu\nu}R_{\mu\nu}\qquad R_{\mu\nu}=R^\lambda_{\mu\lambda\nu} \tag{106}$$

$$R^\lambda_{\mu\lambda\nu}=\partial_\nu\Gamma^\lambda_{\mu\rho}-\partial_\rho\Gamma^\lambda_{\mu\nu}+...$$

$$T^a_{\mu\nu}=\partial_\mu f^a_{\ \nu}-\partial_\nu f^a_{\ \mu}+\varepsilon^a_{bc}f^b_{\ \mu}f^c_{\ \nu}$$

G and Λ are the geometrically induced Newton gravitational constant (as we have been discussed before) and the integration cosmological constant, respectively. From the last equation for the totally antisymmetric torsion 2-form, the potential $f^a_{\ \mu}$ defines the affine connection $\Gamma^\lambda_{\mu\nu}$ .Similarly to the case of Einstein-Yang-Mills systems, for our new UFT model it can be interpreted as a prototype of gauge theories interacting with gravity (*e.g.*, QCD, GUTs, etc.). We stress here the important fact that all the fundamental constants are really geometrically induced as required by the Mach principle. After varying the action, we obtain the gravitational equation (41), namely

$$\begin{aligned}\overset{\circ}{R}_{\alpha\beta}-\frac{g_{\alpha\beta}}{2}\overset{\circ}{R}=&6\left(-h_\alpha h_\beta+\frac{g_{\alpha\beta}}{2}h_\gamma h^\gamma\right)+\\&+\kappa_{geom}\left[F_{\alpha\lambda}F_\beta^{\ \lambda}-F_{\mu\nu}F^{\mu\nu}\frac{g_{\alpha\beta}}{4}\right]+\Lambda g_{\alpha\beta}\end{aligned} \tag{107}$$

with the "gravitational constant" geometrically induced as

$$\kappa_g \equiv \frac{R_s}{2\mathcal{W}} = \left.\frac{8\pi G}{c^4}\right|_{today} \tag{108}$$

and the field equation for the torsion 2-form in differential form

$$d^*T^a + \tfrac{1}{2}\varepsilon^{abc}\left(f_b \wedge^* T_c -^* T_b \wedge f_c\right) \;=\; -\lambda\,^*f^a \tag{109}$$

Notice that 108 κ_g and Λ are not independent, but related by $R_s = 2\Lambda$. In this case $\beta = 0$ we have the simplest expression:

$$\kappa_g \equiv \frac{R_s}{2\left(1+\frac{R_s}{4\lambda}\right)^2} = \frac{\Lambda}{\left(1+\frac{2\Lambda}{4\lambda}\right)^2}$$

in consequence, generalizing the conjecture of Markov in [25], **if Λ is proportional to the energy,** κ goes as Λ if $|\Lambda| \leq 1$, and as Λ^{-1} in other case.

We are going to seek for a classical solution of (107) and (109) with the following ansatz for the metric and gauge connection

$$ds^2 = d\tau^2 + a^2(\tau)\,\sigma^i \otimes \sigma^i \equiv d\tau^2 + e^i \otimes e^i. \tag{110}$$

Here τ is the euclidean time and the dreibein is defined by $e^i \equiv a(\tau)\,\sigma^i$. The gauge connection is

$$f^a \equiv f^a_\mu dx^\mu = f\sigma^a, \tag{111}$$

for $a, b, c = 1, 2, 3$, and for $a, b, c = 0$ we have

$$f^0 \equiv f^0_\mu dx^\mu = s\sigma^0. \tag{112}$$

This choice for the potential torsion is accordingly to the symmetries involved in the problem.

The σ^i 1-form satisfies the $SU(2)$ Maurer-Cartan structure equation

$$d\sigma^a + \varepsilon^a_{\ bc}\sigma^b \wedge \sigma^c = 0 \tag{113}$$

Notice that in the ansatz the frame and $SU(2)$ (isospin-like) indices are identified (as for the case with the non-abelian-Born-Infeld (NBI) Lagrangian of [31]). The torsion 2-form

$$T^\gamma = \frac{1}{2}T^\gamma_{\ \mu\nu}dx^\mu \wedge dx^\nu \tag{114}$$

becomes

$$\begin{aligned} T^a &= df^a + \frac{1}{2}\varepsilon^a_{\ bc}f^b \wedge f^c \\ &= \left(-f + \frac{1}{2}f^2\right)\varepsilon^a_{\ bc}\sigma^b \wedge \sigma^c \end{aligned} \tag{115}$$

$$\begin{aligned} d^*T^a + \tfrac{1}{2}\varepsilon^{abc}\left(f_b \wedge^* T_c -^* T_b \wedge f_c\right) &= -2\lambda\,^*f^a \\ (-2f + f^2)(1-f)d\tau \wedge e^b \wedge e^c &= -2\lambda d\tau \wedge e^b \wedge e^c \end{aligned} \tag{116}$$

$$^{*}T^{a} \equiv h(-2f+f^{2})d\tau\wedge\frac{e^{a}}{a^{2}} \tag{117}$$

$$^{*}f^{a}=-f\frac{d\tau\wedge e^{b}\wedge e^{c}}{a^{3}} \tag{118}$$

Note that to be complete in our description of the possible physical scenarios, we include f^0 as an $U(1)$ component of the torsion potential (although does not belong to the space $SU(2)/U(1)$). Having all the above issues into account, the expression for the torsion is analogous to the non-abelian 2-form strength field of [31].

Inserting T^a from (115) into the dynamic equation (109) we obtain

$$(-2f+f^{2})(1-f)d\tau\wedge e^{b}\wedge e^{c}=-\lambda d\tau\wedge e^{b}\wedge e^{c}, \tag{119}$$

and from expression (119) we have an algebraic cubic equation for f

$$(-2f+f^{2})(1-f)+\lambda=0 \tag{120}$$

We can see that, in contrast with our previous work with a dualistic theory [31] where the NBI energy-momentum tensor of Born-Infeld was considered, there exist three non trivial solutions for f, depending on the cosmological constant λ. In this preliminary analysis of the problem, only the values of f that make the quantity $\left(-f+\frac{1}{2}f^{2}\right)\in\mathbb{R}$. Consequently for $\lambda=2$ we find $f=2.35$ then

$$T^{a}_{bc}=\frac{2}{5}\frac{\varepsilon^{a}_{bc}}{a^{2}}; \qquad T^{a}_{0c}=0 \tag{121}$$

That is, only spatial torsion field is non vanishing while cosmic time torsion field vanishes(an analogous feature with magnetic and electric Yang-Mills can be seen in the solution of Giddings and Strominger and in [31]). Substituting the expression for the torsion 2-form (121) [1] into the symmetric part of the variational equation we reduce the gravitational equations to an ordinary differential equation for the scale factor a,

$$3\left[\left(\frac{\dot{a}}{a}\right)^{2}-\frac{1}{a^{2}}\right]-\Lambda=\frac{3\kappa_{g}}{4a^{2}}\left(f^{2}+s^{2}\right)+\frac{3}{2a^{4}}f^{2}\left(f-2\right)^{2} \tag{122}$$

that in the case for the computed value for $f\sim2.35$ with $s=10$ and $\Lambda\lesssim1$ the scale factor is described in Figure 1 and the scale factor goes as:

$$a\left(\tau\right)=\Lambda^{-1/2}\sqrt{\left(1-\frac{12\kappa_{g}^{2}\Lambda}{\alpha}\right)^{1/2}\sinh\left(\sqrt{\Lambda/3}\left(\tau-\tau_{0}\right)\right)-1+\kappa_{g}\left(f^{2}+s^{2}\right)/4} \tag{123}$$

where we define the geometrically induced fine structure function $\alpha\equiv\kappa_{g}\left(f^{2}+s^{2}\right)/4$

[1] in the tetrad: $\overset{\circ}{R}_{00}=-3\frac{\ddot{a}}{a}$, $\overset{\circ}{R}_{ab}=-\left[\frac{\ddot{a}}{a}+2\left(\frac{\dot{a}}{a}\right)^{2}-\frac{2}{a^{2}}\right]$

10.1. Primordial Symmetries of Standard Model and Torsion Field

In [26] the cross section for neutrino helicity spin flip obtained from this type of $f(R;T)$ model of gravitation with dynamic torsion field introduced by us in [16], was phenomenologically analyzed using the relation with the axion decay constant f_a (Peccei-Quinn parameter) due the energy dependence of the cross section, Consequently, the link with the phenomenological energy/mass window was found from the astrophysical and high energy viewpoints. The important point is that, in relation with the torsion vector interaction Lagrangian, the f_a parameter gives an estimate of the torsion field strength that can variate with time within cosmological scenarios as the described above, potentially capable of modifying the overall leptogenesis picture, the magnetogenesis, the bariogenesis and also to obtain some indication about the primordial (super) symmetry of the early universe.

In FRW scenario given here we saw that the torsion through its dual vector, namely:

$$h^0 = \frac{2}{5}\frac{\delta^0_a C_\tau}{a^2} d\tau \wedge e^a \tag{124}$$

goes as $\sim a^{-2}$ with C_τ a covariantly constant vector field $\left(\text{e.g.,}\overset{\circ}{\nabla} C_\tau = 0\right)$ that we take of the form $C_\tau \sim \left(\dot{\theta} + q_\tau\right)$ (due the Hodge-de Rham decomposition of h_μ, expression(42))where θ is a pseudoscalar field playing the obvious role of axion and q_τ :vector field linking h^0 with the magnetic field via the equation of motion for the torsion. Consequently, the torsion dual vector h^ihas the maximum value when the radius of the universe is $a_{\min}$, e.g., $a_{\min} = a(\tau_0)$ increases to the maximum value the spin-flip neutrino cross section and, for instance, the quantity of right neutrinos compensating consequently the actual (e.g., $a_{today} = a(\tau)$)asymmetry of the electroweak sector of the SM (see the behaviour of a in Fig. 1 of ref.[18]). This fact indicates that the original symmetry group contains naturally $SU_R(2) \times SU_L(2) \times U(1)$ typically inside GUT's structurally based generally in $SO(10)$, $SU(5)$ or some exceptional groups as $E(6)$,$E(7)$, *etc*

Also it is interesting to note that from the FRW line element written in terms of the cosmic time, the Hubble and the flow electromagnetic fields $E_\mu \equiv (0, E_i) = a^{-2}(0, \partial_\tau A_i)$ $B_\mu \equiv (0, B_i) = a^{-2}(0, \varepsilon_{ijk}\partial_j A_k)$ we have

$$\overline{\nabla}\cdot\overline{E} + \left(\frac{\alpha}{f}\overline{\nabla}\theta + \overline{\Pi}\right)\cdot\left(a^2\overline{B}\right) = 0$$

$$\partial_\tau\left(a^2\overline{E}\right) - \overline{\nabla}\times\left(a^2\overline{B}\right) = \left(\frac{\alpha}{f}\partial_\tau\theta + \Pi_0\right)\left(a^2\overline{B}\right) - \left(\frac{\alpha}{f}\overline{\nabla}\theta + \overline{\Pi}\right)\times\overline{E}$$

$$\overline{\nabla}\cdot\overline{B} = 0$$

$$\partial_t\overline{B} = -\overline{\nabla}\times\overline{E}$$

where $\Pi_\mu \equiv f_\mu\left(u_\mu, \gamma^5 b_\mu, eA_\mu.....\right)$ is a vector function of physical entities as potential vector, vorticity, angular velocity, axial vector etc. etc. as described by expression (42). In principle we can suppose that it is zero (low back reaction [29]) then

$$\overline{h} = \frac{\alpha}{f}\overline{\nabla}\theta, \qquad h^0 = \frac{\alpha}{f}\partial_\tau\theta$$

being $\left[\partial_\tau^2 - \overline{\nabla}^2 - \frac{\alpha}{f}\partial_\tau\theta\overline{\nabla}\times\right]\left(a^2\overline{B}\right) = 0$ the second order equation for the magnetic field that shows the chiral character of the plasma particles.

10.2. Magnetogenesis and Cosmic Helicity

Now we pass to see which role plays the torsion field in the magnetic field generation in a FRW cosmology. Taking as the starting point the (hyper) electrodynamic equations [42] and introducing a Fourier mode decomposition $\overline{B}(\overline{x}) = \int d^3\overline{k}\overline{B}(\overline{k})\, e^{-i\overline{k}\cdot\overline{x}}$ with $\overline{B}(\overline{k}) = h_i\overrightarrow{e}_i$ where $i = 1, 2$, $\overrightarrow{e}_i^2 = 1$, $\overrightarrow{e}_i \cdot \overrightarrow{k} = \overrightarrow{e}_1 \cdot \overrightarrow{e}_2 = 0$ the torsion-modified dynamical equations for the expanding FRW become

$$\dot{\overline{z}} + \left[\left(2\dot{a} + \frac{k^2}{\sigma}\right) + \frac{ah^0\,|k|}{\sigma}\right]\overline{z} = 0$$

$$\dot{z} + \left[\left(2\dot{a} + \frac{k^2}{\sigma}\right) - \frac{ah^0\,|k|}{\sigma}\right]z = 0$$

where the magnetic field is written in terms of complex variable $z\ (\overline{z})$ as

$$z = h_1 + ih_2$$
$$\overline{z} = h_1 - ih_2$$

from the above equations we see that the solution for z namely:

$$z = z_0 e^{-\left(2a+\frac{k^2}{\sigma}\tau\right)+\int\frac{ah^0|k|}{\sigma}d\tau}$$

contains the instable mode in the sense of [42] $\frac{k}{\sigma}\tau < \int\frac{ah^0}{\sigma}d\tau$. Consequently a defined polarization of the magnetic field appear and from the dynamical equation for the torsion field: $\nabla_{[\mu}h_{\nu]} = -\lambda\widetilde{F}_{\mu\nu}$ that in this case we have

$$\nabla_{[i}h_{\tau]} = \nabla_i\left(a^{-2}q_\tau\right) = -\lambda B_i \tag{a}$$

that implies a relation between the vector part of the h^0 (namely q_τ) with the vector potential A^k of the magnetic field as follows:

$$\nabla_i q_\tau \approx -\lambda\varepsilon_{ijk}\nabla^j A^k \tag{b}$$

Consequently, the primordial magnetic field (or seed) would be connected in a self-consistent way with the torsion field by means of the dual vector h_0.It in turn, would be connected phenomenologically with the physical fields (matter) of theory through Hodge-de Rham decomposition expression (42). We note from expression for z solution that the pseudo-scalar (axion) controls the stability, growth and dynamo effect but not the generation of the magnetic field (primordial or seed) as is clear from expression (b) where the (pseudo) -vector part of h_0 contributes directly to the generation of the magnetic field as clearly given by eq. (a).

10.3. Magnetogenesis and Cosmic Helicity II

In the case to include the *complete alpha term* given by equations (92) and in the same analytical conditions (e.g., Fourier decomposition) from the previous paragraph, the torsion-modified dynamic equations for the expanding FRW become

$$\dot{\overline{z}} + \left[\left(2\dot{a} + \frac{k^2}{\sigma}\right) + \frac{a\,|k|}{\sigma}\left(h^0 - \frac{\cos\alpha\,\left|\overline{j}_{gen}\right|}{\cos\beta\,|\overline{z}|}\right)\right]\overline{z} = 0$$

$$\dot{z} + \left[\left(2\dot{a} + \frac{k^2}{\sigma}\right) - \frac{a\,|k|}{\sigma}\left(h^0 - \frac{\cos\alpha\,\left|\overline{j}_{gen}\right|}{\cos\beta\,|z|}\right)\right]z = 0$$

where in this case the magnetic field is written (by convenience) in terms of complex variable z $(\overline{z})$ as

$$z = |z|\,e^{i\rho} \rightarrow \dot{z} = \left(\dot{|z|} + i\dot{\rho}\,|z|\right)e^{i\rho} \tag{125}$$

$$\overline{z} = |\overline{z}|\,e^{-i\rho} \rightarrow \dot{\overline{z}} = \left(\dot{|\overline{z}|} - i\dot{\rho}\,|\overline{z}|\right)e^{-i\rho} \tag{126}$$

From the above equation we see that the solution for z namely:

$$z = z_0 \exp\left[-\left(2a + \frac{k^2}{\sigma}\tau\right) + \int \frac{a\,|k|}{\sigma}\left(h^0 - \frac{\cos\alpha\,\left|\overline{j}_{gen}\right|}{\cos\beta\,|z_0|}\right)d\tau\right] \tag{127}$$

$$\text{with } z_0 = |z_0|\,e^{i\rho_0} \qquad (|z_0| = const)$$

contains the instable mode in the sense of [42] for example (117): $\frac{k}{\sigma}\tau < \int \frac{a}{\sigma}\left(h^0 - \frac{\cos\alpha\left|\overline{j}_{gen}\right|}{\cos\beta|z|}\right)d\tau$. But now *there are not a definite polarization for the magnetic field*, but now all depends on the difference:

$$\int \frac{a}{\sigma}\left(h^0 - \frac{\cos\alpha\,\left|\overline{j}_{gen}\right|}{\cos\beta\,|z_0|}\right)d\tau$$

Replacing explicitly h_α from the decomposition (42) we can see in a clear form, the interplay between the physical entities, as the vortical and magnetic helicities for example:

$$\left(\nabla_0\Omega + \frac{4\pi}{3}\left[h_M + q_s n_s \overline{u}_s \cdot \overline{B}\right] + \gamma_1 h_V + \gamma_2 P_0\right) - \frac{\cos\alpha\,\left|\overline{j}_{gen}\right|}{\cos\beta\,|z_0|}$$

Now considering in $\left|\overline{j}_{gen}\right|$ the fermionic current $\sum_f q_f\overline{\psi}_f\gamma_\mu\psi_f$, Ω as the axion a , $|z_0| = \frac{\cos\beta}{\cos\alpha}$ and putting $\gamma_2 = 0$ we have an interesting expression:

$$\nabla_0 a + \frac{4\pi}{3}\left[h_M + q_s n_s \overline{u}_s \cdot \overline{B}\right] + \gamma_1 h_V = \left|\sum_f q_f\overline{\psi}_f\gamma_\mu\psi_f\right|$$

The above expression it is very important because establishes the desired connection between helicities, magnetic field and fermionic fields and axion. We can order it as

$$\nabla_0 a - \left| \sum_f q_f \overline{\psi}_f \gamma_\mu \psi_f \right| = -\left[\frac{4\pi}{3} \left(h_M + q_s n_s \overline{u}_s \cdot \overline{B} \right) + \gamma_1 h_V \right]$$

We now clearly see the link between the axion and the fermionic fields (the dynamics of the interacting fields and the involved currents) in the LHS and the macroscopic physical observables in the RHS giving an indication of the origin of leptogenesis and bariogenesis in the context of this non Riemannian gravitational model.

10.4. Dark Matter, Energy Condition and UFT Model

As is well know, in a wormhole solution energy conditions are always violated in the standard general relativity. In the context of general relativity, this fact is closely related to the necessity to introduce exotic matter through the energy momentum tensor. Physically speaking, the observations of Type Ia Supernova (SNIa), together with the cosmic microwave background radiation (CMB) [and the larger scale structure, suggest that the present universe is in accelerating expansion, which needs something as dark energy with a negative equation of state (e.g., phantom field, non-canonical dynamical terms, etc.). The simplest standard model introduces the cosmological constant term Λ, which has a constant effective equation of state $w = -1$, and drive the acceleration of the universe assuming the effective energy of the Λ term occupies$\sim 73\%$ of the total energy (assuming also $\sim 23\%$ dark matter, $\sim 4\%$ baryon matter and $\sim 10-5\%$ radiation) constituting the ΛCDM model. This simple model satisfies more or less all the cosmological observations but is still a phenomenological one. Also the model suffers 2 important drawbacks: the'fine-tuning' [38] and the 'coincidence' [39] problems. In consequence other candidates as the dark energy (especially the dynamical models) are required. Beyond the general relativity, there are proposals in the literature where wormholes of different kinds [36] and other solutions with torsion were performed. For example, the Einstein-Cartan model ECM is the simplest version of the Poincare gauge theory of gravity (PGTG), in which the torsion is not dynamic because the gravitational action is proportional to the curvature scalar of Riemann-Cartan space-time (the ECT is a degenerate gauge theory in this particular aspect): wormholes were treated for example in [36].

The proposed model presented in this chapter is purely geometric: no energy-momentum tensor (EMT) is introduced. As have been seen, an effective TEM (e.g., eqs. (39-41)) is obtained when, from the general gravitational equations, the standard Einstein tensor is isolated. In our case, the effective TEM of the wormhole solution is diagonal (in the corresponding coordinates) with the isotropic typical structure $\approx \mathbb{R} \otimes SU(2)$ .Consequently, we can proceed analyzing this effective TEM (geometrical) by mean the standard energy conditions expressions [40], namely:

i) *WEC (weak energy condition)*

$$T_{\mu\nu}\big|_{eff} \zeta^\mu \zeta^\nu \geqq 0, \quad (\zeta^\nu : \text{ any timelike vector}) \Rightarrow \rho \geqq 0, \rho + p_k \geqq 0,$$

$$(i, k = 1, 2, 3)$$

guarantees that the energy density as measured by any local observer is nonnegative.

ii) *DEC (dominant energy condition)*

$$T_{00}|_{eff} \geqq |T_{ik}|_{eff} \quad (\zeta^{\nu} : \text{ any timelike vector}) \Rightarrow \rho \geqq 0, \rho + p_k \geqq 0,$$

$$(i, k = 1, 2, 3)$$

includes *WEC* and requires each each principal pressure never exceeds the energy density which guarantees that the speed of sound cannot exceed the light velocity c

iii) *SEC (strong energy condition)*

$$requires \rightarrow \rho + \sum p_k \geqq 0$$

and defines the sign of the acceleration due to gravity. In our case, the wormhole solution presented in Figure 1 of [18]), *the condition iii) is fulfilled jointly with conditions i) and ii).* As we have made mention above, in [36] wormhole solutions with nondynamical torsion were constructed in the context of the standard Einstein-Cartan model (ECM) fulfilling the energy conditions also. The fundamental differences between the model in [36] and here are:

∘In the case of [36] the energy conditions are fulfilled only for particular values (local conditions or windows) of the parameters in the introduced equation.

∘In our case there are not free parameters but geometrically induced functions mutually related. Consequently, there are no parameters that can be freely chosen but geometrically induced and mutually related functions, so that the freedom to choose them independently is restricted: e.g., once one of them is fixed, the others are automatically related to each other by means of expressions of a dynamic character (like the analogue of field b) or by means of the constraint given by Λ, etc (see Sections V and VI). This important fact, which will be treated in a particular way in another work [37], would give an indication that the solutions could have an overall character in our model.

The last important point to highlight is that the dynamic torsion field h_μ acts as mechanism of the reduction of an original (early, primordial) GUT (Grand Unified Theory) symmetry of the universe containing ∼SU(3) ×SU$(2)_R$ ×SU$(2)_L$ ×U(1) to SU(3) ×SU$(2)_L$ ×U(1) today. Consequently, the GUT candidates are $SO(10)$, $SU(5)$ or some exceptional groups as $E(6)$,$E(7)$ for example.

11. Axisymmetric Force-Free Magnetosphere beyond General Relativity

As is well known, a huge amount of the baryonic component of the Universe is found in the form of plasma, e.g., the standard conditions of the interstellar medium. Generally astrophysical plasmas can be (at the macroscopic level) treated as an electric conductive fluid, locally neutral, where currents and magnetic fields play a central role (magnetohydrodynamics, or MHD). This approach can be either employed in its classical version (nonrelativistic), or extended to a relativistic fluid if the physical scenario requires it. In the latter case the theory is the relativistic MHD and is necessary when the presence of strong gravity requires the use of general relativity (e.g., for compact objects like neutron stars and black holes).The magnetic fields involved in astrophysical plasmas may be extremely strong and they are responsible for many steady and variable emission processes, such as the coronal activity and the solar flares driven by subphotospheric motions and flux emergence; [66] or the electromagnetic spindown emission from pulsars [67], etc.

Ideal MHD alone fails at explaining the process of magnetization of a fluid initially with $B = 0$. Moreover, one of the most important problems for astrophysical plasmas is the origin of the magnetic fields (primordial or not), in the various physical scenarios as that described above, and their growth through the cosmic time.

For a long time, attempts have been made to give concrete answers to various astrophysical and cosmological mechanisms. In particular the origin, both of primordial fields of different types, as well as of the stellar and cosmological dynamics. Given that both general relativity (GR) and the standard model of elementary particles (SM) do not give fully conclusive explanations to these questions, the idea of a reformulation of a unified theory beyond RG and MS seems very attractive. Now we give (referring textually to some introductory comments of [61] and [63]) the main motivations of this research.

In previous references [14], we have introduced a unified model based on a non-Riemannian geometry containing a dynamic antisymmetric torsion that admits the same results of GR and SM already proven, but also satisfactorily solving problems for which GR and SM present difficulties or inconsistencies. Some of those problems that were successfully treated in the context of this new formulation were the determination of the mass of the axion [53], violation of CP of the neutrino [55], primordial magnetogenesis [51], [52], etc. In this work we calculate the equations controlling the stellar magnetospheres of compact objects in particular, in this new context. To this end, force free conditions are adopted by deriving the equilibrium conditions depending on a flow function with a pseudoscalar part coming from the torsion.

Actually, a typical example is the axisymmetric force-free magnetosphere in the exterior of a neutron star. Two possibilities are proposed for the energy storage prior to magnetar outbursts to explain the relevant phenomena: storage in the magnetar crust or in the magnetosphere [64] (the case of twisted magnetospheres in the context of standard GR see [63]). The latter model is discussed in terms of similarity with solar flares [56], [57]. In the solar flare model (e.g., [58]), the energy is quasi-statically stored by thermal motion at the surface, and is suddenly released as large-scale eruptive coronal mass ejections. The energy is dissipated via a magnetic reconnection associated with the field reconfiguration. Analogous energy buildup and release processes may be relevant to the magnetar giant flares, although the energy scale differs by many orders. In sum, one must entirely rethink the physics of neutrino cooling, photon emission, and particle emission from a neutron star, when its magnetic field (instead of its rotation) is the main source of free energy. This possibility is completely feasible in the context of the model previously presented in [52] that is based on a geometric (Lagrangian) action that can be considered the non-Riemannian generalization of the Born-Infeld model (see details in [51])

In this model, the torsion $T^{\alpha}_{\beta\gamma}$ has a dynamic character (contrary to other models in the literature [65]) and is totally antisymmetric, which allows it to be related to its dual vector h_{μ}. The other important feature is that the energy-momentum tensor and fundamental constants (really functions of the spacetime) are geometrically induced and not imposed "by hand".

Consequently the simplest mechanism to generate the necessary amount of energy of magnetospheres (even without star rotation) can be described as follows:

1) The axion and other pseudoscalars and pseudovector particles (contained in h_α) plus all helicities increase the original magnetic field B e.g., due the induction (dynamo) linearized expression, as

$$\nabla \times (\alpha_g B) = h \times E - h_0 B - (E \cdot \nabla \omega) \overline{m} \tag{128}$$

2) The increased B, increases the magnetic helicity H_M defined as (g_3 determinant of the absolute space)

$$H_M = \int A \cdot B \sqrt{g_3} d^3 x \tag{129}$$

3) The H_M in turn increases B even more via expressions (3) and (7) through the torsion vector h_α.

4) Consequently, the total energy in the magnetosphere will be increased to a certain limit.

$$E_M = \int \alpha B^2 \sqrt{g_3} d^3 x \tag{130}$$

5) After some limit to be determined, the excess energy in the magnetosphere is ejected and the process is repeated. This limit, as we shall see later, is subject to several topological bounds between the total energy of the system and the helicity. Precisely, in flat space the well-known inequality between energy and helicity namely $E(\xi) \geqslant C. |\mathcal{H}(\xi)|$, (C being a positive constant depending on the shape and size of a compact domain M) is exactly fulfilled in the case treated here, were the metric is fully regular without any singularity. However, the problem of energy conservation in the case of theories based on a basically a non-Riemannian structure will not be discussed here.

12. The Spherically Symmetric Solution: The Axion Star Background

In this section we only give a review of the spherically symmetric solution that describes, in principle, an axion star in the theory with torsion totally antisymmetric of the reference [51], [52]. Details of this type of solutions correspond to a separate publication and will not be given here [69]. The starting point to determine the simple solution that we will give here for the axion star is based on the Lagrangian of [51] and with the following ansatz of the general metric of a spherically symmetric configuration:

$$ds^2 = -e^{2\Lambda} dt^2 + e^{2\Phi} dr^2 + e^{2F(r)} d\theta^2 + e^{2G(r)} \sin^2 \theta \, d\varphi^2 \tag{131}$$

. The gravitational equations (Einstein-like equations) in the case of null cosmological constant λ, take the form

$$\underbrace{\overset{\circ}{R}_{\alpha\beta} - \frac{g_{\alpha\beta}}{2} \overset{\circ}{R}}_{\equiv G_{\alpha\beta}} = \underbrace{6 \left(-h_\alpha h_\beta + \frac{g_{\alpha\beta}}{2} h_\gamma h^\gamma \right)}_{\equiv T^h_{\alpha\beta}} \tag{132}$$

$$\underbrace{\overset{\circ}{R}_{\alpha\beta} - \frac{g_{\alpha\beta}}{2}\overset{\circ}{R}}_{\equiv G_{\alpha\beta}} = 6\left(-\overset{\circ}{\nabla}_{\alpha}a\overset{\circ}{\nabla}_{\beta}a + \frac{g_{\alpha\beta}}{2}\overset{\circ}{\nabla}_{\gamma}a\overset{\circ}{\nabla}^{\gamma}a\right) \tag{133}$$

where the overcircle on operators and tensors indicates that they are defined as in GR and a is a pseudoscalar field: e.g., axion. Note that the fact of having zero cosmological constant in this type of theory implies a constraint of geometric character between energy and geometrically induced matter. In equation (134) the following equation of motion for the torsion field was used, namely

$$\overset{\circ}{\nabla}_{\alpha}T^{\alpha}{}_{\mu\nu} = \lambda^{*}F_{\mu\nu} = 0 \tag{134}$$

$$\nabla_{\mu}h_{\nu} - \nabla_{\nu}h_{\mu} = 0 \Rightarrow h_{\alpha} = \pm\overset{\circ}{\nabla}_{\alpha}\Omega \tag{135}$$

(however $T_{\mu\nu\rho} = \varepsilon_{\mu\nu\rho\sigma}h^{\sigma}$ and the divergence of the G-connection coincides with the Christoffel one due to the full antisymmetry of the torsion tensor).

The conservation law is equivalent to the divergence of the h_{μ} that allows to determine the radial component of the torsion vector

$$h^{1} = \rho e^{-(\Lambda+2F)} = e^{-\Phi}h^{r} \tag{136}$$

Notice that the above result is analog to the standard conservation law for a (pseudo) scalar field in standard GR due the RHS of (133). The remaining components h_{μ}, on the other hand, are determined by equations (135) and are related via Hodge operation to the totally antisymmetric component of the connection, via the formula

$$\partial_{\mu}h_{\nu} = \partial_{\nu}h_{\mu} - 2\Gamma^{\rho}_{[\nu\mu]}h_{\rho} \tag{137}$$

$$= \partial_{\nu}h_{\mu} - \epsilon^{\rho}_{\nu\mu\sigma}h^{\sigma}h_{\rho} \tag{138}$$

h_0 being determined by

$$\partial_r\left(e^{\Lambda}\partial_r\left(e^{-\Lambda}h_t\right)\right) = 0 \rightarrow h_t = e^{\Lambda}\left(c\int dre^{-\Lambda} + d\right) \tag{139}$$

with c, d =const. and $h_2 = h_3 = 0$ in the simplest case.

Consequently, after solving the full system, the metric coefficients and the torsion vector components are given by the following expressions (considering cosmological constant $\lambda = 0$)

$$e^{2\Lambda} = 1 - \frac{2(\rho^2 c_2 + M)r}{(r-r_0)^2 + \rho^2} \tag{140}$$

$$= 1 - \frac{2(\rho^2 + M)r}{\left(r - \frac{\rho^2}{(\rho^2 c_2 + M)}\right)^2 + \rho^2}$$

$$e^{2F} = (r-r_0)^2 + \rho^2 \tag{141}$$

$$= \left(r - \frac{\rho^2}{(\rho^2 c_2 + M)}\right)^2 + \rho^2$$

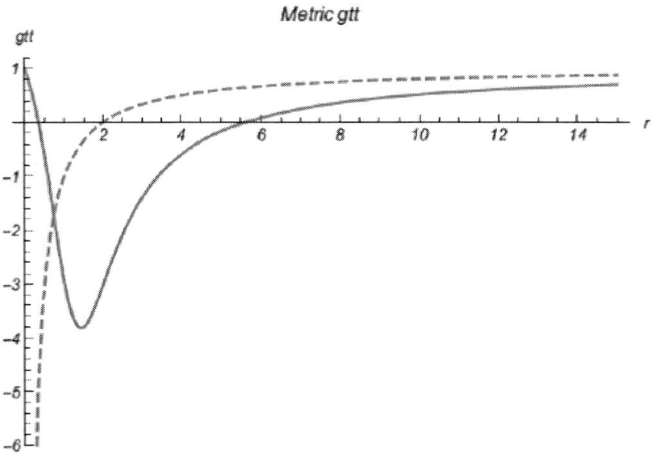

Figure 1. Metric solution, full line $M = 1$, $\rho = 1$ (normalized) and dotted line with $M = 1$ and $\rho = 0$: e.g., without torsion field (that is precisely the Schwarzchild solution).

In the case of SSS configurations the simplest admissible case is $F = G$ and $\Lambda = -\Phi$. Consequently

$$h_r = -\frac{\rho}{(r - r_0)^2 + \rho^2} \tag{142}$$

$$h_t = e^{\Lambda}\left(a \int dr e^{-\Lambda} + b\right) \tag{143}$$

Through the next part of this work, the ρ constant plays the role of v.e.v of the axion field. Another important feature of the solution is that there exists a minimal radius, namely $r_0 = \frac{\rho^2(c_1^2+1)}{2c_1^2(\rho^2 c_2+M)}$,)($c_1$ is a constant that we will put equal to one without loss of generality and c_2 a constant with units of inverse of mass) depending on the torsion field through ρ, that makes spacetime completely regular as described by Fig.1 (e.g., free of singularities). Note that when the torsion field goes to zero, the metric goes to the Schwarzchild one. While the SSS metric depends completely on h_1, with respect to the zero component of the torsion vector, the only nontrivial admissible solution that does not modify the geometrically induced energy-momentum tensor is $h_0 = e^{-\Lambda} h_t$ =constant.

13. Magnetosphere Dynamics and GS Equation

In this case, due the lack of rotation of the compact object, the general equation (197) from Appendix II [61] will be simplified according to the following considerations:

The field h_μ does not generate electromagnetic field directly obeying the equation (137). consequently, $\xi \equiv \psi$

The additional term in the GSE in spherical coordinates below, that contains $|h_0|$, comes from the equation of induction (dynamo) and corresponds to the chiral anomaly

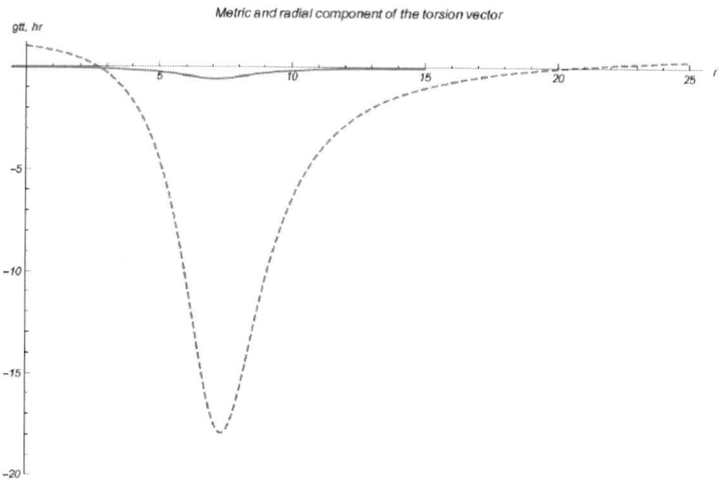

Figure 2. Metric solution g_{tt} (dotted line) and the corresponding radial (full line) component of the torsion vector h_r $(M = 1, \rho = 2)$.

in the generalized current (alpha effect) namely

$$\alpha^2 \frac{\partial}{\partial r}\left(\alpha^2 \frac{\partial \psi}{\partial r}\right) + \frac{\alpha^2 \sin\theta}{r^2}\frac{\partial}{\partial \theta}\left(\frac{1}{\sin\theta}\frac{\partial \psi}{\partial \theta}\right) = -\left(I(\psi) + |h_0|\,\psi/2\right)\frac{dI(\psi)}{d\psi} \tag{144}$$

The simplest way to reduce the above equation to an homogeneous one is with $I(\psi) = const$ or

$$I(\psi) = -h_0 \psi/2 \tag{145}$$

with this condition the solution for (144) is

$$\psi(r,\theta) = \sum_{l \geq 1} f_l(r) \sin\theta \frac{dP_l(\theta)}{d\theta} \tag{146}$$

For the dipole field $l = 1$ the radial function is

$$f_1(r) = -3\mu_\infty \frac{\widetilde{r}^2}{8\widetilde{M}^3}\left[\ln\left(1 - \frac{2\widetilde{M}}{\widetilde{r}}\right) + \frac{2\widetilde{M}}{\widetilde{r}}\left(1 + \frac{2\widetilde{M}}{\widetilde{r}}\right)\right] \tag{147}$$

with the following definitions in order to be compact:

$$\widetilde{M} = (\rho^2 c_2 + M) \tag{148}$$

$$\widetilde{r} = \frac{(r - r_0)^2 + \rho^2}{r} \tag{149}$$

$$\alpha = \left(1 + \frac{2\widetilde{M}}{\widetilde{r}}\right) \tag{150}$$

Consequently

$$\psi(r,\theta) = -3\mu_\infty \frac{\widetilde{r}^2 \sin^2\theta}{8\widetilde{M}^3}\left[\ln\left(1-\frac{2\widetilde{M}}{\widetilde{r}}\right)+\frac{2\widetilde{M}}{\widetilde{r}}\left(1+\frac{2\widetilde{M}}{\widetilde{r}}\right)\right] \tag{151}$$

which in the weak gravity regime is given by

$$= B_0^3(\widetilde{r}^*)\frac{\widetilde{r}^2\sin^2\theta}{\widetilde{r}}\left[1+\frac{3\widetilde{M}}{2\widetilde{r}}+\frac{12\widetilde{M}^2}{5\widetilde{r}^2}+...\right] \tag{152}$$

As can be seen, we have taken the usual boundary conditions for , ie $\psi(r,\theta)$

$$\psi(r,0) = \psi(r,\pi) = 0 \tag{153}$$

$$\psi(r\to\infty,\theta) \sim r^{-1} \tag{154}$$

whereas at the stellar surface $r = r^*$ the called magnetic function $\psi(r,\theta)$ is assumed to be dipolar field $(e.g., l = 1)$

$$\psi(r^*,\theta) = f_1(r^*)\sin^2\theta \tag{155}$$

The components of the magnetic field can be obtained as

$$(B_r, B_\theta, B_\phi) = \left(\frac{\psi_{,\theta}}{r\varpi}, -\frac{\alpha\psi_{,r}}{\varpi}, \frac{I}{\alpha\varpi}\right)$$

then

$$B_r = -3\mu_\infty \frac{\left((r-r_0)^2+\rho^2\right)^2\cos\theta}{8(\rho^2 c_2+M)^3 r^6}\left[\ln\alpha+\frac{2r(\rho^2 c_2+M)}{(r-r_0)^2+\rho^2}\alpha\right] \tag{156}$$

$$B_\theta = -\frac{3}{8}\mu_\infty\alpha\sin\theta\frac{r^2-r_0^2}{r^3(\rho^2c_2+M)^3}\left[2\frac{(r-r_0)^2+\rho^2}{r}\ln\alpha+(\rho^2c_2+M)\left(1+\frac{1}{\alpha}\right)\right] \tag{157}$$

$$B_\phi = 3\mu_\infty\frac{|h_0|}{2r^3}\frac{\left((r-r_0)^2+\rho^2\right)^2\sin\theta}{8(\rho^2c_2+M)^3\alpha}\left[\ln\alpha+\frac{2(\rho^2c_2+M)r}{(r-r_0)^2+\rho^2}\alpha\right] \tag{158}$$

Relating $F_{\mu\nu}$ to the locally measured magnetic field $\mathbf{B}$ and defining $x \equiv \frac{\widetilde{r}}{2\widetilde{M}}$ we have

$$B_r = \frac{2\mu\cos\theta}{r^3}\left[-3x^3\ln\left(1-x^{-1}\right)-3x^2\left(1+x^{-1}/2\right)\right]$$

$$B_\theta = \frac{2\mu\sin\theta}{r^3}\left[6x^3\left(1-x^{-1}\right)^{1/2}\ln\left(1-x^{-1}\right)+\frac{6x^2\left(1+x^{-1}/2\right)}{\left(1-x^{-1}\right)^{1/2}}\right]$$

$$B_\phi = B_\phi = 3\mu_\infty\frac{|h_0|}{2r^2}\frac{x^2\sin\theta}{\left(1-x^{-1}\right)^{1/2}\widetilde{M}}\left[\ln\left(1-x^{-1}\right)+x^{-1}\left(1-x^{-1}\right)\right]$$

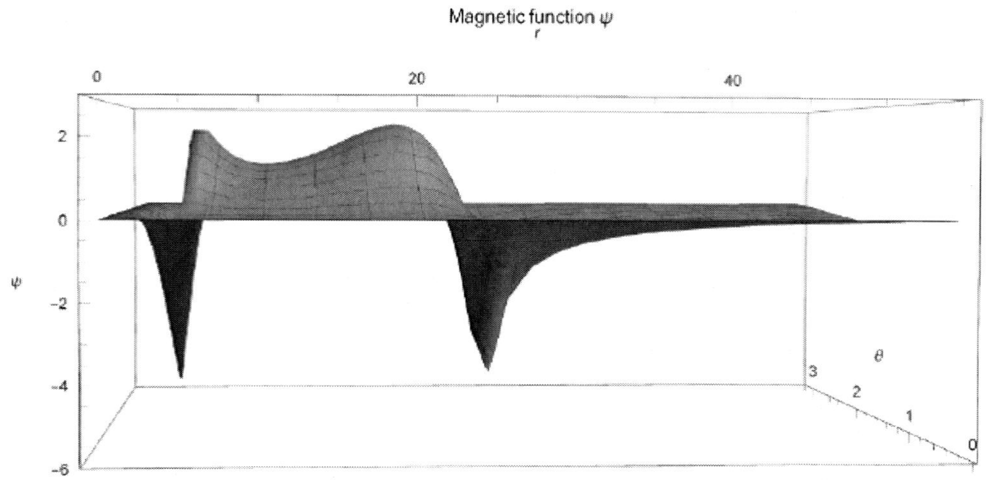

Figure 3. Magnetic function ψ as function of θ and r. Note that the peaks coincide with the horizons of the compact star.

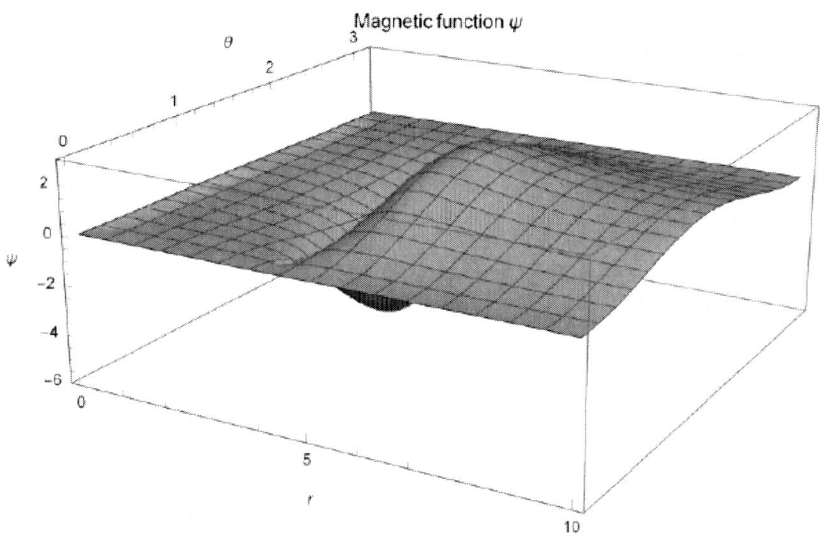

Figure 4. Magnetic function ψ as function of θ and r near the first horizon of the compacat object.

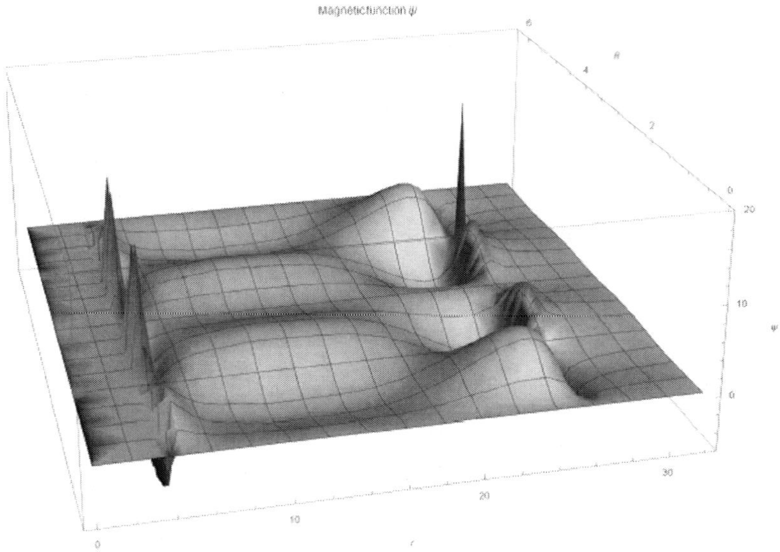

Figure 5. Magnetic function ψ as function of θ and r in the quadrupole case. Note that the peaks coinciding with the horizons of the compact star are sharp.

14. Quadrupole and Quadrupole-Dipole Case

In this case and without going into details that will be given in a subsequent work, $f_2(r)$ in eq. (146) takes the form:

$$f_2(r) = -5\mu_Q \frac{\widetilde{r}^3}{2\widetilde{M}^5}\left[\left(1+\frac{3\widetilde{M}}{2\widetilde{r}}\right)\ln\left(1-\frac{2\widetilde{M}}{\widetilde{r}}\right)+\frac{\widetilde{M}}{\widetilde{r}}\left(2-\frac{\widetilde{M}}{\widetilde{r}}-\frac{\widetilde{M}}{3\widetilde{r}}\right)\right]$$
$$\simeq \frac{\mu_Q}{\widetilde{r}^2}\left[1+\frac{8\widetilde{M}}{3\widetilde{r}}+\frac{40\widetilde{M}^2}{7\widetilde{r}^2}+\cdots\right]$$

the second line in the equation above is given in the weak gravity regime.
As can be seen, we have taken the usual boundary conditions for , ie $\psi(r,\theta)$

$$\psi(r,0) = \psi(r,\pi) = 0 \tag{159}$$
$$\psi(r\to\infty,\theta) \sim r^{-1} \tag{160}$$

whereas at the stellar surface $r = r^*$ the called magnetic function $\psi(r,\theta)$ is assumed now to be dipolar-quadrupolar field ($e.g., l = 1, 2$). Some behaviors have been plotted below, as we see there, the introduction of dipole fields introduces sharp peaks and singularities in the toroidal magnetic field

15. Helicity and Energy: The Mechanism

The helicity in function of I and the function ψ is

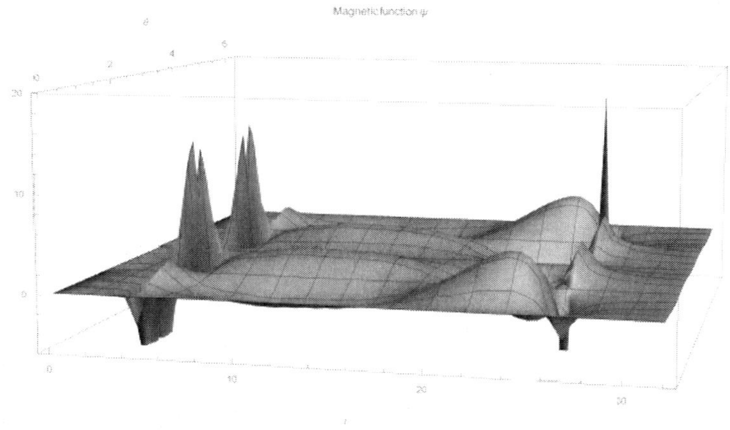

Figure 6. Magnetic function ψ as function of θ and r dipole-quadrupole case Note that the peaks are increased in comparison with quadrupole case.

$$H \equiv \int \sqrt{g_3} A_i B^i d^3x = 4\pi \int \frac{\psi I}{\alpha^2 \sin\theta} dr d\theta \tag{161}$$

Looking at the geometrically induced energy-momentum tensor from (132) and taking into account the results of the previous Section, we have the energy as

$$E_{EM} \equiv \int \sqrt{g_3} \frac{\alpha B_i B^i}{8\pi} d^3x = \frac{1}{4} \int B^2 r^2 \sin\theta dr d\theta \tag{162}$$

$$= \frac{1}{4} \int \left[(\alpha \partial_r \psi)^2 + \left(\frac{1}{r} \partial_\theta \psi \right)^2 + \left(\frac{I}{\alpha} \right)^2 \right] \frac{dr d\theta}{\sin\theta} \tag{163}$$

In the above expressions $\sqrt{g_3}$ is the determinant of the 3-space and generally the integrals are performed in the region $r \geqslant r^*$. As we see from the expression, energy and helicity are linked by h_0 (our alpha term)

At this point we will see the role of the torsion in relation to the energy contained in the magnetospheric region. Since the constant λ was considered null, there is no a priori direct relationship between B_ρ and h_ρ via expression (134). Consequently the relationship will be via h_0 that takes the place of alpha in the dynamo expression (128) and the metric coefficients that form a fundamental part in the stream function ψ(also called magnetic function) that contains h_1. Therefore, h_0 is relevant in the electrodynamics of the pulsar mechanism and h_1 is relevant in the gravitodynamic effects because it is part of the compact object itself by means of g_{tt} (metric solution of the spherical configuration).

As we are going to see now, the torsion is a special ingredient in the increase of the energy contained in the magnetosphere. The helicity and the toroidal component of the energy ($E_{EM}|_T$) are directly related as follows

$$H \equiv 2\pi \int \frac{-h_0 \psi^2}{\alpha^2 \sin\theta} dr d\theta \equiv \int \mathcal{H} dr d\theta \tag{164}$$

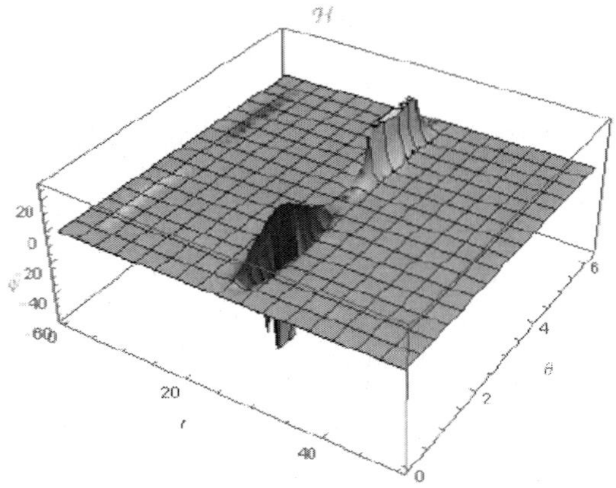

Figure 7. Helicity density $\mathcal{H}$ in the spacetime geometry of the axion star for a h_0 (normalized value) of 1.51

$$E_{EM}|_T = \frac{1}{8}\int\left(\frac{-|h_0|\,\psi}{\alpha}\right)^2 \frac{drd\theta}{\sin\theta} = \frac{-h_0}{4\pi}H \tag{165}$$

$$= \int \mathcal{E}_{EM}|_T \, drd\theta = \int \frac{|h_0|}{4\pi}\,|\mathcal{H}|\, drd\theta \tag{166}$$

Consequently, the inequality suggested by Arnold in [60] is fulfilled.

16. Axion vs. Boson Stars: Qualitative Aspects

We will now briefly discuss some qualitative aspects regarding similarities and differences between axion and boson stars.

16.1. Galactic Center and Supermassive Pseudoscalar Star

The solution presented in this chapter meets partially the conditions described in reference [70] for the stars of scalars. As we see, although our solution have not singularity, the existence of horizons is evident and the interaction with the electromagnetic field is clear. However, at least with respect to the horizons, the problem would be solved for the extreme solution, or that the horizons would be hidden behind the stellar surface. Consequently one could roughly infer in a purely qualitative way, similar arguments to consider a pseudoscalar star as a galactic center. A detailed discussion conveniently with phenomenological and quantitative aspects of this model are given in a subsequent paper [69]

16.2. Cerenkov Radiation and Pseudoscalar Star

In [71] the authors explore the possibility that a charged particle moving in the gravitational field generated by a scalar star could radiate energy via a recently proposed gravitational Čerenkov mechanism. Due that they numerically prove that this is not possible for stable

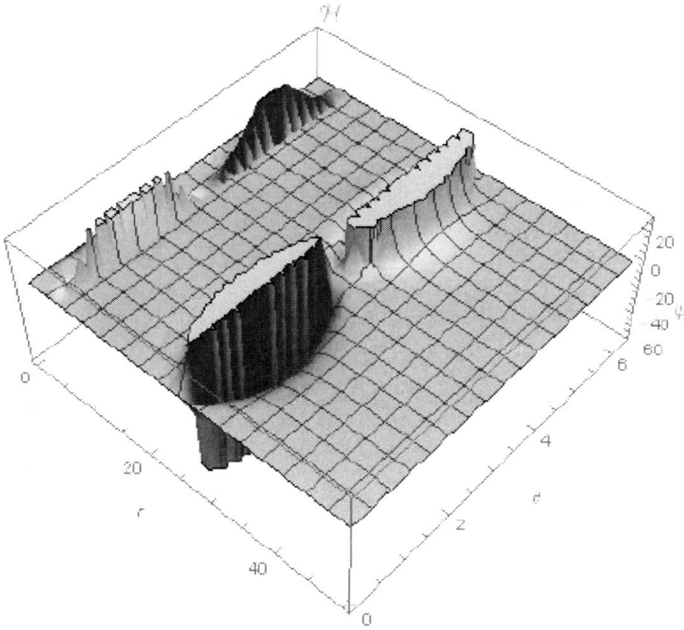

Figure 8. Helicity density $\mathcal{H}$ in the spacetime geometry of the axion star for a h_0 (normalized value) of 0.051

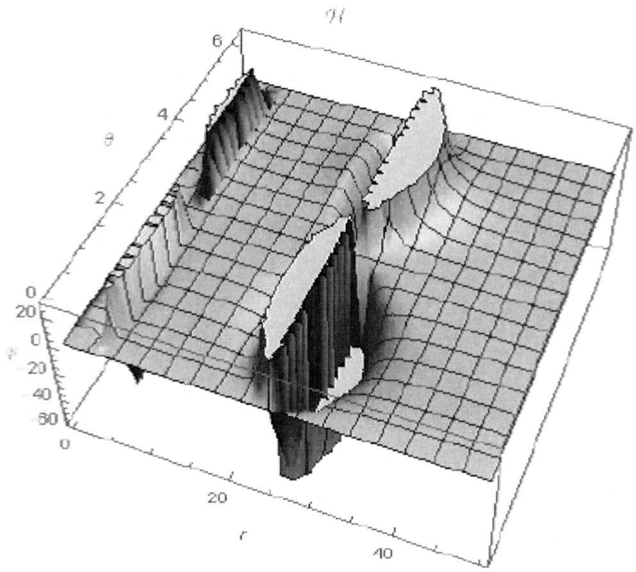

Figure 9. Helicity density $\mathcal{H}$ in the spacetime geometry of the axion star for a h_0 (normalized value) of 0.51

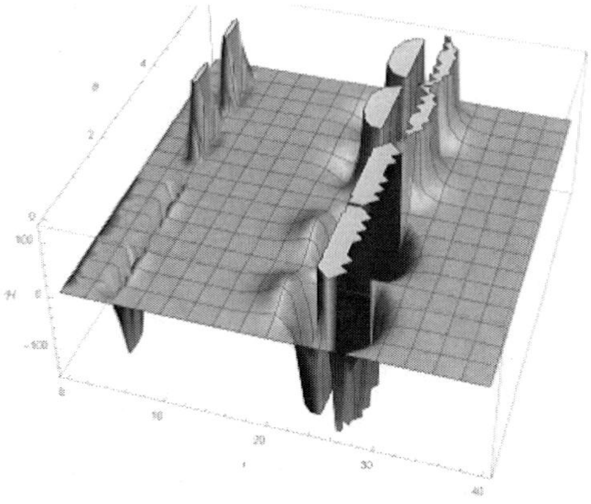

Figure 10. Helicity density $\mathcal{H}$ in the spacetime geometry of the axion star for a h_0 (normalized value) of 1.51 dipole case.

boson stars. From the theoretical viewpoint, the emission of Cerenkov radiation is a kinematically allowed process if one invokes the condition that the spatial variation of the scalar field ψ is greater than three half its potential eq. (21) of [71] that in our case from (132) we see that the energy-momentum tensor takes that same form for the axion field a that for the scalar field ψ but i our case the potential is zero.Then we have simply: $|h_r|^2 = |\partial_r a|^2 > 0$. Consequently from explicit expression for h_r eq. (142) this Cerenkov process is possible allowing in our particular case, a concrete way of distinguishing a star from axions from one of bosons or from another type of star composition.

16.3. Gravitational Collapse and Magnetosphere Structure

From the purely qualitative point of view and in the context of recent work [72] in which it is shown that it is possible to extend Einstein's theory by considering f (R) gravity and inquiring if the further degrees of freedom in the gravitational action can be modeled as perfect fluids sourcing the field equations. In our particular case with dynamic torsion, a similar treatment could be done considering that one has a energy-momentum tensor that allows an interpretation of fluid (exotic or not).

The serious analysis of gravitational collapse in our context with an analogous treatment to work [73] and its relationship with the structure of the magnetosphere requires to go beyond a merely qualitative description that is left out of this work.

Conclusion

As we have shown in other contexts [59], if we initially suppose the fields in a pure toroidal (axial) configuration (possibly confined), the field bundle allows twisted solutions. These configurations appear through the geometrical constraint between the angles of the twist in question, reducing the 3-geometry of the torus to a unilateral helical geometry (surface) isomorphic to the Moebius strip. Fig 11 would be the basic configuration, then Fig. 12 with

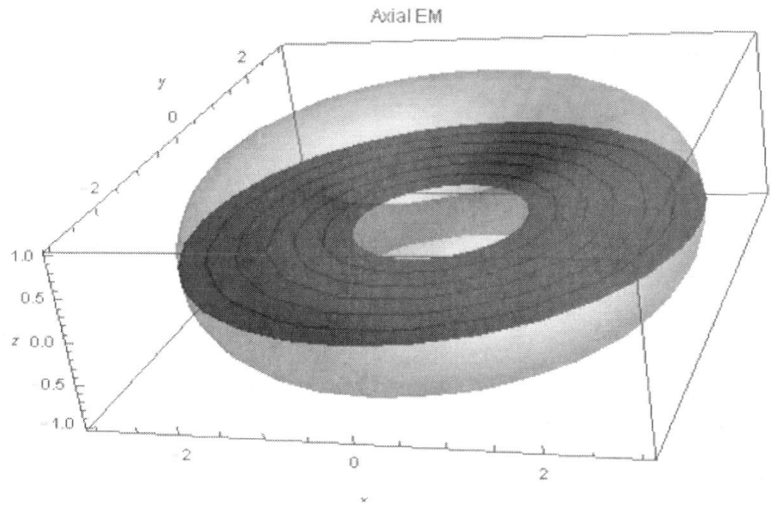

Figure 11. Toroidal field (initial configuration) with no twist (lower magnetosphere energy).

one and Fig. 13 with two twists respectively. Geometrically and topologically speaking, we have here that the energy stored in the magnetosphere of the compact object increases proportional to Tw (twisting number), being the twist number h_0 precisely in our case, fact that is corroborated in well known topological theorems (e.g., [60] in flat space). Note that here the configurations are not necessarily closed on the surface of the star or in the horizons as it is usually assumed, but they close exactly on surfaces of toroidal geometry making the twist (as in the Möbius strip case). The topological/geometric invariant is the cylinder that contains the central ring of the torus. The number of twists(turns)will describe an increase in the energy contained in the configuration (the magnetosphere in our case).

The important remark is that, because in flat space helicity is proportional to the linking number times the square of the flux of the vector field across any section of the flux tubes [60], in our case this translates into the formula (164) namely $H \propto h_0.\psi^2$

Self-Linking of Magnetic Lines: No Reconnection

As we saw in the previous discussion, the number of twist in the toroidal fields increases the helicity of the system. As a result of the expressions (164) and (165) we see that the toroidal component of the energy differs from the helicity by the constant pseudoscalar quantity h_0, therefore the increase in the number of turns (increase of the linking number) increases the energy of the configuration. The number of turns in which the helical field lines close on themselves does not produce the effect of usual reconnection, and for fixed energy of the magnetosphere this number has fixed values as well.

Energy Threshold

For the helicity / toroidal energy relationship discussed in the previous paragraph, the limit value in the force-free magnetosphere will be defined by the topological/geometrical condi-

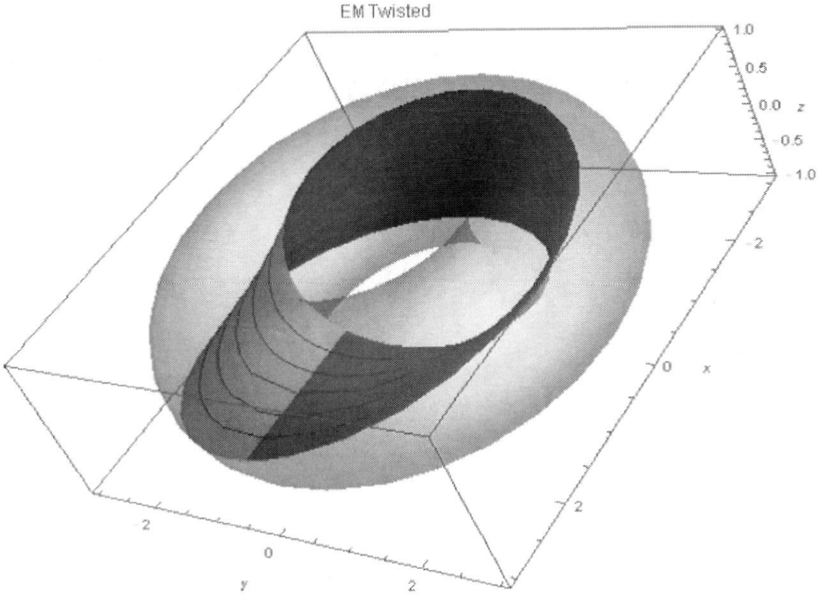

Figure 12. One twist configuration with the magnetic lines in a helical bundle (Moebius strip isomorphic)

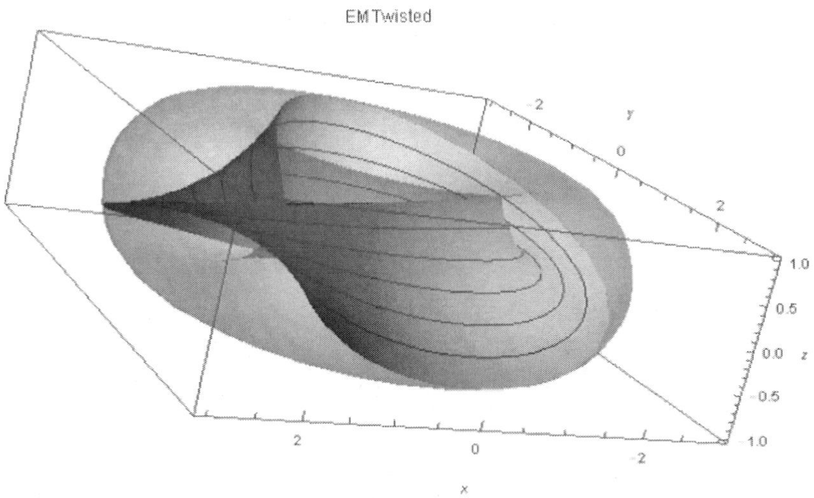

Figure 13. Two twist configuration (higher energy stored in the magnetosphere) with the magnetic lines in a helical bundle (Moebius strip isomorphic)

tions of the helicity and higher invariants. Physically, the excess of energy will be emitted in the vicinity of the horizons of events of the regular black hole as we see in the Fig. 3 and also Fig. 2, since the helicity peaks are at those points.

Seed Magnetic Field

As we have been seen before [52], considering only the terms of interest without the diffusive and advective term in the induction equation (only time-dependence for the magnetic field is preserved) we had the following formula

$$\eta \nabla_j \times \left[\left(\frac{\cos\xi \left| \overline{j}_{gen} \right|}{\cos\varsigma} \right) e_k \right] = -\partial_t B_i \tag{167}$$

being $\{e_k\}$ our vector basis in g_3 (usually in the orthonormal system) giving $\cos\xi$ and $\cos\varsigma$ the relative directions between h and j and between h and the generated B respectively[51], [52]. We can see from expression (167) that the currents given by the fields (related to the geometry via h_α) originate the magnetic field. Considering, for example, fermion currents for different species (e.g., neutrino, electron, positron, etc.) (167) takes the form

$$\eta \nabla_j \times \left[\left(\frac{\cos\xi \left| \sum_f q_f \overline{\psi}_f \gamma_\mu \psi_f \right|}{\cos\varsigma} \right) e_k \right] = -\partial_t B_i$$

As a simple example, one could consider a neutron star with a neutrino dynamics in its interior that under certain conditions (roughly considered in (167)) would generate the initial magnetic field (seed) required.

Axion, Cosmic Magnetic Field and Magnetosphere Dynamics

One point to consider is the possibility of a cosmological origin of the h_0 that would appear as constant in our GS equations due to the differences of temporal scales in the formation and dynamics of the magnetic fields in stars and pulsars and the global cosmological time. Indeed, we have seen in previous papers [51], [52] that h_0 appears as a cosmological solution describing an expanding universe where the h_0 falls as the scale factor $(1/a(\tau))^2$ being h_0 in that case proportional to the temporal derivative of the axion: e.g., $\dot{\theta}$ introducing, in this way, the dynamics of dark matter candidates to large-scale astrophysical phenomena. This fact clearly justifies (in addition to the purely mathematical reasons to obtain a solution) the choice h_0 = constant.

Open Questions

There are many things still to be investigated, one of which is the equilibrium condition beyond the force free condition [62] which also includes a turbulent part in the dynamo mechanism with an original simulation treatment that was used successfully for the solar case in [75].The other interesting topic that we will be working on has its origin in the enormous velocities of the pulsars that can not be explained only in the asymmetric explosions of supernovas during the birth of the pulsar, probably the existence of the dynamic torsion in the model will be the right mechanism to give the pulsar the required "kick" that would take place in magnetic and vortical chiral effects depending also on matter and dark energy.

Acknowledgments

We are very grateful to Professor Alexander Bruno and Professor Alexander Aptekarev for discussions and to the anonymous reviewer who brought to my attention the problem of nonlinear acoustic waves. Also to Conicet and Keldysh Institute of Applied Mathematics for hospitality. This work was supported by Moscow Center of Fundamental and Applied Mathematics, Agreement with the Ministry of Science and Higher Education of the Russian Federation, No. 075-15-2022-283.

Appendix I: Generalized Lorentz Force

As we shown before in [51], [52], the geometrically induced Lorentz force that we have obtained from the model in the linear limit was

$$\left(\overline{h} \cdot B + \rho_e\right) E + J \times B = (E \cdot B)\, \overline{h} \tag{168}$$

consequently, the case of force free condition with non-vanishing torsion field implies: $(E \cdot B) = 0$. General assumptions for 3+1 splitting in axisymmetrical spacetimes can be introduced in standard form (e.g., j, E and B can be treated as 3-vectors in spacelike hypersurfaces). In terms of these 3-vectors the nonlinear eqs. of the original model can be linearized and consequently written in a Maxwellian form as

$$\nabla \cdot \mathbb{E} = -h \cdot \mathbb{B} \equiv 4\pi \rho_e \tag{169}$$

$$\nabla \cdot B = 0 \tag{170}$$

$$\nabla \times (\alpha E) = (B \cdot \nabla\omega)\, \overline{m} \tag{171}$$

$$\nabla \times (\alpha \mathbb{B}) = h \times \mathbb{E} - h_0 \mathbb{B} - (\mathbb{E} \cdot \nabla\omega)\, \overline{m} \tag{172}$$

The derivatives in these equations are covariant derivatives with respect to the metric of the absolute space γ_{ij} being α, β : lapse and shift functions respectively. Notice that we need to replace $\overline{h} \times \overline{E}$ in order to introduce the physical currents as follows. From the above equations in exact form, the geometrical current induced by the non-Riemannian framework is

$$J \equiv +h \times \mathbb{E} - h_0 \mathbb{B} \tag{173}$$

consequently

$$J \times \mathbb{B} \Rightarrow (h \cdot \mathbb{B})\, \mathbb{E} = J \times \mathbb{B} + (\mathbb{B} \cdot \mathbb{E})\, h \tag{174}$$

then $h \times \mathbb{E}$ is

$$h \times \mathbb{E} = h \times \left[\frac{(\mathbb{B} \cdot \mathbb{E})\, h + J \times \mathbb{B}}{(h \cdot \mathbb{B})}\right] = J - \frac{(h \cdot J)\, \mathbb{B}}{(h \cdot \mathbb{B})} \tag{175}$$

consequently, the relation with the physical scenario can be implemented as follows:

$$J-\frac{(h\cdot J)\,\mathbb{B}}{(h\cdot\mathbb{B})}\longrightarrow\alpha_g\left(j_{ph}-\frac{(h\cdot j_{ph})\,\mathbb{B}}{(h\cdot\mathbb{B})}\right)\tag{176}$$

where we define the lapse function α_g(metric coefficient g_{tt}) in order to differentiate it from the alpha-term, transforming the set (169-172) at the linear level, namely $\mathbb{E}\to E$ and $\mathbb{B}\to B$, to

$$\nabla\cdot E=-h\cdot B\equiv 4\pi\rho_e\tag{177}$$

$$\nabla\cdot B=0\tag{178}$$

$$\nabla\times(\alpha_g E)=(B\cdot\nabla\omega)\,\overline{m}\tag{179}$$

$$\nabla\times(\alpha_g B)=\alpha_g\left(j_{ph}-\frac{(h\cdot j_{ph})\,B}{(h\cdot B)}\right)-h_0 B-(E\cdot\nabla\omega)\,\overline{m}\tag{180}$$

$$=\alpha_g j_{ph}-\left[h_0+\frac{(h\cdot j_{ph})}{(h\cdot B)}\right]B-(E\cdot\nabla\omega)\,\overline{m}\tag{181}$$

Similar equations, without considering a fundamental theory, were recently obtained in[45]

Appendix II: Generalized Grad-Shafranov Equation

The consistent theoretical description of gravitational magnetohydrodynamics (MHD) equilibria is of fundamental importance for understanding the phenomenology of accretion disks (AD) around compact objects (black holes, neutron stars, etc.). The very existence of these equilibria is actually suggested by observations, which not only show evidence of quiescent, and essentially non-relativistic, AD plasmas close to compact stars, but also the dynamical interplay with high energy processes involving the magnetospheres of compact objects, in particular pulsars, quasars and magnetars. The electromagnetic (EM) fields involved, in particular the electric field, may locally be extremely intense [46], so several standard processes such as electron positron pair creation occur, but several exotic interactions involving neutrinos with axions and other dark matter candidates must also be taken into account. This suggests therefore that such equilibrium (if it certainly exists) should be described in the framework of unified field theory beyond general relativity (GR) [50]-[53] and beyond the standard model (SM). Extending previous approaches, holding for compact objects/black hole axisymmetric geometries having into account effect of space-time curvature, the purpose of this work is the formulation of a generalized Grad-Shafranov (GGS) [42]-[45] equation based in a non-Riemannian geometry with dynamical torsion field suitable for the investigation of accretion, jets and winds and other astrophysical effects when high energy effects (exotic or not) are present. Now we will calculate the GSE with axisymmetry. Arguments and procedures for calculating GSE in this model are similar in form to works well known in the context of GR [47]-[49] (we use through the notation in work [49]) to have a reasonable comparison parameter with those results.

Magnetic Fields and Currents:

From the electrodynamic equations in 3+1 formulation of curved spacetimes, under axisymmetry, the magnetic field is splitted as $\overline{B} = B_p + B_t$ where

$$\begin{aligned} B_p &= \frac{1}{2\pi\varpi^2}\left(\nabla\psi + \overline{m} \times \nabla h_0\right) \times \overline{m} = \\ &\frac{1}{2\pi\varpi}\left(\nabla\psi \times e_{\widehat{\phi}} + \varpi\nabla h_0\right) = \frac{\nabla\psi \times \overline{m}}{2\pi\varpi^2} + \nabla\chi \\ B_t &= -\frac{2I\left(\psi,\chi\right)}{\alpha\varpi c} e_{\widehat{\phi}} \end{aligned} \tag{182}$$

with $e_{\widehat{\phi}}$ unitary toroidal vector, h_0 pseudoscalar field that we redefine as χ(zero component of the dual of the antisymmetric torsion field) and $\overline{m} \cdot \overline{m} = g_{\phi\phi} = \varpi^2$. The expression for the toroidal magnetic field coming from the Ampere law and the currents enclosed by the surface A , namely I (depending on ψ and χ), are obtained similarly to the magnetic flux assuming the form: $\nabla I = \nabla I\left(\psi\right) + \overline{m} \times \nabla I\left(\chi\right)$. Consequently, since $(E \cdot B) = 0$ the eqs. (182) brings the force free condition as

$$\begin{aligned} j_p &= \frac{1}{2\pi\alpha\varpi}\left(e_{\widehat{\phi}} \times \nabla I\right) = -\frac{\nabla I\left(\psi\right) \times \overline{m}}{2\pi\alpha\varpi^2} + \frac{\nabla I\left(\chi\right)}{2\pi} \\ &= -\frac{1}{\alpha}\frac{dI}{d\zeta}B_p \end{aligned} \tag{183}$$

where we have defined the multi-vector $\zeta \equiv \psi + \overline{m}\chi \left(\overline{m} \equiv \varpi e_{\widehat{\phi}}\right)$ (and consequently under action of exterior derivative: $d\zeta = d\psi + \overline{m} \times d\chi$ and $\nabla\zeta = \nabla\psi + \overline{m} \times \nabla\chi$). Thus:

$$B_p \cdot dA = d\zeta \tag{184}$$

$$\oint \alpha j_p \cdot dA = -\oint \frac{dI}{d\zeta} B_p \cdot dA = -\int_0^{\zeta} \frac{dI}{d\zeta} B_p \cdot dA \tag{185}$$

$$I\left(0\right) = I\left(\zeta\right) \tag{186}$$

$$j_t = -\frac{1}{8\pi}\left[\frac{\varpi c}{\alpha}\nabla \cdot \left(\underbrace{\frac{\alpha}{\varpi^2}\left(\nabla\psi + \overline{m} \times \nabla\chi\right)}_{\equiv\nabla\zeta}\right) + \frac{\varpi\left(\Omega_F - \omega\right)}{\alpha^2 c}\left(\nabla\psi + \overline{m} \times \nabla\chi\right) \cdot \nabla\omega\right] \tag{187}$$

$$v_F = \frac{1}{\alpha}\left(\Omega_F - \omega\right)\varpi \tag{188}$$

$$E_p = -\frac{v_F}{c}\left(e_{\widehat{\phi}} \times B_p\right) = -\frac{1}{2\pi c\alpha}\left(\Omega_F - \omega\right)\nabla\zeta \text{ (force free)} \tag{189}$$

$$\begin{aligned} \rho_e &= \frac{1}{4\pi}\left(\nabla \cdot E_p + \overline{h} \cdot B\right) \\ &= \frac{1}{4\pi}\nabla \cdot E_p \end{aligned} \tag{190}$$

then

$$\rho_e = -\frac{1}{8\pi^2}\frac{\varpi^2(\Omega_F-\omega)}{\alpha^2 c}\left\{\nabla\cdot\left(\frac{\alpha}{\varpi^2}\nabla\zeta\right)+\frac{\alpha}{\varpi^2}\nabla\zeta\cdot\nabla\ln\left[\frac{\varpi^2(\Omega_F-\omega)}{\alpha^2 c}\right]\right\} \tag{191}$$

as usual we can eliminate the first factor: $\nabla\cdot\left(\frac{\alpha}{\varpi^2}\left(\nabla\Psi+\overline{m}\times\nabla\chi\right)\right)$ between expressions (187) and (191), consequently

$$\begin{aligned}\rho_e - \frac{\varpi(\Omega_F-\omega)}{\alpha c}\frac{j_t}{c} = \\ -\frac{1}{8\alpha\pi^2}\left\{\left(\frac{\varpi(\Omega_F-\omega)}{\alpha c}\right)^2\nabla\zeta\cdot\nabla\omega-\frac{c\alpha^2}{\varpi^2}\nabla\zeta\cdot\nabla\left(\frac{\varpi^2(\Omega_F-\omega)}{\alpha^2 c}\right)\right\}\end{aligned} \tag{192}$$

In this case with dynamical torsion field, the transfield component of the momentum equation for the force free case, namely

$$\left(\overline{h}\cdot B+\rho_e\right)E+J\times B=0 \tag{193}$$

becomes

$$\frac{j_t}{c}-\frac{\varpi(\Omega_F-\omega)}{\alpha c}\rho_e=\frac{1}{2\alpha^2\varpi c^2}\frac{IdI}{d\zeta} \tag{194}$$

Solving for ρ_e and j_t

$$\begin{aligned}8\pi^2\rho_e = \\ \frac{\frac{\varpi(\Omega_F-\omega)}{\alpha c}}{1-\left(\frac{\varpi(\Omega_F-\omega)}{\alpha c}\right)^2}\left[\frac{8\pi^2}{2\alpha^2\varpi c}\frac{IdI}{d\zeta}+\frac{\varpi(\Omega_F-\omega)}{\alpha^2 c}\nabla\zeta\cdot\nabla\omega-\frac{c}{\varpi}\nabla\zeta\cdot\nabla\ln\left(\frac{\varpi^2(\Omega_F-\omega)}{\alpha^2 c}\right)\right]\end{aligned} \tag{195}$$

$$\begin{aligned}8\pi^2 j_t = \\ \frac{1}{1-\left(\frac{\varpi(\Omega_F-\omega)}{\alpha c}\right)^2}\left[\frac{8\pi^2}{2\alpha^2\varpi c}\frac{IdI}{d\zeta}+\frac{\varpi(\Omega_F-\omega)}{\alpha^2 c}\nabla\zeta\cdot\nabla\omega-\frac{c}{\varpi}\nabla\zeta\cdot\nabla\ln\left(\frac{\varpi^2(\Omega_F-\omega)}{\alpha^2 c}\right)\right]\end{aligned} \tag{196}$$

(it is due Ampere eq. (fourth Maxwell eq. above) projected toroidally)

Consequently, from(195) and (196) we obtain

$$\nabla\cdot\left(\frac{\alpha}{\varpi^2}\nabla\zeta\right)=\nabla\cdot\left(\frac{(\Omega_F-\omega)^2}{\alpha c^2}\nabla\zeta\right)-\frac{(\Omega_F-\omega)}{\alpha c^2}\frac{d\Omega_F}{d\zeta}|\nabla\zeta|^2-\frac{8\pi^2}{2\alpha(\varpi c)^2}\frac{IdI}{d\zeta} \tag{197}$$

Eqs.(195), (196) and (197) are the Grad Shafranov ones, notice that they can be seen as a 2-dimensional Poisson type equation (axisymmetry) of a complex variable $\zeta\equiv\psi+\overline{m}\chi\rightarrow\psi+i\chi$ with a source term $\propto\frac{IdI}{d\zeta}$

Here we derive the expression for the toroidal magnetic field in the case described in previous sections. Starting from the equation of the current (Ampere), namely

$$\int_A (\nabla \times (\alpha \mathbb{B})) \cdot d\overline{s} = 4\pi \overbrace{\int_A \alpha \overline{j} \cdot d\overline{s}}^{\Sigma_1} - \overbrace{\int_A (\mathbb{E} \cdot \nabla \omega)\, \overline{m} \cdot d\overline{s}}^{\Sigma_2} - \int_A h_0 \mathbb{B} \cdot d\overline{s} + \overbrace{\int_A (h \times \mathbb{E}) \cdot d\overline{s}}^{\Sigma_3} \tag{198}$$

in the linearized case and taking into account that: i) $h_0 = const$,ii) $(h \times E) \propto \alpha_g \left(j_{ph} - \frac{(h \cdot j_{ph})\mathbb{B}}{(h \cdot \mathbb{B})} \right)$, iii) $\nabla_\mu h_\nu - \nabla_\nu h_\mu = 0$ (h_μ does not generate field directly because $\lambda = 0$) y iv) $\Sigma_2 \to \propto 2\pi \overline{m}\,(dr \times \overline{m}) = 0$ we have finally

$$\int_A (\nabla \times (\alpha B)) \cdot d\overline{s} = 4\pi \overbrace{\int_A \alpha \overline{j} \cdot d\overline{s}}^{\Sigma_1} - h_0 \int_A B \cdot d\overline{s} \tag{199}$$

Consequently we obtain

$$B_T = -\left(\frac{2I(\psi) + |h_0|\,\psi}{\alpha \widetilde{\omega}} \right) e_{\widehat{\phi}} \tag{200}$$

Notice that the h_0 enter into the induction equation as alpha term, also it produces the twist into the flux lines (due its pseudoscalar character) as explained in the last part of this chapter.

References

[1] David R. Finkelstein, Andrei A. Galiautdinov, James E. Baugh, *J. Math. Phys.* **42** (2001) 340-346.

[2] George F. R. Ellis, Henk van Elst, Jeff Murugan and Jean-Philippe Uzan, *Class. Quant. Grav.* **28** (2011) 225007.

[3] G. Ellis, *Gen. Rel. Grav.* **46** (2014) 1619.

[4] John Ellis, and Nick E. Mavromatos, *Phys. Rev.* **D88** (2013) no.8, 085029.

[5] Aurelien Barrau and Linda Linsefors, *JCAP* **12** (2014) 037; Carlos Barcelo, Raul Carballo-Rubio and Luis J. Garay, *Phys. Rev.* **D89**, 124019 (2014); Raul Carballo-Rubio, *Phys. Rev.* **D91**, 124071 (2015); Javad T. Firouzjaee, George F. R. Ellis *Phys. Rev.* D. 91.103002 (2015).

[6] K. Thorne and D. Macdonald, *Mon. Not. R. Astron. Soc.* 198, 339 (1982).

[7] D. Macdonald and K. Thorne, *Mon. Not. R. Astron. Soc.* 198, 345 (1982).

[8] D. Kharzeev, *Ann. Phys.* 325, 2015 (2010).

[9] F. Wilzcek, *PRL* **58**, 1799 (1987).

[10] H. Weyl, "*Space-Time-Matter*", Dover (1952).

[11] Yu Xin, 1996, "General Relativity on Spinor-Tensor Manifold", in: "*Quantum Gravity - Int.School on Cosmology & Gravitation", XIV Course.* Eds. P.G. Bergman, V. de.Sabbata & H.J. Treder, pp. 382-411, World Scientific.

[12] D.J. Cirilo-Lombardo, *Int. J. Theor. Phys.* **49**, 1288, (2010).

[13] D.J. Cirilo-Lombardo, *Int. J. Theor. Phys.* **50**, 1699 (2011).

[14] D.J. Cirilo-Lombardo, *Int. J. Theor. Phys.* **50**, 3621 (2011).

[15] D.J. Cirilo-Lombardo, *J. Math. Phys.* **48**, 032301, (2007); Class.Quant.Grav. **22** , 4987 (2005).

[16] D.J. Cirilo-Lombardo, *Astropart. Phys.* **50-52**, 51 (2013).

[17] D.J. Cirilo-Lombardo, *Int. J. Theor. Phys.* 54 (2015) no.10, 3713-3727.

[18] D.J. Cirilo-Lombardo / *Journal of High Energy Astrophysics* 16 (2017) 1–14

[19] Brett McInnes, *Class. Quant. Grav.* 1 (1984) 105-113.

[20] Freydoon Mansouri, *Phys. Rev.* D**13** (1976) 3192.

[21] P.M.S. Blackett, *Nature* **159** (1947), 658.

[22] H. Helmholtz, "Über Integrale der Hydrodynamischen Gleichungen,Welche den Wirbelbewegungen Entsprechen [About integrals of the hydrodynamic equations, which correspond to the eddy movements]," *J. fürdie reine und angewandte Mathematik*, vol. 1858, no. 55, pp. 25-55, Jan. 1858.

[23] H. Helmholtz, "On Integrals of the Hydrodynamical Equations, which Express Vortex-Motion," *Philosophical Magazine and J. Science*, vol. 33, no. 226, pp. 485-512, 1867.

[24] K.-H. Rädler and M. Rheinhardt, *Geophysical and Astrophysical Fluid Dynamics*, Vol. 101, No. 2, April 2007, 117–154; Diego Julio Cirilo-Lombardo, *Int. J. Geom. Meth. Mod. Phys.* 14 (2017) no.07, 1750108.

[25] M.A. Markov, *Annals Phys.* **155** (1984) 333-357.

[26] David Alvarez-Castillo, Diego Julio Cirilo-Lombardo and Jilberto Zamora-Saa, Solar neutrinos, helicity effects and new affine gravity with torsion II, *JHEAp* **13-14** (2017) 10-16.

[27] David Alvarez-Castillo, Diego Julio Cirilo-Lombardo and Jilberto Zamora-Saa, work in progress.

[28] Albert Einstein, *The Meaning of Relativity: Including the Relativistic Theory of the Non-Symmetric Field,* (PUP, Princeton Science Library, 2014).

[29] Edward Kolb and Michael Turner: The Early Universe, *Frontiers in Physics* (Book 69), Westview Press (February 21, 1994).

[30] M. Joyce, M. Shaposhnikov, *Phys. Rev. Lett.*, **79**, 1193 (1997).

[31] Diego Julio Cirilo-Lombardo, *Class. Quant. Grav.* **22** (2005) 4987-5004.

[32] Green, Schwartz and Witten: Superstring Theory I and II, *Cambridge Monogr. Math. Phys.*, (Cambridge University Press, 1987).

[33] B. Carter, *Commun. Math. Phys.* **10**, 280 (1968).

[34] K. Yano, *Ann. Math.* **55**, 328 (1952).

[35] V.I. Ogievetsky and I.V. Polubarinov, "Notoph and photon", preprint JINR P-2330 (1965) (unpublished); V.I. Ogievetsky and I.V. Polubarinov, *Soviet J. Nucl. Phys.* **4**, 156 (1967); M. Kalb and P. Ramond, *Phys. Rev.* D **9**, 2273 (1974).

[36] K.A. Bronnikov and A.M. Galiakhmetov, *Grav. Cosmol.*, **21** (4) 283-288 (2015).

[37] Diego Julio Cirilo-Lombardo, work in preparation.

[38] Pankaj Jain, Gopal Kashyap, and Subhadip Mitra, *Int. J. Mod. Phys.* **A** 30, 1550171 (2015).

[39] H.E.S. Velten, R.F. vom Marttens and W. Zimdahl, *Eur. Phys. J.* **C**74 (2014) 11, 3160.

[40] S.W. Hawking and G.F.R. Ellis, *The large scale structure of spacetime*, (Cambridge University Press, CUP, England, 1973).

[41] Josef Kluson, *Phys. Rev.* D 91, 064058 (2015).

[42] V. D. Shafranov, *Sov. Phys. JETP* 6, 545 (1958).;

[43] H. Grad and H. Rubin, in *Proceedings of the Second United Nations Conference on the Peaceful Uses of Atomic Energy* (United Nations, Geneva, 1958), Vol. 21, p. 190.

[44] L.S. Solov'ev, *Sov. Phys. JETP* 26, 400 (1968);

[45] L.E. Zakharov and V. D. Shafranov, *Reviews of Plasma Physics* (Consultants Bureau, New York, 1986), Vol. 11, p. 153.

[46] Goldreich, P., and Julian, W. H. 1969, *ApJ*, 157, 869.

[47] R.V.E. Lovelace, C. Mehanian, C.M., Mobarry, M.E. Sulkanen, 1986, *ApJS*, 62, 1.

[48] Macdonald, D.A., & Thorne, K.S. 1982, *MNRAS*, 198, 345.

[49] Okamoto, I. 1975, *MNRAS*, 173, 357.

[50] D.J. Cirilo-Lombardo, *Phys. Part. Nucl. Lett.* 14 (2017) no.6, 799-810.

[51] D.J. Cirilo-Lombardo, *JHEAp* 16 (2017) 1-14.

[52] D.J. Cirilo-Lombardo, *Int. J. Geom. Meth. Mod. Phys.* 14 (2017) no.07, 1750108.

[53] David Alvarez-Castillo, D. J. Cirilo-Lombardo, Jilberto Zamora-Saa, *JHEAp* 13-14 (2017) 10-16.

[54] N. Yokoi, Cross helicity and related dynamo, *Geophysical & Astrophysical Fluid Dynamics*, 107:1-2,(2013) 114-184.

[55] D.J. Cirilo-Lombardo, *Astropart. Phys.* 50-52 (2013) 51-56.

[56] Maxim Lyutikov, V. I. Pariev and Roger Blandford, *Astrophys. J.* 597 (2003) 998-1009.

[57] Andrei M. Beloborodov and Christopher Thompson 2007 *ApJ* 657 967.

[58] Aly, J.J. 1984, ApJ, 283, 349; 1991, *ApJ*, 375, L61.

[59] D.J. Cirilo-Lombardo, *J. Phys. A: Math. and Theor.* **45** (2012) 244026.

[60] Vladimir I. Arnold and Boris A. Khesin, Topological Methods in Hydrodynamics, *Applied Mathematical Sciences*, Springer Verlag-New York, 1995.

[61] D.J. Cirilo-Lombardo, *IJGMMP* (online first) doi.org/10.1142/S0219887819500130, arXiv:1812.04481.

[62] D.J. Cirilo-Lombardo and F.O. Minotti, work in progress.

[63] Yasufumi Kojima, *Monthly Notices of the Royal Astronomical Society*, Volume 468, Issue 2, (2017), 2011–2016.

[64] C. Thompson, M. Lyutikov, S.R. Kulkarni, *ApJ*, 574, 332, (2002).

[65] Friedrich W. Hehl, Paul von der Heyde, G. David Kerlick, and James M. Nester, *Rev. Mod. Phys.* 48, 393, (1976).

[66] K. Shibata and T. Magara, *Living Reviews in Solar Physics*, Volume 8, Issue 1, article id. 6, 99 pp.

[67] F. Pacini, *Nature* 219, pages 145–146 (1968).

[68] L Del Zanna and N Bucciantini, *MNRAS* 479, Issue 1, 657 (2018).

[69] D.J. Cirilo-Lombardo work in progress.

[70] Diego F. Torres, S. Capozziello, G. Lambiase, *Phys. Rev.* D62 (2000) 104012.

[71] Diego F. Torres, S. Capozziello, G. Lambiase, *Class. Quant. Grav.* 17 (2000) 3171-3182.

[72] Salvatore Capozziello, Carlo Alberto Mantica and Luca Guido Molinari, *IJGMMP*, Vol. 16, No. 01, 1950008 (2019).

[73] Artyom V. Astashenok, Karim Mosani, Sergey D. Odintsov and Gauranga C. Samanta, online ready: https://doi.org/10.1142/S021988781950035X.

[74] D.J. Cirilo-Lombardo and F.O. Minotti, *IJGMMP*, Vol. 16, No. 04, 1950064 (2019).

[75] Sraibman, L., Minotti, F. Large-scale Model of the Axisymmetric Dynamo with Feedback Effects. *Sol Phys* 294, 14 (2019).

In: Future Relativity, Gravitation, Cosmology
Editors: V. V. Dvoeglazov et al.
ISBN: 979-8-88697-455-3

Chapter 7

NEGATIVE MASSES: A HIGHWAY FOR INTERSTELLAR TRAVELS THROUGH DARK MATTER?

N. Debergh*, J.-P. Petit† and G. D'Agostini‡
Department of Agronomy, Haute Ecole Charlemagne,
Huy, Belgium

Abstract

We take a fresh look on negative masses both in the quantum mechanical as well as cosmological contexts. We show that the concept of negative masses naturally takes place in these two fields provided that slight modifications are made to the traditional theories (i.e., to consider a unitary temporal reversal operator in quantum mechanics and to opt for different laws of attraction in cosmology).

1. Introduction

Until recently, physicists considered states of negative energies as an insane concept. Indeed, the beginning of the twentieth century saw the emergence of two major theories: quantum mechanics that describes the behaviour of elementary particles and general relativity, the paradigm for cosmology. Soon, physicists tried to unify these two theories. Thus, in 1927, Oskar Klein and Walter Gordon, independently, proposed a relativistic equation to describe the electron. Unfortunately, this first trial turned out to have several disadvantages including the absence of spin and the presence of negative energy solutions. A year later, Paul Dirac proposed [1] another equation that overcame the first defect of the Klein-Gordon equation, but states of negative energy remained.

Dirac therefore conjectured at first that Universe consists of both positive and negative energies. This hypothesis was very badly perceived in the scientific community and forced Dirac to indicate, in 1931, that this negative energy actually represented a charge opposed

*Corresponding Author's Email: nathalie.debergh@hech.be
†Corresponding Author's Email: jean-pierre.petit@manaty.net
‡Corresponding Author's Email: gilles.dagostini@manaty.net

to that of the electron charge. Among the famous opponents [2] of these negative energies were Niels Bohr, who said that Dirac's theory is only good at capturing elephants in Africa or Wolfgang Pauli claiming that *any theory proposed by a theoretical physicist should be applied to him.*

It will be said that these were the ancient times but, even much more recently, these negative energies still were a taboo. Take, for instance, Steven Weinberg. In [3], one can read parts of sentences such as *The other two solutions have* $E = -\sqrt{\vec{p}^2c^2 + m^2c^4}$ *and no obvious physical interpretation.* (p. 11) or *the again disastrous conclusion that there is another state of energy* $-E$ (p. 76).

So, let's take a look on the main reason why these energies have been so excluded. The Klein-Gordon equation, as already mentioned, had some unpleasant consequences. Among them, one found a density probability equal to $\frac{E}{m}$ which turns out to be negative if the energies are allowed to take negative values while masses stay positive, as usual.

At that time, it was indeed no way to think of masses taking negative values ! Nearly three decades later, there had been an attempt initiated by Hermann Bondi [4] and revisited by William Bonnor [5] in the context of general relativity. But it soon turned out that these negative masses implied an unmanageable phenomenon called the Runaway effect. What does it stand for ? As well known, general relativity assumes that Universe is a manifold equipped with a metric $g_{\mu\nu}$ according to the Einstein equation

$$R_{\mu\nu} - \frac{1}{2}Rg_{\mu\nu} = \chi T_{\mu\nu}$$

The matter energy tensor $T_{\mu\nu}$ is the source of the gravitational field. The Runaway effect comes from the fact that it is arbitrarily assumed that particles of positive or negative masses immersed in the same gravitational field, react in the same way. Indeed, as there is only one metric, the effect on each particle from that field created by a positive or negative mass ($T \to -T$) is the following:

- Positive masses attract both other positive masses and negative masses
- Negative masses repel both other negative masses and positive masses

This implies that the particle of negative mass pursues, in a uniformly accelerated movement with a conserved kinetic energy, the particle of positive mass.

Does it mean that we have to definitively exclude negative masses? Certainly not. Like all new concepts, it takes time to make them physically meaningful. To do so, two of us have shown [6] that the assumption of positive and negative masses reacting in the same way must be forgotten and replaced by a bimetric model giving rise to two field equations (generalizing the Einstein one) with the corresponding new interaction rules:

- Positive masses attract positive masses and repel negative masses
- Negative masses attract negative masses and repel positive masses

There is then no more Runaway effect and negative masses can be considered. This bimetric model is called the Janus model and it will be shortly introduced with its latest developments in Section 3. Its perfect agreement with experimental observations [7] restores some kind of legitimacy to negative masses. They are actually the material of dark matter. Janus also brings back to light previous theoretical works ([8]- [10]) devoted to the subject.

We thus propose to take a new look on these negative masses, associated with negative energies, in the quantum mechanical context (Section 2) as well as in cosmology (Section 3). We conclude in Section 4.

2. Negative Masses in Quantum Mechanics

Since Eugene Wigner [11], we know that a symmetry operator of a Hamiltonian is necessarily

- either a unitary one:

$$< U\psi|U\phi >=< \psi|\phi >, U(a|\psi > +b|\phi >) = a\,U|\psi > +b\,U|\phi > ,$$
$$a, b \in C$$

  for all states $|\psi >$ and $|\phi >$ of the Hilbert space H

- or an anti-unitary one:

$$< A\psi|A\phi >=< \psi|\phi >^*,\ A(a|\psi > +b|\phi >) = a^*\ A|\psi > +b^*\ A|\phi >$$

  Here, the symbol $*$ denotes the complex conjugation.

If continuous transformations such as rotations are perforce related to unitary operators in order to obey the Lie group requirements, the choice is left free in what concerns the discrete symmetries. However, it is the custom to consider

- the parity operator P to be unitary:

$$\vec{x} \to -\vec{x},\ \vec{p}\,(= -i\bar{h}\vec{\nabla}, i \to i) \to -\vec{p},\ t \to t$$

- the time-reversal operator T to be anti-unitary:

$$\vec{x} \to \vec{x},\ \vec{p}\,(= -i\bar{h}\vec{\nabla}, i \to -i) \to -\vec{p},\ t \to -t$$

Where does this choice come from? It follows from the invariance of the fundamental relation of quantum mechanics

$$[x_j, p_k] = i\bar{h}\delta_{jk}$$

under each of these discrete symmetries as well as the arbitrary option to keep positive energies only through

$$E \leftrightarrow i\bar{h}\frac{\partial}{\partial t}$$

We claim that another choice [12] is also relevant if we allow energies to be negative. This choice corresponds to

- the parity operator P stays unitary:

$$\vec{x} \to -\vec{x},\ \vec{p}\,(= -i\bar{h}\vec{\nabla}, i \to i) \to -\vec{p},\ t \to t$$

- the time-reversal operator T can be unitary, too:

$$\vec{x} \to \vec{x},\ \vec{p}\,(= -i\bar{h}\vec{\nabla}, i \to i) \to \vec{p},\ t \to -t$$

Note that this second choice is the most natural one for mathematicians. For example, in [13], one can read *l'inversion temporelle change le signe de l'énergie* [1] , implying the unitarity of T.

This comes from the fact that physicists often restrict themselves to one of the four components of the Lorentz or Poincaré groups.

More in details, the Lorentz group [14] is the set of matrices Λ such that

$$\Lambda^T\, g\, \Lambda = g,\ g = diag(1, -1, -1, -1)$$

(Λ^T meaning the transpose of Λ) and any of these matrices can be such that

- its determinant is equal to $+1$ and its first element is $\geq 1 \to \{\Lambda\} \equiv L_+^\uparrow$
- its determinant is equal to $+1$ and its first element is $\leq -1 \to \{\Lambda\} \equiv L_+^\downarrow$
- its determinant is equal to -1 and its first element is $\geq 1 \to \{\Lambda\} \equiv L_-^\uparrow$
- its determinant is equal to -1 and its first element is $\leq -1 \to \{\Lambda\} \equiv L_-^\downarrow$

Only a few combinations of these components do form a group and only one of them form a Lie group:

$$L_+^\uparrow \cup L_+^\downarrow \Rightarrow SO(1,3)$$

Usually, physicists restrict themselves to $L_+^\uparrow$. They are not used to consider the second component because this one is related to the first one by applying to it the time-reversal operator as well as the parity operator

$$L_+^\downarrow = PTL_+^\uparrow$$

with

$$P = diag(1, -1, -1, -1)\ ,\ T = diag(-1, 1, 1, 1)\ ,\ PT = -I$$

But this last relationship clearly shows the unitarity of T and, by implication, the existence of negative energies, excluded, let's recall it, by most of physicists...

So, the result of T being unitary has, actually, always been subtended by group theory. It is the possibility associated to the second component of $SO(1,3)$. There is no mathematical reason to let it down.

Adding to these transformations the four translations of the Minkowski space-time, we are led to two components of the Poincaré group, the one, $P_+^\uparrow$, usually considered by physicists and another one, $P_+^\downarrow = PTP_+^\uparrow$, dealing with a unitary time-reversal operator.

[1] time inversion changes the sign of the energy

Limiting ourselves, in a traditional way, to isotropy groups, we can choose a particular time-like momentum 4-vector, say

$$\begin{pmatrix} m \\ 0 \\ 0 \\ 0 \end{pmatrix}, \; m \geq 0$$

This means that this 4-vector belongs to one of the two parts of the light cone. If we apply one of the transformations of $P_+^\uparrow$ on it, it will stay in the same part of the light cone i.e., the image 4-vector is still characterized by a positive first component. This is due to the fact that all matrices of $P_+^\uparrow$ have a first element ≥ 1. Instead, if we apply one of the transformations of $P_+^\downarrow$ on it, it will give a 4-vector on the other part of the light cone or, in other words, a momentum 4-vector with a negative mass. This is due to the fact that, with T being unitary, we have here $PT = -I$.

So, the two components $P_+^\uparrow$ and $P_+^\downarrow$ give rise to two kinds of particles:

- $P_+^\uparrow \rightarrow m \geq 0, \; E \geq 0$
- $P_+^\downarrow = PTP_+^\uparrow = -P_+^\uparrow \rightarrow m \leq 0, \; E \leq 0$

In these relations, the signs of the energies come from the Einstein relation $E = mc^2$.

These two possibilities perfectly agree with the eigenvalues m^2 of the Poincaré Casimir whose unique constraint is to be real, which opens the door to positive as well as negative but also to imaginary masses.

Let us briefly mention the impact of this new option within the Dirac framework. For more details, see [12]. As well known, the free Dirac equation writes [1]

$$(\gamma^\mu p_\mu - m)\psi(t, \vec{x}) = 0$$

The usual unitary parity operator P is realized through

$$\psi(t, \vec{x}) \rightarrow \gamma^0 \psi(t, -\vec{x})$$

The traditional anti-unitary time-reversal operator

$$\psi(t, \vec{x}) \rightarrow i\gamma^0\gamma^1\gamma^3\psi(-t, -\vec{x})^*$$

is now replaced by a unitary one

$$\psi(t, \vec{x}) \rightarrow i\gamma^1\gamma^2\gamma^3\psi(-t, \vec{x})$$

This implies that

$$PT\psi(t, \vec{x}) = i\gamma^0\gamma^1\gamma^2\gamma^3\psi(-t, -\vec{x}) \equiv \gamma^5\psi(-t, -\vec{x})$$

Thus, the PT-symmetry within the Dirac formalism is not just only related to opposite space-time coordinates, it is also subtended by the γ^5 matrix. We are used to interpret this matrix as the one realizing the chirality operator. But it is not just so. Indeed, the γ^5 matrix anticommutes with each of the γ^μ's. Acting on the Dirac equation, this implies that either the sign of each of the p_μ's has to change (which is expected in a unitary PT symmetry) or, in an equivalent way, the mass has to become negative. [2]

As a conclusion of this first part, we claim that group theory as well as quantum mechanical requirements are actually in agreement with the two possibilities

- $(m \geq 0, E \geq 0)$
- $(m \leq 0, E \leq 0)$

despite of the fact that only one of them is familiar to physicists. The second option is due to a unitary time-reversal operator.

Now comes the physical interpretation of the $(m \leq 0, E \leq 0)$ possibility. This is the subject of the next Section.

3. Negative Masses in Cosmology: The Janus Model

As already mentioned in the Introduction, attempts [4], [5] to introduce negative masses in general relativity ran up against the Runaway paradox. It follows immediately from the fact that, by limiting oneself to a description of the Universe based on Einstein's equation, we have, as a solution, only one system of geodesics, resulting from a single metric. As the particles, both of positive and negative masses, use the same geodesics, this generates the Runaway phenomenon, hardly compatible with an acceptable model for a physicist.

To work around this problem, it is necessary to consider that positive and negative masses particles behave differently when subjected to a gravitational field. Realize this requires a drastic change in the geometric paradigm, the Universe being seen as a 4-dimensional manifold with two metrics $g^{(+)}_{\mu\nu}$ and $g^{(-)}_{\mu\nu}$. These metrics generate two Ricci tensors $R^{(+)}_{\mu\nu}$, $R^{(-)}_{\mu\nu}$ as well as two Ricci scalars $R^{(+)}$, $R^{(-)}$, from which one can build an action leading to two coupled field equations [16]

$$R^{(+)}_{\mu\nu} - \frac{1}{2} R^{(+)}\, g^{(+)}_{\mu\nu} = \chi (T^{(+)}_{\mu\nu} + \phi \sqrt{\frac{g^{(-)}}{g^{(+)}}} T^{(-)}_{\mu\nu})$$

$$R^{(-)}_{\mu\nu} - \frac{1}{2} R^{(-)}\, g^{(-)}_{\mu\nu} = -\chi (T^{(-)}_{\mu\nu} + \phi \sqrt{\frac{g^{(+)}}{g^{(-)}}} T^{(+)}_{\mu\nu})$$

with

$$\phi = diag(1, -1, -1, -1)$$

This system provides solutions in agreement with local data, usually attributed to general relativity. In particular, it gives rise to an unsteady solution [17] associated with cosmic expansion, this one being related to a content where negative masses are strongly in the majority. This explains the effect that one usually associate with dark energy. Besides, the model, in its Newtonian approximation, is in agreement with phenomena hitherto attributed

[2]Such developments can also be found in the quantum field theory context [15] where it was quoted that the γ^5 chiral transformation is associated to the fact that the sign of the mass term in the Lagrangian is reversed.

to dark matter [7]. In fact, negative masses completely replace the couple "dark matter-dark energy" by ensuring the two roles.

The interpretation of CMB fluctuations leads to the conclusion that, in the negative masses sector, distances are a hundred times shorter while the speed of light is ten times higher [18]. If, one day, it turns out to be technologically feasible to make the mass of a shuttle and that of its occupants negative, we would thus have a way to bypass the "light barrier". This project is schematically sketched in [6].

Conclusion

It has long been believed that both negative mass and negative energy were science fiction. Physics has been built, until recently, to avoid these signs. However, group theory clearly shows that a relativistic equation can be associated with a Poincaré group which is no longer restricted and, therefore, which predicts the existence of negative mass (and energy) particles. It is simply a question of involving a unitary time-reversal operator instead of an anti-unitary one. Such an operator then not only gives meaning to negative masses but also rehabilitates the negative energies. The key point is that these two formerly excluded concepts must be considered simultaneously to give rise to coherent physics.

These particles of negative mass and energy form conglomerates. Universe would therefore consist of positive mass (and energy) components such as planets, galaxies, ... and of these negative conglomerates, in a much more primitive state. The negative sector is in agreement with experimental data and phenomena generally attributed to dark matter and dark energy, jointly.

Both constituents, positive and negative, have different metrics. Thus, in the negative sector, distances are divided by one hundred while speed of light is multiplied by ten. As a result, traveling times are divided by a thousand. If, from a technological point of view, one manage to make negative a positive mass, interstellar travels will no longer be a utopia.

African elephants have not finished being fascinated...

References

[1] P. A.M. Dirac, *Proc. Roy. Soc.* **117** (1928) 610.

[2] G. Gamow, *Thirty years that shook physics*, Dover, New-York (1985).

[3] Steven Weinberg, *The quantum theory of fields*, vol. I, Cambridge University Press (1995).

[4] H. Bondi, *Rev. Mod. Phys.* **29** (1957) 423.

[5] W. B. Bonnor, *Gen. Rel. and Gravit.* **21** (1989) 1143.

[6] J.-P. Petit and G. D'Agostini, *Mod. Phys. Lett.* **A29** (2014) 1450182.

[7] J.-P. Petit and G. D'Agostini, *Astrophysics and Space Science* **363** (2018) 139.

[8] M.A. Markov, *JETP* **7** (1937) 603.

[9] M.A. Markov, *Nucl. Phys.* **55** (1964) 130.

[10] V.V. Dvoeglazov, Z. Naturforsch. **A71** (2016) 345.

[11] E.P. Wigner, *Gruppentheorie und ihre Anwendung auf die Quanten mechanik der Atomspektren Braunschweig* [*Group theory and its application to the quantum mechanics of atomic spectra Braunschweig*] (Friedrich Vieweg und Sohn) (1931) pages 251-254.

[12] N. Debergh, J.-P. Petit and G. D'Agostini, *J. Phys. Comm.* **2** (2018) 115012.

[13] J.-M. Souriau, *Structure des systèmes dynamiques* [*Structure of dynamical systems*], (Paris, Dunod) (1970) page 199.

[14] K.N. Srinivasa Rao (1988), *The Rotation and Lorentz Groups and Their Representations for Physicists*, John Wiley and Sons, page 213.

[15] V.V. Dvoeglazov, *Acta Physica Slovaca* **50** (2000) 629.

[16] J.-P. Petit, G. D'Agostini and N. Debergh, Physical and mathematical consistency of the Janus cosmological model (JCM) ; to be published in *Progress in Physics* **15** (2019).

[17] J.-P. Petit and G. D'Agostini, *Astrophys. and Space Sci.* **A29** (2014) 145.

[18] J.-P. Petit, *Progress in Physics* **14** (2018) 226.

In: Future Relativity, Gravitation, Cosmology
Editors: V. V. Dvoeglazov et al.
ISBN: 979-8-88697-455-3

Chapter 8

NEGATIVE ENERGIES/MASSES IN THE RELATIVITY THEORY

Valeriy V. Dvoeglazov*
Universidad de Zacatecas
Zacatecas, México

Abstract

It is easy to check that both algebraic equation $Det(\hat{p} - m) = 0$ and $Det(\hat{p} + m) = 0$ for $u-$ and $v-$ 4-spinors have solutions with $p_0 = \pm E_p = \pm\sqrt{\mathbf{p}^2 + m^2}$. The same is true for higher-spin equations. Meanwhile, every book considers the equality $p_0 = E_p$ for both $u-$ and $v-$ spinors of the $(1/2, 0) \oplus (0, 1/2)$ representation only, thus applying the Dirac-Feynman-Stueckelberg procedure for elimination of the negative-energy solutions. The recent Ziino works (and, independently, the articles of several others) show that the Fock space can be doubled. We re-consider this possibility on the quantum field level for both $s = 1/2$ and higher spin particles.

1. Introduction

The recent problems of superluminal neutrinos, e. g., Ref. [1], negative-mass squared neutrinos, e. g. [2], various schemes of oscillations including sterile neutrinos, e. g. [3], require much attention. The problem of the lepton mass splitting (e, μ, τ) has long history [4]. This suggests that something missed in the foundations of relativistic quantum theories. Modifications seem to be necessary in the Dirac sea concept, and in the even more sophisticated Stueckelberg concept of the backward propagation in time. The Dirac sea concept is intrinsically related to the Pauli principle. However, the Pauli principle is intrinsically connected with the Fermi statistics and the anticommutation relations of fermions. Recently, the concept of the *bi-orthonormality* has been proposed; the (anti) commutation relations and statistics are assumed to be different for *neutral* particles [5]. One can speculate that they go in the negative-energy sea, but due to some reasons (interaction?) they do not live there (from our viewpoint), but return back (been expelled), thus showing us the new kind

*Corresponding Author's Email: valeri@fisica.uaz.edu.mx

of oscillations on the Planck scale $\omega \sim E/\hbar$, Ref. [6]. Perhaps, some of the neutrinos remain *sterile* even in our world.

We propose the relevant modifications in the basics of the relativistic quantum theory for *neutral* particles below. However much work is still needed.

2. The General Framework and Connections with Previous Models

The Dirac equation is:

$$[i\gamma^\mu\partial_\mu - m]\Psi(x) = 0\,. \tag{1}$$

At least, 3 methods of its derivation exist [7, 8, 9]:

- the Dirac one (the Hamiltonian should be linear in $\partial/\partial x^i$, and be compatible with $E_p^2 - \mathbf{p}^2c^2 = m^2c^4$);
- the Sakurai one (based on the equation $(E_p - \sigma\cdot\mathbf{p})(E_p + \sigma\cdot\mathbf{p})\phi = m^2\phi$);
- the Ryder one (the relation between 2-spinors at rest is $\phi_R(\mathbf{0}) = \pm\phi_L(\mathbf{0})$, and boosts).

The γ^μ are the Clifford algebra matrices

$$\gamma^\mu\gamma^\nu + \gamma^\nu\gamma^\mu = 2g^{\mu\nu}\,. \tag{2}$$

Usually, everybody uses the following definition of the field operator [10] in the pseudo-Euclidean metrics:

$$\Psi(x) = \frac{1}{(2\pi)^3}\sum_h \int \frac{d^3\mathbf{p}}{2E_p}[u_h(\mathbf{p})a_h(\mathbf{p})e^{-ip\cdot x} + v_h(\mathbf{p})b_h^\dagger(\mathbf{p})]e^{+ip\cdot x}]\,, \tag{3}$$

as given *ab initio*. After actions of the Dirac operator at $\exp(\mp ip_\mu x^\mu)$ the 4-spinors ($u-$ and $v-$) satisfy the momentum-space equations: $(\hat{p} - m)u_h(p) = 0$ and $(\hat{p} + m)v_h(p) = 0$, respectively; the h is the polarization index. It is easy to prove from the characteristic equations $Det(\hat{p} \mp m) = (p_0^2 - \mathbf{p}^2 - m^2)^2 = 0$ that the solutions should satisfy the energy-momentum relation $p_0 = \pm E_p = \pm\sqrt{\mathbf{p}^2 + m^2}$.

The general scheme of construction of the field operator has been presented in [11]. In the case of the $(1/2, 0) \oplus (0, 1/2)$ representation we have:

$$\begin{aligned}
\Psi(x) &= \frac{1}{(2\pi)^3}\int d^4p\,\delta(p^2 - m^2)e^{-ip\cdot x}\Psi(p) = \\
&= \frac{1}{(2\pi)^3}\sum_h \int d^4p\,\delta(p_0^2 - E_p^2)e^{-ip\cdot x}u_h(p_0,\mathbf{p})a_h(p_0,\mathbf{p}) = \\
&= \frac{1}{(2\pi)^3}\int \frac{d^4p}{2E_p}[\delta(p_0 - E_p) + \delta(p_0 + E_p)][\theta(p_0) + \theta(-p_0)]e^{-ip\cdot x} \\
&\sum_h u_h(p)a_h(p) = \frac{1}{(2\pi)^3}\sum_h \int \frac{d^4p}{2E_p}[\delta(p_0 - E_p) + \delta(p_0 + E_p)]
\end{aligned} \tag{4}$$

$$\left[\theta(p_0)u_h(p)a_h(p)e^{-ip\cdot x} + \theta(p_0)u_h(-p)a_h(-p)e^{+ip\cdot x}\right]$$
$$= \frac{1}{(2\pi)^3}\sum_h \int \frac{d^3\mathbf{p}}{2E_p}\theta(p_0)\left[u_h(p)a_h(p)|_{p_0=E_p}e^{-i(E_pt-\mathbf{p}\cdot\mathbf{x})}+\right.$$
$$+ \quad \left. u_h(-p)a_h(-p)|_{p_0=E_p}e^{+i(E_pt-\mathbf{p}\cdot\mathbf{x})}\right]$$

During the calculations above we had to represent $1 = \theta(p_0) + \theta(-p_0)$ in order to get positive- and negative-frequency parts.[1] Moreover, during these calculations we did not yet assumed, which equation this field operator (namely, the $u-$ spinor) satisfies, with negative- or positive- mass?

In general we should transform $u_h(-p)$ to the $v(p)$. The procedure is the following one [13]. In the Dirac case we should assume the following relation in the field operator:

$$\sum_h v_h(p)b_h^\dagger(p) = \sum_h u_h(-p)a_h(-p)\,. \tag{5}$$

We know that [9]

$$\bar{u}_\mu(p)u_\lambda(p) = +m\delta_{\mu\lambda}\,, \tag{6}$$
$$\bar{u}_\mu(p)u_\lambda(-p) = 0\,, \tag{7}$$
$$\bar{v}_\mu(p)v_\lambda(p) = -m\delta_{\mu\lambda}\,, \tag{8}$$
$$\bar{v}_\mu(p)u_\lambda(p) = 0\,, \tag{9}$$

but we need $\Lambda_{\mu\lambda}(p) = \bar{v}_\mu(p)u_\lambda(-p)$. By direct calculations, we find

$$-mb_\mu^\dagger(p) = \sum_\lambda \Lambda_{\mu\lambda}(p)a_\lambda(-p)\,. \tag{10}$$

Hence, $\Lambda_{\mu\lambda} = -im(\boldsymbol{\sigma}\cdot\mathbf{n})_{\mu\lambda}$, $\mathbf{n} = \mathbf{p}/|\mathbf{p}|$, and

$$b_\mu^\dagger(p) = i\sum_\lambda(\boldsymbol{\sigma}\cdot\mathbf{n})_{\mu\lambda}a_\lambda(-p)\,. \tag{11}$$

Multiplying (5) by $\bar{u}_\mu(-p)$ we obtain

$$a_\mu(-p) = -i\sum_\lambda(\boldsymbol{\sigma}\cdot\mathbf{n})_{\mu\lambda}b_\lambda^\dagger(p)\,. \tag{12}$$

The equations are self-consistent. In the $(1,0)\oplus(0,1)$ representation the similar procedure leads to somewhat different situation:

$$a_\mu(p) = [1 - 2(\mathbf{S}\cdot\mathbf{n})^2]_{\mu\lambda}a_\lambda(-p)\,. \tag{13}$$

This signifies that in order to construct the Sankaranarayanan-Good field operator (which was used by Ahluwalia, Johnson and Goldman [Phys. Lett. B (1993)], it satisfies $[\gamma_{\mu\nu}\partial_\mu\partial_\nu - \frac{(i\partial/\partial t)}{E}m^2]\Psi(x) = 0$, we need additional postulates. For instance, one can try to construct the left- and the right-hand side of the field operator separately each other [12].

[1] See Ref. [12] for some discussion.

However, other ways of thinking are possible. First of all to mention, we have, in fact, $u_h(E_p, \mathbf{p})$ and $u_h(-E_p, \mathbf{p})$ originally, which may satisfy the equations:[2]

$$\left[E_p(\pm\gamma^0) - \boldsymbol{\gamma}\cdot\mathbf{p} - m\right] u_h(\pm E_p, \mathbf{p}) = 0\,. \tag{14}$$

Due to the properties $U^\dagger \gamma^0 U = -\gamma^0$, $U^\dagger \gamma^i U = +\gamma^i$ with the unitary matrix $U = \begin{pmatrix} 0 & -1 \\ 1 & 0 \end{pmatrix} = \gamma^0\gamma^5$ in the Weyl basis,[3] we have

$$\left[E_p\gamma^0 - \boldsymbol{\gamma}\cdot\mathbf{p} - m\right] U^\dagger u_h(-E_p, \mathbf{p}) = 0\,. \tag{15}$$

Thus, unless the unitary transformations do not change the physical content, we have that the negative-energy spinors $\gamma^5\gamma^0 u^-$ (see (15)) satisfy the accustomed "positive-energy" Dirac equation. Their explicite forms $\gamma^5\gamma^0 u^-$ are different from the textbook "positive-energy" Dirac spinors. They are the following ones (the linear combination of the positive-energy spinors):[4]

$$\tilde{u}(p) = \frac{N}{\sqrt{2m(-E_p + m)}} \begin{pmatrix} -p^+ + m \\ -p_r \\ p^- - m \\ -p_r \end{pmatrix}, \tag{16}$$

$$\tilde{\tilde{u}}(p) = \frac{N}{\sqrt{2m(-E_p + m)}} \begin{pmatrix} -p_l \\ -p^- + m \\ -p_l \\ p^+ - m \end{pmatrix}. \tag{17}$$

$E_p = \sqrt{\mathbf{p}^2 + m^2} > 0$, $p_0 = \pm E_p$, $p^\pm = E \pm p_z$, $p_{r,l} = p_x \pm i p_y$. Their normalization is to $(-2N^2)$. What about the $\tilde{v}(p) = \gamma^0 u^-$ transformed with the γ_0 matrix? Are they equal to $v_h(p) = \gamma^5 u_h(p)$? Our answer is '*no*'. Obviously, they also do not have well-known forms of the usual $v-$ spinors in the Weyl basis, differing by phase factor and in the sign at the mass term (!)

Next, one can prove that the matrix

$$P = e^{i\theta}\gamma^0 = e^{i\theta} \begin{pmatrix} 0 & 1_{2\times 2} \\ 1_{2\times 2} & 0 \end{pmatrix} \tag{18}$$

can be used in the parity operator as well as in the original Weyl basis. The parity-transformed function $\Psi'(t, -\mathbf{x}) = P\Psi(t, \mathbf{x})$ must satisfy

$$[i\gamma^\mu \partial'_\mu - m]\Psi'(t, -\mathbf{x}) = 0\,, \tag{19}$$

with $\partial'_\mu = (\partial/\partial t, -\nabla_i)$. This is possible when $P^{-1}\gamma^0 P = \gamma^0$ and $P^{-1}\gamma^i P = -\gamma^i$. The matrix (18) satisfies these requirements, as in the textbook case. However, if we would

[2] Remember that, as before, we can always make the substitution $\mathbf{p} \to -\mathbf{p}$ in any of the integrands of (4).

[3] The properties of the $U-$ matrix are opposite to those of $P^\dagger\gamma^0 P = +\gamma^0$, $P^\dagger\gamma^i P = -\gamma^i$ with the usual $P = \gamma^0$, thus giving $\left[-E_p\gamma^0 + \boldsymbol{\gamma}\cdot\mathbf{p} - m\right] P u_h(-E_p, \mathbf{p}) = -\left[\hat{p} + m\right]\tilde{v}_?(E_p, \mathbf{p}) = 0$. While, the relations of the spinors $v_h(E_p, \mathbf{p}) = \gamma_5 u_h(E_p, \mathbf{p})$ are well-known, it seems that the relations of the $v-$ spinors of the positive energy to $u-$ spinors of the negative energy are frequently forgotten, $\tilde{v}_?(E_p, \mathbf{p}) = \gamma^0 u_h(-E_p, \mathbf{p})$.

[4] We use tildes because we do not yet know their polarization properties.

take the phase factor to be zero we obtain that while $u_h(p)$ have the eigenvalue $+1$, but $(R = (\mathbf{x} \to -\mathbf{x}, \mathbf{p} \to -\mathbf{p}))$

$$PR\tilde{u}(p) = PR\gamma^5\gamma^0 u(-E_p, \mathbf{p}) = -\tilde{u}(p)\,, \tag{20}$$
$$PR\tilde{\tilde{u}}(p) = PR\gamma^5\gamma^0 u(-E_p, \mathbf{p}) = -\tilde{\tilde{u}}(p)\,. \tag{21}$$

Perhaps, one should choose the phase factor $\theta = \pi$. Thus, we again confirmed that the relative (particle-antiparticle) intrinsic parity has physical significance only.

Similar formulations have been presented in Refs. [14], and [15]. The group-theoretical basis for such doubling has been given in the papers by Gelfand, Tsetlin and Sokolik [16], who first presented the theory in the 2-dimensional representation of the inversion group in 1956 (later called as "the Bargmann-Wightman-Wigner-type quantum field theory" in 1993).

M. Markov wrote long ago *two* Dirac equations with the opposite signs at the mass term [14].

$$[i\gamma^\mu\partial_\mu - m]\,\Psi_1(x) = 0\,, \tag{22}$$
$$[i\gamma^\mu\partial_\mu + m]\,\Psi_2(x) = 0\,. \tag{23}$$

In fact, he studied all properties of this relativistic quantum model (while he did not know yet the quantum field theory in 1937). Next, he added and subtracted these equations. What did he obtain?

$$i\gamma^\mu\partial_\mu\varphi(x) - m\chi(x) = 0\,, \tag{24}$$
$$i\gamma^\mu\partial_\mu\chi(x) - m\varphi(x) = 0\,. \tag{25}$$

Thus, φ and χ solutions can be presented as some superpositions of the Dirac 4-spinors $u-$ and $v-$. These equations, of course, can be identified with the equations for the Majorana-like $\lambda-$ and $\rho-$, which we presented in Ref. [17].[5]

$$i\gamma^\mu\partial_\mu\lambda^S(x) - m\rho^A(x) = 0\,, \tag{26}$$
$$i\gamma^\mu\partial_\mu\rho^A(x) - m\lambda^S(x) = 0\,, \tag{27}$$
$$i\gamma^\mu\partial_\mu\lambda^A(x) + m\rho^S(x) = 0\,, \tag{28}$$
$$i\gamma^\mu\partial_\mu\rho^S(x) + m\lambda^A(x) = 0\,. \tag{29}$$

Neither of them can be regarded as the Dirac equation. However, they can be written in the 8-component form as follows:

$$[i\Gamma^\mu\partial_\mu - m]\,\Psi_{(+)}(x) = 0\,, \tag{30}$$
$$[i\Gamma^\mu\partial_\mu + m]\,\Psi_{(-)}(x) = 0\,, \tag{31}$$

with

$$\Psi_{(+)}(x) = \begin{pmatrix} \rho^A(x) \\ \lambda^S(x) \end{pmatrix}, \Psi_{(-)}(x) = \begin{pmatrix} \rho^S(x) \\ \lambda^A(x) \end{pmatrix}, \tag{32}$$
$$\text{and}\ \Gamma^\mu = \begin{pmatrix} 0 & \gamma^\mu \\ \gamma^\mu & 0 \end{pmatrix}. \tag{33}$$

[5]Of course, the signs at the mass terms depend on, how do we associate the positive- or negative- frequency solutions with λ and ρ.

It is easy to find the corresponding projection operators, and the Feynman-Stueckelberg propagator.

You may say that all this is just related to the spin-parity basis rotation (unitary transformations). However, in the previous papers I explained: the connection with the Dirac spinors has been found [17, 19].[6] For instance,

$$\begin{pmatrix} \lambda^S_\uparrow(\mathbf{p}) \\ \lambda^S_\downarrow(\mathbf{p}) \\ \lambda^A_\uparrow(\mathbf{p}) \\ \lambda^A_\downarrow(\mathbf{p}) \end{pmatrix} = \frac{1}{2}\begin{pmatrix} 1 & i & -1 & i \\ -i & 1 & -i & -1 \\ 1 & -i & -1 & -i \\ i & 1 & i & -1 \end{pmatrix}\begin{pmatrix} u_{+1/2}(\mathbf{p}) \\ u_{-1/2}(\mathbf{p}) \\ v_{+1/2}(\mathbf{p}) \\ v_{-1/2}(\mathbf{p}) \end{pmatrix}, \tag{34}$$

provided that the 4-spinors have the same physical dimension. Thus, we can see that the two 4-spinor systems are connected by the unitary transformations, and this represents itself the rotation of the spin-parity basis. However, it is usually assumed that the $\lambda-$ and $\rho-$ spinors describe the neutral particles, meanwhile $u-$ and $v-$ spinors describe the charged particles. Kirchbach [19] found the amplitudes for neutrinoless double beta decay ($00\nu\beta$) in this scheme. It is obvious from (34) that there are some additional terms comparing with the standard formulation.

One can also re-write the above equations into the two-component forms. Thus, one obtains the Feynman-Gell-Mann [18] equations. As Markov wrote himself, he was expecting "new physics" from these equations.

Barut and Ziino [15] proposed yet another model. They considered γ^5 operator as the operator of the charge conjugation. Thus, the charge-conjugated Dirac equation has the different sign comparing with the ordinary formulation:

$$[i\gamma^\mu\partial_\mu + m]\Psi^c_{BZ} = 0\,, \tag{35}$$

and the so-defined charge conjugation applies to the whole system, fermion+electromagnetic field, $e \to -e$ in the covariant derivative. The superpositions of the Ψ_{BZ} and Ψ^c_{BZ} also give us the "doubled Dirac equation", as the equations for $\lambda-$ and $\rho-$ spinors. The concept of the doubling of the Fock space has been developed in the Ziino works (cf. [16, 20]) in the framework of the quantum field theory. In their case the charge conjugate states are simultaneously the eigenstates of the chirality. Next, it is interesting to note that for the Majorana-like field operators we have

$$\left[\nu^{^{ML}}(x^\mu) + \mathcal{C}\nu^{^{ML\,\dagger}}(x^\mu)\right]/2 = \int \frac{d^3\mathbf{p}}{(2\pi)^3}\frac{1}{2E_p} \tag{36}$$

$$\sum_\eta \left[\begin{pmatrix} i\Theta\phi^{*\,\eta}_L(p^\mu) \\ 0 \end{pmatrix} a_\eta(p^\mu)e^{-ip\cdot x} + \begin{pmatrix} 0 \\ \phi^\eta_L(p^\mu) \end{pmatrix} a^\dagger_\eta(p^\mu)e^{ip\cdot x}\right],$$

$$\left[\nu^{^{ML}}(x^\mu) - \mathcal{C}\nu^{^{ML\,\dagger}}(x^\mu)\right]/2 = \int \frac{d^3\mathbf{p}}{(2\pi)^3}\frac{1}{2E_p} \tag{37}$$

$$\sum_\eta \left[\begin{pmatrix} 0 \\ \phi^\eta_L(p^\mu) \end{pmatrix} a_\eta(p^\mu)e^{-ip\cdot x} + \begin{pmatrix} -i\Theta\phi^{*\,\eta}_L(p^\mu) \\ 0 \end{pmatrix} a^\dagger_\eta(p^\mu)e^{ip\cdot x}\right],$$

which, thus, naturally lead to the Ziino-Barut scheme of massive chiral fields, Ref. [15].

---

[6] I also acknowledge personal communications from D. V. Ahluwalia on these matters.

Finally, I would like to mention that, in general, in the Weyl basis the $\gamma-$ matrices are *not* Hermitian, $\gamma^{\mu^\dagger} = \gamma^0\gamma^\mu\gamma^0$. So, $\gamma^{i^\dagger} = -\gamma^i$, $i = 1, 2, 3$, the pseudo-Hermitian matrix. The energy-momentum operator $i\partial_\mu$ is obviously Hermitian. So, the question, if the eigenvalues of the Dirac operator (the mass, in fact) would be always real? The question of the complete system of the eigenvectors of the *non*-Hermitian operator deserve careful consideration [21]. As mentioned before, Bogoliubov and Shirkov [11, p.55-56] used the scheme to construct the complete set of solutions of the relativistic equations, fixing the sign of $p_0 = +E_p$.

Conclusion

The main points of my paper are: there are "negative-energy solutions" in that is previously considered as "positive-energy solutions" of relativistic wave equations, and vice versa. Their explicit forms have been presented in the case of spin-1/2. Next, the relations to the previous works have been found. For instance, the doubling of the Fock space and the corresponding solutions of the Dirac equation obtained additional mathematical bases in this paper. Similar conclusion can be deduced for the higher-spin equations.

I appreciate the discussions with participants of several recent Conferences.

References

[1] Guang-jiong Ni, In *Relativity, Gravitation, Cosmology.* (Nova Science Pub., Huntington, NY, USA, 2004), hep-ph/0306028.

[2] N. Simicevic, *Implications of Weak-Interaction Space Deformation for Neutrino Mass Measurements*, nucl-th/9909034.

[3] S. M. Bilenky and C. Giunti, *Nucl. Phys. Proc. Suppl.* **48**, 381 (1996).

[4] A. O. Barut, *Phys. Lett.* B**73**, 310 (1978); *Phys. Rev. Lett.* **42**, 1251 (1979).

[5] D. V. Ahluwalia and A. Ch. Nayak, *Int. J. Mod. Phys.* D**23**, 1430026 (2014).

[6] V. V. Dvoeglazov, *Adv. Appl. Clifford Algebras* (2016), accepted.

[7] P. A. M. Dirac, *Proc. Roy. Soc. Lond.* A**117**, 610 (1928).

[8] J. J. Sakurai, *Advanced Quantum Mechanics.* (Addison-Wesley, 1967).

[9] L. H. Ryder, *Quantum Field Theory.* (Cambridge University Press, Cambridge, 1985).

[10] C. Itzykson and J.-B. Zuber, *Quantum Field Theory.* (McGraw-Hill Book Co., 1980), p. 156.

[11] N. N. Bogoliubov and D. V. Shirkov, *Introduction to the Theory of Quantized Fields.* 2nd Edition. (Nauka, Moscow, 1973).

[12] V. V. Dvoeglazov, *J. Phys. Conf. Ser.* **284**, 012024 (2011), arXiv:1008.2242.

[13] V. V. Dvoeglazov, *Hadronic J. Suppl.* **18**, 239 (2003), physics/0402094; *Int. J. Mod. Phys.* B**20**, 1317 (2006).

[14] M. Markov, *ZhETF* **7**, 579 (1937); ibid. 603; *Nucl. Phys.* **55**, 130 (1964).

[15] A. Barut and G. Ziino, *Mod. Phys. Lett.* A**8**, 1099 (1993); G. Ziino, *Int. J. Mod. Phys.* A **11**, 2081 (1996).

[16] I. M. Gelfand and M. L. Tsetlin, *ZhETF* **31** (1956) 1107; G. A. Sokolik, *ZhETF* **33** (1957) 1515.

[17] V. V. Dvoeglazov, *Int. J. Theor. Phys.* **34** (1995) 2467; *Nuovo Cim.* **108**A (1995) 1467; *Hadronic J.* **20** (1997) 435; *Acta Phys. Polon.* B**29** (1998) 619.

[18] R. P. Feynman and M. Gell-Mann, *Phys. Rev.* **109** (1958) 193.

[19] M. Kirchbach, C. Compean and L. Noriega, *Eur. Phys. J.* A**22** (2004) 149.

[20] V. V. Dvoeglazov, *Int. J. Theor. Phys.* **37** (1998) 1915.

[21] V. A. Ilyin, *Spektralnaya Teoriya Differencialnyh Operatorov.* [*Spectral Theory of Differential Operators.]* (Nauka, Moscow, 1991); V. D. Budaev, *Osnovy Teorii Nesamosopryazhennyh Differencialnyh Operatorov.* [*Fundamentals of the Theory of Non-Self-Adjoint Differential Operators]* (SGMA, Smolensk, 1997).

In: Future Relativity, Gravitation, Cosmology
Editors: V. V. Dvoeglazov et al.
ISBN: 979-8-88697-455-3

Chapter 9

ALGEBRA OF DISCRETE SYMMETRIES IN THE EXTENDED POINCARÉ GROUP*

Valeriy V. Dvoeglazov†
UAF, Universidad Autónoma de Zacatecas
Zacatecas, México

Abstract

We begin with the comprehensible review of the basics of the Lorentz, (extended) Poincare Groups and O(3,2) and O(4,1). On the basis of the Gelfand-Tsetlin-Sokolik-Silagadze research [1-3], we investigate the definitions of the discrete symmetry operators both on the classical level, and in the secondary-quantization scheme. We studied the physical content within several bases: light-front form formulation, helicity basis, angular momentum basis, on several practical examples. The conclusion is that we have ambiguities in the definitions of the the corresponding operators P, C; T, which lead to different physical consequences.

1. The Standard Definitions

The Lorentz Group conserves the interval $ds^2 = dx^\mu dx_\mu$ in the Minkowski space with respect to (pseudo)Euclidean rotations. The Poincaré Group includes translations in the Minkowski space. The extended Poincaré Group includes discrete transformations, the unitary C, P, and the antiunitary T in the quantum field theory (QFT). The P is the space inversion: $x^0 \to x^0$, $\mathbf{x} \to -\mathbf{x}$. The T is the time reversal: $x^0 \to -x^0$, $\mathbf{x} \to \mathbf{x}$. The C is the electric charge conjugation. It is related to the PT operation: $x^0 \to -x^0$, $\mathbf{x} \to -\mathbf{x}$. The interval is also conserved under these operations. In the QFT the eigenvalues of the combined CPT are also invariants.

---

*Talk at the LXII Congreso Nacional de Física. 6-11/10/2019. Villahermosa, Tab., México. First published in Progress in Physics, Volume 17, Issue 1, 2021. Reprinted by permission.

†Corresponding Author's Email: valeri@fisica.uaz.edu.mx; URL: http://fisica.uaz.edu.mx/~valeri/.

While [4] presented the derivation method to obtain the field operator *ab inition*, we define the field operator [5, 6] in the pseudo-Euclidean metrics as follows:

$$\Psi(x) = \frac{1}{(2\pi)^3}\sum_h \int \frac{d^3\mathbf{p}}{2E_p}\left[u_h(\mathbf{p})a_h(\mathbf{p})e^{-ip\cdot x} + v_h(\mathbf{p})b_h^\dagger(\mathbf{p})e^{+ip\cdot x}\right]. \quad (1)$$

Hence, the Dirac equation is:

$$[i\gamma^\mu\partial_\mu - m]\,\Psi(x) = 0\,. \quad (2)$$

At least, 3 methods of its derivation exist [7, 8, 9]:

- the Dirac one (the Hamiltonian should be linear in $\partial/\partial x^i$, and be compatible with $E_p^2 - \mathbf{p}^2c^2 = m^2c^4$);
- the Sakurai one (based on the equation $(E_p - \sigma\cdot\mathbf{p})(E_p + \sigma\cdot\mathbf{p})\phi = m^2\phi$);
- the Ryder one (the relation between 2-spinors at rest is $\phi_R(\mathbf{0}) = \pm\phi_L(\mathbf{0})$ and boosts).

It has solutions of the positive energies and the negative energies. The latter are reinterpreted as the antiparticles. $E_p = \sqrt{\mathbf{p}^2 + m^2}$, $c = \hbar = 1$, $g^{\mu\nu} = diag\{1, -1, -1, -1\}$. The solutions in the momentum representation are: $u_h(\mathbf{p}) = column(\phi_R^h(\mathbf{p})\ \phi_L^h(\mathbf{p}))$. Next,

$$u_h = \begin{pmatrix} \exp(+\boldsymbol{\sigma}\cdot\boldsymbol{\varphi})\phi_R^h(\mathbf{0}) \\ \exp(-\boldsymbol{\sigma}\cdot\boldsymbol{\varphi})\phi_L^h(\mathbf{0}) \end{pmatrix}, \quad v_h(\mathbf{p}) = \gamma^5 u_h(\mathbf{p})\,, \quad (3)$$

where $cosh(\varphi) = E_p/m$, $sinh(\varphi) = |\mathbf{p}|/m$, $\hat{\varphi} = \mathbf{p}/|\mathbf{p}|$, h is the polarization index. It is shown that the parity operator can be chosen as

$$P = e^{i\alpha_s}\gamma^0 R,\ \gamma^0 = \begin{pmatrix} 0 & 1 \\ 1 & 0 \end{pmatrix}, \quad (4)$$

because

$$\left[i\gamma^\mu\partial'_\mu - m\right]\Psi(x^{\mu'}) = 0\,, (change\ of\ variables)\,, \quad (5)$$

where

$$\Psi(x^{\mu'}) = \mathcal{A}\Psi(x^\mu)\,, (lineality)\,. \quad (6)$$

These conditions can be satisfied by γ^0 matrix in the Weyl basis. R can be chosen $R \equiv (\theta \to \pi - \theta, \phi \to \pi + \phi, r \to r)$. For fermions it is well known that a particle and an antiparticle have opposite eigenvalues of the parity operator in this $(1/2, 0) \oplus (0, 1/2)$ representation of the Lorentz Group. In the QFT we should have:

$$U_P\psi(x)U_P^\dagger = e^{i\alpha_s}\gamma^0\psi(x')\,. \quad (7)$$

So,

$$U_P a_h(\mathbf{p})U_P^\dagger = e^{+i\alpha_s}a_h(\mathbf{p}')\,,\ U_P b_h(\mathbf{p})U_P^\dagger = -e^{-i\alpha_s}b_h(\mathbf{p}')\,. \quad (8)$$

The operator U_P can be constructed in the usual way, see [5] and [6]. The charge operator interchange the particle and the antiparticle. For example, in the Dirac case on the classical level:

$$u_\uparrow \to -v_\downarrow,\ u_\downarrow \to +v_\uparrow, v_\uparrow \to +u_\downarrow,\ v_\downarrow \to -u_\uparrow, \quad (9)$$

Thus, we can write thanks to E. Wigner:

$$\mathcal{C}_{1/2} = e^{i\alpha_c} \begin{pmatrix} 0 & i\Theta \\ -i\Theta & 0 \end{pmatrix} \mathcal{K}, \; \Theta = \begin{pmatrix} 0 & -1 \\ 1 & 0 \end{pmatrix} = -i\sigma_2 \,. \tag{10}$$

In the QFT we should have:

$$U_C \psi(x) U_C^\dagger = e^{i\alpha_c} C \psi^\dagger(x) \,. \tag{11}$$

So [5],

$$U_C a_h(\mathbf{p}) U_C^\dagger = e^{+i\alpha_c} b_h(\mathbf{p}) \,, \; U_C b_h(\mathbf{p}) U_C^\dagger = e^{-i\alpha_c} a_h(\mathbf{p}) \,. \tag{12}$$

See, however, Ref. [11], where **two** possibilities for the charge conjugation operator have been proposed.

The time reversal operator is antiunitary (see Wigner and [4]). Let us remind that the operator of hermitian conjugation does not act on c-numbers on the left side of the equation (13). This fact is conected with the properties of an antiunitary operator: $\left[V^T \lambda A (V^T)^{-1}\right]^\dagger = \left[\lambda^* V^T A (V^T)^{-1}\right]^\dagger = \lambda \left[V^T A^\dagger (V^T)^{-1}\right]$.

$$\left[V^T_{[1/2]} \Psi(x^\mu) (V^T_{[1/2]})^{-1}\right]^\dagger = S(T) \Psi^\dagger(x''^\mu) \,. \tag{13}$$

We can see that C and P anticommutes in the Dirac case:

$$\{C, P\}_+ = 0 \,, P^2 = 1, \,, C^2 = 1 \,, \tag{14}$$

and $(CPT) = \pm 1$. However, we present the opposite case later, where $(CPT) = \pm i$, which is related to the commutation (anticommutation) of the C and P operators.

The Table in p. 157 of Ref. [5] gives us the properties of the scalar, 4-vector, tensor, axial-vector and pseudoscalar under these transformations in the case of the "Dirac-like-parity" definitions. However, see the next Section.

2. Anomalous Representations of the Inversion Group

The previous Section perfectly describes the CPT properties of the charged fermions. Nevertheless, the authors of [1, 2, 10] proposed another class of representations of the full Lorentz Group long ago . As it was shown recently, it may be applied to the (anti)bosons of the opposite parities, and to the (anti)fermions of undefined parities. The latter are not the eigenstates of the parity operator, but they are the eigenstates of the charge-conjugate operator. Gelfand, Tsetlin and Sokolik noted that there exist representations of the full Lorentz Group of the anomalous parity. Originally, this concept was intended to be applied to explanation of the decay of $K-$ mesons.

Briefly, the examples are: one can note that in the $(1/2, 1/2)$ representation (or for x^μ) the operators of the space inversion (t_{01}), the time reversal (t_{10}) and the combined space-time inversion (t_{11}) are commutative. They form the inversion group together with the unit element. Let us search the projective representations of the Lorentz group combined with the discrete group. As opposed to the usual case, $t_{01}t_{10} = t_{11}$, $t_{10}t_{11} = t_{01}$, $t_{01}t_{11} = t_{10}$, for instance, one can drop the condition of the commutativity, and one can form the projective rep. with $T_{10}T_{01} = -T_{11}$, or $T_{11}T_{11} = -1$, see the full Table in [1]. They noted that there are **two** normal-parity (in their notation) and **two** anomalous parity reps. for (bi)spinors. Then, they extended the concept of the anomalous parity to any rep. of the proper Lorentz Group characterized by the numbers (k_0, k_1) and $(-k_0, k_1)^*$. When

$$[T_{i'k'}, T_{i''k''}]_+ = 0\,, \tag{15}$$

this is the case of the anomalous parity (later, this was confirmed by Nigam and Foldy [12]). G. Sokolik noted that this concept is related to the concept of the 5-dim representations of the proper orthogonal group with pseudo-Euclidean metrics. For example,

$$T_{10} \sim H_{54} = \exp(i\pi I_{54}/2)\;,\; T_{11} \sim H_{43}H_{21} = \exp(i\pi I_{43})\exp(i\pi I_{21})\,, \tag{16}$$

$$T_{01} = T_{11}T_{10}\,. \tag{17}$$

T_{10}, T_{01}, T_{11} leave invariant the extended 8-component Dirac equation only (compare with [13] and [14]):

$$\Gamma_\mu \partial^\mu \psi + m\psi = 0\,,\; \Gamma_\mu = \begin{pmatrix} \gamma_\mu & 0\\ 0 & -\gamma_\mu \end{pmatrix} \tag{18}$$

They claimed relations to the concepts (known in that time):

- istopic spin;
- fusion theory;
- the non-linear Heisenberg theory

were mentioned. The corresponding matrix representations of the anomalous-parity representations have been presented:

$$T_{01} = \begin{pmatrix} 0 & I\\ I & 0 \end{pmatrix}, T_{10} = \begin{pmatrix} 0 & -I\\ I & 0 \end{pmatrix}, T_{11} = \begin{pmatrix} I & 0\\ 0 & -I \end{pmatrix}, \tag{19}$$

and

$$T_{01} = \begin{pmatrix} 0 & -iI\\ iI & 0 \end{pmatrix}, T_{10} = \begin{pmatrix} 0 & iI\\ iI & 0 \end{pmatrix}, T_{11} = \begin{pmatrix} I & 0\\ 0 & -I \end{pmatrix}. \tag{20}$$

Later Wigner [10] repeated their results at the Istanbul School lectures (1962). And Silagadze [3] rediscovered and applied this possibility in 1992. The conclusion of these papers is: we noted that both new versions of the reps. of the full Lorentz Group (commuting spinor and anticommuting boson representations) lead to the doubling of the dimensionality of the $\psi-$ function.

3. The Self/Anti-Self Charge Conjugate States

The content of this Section contains the material of [11]. The conclusions are: we have constructed another explicit example of the BWW-GTS theory. The matter of physical dynamics connected with this mathematical construct should be solved in future as depended on what gauge interactions with potential fields do we introduce [14] and what experimental setup do we choose.

References

[1] I. M. Gelfand and M. L. Tsetlin, *Sov. Phys. JETP* **4** (1957) 947.

[2] G. A. Sokolik, Sov. Phys. JETP **6** (1958) 1170; ibid. **9** (1959) 781; *Dokl. Akad. Nauk SSSR* **114** (1957) 1206.

[3] Z. K. Silagadze, *Sov. J. Nucl. Phys.* **55** (1992) 392.

[4] N. N. Bogoliubov and D. V. Shirkov, *"Introduction to the Theory of Quantized Fields"* (John Wiley & Sons, NY, USA, 1980).

[5] C. Itzykson and J.-B. Zuber, *"Quantum Field Theory"* (McGraw-Hill International Book Co., 1980).

[6] W. Greiner, *"Field Quantization"* (Springer, Berlin-Heidelberg, 1996).

[7] P. A. M. Dirac, *Proc. Roy. Soc. Lond.* A**117** (1928) 610.

[8] J. J. Sakurai, *"Advanced Quantum Mechanics"* (Addison-Wesley, 1967), §3.2.

[9] L. H. Ryder, *"Quantum Field Theory"* (Cambridge University Press, Cambridge, 1985).

[10] E. P. Wigner, in *"Group Theoretical Concepts and Methods in Elementary Particle Physics"*, ed. F. Gursey (Gordon and Breach, 1964).

[11] V.V. Dvoeglazov, *Mod. Phys. Lett* A**12** (1997) 2741, hep-th/9609142.

[12] B. P. Nigam and L. L. Foldy, *Phys. Rev.* **102** (1956) 1410.

[13] M. A. Markov, *ZhETF* **7** (1937) 579; ibid. 603.

[14] V. V. Dvoeglazov, *Nuovo Cim.* **108**A (1995) 1467.

In: Future Relativity, Gravitation, Cosmology
Editors: V. V. Dvoeglazov et al.

ISBN: 979-8-88697-455-3
2023 Nova Science Publishers, Inc.

Chapter 10

EMPIRICAL EVIDENCE FOR THE RELATIVISTIC THEORIES OF LORENTZ, POINCARÉ AND EINSTEIN: THE 1881 AND 1887 MICHELSON EXPERIMENTS, REVISITED*

Hector A. Munera†
International Centre for Physics, (CIF, Centro Internacional de Física), Bogotá, Colombia
Department of Physics, National University, Bogotá, Colombia

ABSTRACT

The present paper begins with a quick revision of the undoubted connection between the inception of the theory of relativity and the "null" interpretation given to the experiments carried out by Michelson in 1881 and 1887. Lorentz always hoped that there would be a loophole in Michelson's analysis that would make his length-contraction hypothesis unnecessary. In this context, section 2 presents for the first time my detailed analysis of the 1881 experiment, and with this new information, revisits the 1887 Michelson-Morley experiment in section 3. The overall result is that the pioneering experiments were designed to test only a particular value of solar motion, a hypotheses not supported by the empirical observations. However, the conclusions from both experiments overstepped the range of validity imposed by the experimental design itself. Thus, the crucial 1881 and 1887 experiments *never* were "null" and consequently there is no contradiction with the "positive" outcome of our own 2002-2005 MMMM experiment at CIF, in Bogota (Colombia): a successful first-time determination by optical means of velocity of Earth relative to a Newtonian frame of reference.

---

* First published in Apeiron, Volume 20, Hors série 4, June 2018. Reprinted by permission.
† Corresponding Author's Email: hmunera@hotmail.com.

DEDICATION

My long-standing connection to Professor Valeri Dvoeglazov

In the editorial introduction to the timely book entitled *Einstein and Poincaré: the Physical Vacuum,* edited by my friend professor Valeri Dvoeglazov in 2006 [1], he noted that "unfortunately, there are almost no insights into the obvious relations between the ether concept and the recently discovered 'dark matter' and 'dark energy.'" The present writer has been working in that direction for the past twenty years, and has postulated an energy-like dark matter (DM) fluid (or aether, if you wish) formed by discrete DM entities called sagions, which are the carriers of all energy and linear momentum in the universe; the DM fluid obeys the Lorentz invariant homogeneous Klein-Gordon equation (HKGE) proposed by Louis de Broglie as a basis for a quantum theory compatible with Einstein's general theory of relativity. Our novel solutions in spherical coordinates (r,θ,ϕ) for the HKGE are formed by a background time-independent scalar field $B(r, \theta,\phi)$ and an entangled scalar field $E(Ct/r,\theta,\phi)$ [2], whose value is the same for all time-distance transformations between coordinate systems in relative motion (Doppler, Lorentz, Poincaré, or Einstein), thus side-stepping many theoretical controversies about the various approaches to relativity theory. Nonetheless, the present writer feels it is very important to clarify some issues that are still open at the empirical foundations of relativity, especially after his own empirical research [11-13] produced results opposite to the claimed "null" results of the 1887 Michelson-Morley (MM) experiment [3].

In 1997 at York University in Toronto, Canada, in one of the earlier symposia to honor Jean-Pierre Vigier —formerly the scientific assistant to Louis de Broglie, and at that time the last living connection to the great physicists of the early 20th century—, I had the pleasure of meeting and exchanging ideas with Valeri Dvoeglazov for the first time. One of my presentations there, a critical review of all MM-type experiments up to 1930 [4], is very relevant for present paper. Two years later, in 1999, Valeri and his colleagues at the then incipient School of Physics of the Autonomous University of Zacatecas (Mexico) organized a successful international workshop on CPT, neutrinos, and Lorentz invariance. They were kind enough to invite me to participate with a couple of papers, one of them my first approach towards a unified fluid and field theory [5], that I continued developing at recent Vigier Symposia [6-8]. My present view is that the discrete energy-like DM aether described at the Vigier Symposia may be identified with the fluid underlying Louis de Broglie's double-solution theory.

Last time that I met Valeri in person was at the Vigier Symposium held in 2000 at the University of California in Berkeley—my doctoral alma mater. On that occasion we had lunch on or near Telegraph Avenue and talked about my incipient photon model [9] that arose from ideas sketched at Zacatecas [5].

During the past twenty years Valeri has been kind enough to include contributions of mine in several of his books and journals on which he has been an invited editor. For instance, for the *Annals of the Louis de Broglie Foundation* in Paris I contributed a paper [10] calculating the effect of solar motion on the fringe-shifts in a MM interferometer, which provided the theoretical basis for the design of my repetition [11-15] of the MM experiment. Many thanks Valeri for your friendship and for that long-standing and helpful connection.

1. Introduction: The "Null" Interpretation of the MM Experiment

An unprecedented change in the way scientists view Nature occurred over the short lapse of forty years from mid-1880s to mid-1920s; here we concentrate on the empirical foundations of relativity [3, 16]. In the general overview of the MM-type experiments [4] written in 1998, some issues were not identified, and some were not stressed enough. Their importance became obvious to this author during the design and implementation of our own MM experiment from 2002 to 2005. The present paper describes for the first time my detailed evaluation of the 1881 Michelson experiment [16]. A remarkable aspect is that several weak features regarding data collection and data reduction from the 1881 experiment were re-used in the 1887 MM experiment.

1.1. Lorentz and Michelson's 1881 and 1887 Experiments

The Dutch professor of physics Hendrik Antoon Lorentz followed Michelson's research closely from the beginning, and in 1886 sent him a correction pertaining to the analysis of the optical path along the arm that was *presumably* transversal to the direction of motion in the 1881 experiment at Postdam. In the process of designing the 1887 experiment, Michelson acknowledged the validity of the criticism by Lorentz and modified his analysis accordingly [3, p. 335].

To avoid, or at least to minimize misunderstandings, the reader will kindly forgive me for being explicit and direct in my comments, and for literally emphasizing the most significant issues, and even words as above.

Lorentz postulated the now famous length-contraction as the basis of his 1895 theory of the electron [17]; but he had doubts, as attested by a letter to Lord Rayleigh dated 18 August 1892 in which Lorentz asked: *"Can there be some point in the theory of Mr. Michelson's experiment which has as yet been overlooked?"* [18, p. 32]. As argued here, our answer is positive: Yes professor Lorentz, indeed. MM only recorded the *residual fraction* of a fringe-shift, instead of the *total* fringe-shift as it should have!

1.2. Poincaré and Michelson's Experiments

The French mathematical physicist Henri Poincaré does not mention Michelson by name in his 1905 book; he writes simply: "many *experiments* have been made *on the influence* of the *motion of the Earth*. The results have always been *negative*. But if these experiments have been undertaken, it is because we have not been certain beforehand... we might expect to find accurate methods giving positive results. I think that such a hope is illusory... I do not believe, in *spite of Lorentz*, that more exact observations will ever make evident anything else but the *relative displacements of material bodies*. *Experiments* have been made that should have disclosed the terms of the first order; the results were *nugatory* ... Then *more exact experiments* were made, which were *also negative*" [19, p. 171-172], emphases added. Poincaré often mentions Lorentz, and at the end of the book he states that "Lorentz's theory is very attractive" and that according to Lorentz it is not possible to measure "absolute velocity, but their relative velocity *with respect to the ether*, so that the principle of relativity is safe" [19, p. 243-244], emphasis in the original.

Comparing the two quotations above, it follows that, regarding motion, for Poincaré the aether had the status of a "*material body*." This is confirmed when, in his Socratic style, Poincaré asks: "Does our ether actually exist? We know the origin of our belief in the ether. If light takes several years to reach us from a distant star, it is no longer on the star, nor is it on the Earth. It must be somewhere, and *supported*, so to speak, by *some material agency*" [19, p. 169], emphases added.

A short digression, there is an immediate counter-question back to Poincaré: where does ether exist? Or, where is it contained? For Newton the answer was clear: in absolute space. In the context of propagation of gravity, aether had for Newton *both active* and *passive* roles as evidenced by the phrases "by and through" and "agent acting constantly" in the third letter to Bentley dated 25 February 1692/3: "that one *body* may *act* upon another *at a distance through a vacuum, without the mediation of anything else, by and through* which their action and force may be conveyed from one to another, is to me so great an absurdity that I believe no man who has in philosophical matters a competent faculty of thinking can ever fall into it. *Gravity* must be *caused by* an *agent acting constantly* according to certain laws, but *whether this agent is material or immaterial* I have left to the consideration of my readers" [20, p. 54], emphases added. Note also that Newton did not decide between a material or immaterial ether.

Returning to Poincaré and Michelson, in *Science and Method* published in French in 1908 and translated into English in 1914 [21] Poincaré mentions Michelson several times, including a repetition of Fresnel's experiment (page 216), a subject not directly relevant here. The section entitled The Principle of Relativity begins thus: "But if the *ether* is *not displaced by the Earth's motion*, is it possible by means of optical phenomena to demonstrate the absolute velocity of the Earth, or rather its velocity in relation to the *motionless* ether? *Experience* has given a *negative reply*, and ..." [21, p. 217], emphases added. What is a "*motionless*" ether? This strange object is mentioned again on page 218. What kind of aether had Poincaré in mind? It cannot be a fluid that *certainly* would be "*displaced*" by motion of Earth, and whose elementary particles would be in permanent motion, similarly to the molecules of oxygen and nitrogen in air. Was Poincaré a supporter of the far-fetched notions of solid and crystal aethers [22] that proliferated in the second half of the 19th century? Be it as it may, let us turn to the "*negative reply*" from experience.

Poincaré continues thus: "Whatever be the method employed, we shall never succeed in disclosing any but relative velocities; I mean the velocities of certain material bodies in relation to other material bodies." [21, p. 217] Yes, agreed. But nothing forbids us from choosing the far away Newtonian "fixed stars" as an anchor to define a reference or preferred frame. Accordingly, at the beginning of this century the present writer succeeded in measuring, by optical means, the velocity of Earth relative to the said preferred frame [11-15].

Then, Poincaré finally mentions an experiment: "Indeed, when the source of the light and the apparatus for observation are both on the Earth and participate in its motion, the experimental results have always been the same, whatever be the direction of the apparatus in relation to the *direction of the Earths's orbital motion*" [21, p. 217], emphasis added. This description may refer either to the 1881 Michelson experiment or to the 1887 Michelson-Morley experiment. The phrase in italics pinpoints the weakest part in those experiments, and will be discussed at length in the present paper.

Next, Poincaré discusses aberration and length-contraction, which may be "easy to explain so long as we neglect the square of aberration... But one day Michelson thought out a much more delicate process. He introduced rays ...reflected by mirrors. Each of the distances being about a yard..." [21, p. 218-219]. No doubt, this apparatus is the interferometer used by Michelson in the 1881 experiment, whose arms were one meter long (see section 2 below for previously unnoticed shortcomings in that experiment). The name of Michelson is mentioned approvingly three more times in that section [21, p. 220-221].

1.3. Einstein and the Michelson Experiments

Einstein's abundant correspondence became available in English translations by the end of past century [23-25] and clearly shows that he had been aware of the "null" interpretation of the MM experiment since his time at ETH-Zurich [26]. Einstein mentioned Lorentz for the first time in a letter to his girlfriend Mileva Marić dated 28 December 1901: "I now want to buckle down and study what Lorentz and Drude have written on the electrodynamics of moving bodies" [23, p. 189-190], [24, p. 72 and 100]. Einstein was talking about Paul Drude's recent 1900 book entitled *Lehrbuch der Optick*, translated into English as The Theory of Optics, chapter VIII of which in the Dover edition begins thus: "The assumption which will be adopted here is that the *ether always remains completely at rest.* Upon this basis H. A. Lorentz has developed a complete and elegant theory" [27, p. 457-482], emphasis in the original. Drude discusses in detail Michelson's 1881 experiment (pp. 478-481), and Michelson and Morley 1887 experiment in somewhat less detail (pp. 481-482).

Strangely enough, in his public papers Einstein always downplayed the influence of Michelson experiments, even in August 19/1952 at Michelson's birthday centennial: "The influence of the crucial *Michelson-Morley experiment* upon my own efforts has been *rather indirect.* I learned of it through H. A. *Lorentz's decisive investigation* of the electrodynamics of moving bodies (1895) [17] *with which I was acquainted before developing the special theory of relativity.* Lorentz's basic assumption of an *ether at rest* seemed to me *not convincing* in itself and also for the reason that it was leading to an *interpretation* of the result of the *Michelson-Morley* experiment which seemed to me *artificial*" [18, p. 35], emphases added.

1.4. Miller and the "Null" Interpretation of the MM Experiment

After Michelson moved to Chicago, Dayton C. Miller continued the interferometer experiments with Morley from 1902 onwards, and had thus first hand knowledge on the empirical details of the MM experiment; he explicitly stated that "the effect did not have the anticipated magnitude. However, and this fact must be emphasized, *the indicated effect was not zero*; ... This is quite different from a null effect now so frequently imputed to this [MM] experiment" [28, p. 206], emphasis in the original. This is consistent with my 1998 review of all MM-type experiments up to 1930 [4], in the sense that all classical experiments observed a small, but non-zero, motion of Earth relative to an aether presumably at rest in Newton's absolute space. Surprisingly, in the conclusions many papers stated that they obtained a "null" result.

2. The 1881 Michelson Experiment Critically Revisited

Albert Abraham Michelson, an officer of the U.S. Army, very early demonstrated a great talent for experimental research in physics, and for further training he went to the laboratory of Hermann von Helmholtz in Berlin (Germany), where Michelson designed and carried out an experiment to measure the relative velocity between Earth and aether, assumed at rest relative to Newton's absolute space (see Figure 1A). Measurements were initially in Berlin, but due to tram vibrations, the apparatus was moved to Postdam, a quiet location in the outskirts of the city [16].

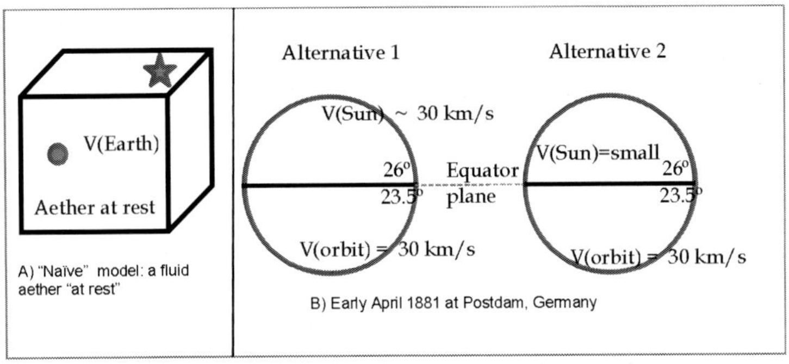

Figure 1. Michelson's analysis to design his experiment at Postdam in early April 1881 to test his hypothesis H-1881 (see text for details).

2.1. Design of the 1881 Experiment

Michelson treated the velocity of Earth as the vector addition of two components: (a) orbital motion with an approximate speed of 30 km/s along the plane of the ecliptic, plus (b) solar motion towards Hercules constellation, that for early April 1881 was at an angle of 26º relative to the terrestrial equatorial plane (see Figure 1B). Being a naval officer, Michelson was quite aware that azimuth associated with daily rotation of Earth should be taken into account, and he explicitly noted that "if the apparatus be so placed that the arms point north and east *at noon*, the arm pointing east would coincide with the resultant motion, and the other would be at right angles" [16, p. 125], emphasis added. Please note that this result holds if, and only if, the Sun *only moves* towards Hercules with a speed of 30 km/s. If Sun moves with a larger speed, or if it moves in a different direction, then *Michelson's expectations do not hold*. As it turned out, the fringe-shift in the NE-SW direction was +0.034, which according to Michelson "should have been zero. The numbers are simply outstanding errors of the experiment" [16, p. 127]. This large unexpected shift, which was one-third the size of the predicted 0.1 fringe-shift (see next paragraph), was eliminated as unwanted linear drift (page 128).

Since the speed of solar motion was unknown at that time, Michelson considered two alternatives: (1) magnitude similar to orbital motion, and (2) small magnitude relative to orbital motion. Michelson calculated the expected fringe-shift for his short one-meter arms interferometer at noon in early April at Postdam for each of the two alternative speeds and decided to take "the *mean of these two numbers* as the most probable … the displacement to be looked for is not far from one-tenth the distance between the fringes" [16, p. 125], emphasis added.

From the foregoing quotations, it is evident that the 1881 Michelson experiment was designed to test *only* the following hypothesis that I will label *H-1881: Is it possible to measure with Michelson's interferometer the velocity of Earth defined as orbital motion around Sun plus solar motion towards Hercules at 30 km/s maximum speed?*

From a logical viewpoint there is a third alternative to Figure 1, namely: a more complex solar motion having, in addition to velocity towards Hercules, other velocity components yielding a *resultant solar speed larger than 30 km/s, in a direction other than toward Hercules*. As we know today, solar speed is indeed quite large [29], at least a factor of ten higher than orbital motion. Michelson cannot be blamed too much for overlooking this third alternative in 1881, because at the time solar speed was considered small. However, since Michelson's mind was anchored to the notion that fringe-shifts were smaller than one fringe he could not correctly interpret some "apparent" drifts of fringe-shift, which he summarily eliminated (see subsection 2.5 for further comments).

2.2. Raw Data from the 1881 Experiment

In a table [16, p. 126], Michelson reported the raw data he obtained in four series of measurements, each one formed by five revolutions of his apparatus relative to the laboratory, i.e., relative to the surface of Earth. Michelson explicitly noted that "the numbers in the columns are the positions of the center of the dark fringe in *twelfths* of the distance between the fringes" [16, p. 127], emphasis in the original. In the first position the reference arm is oriented toward north (N), in the second same arm points northeast (NE), and so on in 45° steps clockwise until

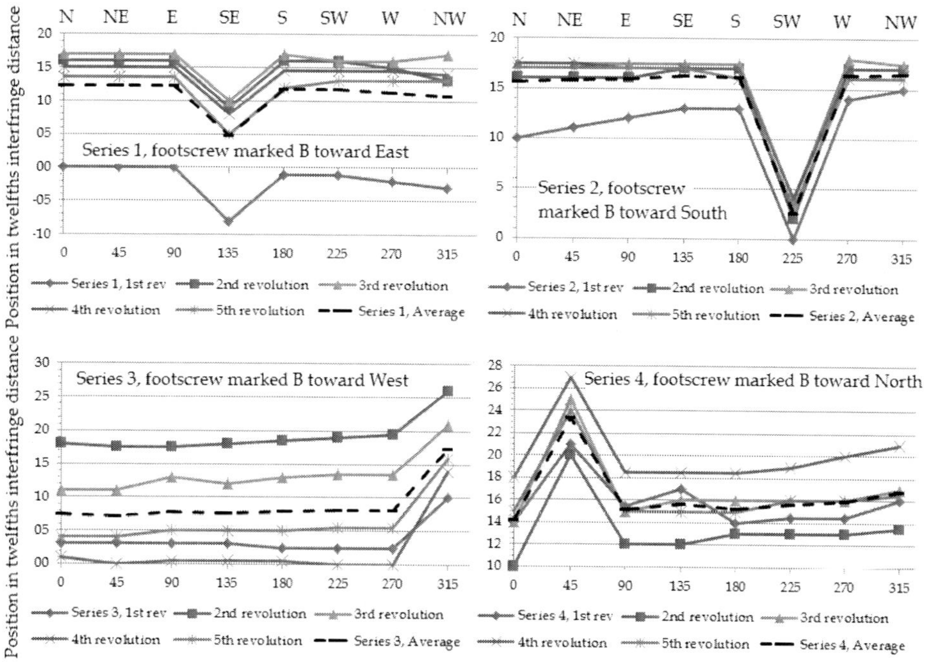

Figure 2. Average raw data curves (dashed lines) in sessions 1 and 4 with one-SD error band above and below each session mean. Average SD in sessions 1 and 4 respectively are 0.61 and 0.21. Dotted curves show Michelson's expectations.

reaching last position with reference arm oriented toward northwest (NW). For each session, Figure 2 graphically shows Michelson's raw data, plus the mean values for each orientation over the five revolutions. In each session there is one outlier that Michelson (plausibly at first sight) attributed to a mechanical asymmetry or peculiarities of his interferometer (page 125).

For each orientation and series of data, Table 1 shows the mean of the raw data over the five revolutions *without removing outliers*. For each mean we have calculated its standard deviation (SD), also included in Table 1. The last row shows overall averages for each series.

Table 1. Mean values and standard deviations for each orientation in the four series (readings in twelfths of the distance between fringes)

		Series 1		Series 2		Series 3		Series 4	
Orientation	o	Mean	SD	Mean	SD	Mean	SD	Mean	SD
N	0	12.3	7.0	15.6	3.2	7.4	7.0	14.2	2.9
NE	45	12.3	7.0	15.8	2.8	7.1	7.1	23.4	2.9
E	90	12.3	7.0	15.9	2.2	7.8	7.2	15.2	2.3
SE	135	4.8	7.4	16.3	1.9	7.7	7.2	15.7	2.4
S	180	11.7	7.3	16.1	1.8	7.9	7.6	15.3	2.1
SW	225	11.7	7.2	2.6	1.7	8.1	7.9	15.7	2.2
W	270	11.3	7.5	16.4	1.5	8.2	8.1	15.9	2.6
NW	315	10.8	7.9	16.5	1.0	17.4	6.2	16.8	2.7
Averages		10.9	7.3	14.4	2.0	9.0	7.3	16.5	2.5

It is quite remarkable that Michelson did NOT calculate standard deviations for his raw data. As a consequence, in the whole paper there is a notorious absence of an analysis of the experimental error associated with each stage of his data reduction process. This is most surprising for an experiment carried out in Germany, homeland of Gauss —the creator of the theory of errors.

To illustrate the importance of this matter, Figure 3 shows the means and the associated standard deviations in fringe-widths; note that values in Table 1 are divided by 12 to convert raw readings into fringe-widths (page 127). Michelson expected to observe in this 1881 experiment a harmonic curve with an amplitude of 0.05 of a fringe-width. This is the dotted curve shown in Figure 3, immersed within a very large experimental dispersion.

For sessions 1 and 3 last row in Table 1 gives an average SD around 0.6 fringe-widths. The experimental dispersion shown on the left side of Figure 3 is *twelve* times larger than Michelson's expected amplitude of 0.05 fringe-widths for his harmonic variations. Likewise, for sessions 2 and 4, the last row in Table 1 gives an average SD of 0.2 fringe-widths, which is FOUR times larger than Michelson's expectations as seen on the right side of Figure 3. For this author the results shown in Figure 3 constitute a *strong indication* that *something is not right with the experiment*, and that the observations summarized in Table 1 above are not useful to test hypothesis H-1881, the objective of the experiment. However, as Michelson did not calculate his standard deviations, he continued his data reduction process, briefly described in the next subsection.

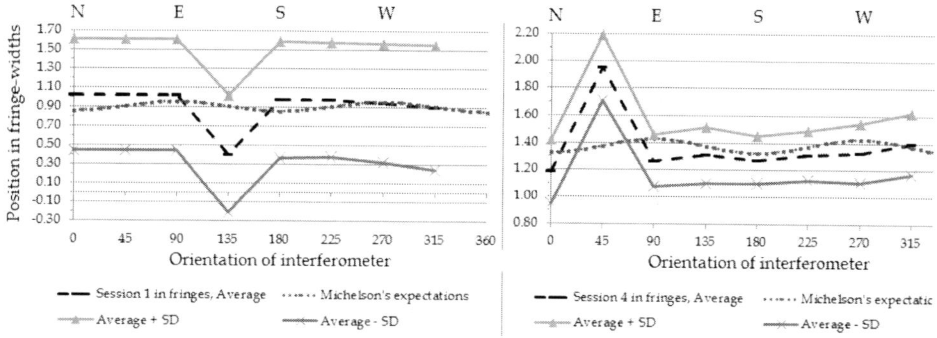

Figure 3. Average raw data curves (dashed lines) in sessions 1 and 4 with one-SD error band above and below each session mean. Average SD in sessions 1 and 4 respectively are 0.61 and 0.21. Dotted curves show Michelson's expectations.

For completeness, let us note that SD values are very stable within a given series, and are much larger for series 1 and 3 corresponding to an East-West orientation of a footscrew at the base of the apparatus, and three times smaller for series 2 and 4 corresponding to a South-North orientation of the footscrew. Michelson noted that the outliers displace 90° together with the 90° rotation of the footscrew at the base of the apparatus between sessions. He thus attributed the outliers to some idiosyncracy of the interferometer and eliminated them by using in his calculations the average of the two adjacent points. Surprisingly, in Table 1 above, which did include the outliers, the SDs associated with all outliers seem to be immune to the peculiarities of the magic footscrew. I leave this fact as an open question.

2.3. Michelson's Analysis and Conclusion

As just stated, for each session Michelson calculated mean fringe-shifts without outliers (see table in [16, p. 126]). *Mutatis mutandis*, Michelson's means are the same dashed curves shown above in Figures 2 and 3. Since he expected two harmonic cycles over a 360° rotation of the interferometer, Michelson averaged the four data for N to SE with the four data for S to NW. Michelson then proceeded to inter-session averaging of the means in the four sessions. As noted in his page 127, the average fringe-shift associated with N-S direction (=0.022) was smaller than the average fringe-shift associated with NE-SW direction (=0.034), which he expected to be zero under the assumption that Sun actually moves towards Hercules *only*.

Instead of considering the possibility that one of his design assumptions might be incomplete, and that his own data hinted at a real fringe-shift along the NE-SW direction (recall our comments in subsection 2.1), Michelson applied a *linear correction* to eliminate the apparent drift along the NE-SW direction because "the numbers are simply outstanding errors of the experiment" (page 127), obtaining values which "represent the *displacements observed, freed from the error in question*" (page 128), emphasis added. The simulation in the next subsection 2.4 clearly demonstrates that, *in some cases*, a linear correction not only distorts the empirical data, but may also be unnecessary and incorrect.

Michelson used the tiny residuals after the linear correction to plot his Figure 4, *without error bars*, showing "the actual curve together with the curve that should have been found if the theory had been correct" (page 128). Then, Michelson concluded that "the interpretation of these results is that there is no displacement of the interference bands. The result of the *hypothesis of a stationary ether is thus shown to be incorrect*, and the necessary conclusion follows that the hypothesis is erroneous" (next to last paragraph in [16, p. 128], emphasis added).

Regarding Michelson's Figure 4 and his conclusions, several comments are in order. In the first place, for a modern reader a curve without error bands is rather strange. Recalling Figure 3 above, it may be expected that the statistical errors associated with Michelson's "actual curve" are much larger than the 0.05 amplitude of his expected dotted curve "drawn on the supposition that the displacement to be expected is one-tenth of the distance between the fringes" (page 128). Hence, the "actual curve" with small amplitude close to zero and *without error bars,* shown as Figure 4 in Michelson's paper [16], is simply meaningless.

Secondly, let us grant for the sake of the discussion that observed fringe-shifts had a very small experimental error. In that case Michelson was right in asserting that there was "no displacement of the interference bands". Since such an experimental outcome was contrary to the expected displacement of "one-tenth of the distance between the fringes," the correct conclusion is that hypothesis H-1881 was not proven. In other words the conclusion should be: *Michelson's interferometer could not measure the velocity of Earth defined as orbital motion around Sun plus solar motion toward Hercules at 30 km/s maximum speed.* This is quite different from Michelson's conclusion that "*the hypothesis of a stationary ether is thus shown to be incorrect*", and significantly oversteps the design objectives of the 1881 experiment at Postdam.

For additional remarks see the next subsection where we try to answer a more general question: was the 1881 experiment capable of measuring the whole speed of Earth's motion?

To end this subsection fairly, it must be acknowledged that the present writer has the benefit of hindsight in two senses: (1) the current evidence that net solar motion has a high speed in a direction other than the Hercules constellation [29], and (2) our experiment [11-13] demonstrated that a Michelson interferometer *operated at small angular steps* does yield large fringe-shifts—really, and beyond any doubt.

2.4. Simplified Simulation of the 1881 Experiment

Let us consider a simplified example that contains just the basic ingredients of Michelson's experiment and his data reduction process. Firstly, let the objective be to measure $y(\varphi)$, which is a sinusoidal curve of unknown amplitude A. Secondly, let the experimental procedure be to collect $y_{obs}(\varphi)$, which are the residuals of $y(\varphi)$ after subtracting the integer component of $y(\varphi)$; see eqs. (1). And, thirdly, in the data reduction process let a linear correction $y_{corr}(\varphi)$ be applied to cancel the value of observed data at $\varphi = 45°$ and 315°, that is along the NE-NW direction, see eq. (2). The final output of Michelson's experiment is $y_M(\varphi)$ defined by eq. (3). To avoid misunderstandings, please note that Michelson used equation (3) below with a negative sign, and that the origin for the horizontal coordinate x in his equation in page 128 is on the right side.

$$y(\phi) = A\sin\phi, y_{obs}(\phi) = y(\phi) - \text{int}\{y(\phi)\}, \quad \phi = 0°, 45°, 90°, \ldots, 360° \tag{1}$$

$$y_{corr}(\phi) = b + m\phi, b = y_{obs}(45) - 45m, \quad m = \frac{y_{obs}(315) - y_{obs}(45)}{315 - 45} \tag{2}$$

$$y_M(\phi) = y_{obs}(\phi) - y_{corr}(\phi), \quad \phi = 0^o, 45^o, 90^o, ..., 360^o \tag{3}$$

The linear correction was implemented by Michelson, who argued that elimination of "this gradual change, which should *not in the least* affect the periodic variation for which we are searching, would of itself necessitate an outstanding error … If therefore, we can eliminate this gradual change, we may expect a much smaller error" [16, p. 127], emphasis added. Of course, the periodic variation is still there. But, the linear correction projects the observed data upon a new horizontal thus distorting the shape and amplitude of observations—this is a sort of rotation as seen in Figures 4 and 5.

Figure 4 shows three examples for different values of the unknown amplitude $A = 0.25$, 1.25, 3.25. The left side shows $y(\varphi)$ the curve to be found by recording succesive values of $y_{obs}(\varphi)$ for $\varphi = 0^o, 45^o, 90^o, ..., 360^o$, as shown on the right side of Figure 4, together with the linear correction defined in eqs. (2).

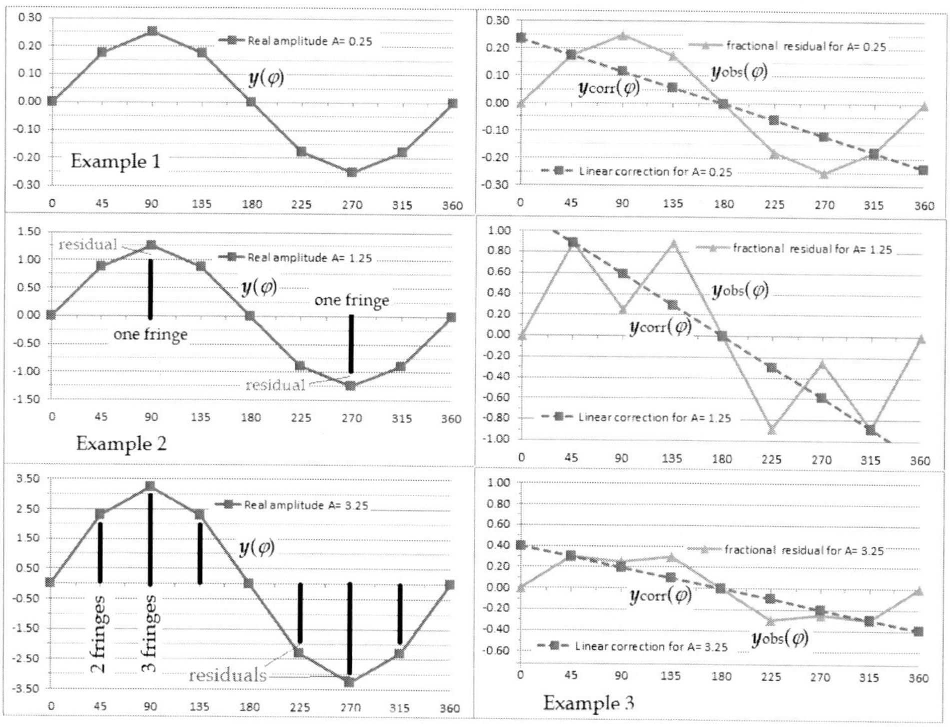

Figure 4. Determination of the sinusoidal curve by measuring residuals y_{obs}, which are graphically defined on the left side of the figure. The resulting observed curves are shown at right, together with a linear correction to cancel *apparent* linear drifts.

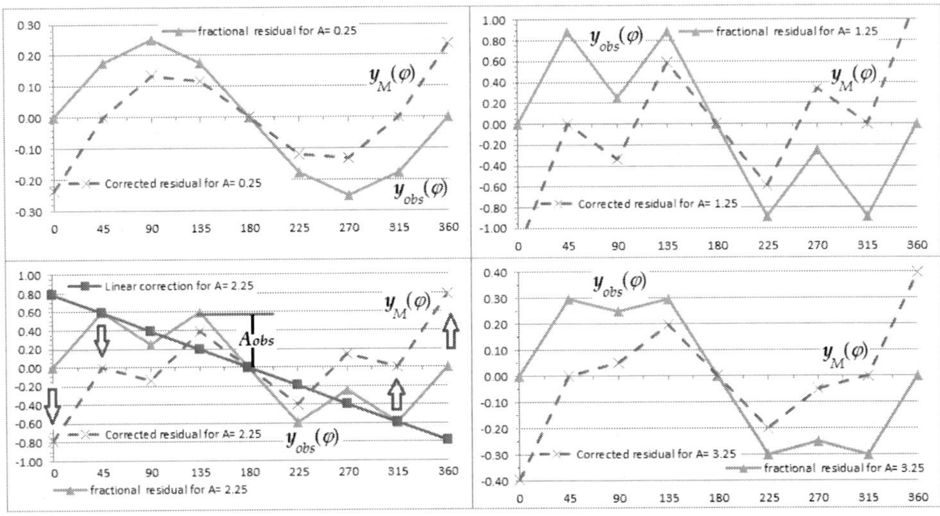

Figure 5. Final curve $y_M(\varphi)$ after applying linear corrections to the three examples in Figure 4. Details of the process are self-explanatory in the observed and final curves for $A = 2.25$ shown at the bottom on left side.

Example 1 has $y(\varphi) < 1$ for all φ ; it is evident that $y_{obs}(\varphi) = y(\varphi)$ for all φ. This means that, in such cases, Michelson experiment is correctly defined. On the contrary, in examples 2 and 3 there are one or more $y(\varphi) > 1$. In such cases the observed data $y_{obs}(\varphi) < 1$ is very different from the original sinusoidal curve, and there is no obvious relationship between the shape of the curve and the value of the original amplitude A. In some cases, such as example 3, one may be tempted to interpret variations in $y_{obs}(\varphi)$ as experimental error and fit a smooth curve to $y_{obs}(\varphi)$. The resulting amplitude $A_{obs} < 1$ will be with certainty smaller than the original amplitude, which by necessity is $A > 1$.

In example 3 the amplitude of the smoothed curve will be $A_{obs} \sim 0.3$, which is ten times smaller than the real amplitude $A = 3.25$. In this case it would be ludicrous to claim that the *real* amplitude is 0.3, and even more absurd to claim that due to experimental errors the amplitude may be considered to be zero (i.e., a "null" result). No! The only logically reasonable conclusion is that the design of the experiment (i.e., the choice of $y_{obs}(\varphi)$) is not appropriate to measure sinusoidal curves having one or more $y(\varphi) > 1$. This is precisely the case in Michelson's experiment when total solar motion has a speed much larger than 30 km/s.

In examples 2 and 3 the scatter of points around Michelson's "linear correction" *apparently* shows a linear drift of data of unknown origin. Michelson interpreted the apparent drift as caused by an unidentified real physical agent and subtracted it. However, in the present simulation there are no unidentified physical agents, and Michelson's interpretation is not appropriate.

Just consider example 1 in Figure 4. The curve for $y_{obs}(\varphi)$ on the right side of the figure is the correct representation of the sought after $y(\varphi)$. Any further correction is not only unnecessary, but also wrong, as can be seen in Figure 5 for the case $A = 0.25$. Indeed, the final curve $y_M(\varphi)$ has a different shape and a smaller amplitude.

For the other three cases in Figure 5 having $A > 1$, the resulting $y_M(\varphi)$ curves depict unexpected variations between successive measurements, which in an experiment would be interpreted as observational errors. However, such variability arises from an incorrect choice of the method of observation $y_{obs}(\varphi)$, which in Michelson's case was recording only the fractional component of a total fringe-shift containing both an integer plus a residual.

2.5. Stationary Aether and Newton's Preferred Frame

The assumption of an aether introduces an unnecessary complication in the experimental design. Consider the cube shown in Figure 1A which may represent an isolated room filled with air, or an isolated swimming pool filled with water, either of them rigidly attached to a larger building; in this sense, the walls of the cube are at rest relative to the larger building. The velocity of water in the pool is defined in equation (4) as the average velocity <**V**> of the *N* molecules of water contained in the pool, where each molecule *j* moves with *velocity* $\mathbf{V}_j$ *relative to the walls of the pool.* The water in the pool is at rest if the average velocity is zero, and the water is in motion otherwise:

$$< \mathbf{V} > \equiv \sum_{j=1}^{N} \mathbf{V}_j, \; < \mathbf{V} > = 0 \text{ fluid at rest}, \; < \mathbf{V} > \neq 0 \text{ fluid in motion} \tag{4}$$

Please note that the fluid may be at rest, while concomitantly the average speed of water molecules relative to the walls of the pool may be $C > 0$. Then, it follows that Poincaré's "motionless" aether cannot be a fluid. For convenience, one may choose a preferred frame of coordinates at rest relative to the walls of the pool, with one axis pointing toward a fixed mark on one of the walls, say the star shown in Figure 1A, which represents Newton's fixed stars. A similar definition holds for air, or for a more tenuous classical aether.

Consider now a material particle in motion, say a swimmer in the pool, or a tennis ball traveling in air. There is no difficulty in defining the velocity of the particle relative to the walls of the cubic room or of the swimming pool, which always are at rest by definition. Likewise, it is easy to define the dual problem where the material particle is at rest and the fluid is in motion relative to the walls of the cube, as in a wind tunnel.

On the contrary, one introduces unnecessary complications if motion of a material particle is defined in relation to a moving fluid, with average velocity given by equation (4).

Some proponents of aether treated in [30] talk about aether entrainment. This notion is reminiscent of a well known but rather complex classical phenomenon in the theory of *viscous* fluids: the formation of boundary layers (see any textbook on fluid theory, say [31, pp. 140-146]). Simplifying many caveats and issues, if relative speed between a material particle and a viscous fluid is small, a *thin* layer of fluid —called the boundary layer— always remains attached to the material particle. In that case the relative speed between the particle and the boundary layer is zero. In a Michelson experiment on the surface of Earth one may assume as a reasonable first approximation that the laboratory is within the boundary layer of a fluid aether, so that the apparatus is at rest relative to the boundary layer. But beyond that thin layer (how many meters?) the fluid has a speed profile that smoothly increases up to some limiting speed associated with the motion of Earth relative to some local bulk of aether. So a new question appears, at what distance from Earth is attained the limiting speed? In such models the only known thing is that, beyond the boundary layer, the relative speed between particle and fluid is always non-zero.

However, what happens if the Reynolds number associated with aether is large? This may occur if the relative speed between aether and Earth is large in some undefined sense. By analogy with classical fluids, in those cases the aether flow would be turbulent, and wakes may appear, as in airplanes, and so on. In entrainment models there is also the implicit assumption that the aether is viscous. Is there any evidence regarding aether's viscosity? In my humble opinion, aether entrainment introduces too many additional and unnecessary hypotheses.

For previous reasons, in the analysis of the MMX the present writer avoids aether as an unnecessary intermediary, and strictly adheres to Newton's approach, namely: refer all velocities to a preferred frame. For instance, in the calculations for the design of our experiment [10] and in our concluding analysis [13] the words "aether" and "aether wind" *never* appear.

3. The 1887 MM Experiment Critically Revisited, Once Again

For the 1887 experiment at Cleveland (USA), Michelson, then teaching at Case School of Applied Science, joined efforts with the chemist Edward W Morley, from nearby Western Reserve University. The Michelson-Morley experiment (MMX) [3] improved the hardware relative to the 1881 experiment, in particular the rigidity of the apparatus and the stability of rotation. However, for the collection of data MM used the same 1881 protocol, consistent–let me repeat once again–in recording *only* the residual fraction of a fringe-shift rather than the total number of fringe-widths produced in each successive change of orientation of the apparatus through 22.5° relative to the laboratory.

In present paper I want to *explicitly* argue that the crucial 1887 MM experiment was designed to test *only one component* of the motion of Earth relative to Newton's preferred frame. Since in Miller's words [28, p. 206] the observed "*effect did not have the anticipated magnitude*" of 30 km/s, the only possible conclusion was that the *motion tested for was not measurable* with MM experimental setup. Any interpretation or extension by MM, or by anybody else, beyond that clear and summary answer must be justified, and most likely would require additional assumptions that must be clearly stated.

3.1. Design of the 1887 Experiment

As already mentioned in section 1, the design for the 1887 experiment incorporated a correction suggested by the Parisian M. A. Potier in 1881, and by Lorentz in his 1886 paper [32]. It also explicitly supposed "*the ether being at rest*" [3, pp.334-335], see Figure 1A.

Surprisingly, MM's analysis for the expected fringe-shifts did *not* include solar motion, which was taken into account in the earlier 1881 experiment at Postdam. For this more sensitive apparatus with eleven-meter arms, MM predicted a 0.4 fringe-shift—as in the 1881 experiment, the *expected* fringe-shift was *smaller than one fringe*. In MM's words "in what precedes, *only the orbital motion of the Earth* is considered. If this is combined with the *motion of the solar system*, concerning which but little is known with certainty, the *result* would *have to be modified*; and it is *just possible that the resultant velocity at the time of the observations was small* though the chances are much against it. The *experiment* will therefore be *repeated at intervals of three months*, and thus all uncertainty will be avoided" [3, p. 341], emphases added. This quotation merits several comments, made in the same sequence as the phrases in italics.

1. Since MM acknowledged that "only the orbital motion of the Earth is considered" it follows that the hypothesis tested in the MM experiment was H-1887: *Is it possible to measure orbital speed of Earth around the Sun with Michelson's interferometer?* Indeed, the expected 0.4 fringe-shift was calculated for an orbital motion of 30 km/s.
2. Regarding "motion of the solar system… little is known with certainty." This attitude is not the same as in 1881. MM could have assumed some values of solar speed as in 1881; why didn't they? Were MM, perhaps, concerned that a non-zero solar speed might increase expected fringe-shift beyond one fringe? In that case the expected sinusoidal variations would have amplitude $A > 1$, and as previously argued in Section 2.4, a correct experimental design would require recording the full fringe-shift. This prompts a different question, did MM have, in their day, the technical capability to measure the full fringe-shift? This latter issue is tackled in next section 3.2.
3. At any rate, MM were quite aware that "the result would have to be modified," i.e., that the expected fringe-shift would *certainly* be different. Indeed, the motion of Earth relative to a Newtonian preferred frame is a unique *indivisible* physical concept: a time-dependent net speed along some direction. A completely different matter is that for calculations, or for expository reasons, the motion of Earth may be considered the vector addition of several components: orbital velocity around the Sun, plus solar velocity relative to the center of our galaxy, plus the velocity of Milky Way relative to the local group of galaxies, and so forth. One may naively posit that Nature does not identify a particular experiment on Earth, and that Nature does not adjust herself to react only to the component of velocity that such experiment is trying to measure! It follows that experiments seeking to measure the motion of Earth necessarily refer to total net motion. In that case the amplitude of the expected sinusoidal variations is approximately A > 10. Taking into account the simulation discussed in 2.4, one may conclude that the 1887 MM experiment was not designed to measure A > 10, and thus was doomed *ab initio*.
4. MM considered a *small* "resultant velocity at the time of the observations" quite unlikely. Why did MM omit the obvious alternative: a *large* resultant velocity of Earth at the time of the observations? This omission prompted us to repeat the experiment; for a brief description see Section 3.4.
5. MM suggested repeating the experiment "at intervals of three months." MM never did it. It was Dayton C. Miller who, almost forty years later, finally performed the experiments between April 1925 and February 1926 [28, pp. 228-231].

Be that as it may, the scope of the MM experiment was restricted by the *very* design of the experiment to test for H-1887 only.

3.2. The Execution of the 1887 MM Experiment, and the Continuation by Miller

When I prepared the overview of MM-type experiments in 1998 [4], I had never carried out the experiment, so that some aspects were missed or did not receive the attention they deserve. Let us stress two issues.

3.2.1. The Number of Fringes between Two Consecutive Orientations

Or, from the observer's viewpoint, how may an experimenter count the total number of fringes shifting from one orientation of the apparatus to the next?

The 1881 Michelson experiment. Let us start with the pioneering 1881 experiment. In the first position (P1), one of the arms called the reference arm is parallel to a South-North line, the interferometer is calibrated to have a *reference* fringe (RF) close to a fiducial point, say an arrow in the field of view of the telescope. Michelson measured the position of the center of that RF relative to the arrow, and wrote it down, this is the first register or R1. Next the apparatus was turned 45° to position 2 (P2) so that the reference arm was parallel to a line from SW to NE, Michelson checked the horizontality of the apparatus and waited until vibrations faded away; he looked at the interference pattern, identified the reference fringe that he expected to be near the fiducial arrow, measured the position of the center of that RF relative to the arrow, and wrote it down; this is the second register or R2. And so on every 45° until five revolutions of the apparatus were completed.

First question: what is a reference fringe? My own experience is with laser light where all fringes are alike, but I have been told by my colleagues from optics that in experiments with non-coherent light there is always a brighter fringe that is easy to identify. So, let that bright line be the reference line for Michelson in 1881. Incidentally, that bright line does not help much for my own experiment.

Second (much harder) question: how many fringes shifted from the S-N position P1 to the SW-NE position P2? For Michelson the answer was simple: the difference between the two readings, i.e., R2-R1. But, was that right? How could he possibly know? Was that mere wishful thinking? Just remember that Michelson was conditioned by the scientific knowledge at the end of the 19th century, whereby the Sun moved at most at 30 km/s—hence the expected small shift (see section 2).

In my humble opinion, in the 1881 experiment with separate or independent measurements at each orientation it was *impossible* to know how many fringes shifted between the two separate and independent orientations of the apparatus.

The 1887 MM experiment: For the 1887 experiment MM introduced a conceptual change in design. Rather than a succession of separate measurements, each one at a different orientation, the apparatus was placed on a stable base so that there was only one calibration per session. The stone base with the apparatus on top was floated in mercury, and slowly rotated at a rate of one revolution every six minutes, i.e., every 360 s. Relative to the 1881 experiment, the spatial resolution in 1887 improved by a factor of two: the interference pattern was checked every 22.5° (instead of 45°) through a telescope also placed on the rotating platform. In a revolution there were 16 readings (=360°/22.5°), so that observer had to walk around the mercury bath, and quickly look through the telescope eyepiece every 22.5 seconds (= 360 s/16).

In their paper MM stated that "the cross wire of the micrometer was set on the *clearest* of the interference fringes at the instant of passing one of the marks ... the reading of the screw-head on the micrometer was noted ... on passing the second mark, the same process was repeated, and this was continued till the apparatus had completed six revolutions" [3, p. 339], emphasis added. So, MM measured the position of *the* "clearest" fringe every 22 seconds at each mark, but they did not know what happened during the 22-second interval between successive marks. For instance, did the clearest fringe drift several fringes before arriving at its current position? Or, is *the* "clearest" fringe at the second mark the *very same* fringe observed at the first position? In summary, in the MM experiment it is not clear whether it was possible to measure, at each orientation, the accumulated fringe-shift over all previous orientations.

Miller experiments: For his own experiments Miller [28, pp. 208-214] built in 1904 a more massive apparatus with higher resolution: effective length of arms was three times the arm length in the MM experiment. He continued Michelson's practice of taking readings every 22.5°. The rotation rate was increased to one revolution every 50 seconds, i.e., the observer shouted a reading every *three seconds*! The values in units of a tenth of a fringe-width were recorded by an assistant on a prepared form (page 213). In Miller's words: "the observer has to walk around a circle about twenty feet in diameter, *keeping his eye at the moving eyepiece* of the telescope attached to the interferometer which is turning on its axis steadily, at the rate of about one turn in fifty seconds; the observer must not touch the interferometer in any way and yet *he must never lose sight of the interference fringes*, ... The string attached to the float ... used as a sensitive guide to assist the observer in maintaining the proper circular path" [28, p. 211], emphases added. So, the arc distance between consecutive marks was about four feet ($20\pi/16$) that the observer had to walk with his head attached to an eyepiece! The meticulous and continuous attention maintained for hours by the observer and his assistant are awe inspiring. Since Miller required keeping the eye "at the moving eyepiece" and never "lose sight of the interference fringes," it is evident that he was aware of the issues under consideration in this subsection, namely, the need to *record the real* fringe-shift. To attain such an impressive goal Miller implemented *continuous monitoring* of interference fringes by the naked eye. It is very sad that an improper operational protocol implemented by Miller himself destroyed the continuity of his monitoring (see the final part of next subsection).

Present writer completely agrees with Miller that the only way to count the total number of fringe-shifts is by *continuous* monitoring of interference fringes. In our case, during 2002 in the preliminary stage of our experiment, we carried out three-day long sessions over several weekends, *day and night*, with an observer *permanently* looking at the pattern of interference-fringes appearing over a frost-glass screen. The objective was to write down the succession of positions and times (hour, minutes and seconds) when the observer could discern by the naked eye a motion towards left or right of a *selected* fringe, *the same* fringe was followed throughout the whole weekend session. Let me once again thank the several students of my Newtonian Mechanics course, and my daughter Natalia and her friends, for being the observers during night-shifts in the basement of CIF.

It was found that the interference-pattern in our interferometer, at rest on top of a pneumatic thirteen metric ton concrete table, was stable, and that the eye could only distinguish changes at the scale of a few minutes, typically five to ten. With that empirical observation we benefited from modern technology, and installed a video camera for the 2003-2005 experiment. To optimize data handling, only a photograph of the pattern was recorded and stored every minute (see section 3.4).

3.2.2. The Standard Deviations in Michelson's and Miller's Raw Data

Regarding the 1881 Michelson's raw data, it was noted in Section 2 that he did not calculate the standard deviation (SD) associated with each orientation, but SDs could be easily calculated from the data for the five individual turns of the apparatus (see Table 1 above). In general, the paper reporting the 1887 experiment [3] was less detailed than the pioneering paper reporting the 1881 experiment [16]; for instance, raw data is not reported in 1887. In the latter, MM only included a table showing, for each session, the means over the six turns of the apparatus [3, p. 340], *without* individual standard deviations associated with each mean value. Thus, we do not have the slightest clue whether the dispersion of raw data in 1887 was smaller or larger than in 1881. As might be expected from this omission, MM do not include error bars in the concluding Figure 6 [3, p. 340]. Consequently, the same comments made regarding the 1881 experiment are valid here: the absence of error bars seriously calls into question the validity of the conclusions offered by MM.

Returning to Miller's thousands of experimental sessions, a run in his experiment was the average of twenty consecutive turns during some 1,000 seconds (20 turns x 50s/turn), i.e., about 16 minutes. A facsimile of the data sheet for a run at Mount Wilson on September 23/925 is shown as Figure 8 in [28, p. 213]. On the right side of that sheet the ominous word "adjust" appears three times. What is it?

Like Michelson and MM, Miller also was psychologically anchored to the belief that fringe-shifts should be small. As a result during the execution of his experiments Miller implemented a procedure that, in retrospect, was *not* appropriate. In Miller's words: "the adjustments are maintained so that the *central fringe* of the field of view ... is *never more than two fringe widths* from the fiducial point. Often the *temperature drift* is such that the fringes drift more than this before a set of twenty turns is completed. When this occurs, the *fringe system* is *restored* to its central position simply by *placing a small weight* of two or three hundred grams on the end of the arm or by *removing* a weight from the arm" [28, p. 212], emphases added.

In the first part of the above quotation, Miller interpreted observed shifts larger than two fringes as caused by an unwanted "temperature drift" that should be eliminated. This *might* be *partially* correct, but if a correction is actually required, it is done during the process of data reduction—*never* by modifying the apparatus during a run. Furthermore, recalling our simulation in 2.4 above, at least part of the observed apparent drift may be a real effect caused by the *real high speed* of solar motion. So, "adjustments" were *not* needed at all.

Present writer is a neutral outsider without emotional ties (one way or another) to Miller, and besides, he is knowledgeable in the field of experimentation. For him it was almost unbelievable to read the final sentence in the above quotation. Indeed, the "*placing*" or "*removing*" of weights shortens or enlarges via (un)bending the lengths of the arms, thus violating a fundamental rule in experimentation: use the *same* apparatus in a run. Thus, in the case shown in Figure 8, Miller used one apparatus for revolutions 1 to 5, a second different apparatus for revolutions 6 to 9, a third different apparatus for revolutions 10 to 19, and a fourth apparatus for the last, twentieth, revolution. Hence, the twenty turns cannot be averaged, as Miller did. Also, it is surprising that Miller did not calculate standard deviations associated with each mean over twenty turns. At any rate, neither the mean nor the SD would have any value in cases where adjustments are made during a run.

Another unwanted by-product of Miller's adjustments is that the *continuous* observation of cumulative fringe-shift was spoiled. Let us peruse Miller's data sheet in Figure 8 [28, p.

213]. At position 1 in the first turn the reading in tenths of a fringe-width was +10, and at position 17 at end of fifth turn the reading was –15; this is a net change of 25 = 2.5 fringe-widths toward the left (the latter from the minus sign). The arm of the interferometer was modified, i.e., "adjusted," and in the sixth turn the reading at position 1 is now 0. The fringes continued moving to the left, and at position 17 at the end of the ninth turn the reading was –10; this corresponds to one additional fringe-width to the left. The arm of the interferometer was modified for a second time, i.e., "adjust," and in the tenth turn the reading at position 1 was now +8. The fringes continued moving to the left, and at position 17 at end of 19th turn the reading was –21; this is a net change of 29 = 2.9 fringe-widths toward the left. Altogether a total of 2.5 + 1.0 + 2.9 = 6.4 fringe-widths to the left. The arm of the interferometer was modified for a third time, i.e., "adjust," and in the 20th turn the reading at position 1 was now +1. After this modification of the apparatus the fringes started to move to the right, and at position 17 at end of 20th turn the reading was +4; this is a net change of 3 = 0.3 fringe-widths towards the right. Altogether for the twenty turns of the interferometer 6.4-0.3 = 6.1 fringe-widths to the left in 16 minutes. This is a significantly large fringe-shift!

Contrary to some opinions, the previous exercise clearly shows that large fringe-shifts were present in Miller's data. Unfortunately, Miller threw away that information by introducing the adjustments.

At the beginning of this 21st century James De Meo went to Cleveland to unearth Miller's data sheets. He succeeded, and he was extremely generous in letting this author have a copy of that valuable information. The present writer considered the possibility of recovering the hidden total fringe-shifts from Miller's data sheets, using the procedure described above. However, it was realized that such an exercise would only be of *qualitative* value, because due to *the* "adjustments" Miller's interferometer kept varying over one run, as well as from run to run. So, this writer finally decided to go ahead with his own experiment.

3.3. Comments on the Conclusions of the 1887 MM Experiment

In the context of the foregoing discussion, the conclusions offered by MM in their crucial 1887 paper were quite misleading: "the *relative velocity of the Earth and the ether* is probably less than one-sixth the Earth's orbital velocity, and certainly less than one-fourth… It appears, from all that precedes, reasonably certain that if there be any *relative motion* between the *Earth and* the luminiferous *ether*, it must be *small*; … If now it were legitimate to conclude from the present work that *the ether is at rest with regard to the Earth's surface*, according to Lorentz there could not be a velocity potential, and his own theory also fails" [3, p. 341], italics added. Additionally, as already stated above, the concluding Figure 6 [3, p. 340] *without* error bars is not meaningful. In 1887 the situation is worse than in 1881, where at least it is possible to recover the SD associated with the raw data.

Strictly speaking, MM's correct conclusion should have been something like this: our experiment only measured "less than one-sixth the Earth's orbital velocity, and certainly less than one-fourth" of the said expected orbital velocity. The experimental error was such and such. However, solar motion was not included in the design of our experiment, and it is unknown at the present time whether an experiment designed to test, by optical means, for a non-zero solar velocity may be successful.

Many people focus on the conclusions, and *may understand* that MM *actually* measured "*the relative velocity of the Earth and the (luminiferous) ether,*" under the most general conditions. No, not quite! The MM claim significantly overstepped the range of validity of their experiment, which ignored solar motion and only registered residual fractions of a fringe-shift, instead of total fringe-shift.

Since in the concluding sentences MM did not emphasize the limitations of their experiment, Lorentz, Poincaré, Einstein, and *almost* everybody else took their concluding remarks at face value (just recall the quotations in the Introduction).

3.4. Our 2003-2005 MMMM Experiment at CIF, Bogota (Colombia)

Of course, this author is well aware that MM actually wanted to test a more ambitious hypothesis that may be called *H-aim: Is it possible to measure the motion of Earth relative to a Newtonian preferred frame with Michelson's interferometer?* Our answer is: yes, it is possible, but it requires a slight modification in the collection of data. To identify an appropriate experimental procedure this writer calculated the response of the MM apparatus when solar velocity is also included; the ensuing theoretical predictions [10, 33] were kindly published by Professor Dvoeglazov.

Let us focus on our *realistic* calculation of the expected fringe-shifts in the MM experiment at Cleveland. The word "realistic" is stressed to indicate that solar motion was included, contrary to MM's toy calculation *without* solar motion.

Our calculations are for the same apparatus used by MM in 1886, placed at the same geographical coordinates in Cleveland, for the same dates and hours of day, and use a modern value of solar velocity taken from [29]. The results are in Figure 4A [33, p. 97]. Amplitude of the oscillation exceeds fifteen fringe-shifts! This is in stark contrast to MM's expected 0.4 fringe-shift *without* solar motion. Thus a correct design of any modern repetition of the MM experiment must be capable of measuring large fringe-shifts (this is a large amplitude A in the simulation in 2.4).

Our calculation also explains why MM observed what they did. Figure 4.B [33, p. 97] shows the expected observations when only the residual fraction of the fringe-shift is recorded, as in the MM experiment [3] (recall the simulation in II.4).

To empirically check the foregoing theoretical analysis, the present writer decided to repeat the MM experiment, with important improvements, from early 2002 to February 2005; this is the Michelson-Morley-Miller-Munera (MMMM) experiment. The first year, 2002, was dedicated to design and preliminary measurements. The experiment itself ran over a period of 26 months from January 2003 to February 2005 at the International Centre for Physics (CIF, for the initials in Spanish) located on the campus of National University in Bogota. The observed data [11-15] support our theoretical predictions.

We used a stationary interferometer (i.e., rotation period of 24 solar hours), laser green light, automatic video recording of the fringe-pattern every minute for a total of 1,440 frames per daily rotation, and during 2002, developed software to convert video images to digital fringe-shifts. It is worthwhile to stress a *very significant difference* between our MMMM experiment and all classical MM-type experiments: *our high* orientational *resolution.*

Let us be explicit. The MM and Miller experiments used an *angular distance of* 22.5^o between two consecutive readings. In our stationary interferometer this would be the same as taking a reading every 90 minutes, or every one and a half hours. Our preliminary observations during 2002, mentioned in 3.2.1. above, clearly indicated that such an interval *was too long* because a reference fringe could move over several fringe-widths during that 90 minute interval. As already explained in 3.2.2, Miller himself did observe a fringe shift of 6.1 fringe-widths in sixteen minutes, but unfortunately his "adjustments" hid that information. For a comparison recall that Miller's apparatus, with its thirty-three meter long arms, could produce fringe shifts 16.5 times larger than our interferometer with two meter long arms, so Miller's observations translated to our apparatus mean a fringe-shift of 0.37 fringe-widths (= 6.1/16.5) in sixteen minutes, completely compatible with our 2002 preliminary observations.

Next, we decided to record one image of the interference-pattern every minute, that is, every 0.25° (=360° per day/1440 minutes per day) [11-15]. Thus, our angular resolution was *ninety times better* than the 22.5° in MM and in Miller experiments. As a result, we *succeeded where Michelson-Morley and Miller failed*, and we were able to witness the slow drift of the reference fringe, and *count* the total number of *fringes drifting forward and backward* during one turn of our apparatus, which, as already stated, was one solar day.

There is an additional aspect that also deserves attention. In designing our process for data reduction we were very concerned about the unfit handling of the effect of ambient temperature upon fringe-shift in the earlier experiments (just recall comments above in 3.2.2). Since in optics it is well-known that speed of light propagation is affected by ambient temperature, humidity and pressure, we acquired commercial data loggers to measure humidity and temperature, but at that time in the local market there were no loggers for ambient pressure. To have an idea of the magnitude of the possible effect, professor German Arenas mounted a miniature interferometer inside a vacuum chamber. We used both the naked eye and a video camera to witness fringes passing by as pressure increased from a moderate vacuum up to atmospheric pressure. Two lessons: (1) The need to include during data reduction a correction for fringe drifts associated with variations of atmospheric pressure, leading to our implementation of stochastic corrections in our data reduction process [34]. In contrast, the subject of *ambient pressure* is notoriously *absent* in the crucial experiments by Michelson, MM, and Miller. (2) A video camera was indeed suitable to register fringe shifts in our experiment, as effectively implemented in our MMMM experiment.

From our data reduction process we obtained significant periodical diurnal and annual fringe-shift variations correlated with terrestrial motion. In the classical MM-type experiments the approach was to suppose a solar velocity and predict the expected fringe-shift. We solved the inverse problem, i.e., from the observed fringe-shifts we calculated solar motion relative to a Newtonian frame attached to the fixed stars [12, 13]. Our solar velocity is compatible with other values of solar velocity obtained with different methods by other investigators [14].

Our recent presentation in Moscow at PIRT-2017 [15] reports an *astonishingly high correlation* between our calculated terrestrial velocity and the observed periodical variations in the *raw* data of a hi-tech experiment at Stanford University carried out with microwave cavities near 0 K, and thermally controlled within $\pm 5 \times 10^{-6}$ K. However, in the process of data reduction the Stanford group interpreted the said periodical variations as unwanted "*mechanical perturbations*", subtracted them, and interpreted the white-noise residual as support for the conventional interpretation of the MM experiment [35]. On the contrary, in our view the Stanford experiment confirms our MMMM experiment: their unwanted "mechanical perturbations" are highly correlated to velocity of Earth projected onto a *vertical plane* containing the two microwave cavities at their laboratory in Palo Alto (California), where velocity of Earth is defined as vector addition of orbital motion *plus our solar velocity* reported in [13] relative to a Newtonian preferred frame.

Conclusion: The Indubitable Validity of Our "Positive" Experiment

Since the foregoing claims cast shadows on the very empirical foundations of contemporary science, it is not a great surprise that our "positive" results are mostly ignored, or at best labeled as "anti-relativistic". The latter is very far from our intentions: in short, this author still *believes* in the scientific method, where experiment and/or observation is the final arbiter. Besides, the present writer is neither anti-Einsteinian, nor anti-Newtonian, nor anti-Cartesian …, nor anti-anybody. Quite the opposite, I greatly admire and respect the scientific work of all of them. Our business is to fulfill Einstein's dream for a unified theory of Nature, based on a fluid aether similar to the classical aethers of Newton and Descartes [22], but using Einstein's field approach where force is not a primitive notion. Granted, such an inclusive process of synthesis requires abandoning, or rejecting, or completing, or reformulating Einsteinian, Newtonian and Cartesian ideas, but always in a constructive manner as manifested throughout present paper.

Please also note that the conclusions from our MMMM experiment do not result from "single experimenters, working unseen, based on relatively few tests" as was the case with Michelson's twenty turns of his interferometer in 1881 to collect 160 readings [16, p. 126] (see also Table 1 above), or with the scanty thirty six turns in the MM experiment in 1887 to collect 576 data points summarized in a laconic six line table of their paper [3, p. 340]. No, our work consisted of 1,440 data per day over hundreds of days covering a two-year span; that is, in a single day we doubled the 736 data collected by both Michelson in 1881 and MM in 1887 together. Furthermore, our ideas and results have been presented at several small international symposia and three large international conferences on different continents—SPIE-2007 in San Diego, PIERS-2009 in Beijing, and PIRT-2017 in Moscow—and published in journals and outlets available to us, for instance [4-15]. Needless to say, the raw data from the MMMM experiment is available for any interested person to check: just come to Bogota, copy the more than 300 CDs, and take them with you.

The overall conclusion in this paper is that the 1881 and 1887 experiments were designed to test only a particular value of solar motion. But the empirical observations did not support those limited hypotheses. However, in both cases the conclusions offered by Michelson in 1881 and by Michelson and Morley in 1887 significantly overstepped the range of validity imposed by the design of each experiment. Thus, the crucial 1881 and 1887 experiments *never* were "null." Consequently, there is no contradiction with the, indubitable in our opinion, "positive" outcome of our MMMM experiment at CIF: a successful *first time ever* determination by optical means of the velocity of Earth relative to a Newtonian frame of reference.

Our MMMM experiment is *à la Miller* in several senses: (a) Measurements were taken throughout the day, and not merely at noon and 6 pm, where Michelson had *assumed* some convenient conditions, as cancellation of orbital speed by solar motion towards Hercules. (b) The experiment was repeated at several epochs of the year, not a scanty observational period of two hours per day during three almost consecutive days, for a total of six hours. Actually, our experiment had many long sessions from several days up to a couple of weeks in a row during 26 months from January 2003 to March 2005. (c) The interference pattern was continuously monitored to count the *whole* fringe-shift (an integer plus a fraction). However, in order to avoid confusion [30, p. 182], it is explicitly stated here that in our analysis corrections by hand are completely absent; in particular we did not apply Miller's multiplication of final speed by some constant, as in his Table V [28, p. 235].

REFERENCES

[1] Dvoeglazov, Valeri V (ed.) (2006) *Einstein and Poincaré: The Physical Vacuum* (Montreal, Canada: Apeiron Press) 184 pp. ISBN: 0-9732911-3-3.

[2] Munera, Hector A and Guzmán Octavio (1997) "New explicit nonperiodic solutions of the homogeneous wave equation," *Foundations of Physics Letters* **10** (1), 31-41.

[3] Michelson, Albert A and Edward W Morley (1887) "On the relative motion of the Earth and the luminiferous ether," *American Journal of Science (third series)* **34** (203), 333-345. Same paper in *London, Edinburgh and Dublin Philosophical Magazine (fifth series)* **24**, 449-463.

[4] Munera, Hector A (1998) "Michelson-Morley experiments revisited: systematic errors, consistency among different experiments, and compatibility with absolute space," *Apeiron* **5** (Nos. 1-2), 371-376.

[5] Munera, Hector A (2000) "A realistic four-dimensional hydrodynamic aether interpreted as a unified field equation" 425-433 in Andrew E. Chubykalo, Valeri V. Dvoeglazov, David J. Ernst, Vladimir G. Kadyshevsky, and Young S. Kim (eds.), Lorentz Group, CPT and Neutrinos, *Proceedings of an International Workshop at the University of Zacatecas,* Zacatecas, Mexico, 23-26 June 1999 (Singapore: World Scientific Publishing) 466 pp. ISBN: 981-02-4062-7.

[6] Munera, Hector A (2016) "Interconnection of all forces of nature via the energy and momentum equations for a fluid aether" 247-267 in Richard L Amoroso, Louis H Kaufman and Paul Rowlands (eds.), *Unified Field Mechanics: Natural Science Beyond the Veil of Spacetime, Proceedings of the IX Symposium Honoring Noted French Mathematical Physicist Jean-Pierre Vigier, Morgan State University, 16-19 Nov 2014* (Singapore: World Scientific) 598 pp. ISBN 978-981-4719-05-6.

[7] Munera, Hector A (2019) "Neo-Cartesian unified fluid and field theory: from Klein-Gordon equation to de Broglie's Lorentz invariant mechanics" presented at *the X Vigier Symposium, Porto Novo, Italy, July 2016*, published as *IOP Conf. Series: Journal of Physics: Conf. Series* **1251** (2019) 012038, 22 pp. doi: 10.1088/1742-6596/1251/1/012038.

[8] Munera, Hector A (2019) "Symmetry in nature: discreteness, three-dimensional extension, motion, and minimum potential energy" presented at *XI Vigier Symposium, Advances in Fundamental Physics: Prelude to Paradigm Shift*, Liege, Belgium August 6-9, 2018, published as *IOP Conf. Series: Journal of Physics: Conf. Series* **1251** (2019) 012037, 20 pp. doi: 10.1088/1742-6596/1251/1/012037.

[9] Munera, Hector A (2002) "The photon as a charge-neutral and mass-neutral composite particle" 469-476 in Richard L Amoroso, Geoffrey Hunter, Menas Kafatos and Jean-

Pierre Vigier (eds.) *Gravitation and Cosmology: From the Hubble Radius to the Planck Scale* (Dordrecht, The Netherlands: Kluwer Academic Publishers).

[10] Munera, Hector A (2002) "The effect of solar motion upon the fringe-shifts in a Michelson-Morley interferometer à la Miller," *Annales de la Fondation Louis de Broglie* **27** (3), 463-484.

[11] Munera, Hector A, Hernández-Deckers, Daniel, Arenas, Germán and Alfonso Edgar. (2006) "Observation during 2004 of periodic fringe-shifts in an adialeiptometric stationary Michelson-Morley experiment," *Electromagnetic Phenomena (Kharkov, Ukraine)* **6**, No. 1-16, 70-92.

[12] Munera, Hector A, Hernández-Deckers, Daniel, Arenas, Germán and Alfonso Edgar. (2007) "Observation of a significant influence of Earth's motion on the velocity of photons in our terrestrial laboratory," presented at SPIE-2007, San Diego, California. Published in Chandrasekhar Roychoudhuri, Aloysius F Kracklauer, and Katherine Creath (eds.), *The Nature of Light: What Are Photons?*, *Proceedings of SPIE* **6664**, 66640K, 8 pages.

[13] Munera, Hector A, Hernández-Deckers, Daniel, Arenas, Germán, Alfonso, Edgar and López Iván. (2009) "Observation of a non-conventional influence of Earth's motion on the velocity of photons, and calculation of the velocity of our galaxy," 113-119 in preliminary proceedings of *Progress in Electromagnetics Research Symposium, PIERS 2009*, Beijing, March 23-27/2009. Final proccedings at www.piers.org.

[14] Munera, Hector A (2009) "Towards the reinstatement of absolute space, and possible cosmological implications," *ICFAI University Journal of Physics* **2** (2-3), 9-24.

[15] Munera, Hector A (2017) "Absolute velocity of Earth from our stationary Michelson-Morley-Miller experiment at CIF, Bogota, Colombia" 109-110 in *Abstracts International Scientific Conference Physical Interpretations of Relativity Theory PIRT2017*, Bauman Moscow State Technical University, July 3-6/2017 Moscow, Russia. Presentation not selected for inclusion in proceedings. See full text at Academia.edu and ResearchGate.

[16] Michelson, Albert Abraham (1881) "The relative motion of the Earth and the luminiferous ether," *American Journal of Science (third series)* **22** (122), 120-129.

[17] Lorentz, Hendrik Antoon (1895) *Versuch einer Theorie der elektrischen und optischen Erscheinungen in bewegten Körpern* [*Attempt at a theory of electrical and optical phenomena in moving bodies*] (Leiden: Brill).

[18] Shankland, Robert S (1964) "Michelson-Morley experiment," *American Journal of Physics* **32** (1), 16-35.

[19] Poincaré Henri (1905) *Science and Hypothesis* (Walter Scott Publishing Company). Republished in 1952 (New York, NY, USA: Dover) 244 pp. ISBN 0-486-60221-4.

[20] Thayer H S (editor and notes) (1974) *Newton's Philosophy of Nature — Selections from his Writings*, fourth printing 1974 (New York, NY, USA: Hafner Press, 1953) 207 pp. Library of Congress card number: 53-12509.

[21] Poincaré Henri (1914) *Science and Method*, translated from the 1908 French edition by Francis Maitland (London and New York: T. Nelson and Sons). Republished in 1952 (Mineola, NY, USA: Dover) 288 pp. ISBN 0-486-43269-6.

[22] Munera, Hector A (2015) "From the classical ethers of Descartes and Newton to cosmons and sagions" *Apeiron* **20** (2) 1-67. Same paper as pp 243-292 in Christopher C Fulton and Martin Kokus (eds.) (2017) *The Galileo of Palomar: Essays in Memory of Halton Arp* (Montreal, Canada: Apeiron) 358 pp. ISBN 978-1-987980-08-0 (PDF).

[23] Beck Anna (translator) and Havas Peter (consultant) (1987) *The Collected Papers of Albert Einstein, Volume 1, The Early Years, 1879-1902* (Princeton, NJ, USA: Princeton University Press) 196 pp. ISBN 0-691-08475-0.

[24] Jürgen, Renn, Schulmann Robert (editors), and Smith Shawn (translator) (1992) *Albert Einstein-Mileva Marić: The Love Letters* (Princeton, NJ, USA: Princeton University Press) 107 pp. ISBN 0-691-08886-1.

[25] Beck Anna (translator) and Howard Don (consultant) (1995) *The Collected Papers of Albert Einstein, Volume 5, The Swiss Years: Correspondence, 1902-1914* (Princeton, NJ, USA: Princeton University Press) 384 pp. ISBN 0-691-00099-9.

[26] Munera, Hector A (2010) "The Michelson-Morley experiment and Einstein's second postulate: inextricably linked" (in Spanish: "El experimento de Michelson-Morley y el segundo postulado de Einstein: inextricablemente unidos") 245-290 in Germán Guerrero Pino (editor), *Einstein: Científico y Filósofo* (Cali, Valle, Colombia: Editorial Universidad del Valle) 411 pp. ISBN 978-958-670-835-7.

[27] Drude Paul (1900) *Lehrbuch der Optik* (Leipzig: Hirzel), translated into English by C. Riborg Mann and Robert A Millikan (1902) as *The Theory of Optics*. Republished in 1959 (New York: Dover Publications).

[28] Miller Dayton C (1933) "The ether-drift experiment and the determination of the absolute motion of the Earth," *Reviews of Modern Physics* **5**, 203-242.

[29] Smoot, G F, Gorenstein M V and Miller R A (1977) "Detection of anisotropy in the cosmic background radiation," *Physical Review Letters* **39** (14), 898-901.

[30] James De Meo (2019) *The Dynamic Ether of Cosmic Space: Correcting a Major Error in Modern Science* (Ashland, Oregon, USA: Natural Energy Works) 389 pp. ISBN 978-0-9974057-1-2.

[31] Byron, Bird R, Warren, Stewart, E and Lightfoot, Edwin N (1960) *Transport Phenomena* (New York and London: John Wiley & Sons) 780 pp. Library of Congress card number: 60-11717.

[32] Lorentz, Hendrik A (1886) "On the influence of Earth's motion on light phenomena" (in French: "De l'influence du mouvement de la Terre sur les Phenomenes Lumineux"), *Archives Néerlandaises* XXI, book 2.

[33] Munera, Hector A (2006) "The evidence for length contraction at the turn of the 20th century: Non-existent" 87-102 in reference [1].

[34] Hernández-Deckers, Daniel (2005) *Data reduction in the Michelson-Morley-Miller experiment with stationary interferometer* (in Spanish: Reducción de datos del experimento con interferómetro estacionario de Michelson-Morley-Miller), B.Sc. thesis, Department of Physics, National University, Bogotá, Colombia, 77 pp.

[35] Lipa, J A, Nissen, J A, Wang, S, Stricker D A and Avaloff D (2003) "A new limit on signals of Lorentz violation in electrodynamics," *Physical Review Letters* **90**, 060403.

In: Future Relativity, Gravitation, Cosmology
Editors: V. V. Dvoeglazov et al.

ISBN: 979-8-88697-455-3
2023 Nova Science Publishers, Inc.

Chapter 11

DARK ENERGY AS ZERO ENERGY OF GRAVITY FIELD*

V. B. Morozov†
P. N. Lebedev Physical Institute, Russian Academy of Sciences, Moscow, Russia

ABSTRACT

Homogeneous isotropic empty space is described here based on its metric. The only parameter of this metric is taken to be equal to the Hubble constant. The tensor of energy-momentum of empty space is calculated. Upon this, the density of the empty homogeneous space turned out to be equal to the expected density of the Universe. In other words, the density of empty space is "dark energy." Of not less importance is the fact that the same metric imposes the exponentially increasing law of redshift asymptotically close to the Hubble law at low redshift values. The solution has no singularities.

Keywords: dark energy, metric of Universe, redshift, accelerated expansion, density of Universe

1. INTRODUCTION

There is a plain and simple kind of energy and this is the energy of the gravity field. In other words, this is the energy of an empty space, in which both substance and electromagnetic fields are absent.

Is this energy equal to zero in absence of a gravity field? There are no reasons for the energy of the empty space being equal to zero.

For calculating of the (allegedly equal to zero) energy of the empty space, the metric of the homogeneous isotropic space assists. This means that the dark energy is nothing but the energy

* First published in Apeiron, Volume 20, Hors série 4, June 2018. Reprinted by permission.

† Corresponding Author's Email: morozov@lebedev.ru.

of the empty space. Notably, it is possible that the dark matter is nothing but thickening (concentration) of the same dark energy, which - as any other energy - possesses a weight.

2. Homogeneous Universe

A. A. Friedman and A. Einstein considered the Universe filled up with substance. The paper [1] laid the foundations to the concept of the non-stationary Universe. The discovery of the cosmological red shift and A. Einstein's support [2] determined the development of cosmology for many years to come.

The effects of the zero fluctuations (of physical vacuum) have been confirmed very precisely by the well-known experiments. This is one of the greatest successes of the scientific physics. However, the attempts of a vacuum energy density determination failed both in calculations and experiments. Any estimations result in incredibly large values of the vacuum density. To be successful in connecting of the vacuum energy with the Einstein's "empty space" is possible only with aid of the "imagination play," though often some authors consider this imagination to be rather ill.

Here we shall make an attempt to describe the Universe background space having no substance based only on the relativity principle and the Riemannian geometry.

The constancy of the speed of light is the necessary condition of the relativity principle. For the metric

$$ds^2 = dx_0^2 - dx_1^2 - dx_2^2 - dx_3^2,$$

the speed constancy follows from the fact that the equation x_0 describes the wave front having the fixed speed c.

Einstein and Fokker [3] noted that the reference system with the constant (local) speed of light is described with the metric of the following kind:

$$ds^2 = v^2\left(dx_0^2 - dx_1^2 - dx_2^2 - dx_3^2\right), \tag{1}$$

notably, v is function of position coordinates[1].

The space with the metric (1) can be qualified as conformally-Galilean [4].

Let us give a consideration to the space with the homogeneous acceleration of the gravity field, i.e., with the homogeneous derivative coordinates on time of second order. In the coordinates system, in which a point has the zero speed, the components of the point acceleration look like as follows [5]:

$$\frac{d^2x^i}{dt^2} - c^2\Gamma^i_{00} = c^2 k_i.$$

---

[1] The authors [3] came from this result to the generally covariant Nordstrom equation. However, the Nordstrom scalar theory failed in the competition with the general relativity theory.

Let us require that the metric (1) is homogeneous on acceleration, i.e., that the acceleration $c^2 k_i$ does not depend on coordinates. The following metric meets this condition[2]:

$$ds^2 = \exp(2k_i x_i)(dx_0^2 - dx_1^2 - dx_2^2 - dx_3^2).$$

In the case that $k_i = 0$ is the metric of the Minkowski space, if at least one of $k_i \neq 0$, anyway, $k_0 = 0$ is the metric of the homogeneous accelerated system [8]. The assumption that the Universe as a whole has non-zero acceleration in one direction, is most unbelievable. The only possible metric left is one with the non-zero component of the vector k_0, which takes the following form:

$$ds^2 = \exp(2k_0 x_0)(dx_0^2 - dx_1^2 - dx_2^2 - dx_3^2). \tag{2}$$

The constant k_0 has the same dimensionality as the Hubble constant H_0. Now let us make the most reasonable step and replace the constant in the equation (2) with the Hubble constant.

$$ds^2 = \exp(2H_0 x_0)(dx_0^2 - dx_1^2 - dx_2^2 - dx_3^2). \tag{3}$$

The metric (3) describes the space-time continuum with the monotonously increasing scale factor (see Figure). It is essential that the Einstein's tensor of the metric (3) is non-zero.

3. Energy of Empty Space

It is well-known that the covariant divergence of the Einstein's tensor is equal to zero. This allows assuming that, following Einstein, it is proportional to the tensor of energy-momentum. The proportionality rate between these values is known from the asymptotical behavior of these tensors in the Newtonian approximation[3]. These considerations allow calculating the tensor of energy-momentum as follows:

$$t_{ik} = \frac{c^4}{8\pi G}\left(R_{ik} - \frac{1}{2}g_{ik}R\right) = \begin{pmatrix} 3p & 0 & 0 & 0 \\ 0 & -p & 0 & 0 \\ 0 & 0 & -p & 0 \\ 0 & 0 & 0 & -p \end{pmatrix}, \tag{4}$$

---

[2] This metric can be considered as a generalization of the exact result obtained by Einstein as early as in 1907 [6] for the homogeneous field of accelerations $d\tau = \exp(ax/c^2)dt$. Besides, it is compatible with the homogeneous one-dimensional metric by Harry Lass [7].

[3] The reversed Einstein equation.

notably $p = \frac{c^2 H_0^2}{8\pi G}$. The expression (4) does not depend on coordinates and thus indeed the metric (2) describes the homogeneous space-time continuum having a constant positive density and a negative pressure in every point of the space. The metric (2) has a negative scalar curvature $R = -6k_0^2 \exp(-2k_0 x)$.

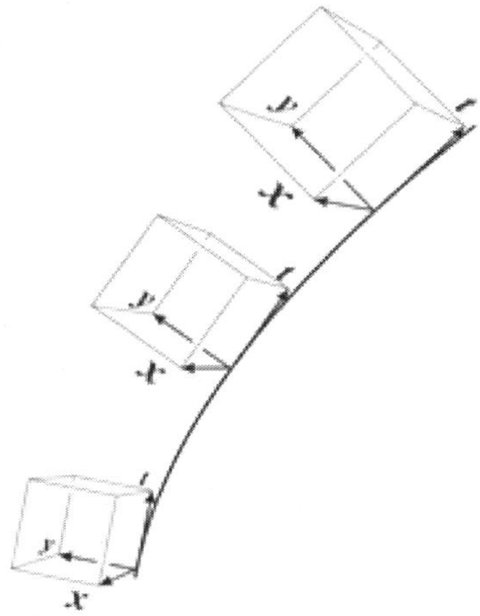

Figure 1. Attempt to depict space evolution with growing scale factor.

Surprisingly, the space density of the homogeneous metric $\rho = 3p / c^2$ coincides with the Friedman's critical density [1]

$$\rho = \frac{3H_0^2}{8\pi G}.$$

This confirms the ideas by A. A. Friedman. The value ρ coincides with the expected value of sum of densities of the dark energy and the dark matter in our Universe. Hence, ρ is the natural contender for the place of the *dark energy* of a homogeneous Universe.

In the real Universe, the tensor of energy-momentum (4) must acquire a non-zero energy flow, for example, at action of a gravity field. This way, a non-homogeneous density ρ distribution in the space is induced, mainly in massive galaxies or clusters of galaxies. This is already quite plausible model of the *dark matter*.

4. Cosmological Red Shift

From the same metric of empty space (3), it follows that for a fixed in space observer, the source own time is connected with the observer's time on the distance x_0 via the expression:

$$d\tau = \exp(k_0 x_0)\,dt.$$

If $T = x_0 / c$ is the signal delay time, then if to assume that $ck_0 = H_0$, we shall obtain the red shift being close to the Hubble law:

$$z = \exp(H_0 T) - 1 \approx H_0 T. \tag{5}$$

However, with the distance increase to a source, the red shift grows faster than it is predicted by the Hubble law, which - in fact - corresponds to the observations.

CONCLUSION

It is readily expected that the tensor of energy-momentum (3) of a homogeneous field is constant through the entire space. From this result, it follows that an observer - *in any space point and at any instant of time* - would find himself in the same conditions and would observe the identical red shift (5). This shift is equivalent to the expanding at a growing rate Universe. We see that this is not necessary that the Hubble law extrapolation to the past makes the Universe shrinking to a point.

Let us pay attention that this result was obtained with relying on neither the Friedman equation nor any hypotheses with use of the lambda member.

Perhaps this is the beginning of a new point of view on the theory of relativity [8, 9].

REFERENCES

[1] A. Friedman, *Z. Phys.* 10 (1922), pp. 377-386.

[2] Einstein, Sitzungsber. Preuss [Einstein, session report. Prussia]. *Akad. Wiss., phys.-math. Kl.*, 1931, 235-237.

[3] Einstein, A. D. Fokker. *Ann. Phys.*, 1914, 44, 321-328.

[4] V. A. Fock *Theory of space, time and gravity*, Pergamon Press 1964.

[5] M. V. Sazhin, *General relativity theory for astronomers. URL:* http://www.astronet.ru/db/msg/1170927.

[6] A. Einstein, *Jahrb. d. Radioaktivitat u. Elektronik*, 4, 411-462 (1907).

[7] H. Lass, *American Journal of Physics,* 31, pp. 274-276, 1963.

[8] V. B. Morozov, arXiv:1305.5412v4 [gr-qc].

[9] Morozov V. B. *The General Theory of Relativity, a New Iteration.* 2019. *URL:* https://www.researchgate.net/publication/331262328_The_General_Theory_of_Rclativity_a_New_Iteration/download.

In: Future Relativity, Gravitation, Cosmology
Editors: V. V. Dvoeglazov et al.

ISBN: 979-8-88697-455-3

Chapter 12

First-Order Phase Transitions of a Generalized Newman–Unti–Tamburino Cosmological Model: A Pseudo-Finsler Interfacial Geometrothermodynamics

H. V. Grushevskaya * ***and N. G. Krylova*** †
Belarusian State University, Minsk, Belarus

Abstract

While within the minimal Standard Model the electroweak phase transition is considered to be the second order, the first-order electroweak phase transitions proposed in the background of beyond Standard Models may solve some cosmological problems, like the generation of the baryon asymmetry of the universe. We develop a geometrothermodynamic model of cosmological first-order vacuum-phase transition based on a theory of the first-order phase transition in a contact statistical manifold. Nucleation of true-vacuum bubbles with axially symmetric generalized Newman–Unti–Tamburino (NUT)-like metrics have been considered. A manifold of evolving bubbles is a pseudo-Finsler statistical manifold of such thermodynamic system. Finsler-Lagrange dynamics has been studied taking into account the heterogeneity of nucleation processes, notably a relaxation times distribution for bubbles. We has shown that a NUT-like variable parameter n in the theory is the gauge parameter of the scalar field which plays a role of fifth dimension and the transition of which from the steady state into strongly oscillating state is accompanied by the first-order phase transition between true and false vacua.

PACS 98.70.Bp,98.80.-K, 05.70.Fh.

Keywords: cosmological first-order phase transitions, Newman–Unti–Tamburino metric, geometrothermodynamics, pseudo-Finsler statistical manifold

* Corresponding Author's Email: grushevskaja@bsu.by.
† Corresponding Author's Email: krylovang@bsu.by.

1. Introduction

According to standard cosmology, an electroweak phase transition associated with the spontaneous symmetry breaking for a scalar field Φ has occurred at the electroweak scale $t \sim 10^{-11}s$ [1, 2]. Within the minimal Standard Model the electroweak phase transition proceeds through crossover [3, 4]. The first-order electroweak phase transitions beyond the Standard Model [5, 6] may implement the observed baryon asymmetry through baryogenesis. One of the simplest extensions of the Standard Model includes a real singlet scalar field v as the additional term [7, 8] $L_v = \frac{1}{2}(\partial_\mu v)^2 - \frac{\lambda}{4}(v^2 - m^2)^2 - \chi H^\dagger H v^2$ in Lagrangian where H is the Brout–Englert–Higgs field, λ and χ are dimensionless coupling constants, m is a mass parameter.

Theories of cosmological electroweak phase transitions are based on the cosmological scenario that a spacetime of early universe is a void spacetime. The void space is populated domains of different vacua, which are called true and false vacua and characterized by a non-zero and a zero vacuum averages $\langle\Phi\rangle$ of the scalar field Φ, respectively [9]. This vacuum-like state of early universe is unstable or metastable one. The phase transition (inflation, the inflationary epoch) is terminated by decay of the unstable false-vacuum with energy release expended on universe expansion. The theory of the cosmological vacuum transition of crossover type allows the existence of large number of instanton configurations (instantons, domain walls) for the scalar field Φ with a very shallow Mexican hat potential in order that an energy density for the scalar-field vacuum in domains does not dominate the total energy density of the universe in agreement with the present observational results [10, 11]. If the crossover is the case then a cosmological constant Λ is the vacuum energy. Relic gravitational waves (GWs) in the vacuum transition of crossover type could be produced by the collisions between light particles mass-acquired by Φ and the walls of neighbouring touching domains. On the other hand, Λ-term within Einstein equations can describe the dark energy (see, for example, [12] and references therein) the latter can be interpreted as a fluid with negative pressure [13]. The pressure takes on a negative value at the first-order phase transition while the matter is in a metastable state. ΛCDM (Lambda cold dark matter) cosmological models [14] and electrically charged AdS (anti–de Sitter) black holes [15, 16] exhibit a liquid-gas-like first-order phase transition culminating in critical points that resemble the phase diagram of a Van-der-Waals fluid [17].

In the phenomenological Van-der-Waals–Maxwell theory of first-order three-dimensional (3D) phase transition the principal indistinguishability of supercooled vapor and superheated liquid is appreciated as the phase coexistence in a given temperature range. Because the phase coexistence, the instanton configurations are absent and, accordingly, the scalar field Φ undergoes a jump only after completion of the first-order phase transition with formation of a domain of size more than a radius of event horizon, to do not conflict with the Universe homogeneity observed experimentally. In the first-order 3D-phase transition the domains were not visualized experimentally owing to the arbitrariness of domains orientation. Therefore, within the framework of the Kolmogorov–Mehl–Johnson–Avrami approach [18] models of phase transition between true and false vacua are limited by nucleation stage with a very small rate v_w of the movement of nuclei (bubbles, domain wall). Since v_w is very small the domain growth up to the sizes exceeded the event horizon radius is time-expanded. Accordingly, the cosmological phase transition of Van-der-Waals type would be too slow. In the approximation of small rates v_w the domain growth stage is modeled by overlapping with the following coalescence of true vacuum bubbles due to that a nucleation rate deflects from an exponential form (the nucleation with one relaxation time) [19]. A shock wave at deflagration type of the transition (shock-wave regime) [20] or a friction force of plasma of light particles from false-vacuum domain acting on the wall for detonation bubbles after the transient process of deflagration (see [21] and references therein) could cause relic GWs from the slow cosmological electroweak phase transition [22, 23]. The model rates of detonation and deflagration bubbles are $v_w < 1$ and $v_w < 0.01$ respectively. Owing to non-relativistic values of rates v_w, collisions of the bubbles would chaotize sources of GWs only, and a large number of GWs would be observed as a stochastic background of gravitational potential. The production of gravitational waves in the cosmological phase transitions was simulated as a perturbation of symmetric Friedmann–Lemaître–Robertson–Walker metrics [24, 25]. But, a metric in the cosmological first-order phase transition should be referred to an evolved domain surrounded by false-vacuum domains. Because of that the using of axially symmetric metrics such as axially symmetric Newman–Unti–Tamburino (NUT)-like metrics is much more suitable.

A surface energy (energy of domain-wall surface tension) σ is minimized and a vacuum expectation $\langle\Phi\rangle$ undergoes a jump with domain growth. However, $\langle\Phi\rangle$ fluctuates strongly at first-order phase transition. The existence of the strong oscillating field Φ can be revealed as the gravitational field strength variations near an average value (see [26] and references therein), that will produce a stochastic background of gravitational potential. The decrease of $\langle\Phi\rangle$ is equivalent to the increase of surface tension force σ with following slowing-down of domain wall growth and collapse of the domain. The case of the fluctuations of Φ owing to the Hubble parameter variation during inflation with the increase of σ to the large values was considered in [27, 28]. At large σ the gravitational waves emission proceeds due to collision between particles acquired mass and false-vacuum domain walls (see [29] and references therein). The influence of the high value of domain growth rate on the GW-production is the open problem.

The geometrothermodynamics has been proposed to model the gravitating systems [30, 31]. Based on the assumption that there exists a correspondence between AdS models and Conformal Field Theory Models (CFT), cosmological models are analyzed as thermodynamic objects with a Bekenstein–Hawking entropy and a Hawking temperature [32, 33, 34]. Recently, series of modern cosmological models are constructed in the framework of thermodynamic approaches, which are developed on the grounds of standard thermodynamics of the first-order phase transitions. In particular, there are examined the large class of asymptotically AdS black holes [16], charged de Sitter black holes [35], black hole in dark matter background [36]. But, the cosmological models which are based on Van-der-Waals–Maxwell theory of the liquid–gas first-order phase transitions, should meet a demand of the agreement of different thermodynamic representations (for example, between energy (black-hole mass M) and Gibbs free energy G representations) to provide comprehensive description of the model. This demand extremely complicates the calculations because of that a thermodynamic geometry of the cosmological model has been constructed without taking into account the cosmological constant Λ or scalar fields [16]. To describe a model of the universe filled by barotropic fluid (dark energy or a scalar field named quintessence) with the Weyl's canonical coordinates and negative cosmological constant Λ, the static solutions for axially symmetric metric for black holes in vacuum have been acquired [37, 38]. Axially symmetric NUT metrics pretend to play a role of such metrics. Although some their thermodynamic properties have been studied in $f(R)$ gravity [39] and within Rastall theory of gravity [40], due to a complexity of the NUT-models a thermodynamic geometry has not been constructed yet. It has been proposed in [41] that the features of behaviour of galaxy rotation curves can be considered as effects of velocity on matter. Because of the above, the generalized NUT metric can be considered as an appropriate metric for cosmologic phase transition description.

However, the Van-der-Waals–Maxwell gas model is not invariant with respect to the physically meaningful Legendre transformations [42, 43]. Moreover, Einstein equations are not conformally invariant because of the Weyl tensor (anisotropic pressure) does not vanish. The thermodynamics of cosmological models, such as Friedmann–Lemaître–Robertson–Walker ones, includes effects of the non-null Weyl tensor by introducing a viscosity parameter and relaxation time (inverse expansion coefficient) [44, 45]. But, relaxation processes are absent in the Van-der-Waals-Maxwell gas model. Thus, a more realistic theory of the first-order phase transition has to be used to develop geometrothermodynamic approaches to cosmology.

In [46] a theory of the two-dimensional (2D) first-order phase transition in a contact statistical manifold that describes interfacial systems with an electrocapillary mechanism of the energy dissipation and with relaxation time distributions was proposed. This theory has not limitations which are in the Van der Waals-like phase transition theories. Velocity effects are typical for 2D first-order phase transition with the distribution of relaxation times that can be described in a configuration space $\left(\vec{r},\ \xi,\ \vec{r}_\xi,\ \dot{\xi}_s,\ s\right)$ possessing the properties of contact manifolds which are invariant with respect to Legendre transformations. Here $\vec{r}$ is a 2D radius-vector, ξ is a time-like coordinate, dots are defined the derivatives with respect to an evolution parameter s. The triad $\left\{\vec{r}, \dot{\vec{r}}, \dot{\vec{\xi}}\right\}$ for spacetime (r, ξ) defines our contact manifold which is called the thermodynamic phase space $\mathcal{T} = \left\{\vec{r}, \dot{\vec{r}}, \dot{\vec{\xi}}\right\}$, $\dot{\vec{r}} = \vec{r}_\xi\left(\frac{ds}{d\xi}\right)^{-1}$ [47, 48, 49]. The tangent space $\mathcal{T}_T = \left\{\dot{\vec{r}}, \dot{\vec{\xi}}\right\}$ is a subspace of the bundle $\mathcal{T}$. The tangent space $\mathcal{T}_T$ has the vertical (hidden) $\mathcal{V}_\xi$ and horizontal $\mathcal{H}$ direct sum components: $\mathcal{T}_T = \mathcal{V}_\xi \oplus \mathcal{H}$. Thus, $\mathcal{V}_\xi$ is the vertical subspace generated by $\dot{\vec{\xi}}$, and $\mathcal{H}$ is the horizontal distribution. The bundle $\mathcal{T}$ of our problem can be redefined by the dynamical-system phase-space $\left\{\vec{r}, \vec{r}_\xi\right\}$, which is augmented by the subspace spanned by the tangent vector $\frac{ds}{d\xi}$: $\mathcal{T} = \left\{\vec{r}, \vec{r}_\xi, \frac{ds}{d\xi}\right\}$. The bundle $\mathcal{T}$ includes the four-dimensional (4D) space of the two pairs $\left\{\vec{r}, \vec{r}_\xi\right\}$ and admits a non-trivial connection determined by the vector $\frac{\partial}{\partial s}$. Here $\frac{\partial}{\partial s}$ is a Reeb vector field of the contact manifold (see [50, 51, 52] and references therein). Therefore, $\mathcal{T}$ is a fifth-dimension (5D) contact manifold. Since the first-order phase transition proceeds on the interface, the metric is an axially symmetric one.

In this paper the interfacial geometrothermodynamics of cosmological first-order vacuum phase transition in a pseudo-Finsler contact statistical manifold is studied. With this purpose in mind, nucleation of true-vacuum bubbles with axially symmetric generalized Newman–Unti–Tamburino-like metrics are considered.

2. A Geometrothermodynamical Background of the Problem

Dynamics of the first-order phase transition is considered as evolution of a system from the metastable state to the stable one (true vacuum). In the classical nucleation theory the free energy of a nucleus (bubble) is determined by two factors: the volume term promoting the nucleation at the supersaturation and the surface term that is proportional to the surface tension and facilitates new-phase nucleus decays. Minimizing of the free-energy, one can obtain a critical radius so that the nuclei larger than the critical radius will grow up, while the smaller ones will decay. Instead of the parameter of nucleus size we will utilize notion of bubble relaxation time (nucleus life time) and will suppose that each nucleus is characterized by own individual relaxation time.

It is useful to choose the relaxation time τ_c of the critical nucleus as the unit measure of relaxation time. Because of this we will utilize further the dimensionless parameter $\tau_i/\tau_c \rightarrow \tau_i$. Let define the evolution parameter s by the following relation: $\tau_i = \Delta t_i/\Delta s$. Evolution of the system in the limit of infinite number of nuclei relaxation times can be described by the differential [53, 54]

$$\dot{\xi} = \frac{dt}{ds}. \tag{1}$$

It has been shown that transition of a system in the metastable state to the stable one is associated with narrowing of nuclei relaxation times $\dot{\xi}$ distribution [55]. There are two assumption in the nucleation theory: homogeneous and heterogeneous nucleations [56]. In homogeneous nucleation one assumes that all nuclei are produced simultaneously and evolve in identical manner; at that, the time-like coordinate t can be considered as the evolution parameter of the such type system. The phase space $\left(\vec{r},\ \dot{\vec{r}}\right)$ consists of individual phase elements described by coordinates and velocities. The statistical distribution of such macroscopic system having N new phase nuclei can be defined with the partition function $p\left(\vec{r}_1, \dot{\vec{r}}_1;\ \vec{r}_2, \dot{\vec{r}}_2;\ \ldots;\ \vec{r}_N, \dot{\vec{r}}_N;\ t\right)$. In heterogeneous nucleation theory one has to consider the production/decay of new-phase nuclei during long-standing nucleation process. In this case the statistical distribution of the macroscopic system has to take into account the moments of nuclei production t_i and the nuclei relaxation times τ_i such that the partition function depends on relaxation time distribution: $p\left(\vec{r}_1, \dot{\vec{r}}_1, t_1, \tau_1;\ \vec{r}_2, \dot{\vec{r}}_2, t_2, \tau_2;\ \ldots;\ \vec{r}_N, \dot{\vec{r}}_N, t_N, \tau_N\right)$. In macroscopic system when N is large enough ($N \rightarrow \infty$), one can pass to the continuous limit $p\left(\vec{r}(s), \dot{\vec{r}}(s), t(s), \tau(s)\right)$. Taking into consideration the parameterization (1) of relaxation times the partition function can be represented as $p\left(\vec{r}(s), \dot{\vec{r}}(s), t(s), \dot{\xi}(s)\right)$. Then, from the physical point of view the parameterization (1) can be interpreted as follows. The evolution parameter s determines the number of produced nuclei (without subtraction of decayed ones). The relation (1) implies that during the period Δt in the system either large number of small nuclei (which decay quickly) or smaller number of long-lived nuclei (growth of which will results in phase transition progression) can be produced.

So, in the continuous medium limit $\left\{\vec{r}, \dot{\vec{r}}, t, \dot{\xi}\right\}$ is the extended configuration space, the transition from micro- to macrostate of the system is described with the parameterization

(1). It should be noted that the scale invariance is a feature of the phase transitions due to their self-similarity [57], the introduced parameterization (1) is in agreement with this behaviour.

Let consider the entropy production ΔS at the heterogenous first-order phase transition [46, 58]:

$$\Delta S = \sum_a \Delta S_a = -\sum_a F_a \Delta t_a + \sum_a E_a \Delta t_a, \tag{2}$$

where ΔS_a is the entropy production, F_a and E_a are the Helmholtz free energy and the internal energy, respectively, for the a-th relaxation process:

$$-\sum_{a=1}^{N} F_a \Delta t_a = \Delta s \sum_{a=1}^{N} \tau_a F_a = \Delta s F = \ln Z, \tag{3}$$

$$\sum_{a=1}^{N} E_a \Delta t_a = -\Delta s \sum_{a=1}^{N} \tau_a E_a = -\sum_{a=1}^{N} \frac{\partial \ln Z}{\partial \Delta t_a} \Delta t_a = -N \Delta s \frac{\partial \ln Z}{\partial \Delta s}. \tag{4}$$

Here Z is the statistical sum

$$Z = \int \prod_{a=1}^{N} DX_a p(\vec{r}_1, \ldots, \vec{r}_N, t) \exp\left\{ \Delta s \sum_a \tau_a \left[H(X_a) + i\Gamma(X_a) \right] \right\}, \tag{5}$$

$\int \prod_{a=1}^{N} DX_a$ represents the path integral $\int DX = \int \prod_{a=1}^{N} DX_a$ over configurations X_a, $X_a = \{\vec{r}_1(t_1); \ldots; \vec{r}_{a-1}(t_{a-1}); \vec{r}_a(t_a + \Delta t_a); \vec{r}_{a+1}(t_{a+1}); \ldots; \vec{r}_N(t_N)\}$; after changing $s \to s$ one gets $\Delta s \tau_a = -\Delta t_a$ being an inverse temperature for the a-th relaxation process. The Hamiltonian $H(X_a)$ and the decay rate $\Gamma(X_a)$ of each individual relaxation process are renormalized by its relaxation time τ_a, the role of inverse temperature β in the transition state would have been played by the parameter $(-\Delta s)$. The formulas (3, 4) are determined the Helmholtz free energy F and the internal energy E for the first-order phase transition as

$$F = \sum_{a=1}^{N} \tau_a F_a = \frac{1}{\Delta s} \ln Z, \quad E = \sum_{a=1}^{N} \tau_a E_a / N = \frac{\partial \ln Z}{\partial \Delta s}. \tag{6}$$

Taking into consideration the expression (5), we expand the expressions (3–4) in series over small parameter Δs as

$$\begin{aligned} -\sum_a F_a \Delta t_a &= \ln Z|_{\Delta s=0} + \sum_a \Delta t_a \\ \times \left[\frac{-\int p\{X_a\}(H(X_a) + i\Gamma(X_a)) DX_a}{Z} \right] \Bigg|_{\Delta t_a = 0} &= -\sum_a \Delta t_a\, C, \end{aligned} \tag{7}$$

$$\begin{aligned} \sum_a E_a \Delta t_a &= -\sum_a \frac{\partial \ln Z}{\partial \Delta t_a} \Delta t_a \\ &= \frac{1}{Z} \sum_a \Delta s \int p\{X_a\}[\tau_a (H(X_a) + i\Gamma(X_a))] DX_a \end{aligned} \tag{8}$$

and passing to the limit $\dot{\xi} = -\frac{\partial t}{\partial s} = -\lim\limits_{\Delta s\to 0}\frac{\Delta t}{\Delta s}$ one can find an action on the statistical manifold of the first-order phase transition:

$$\Delta S = \int \left[\dot{\xi}C - L(\dot{\xi})\right] ds. \tag{9}$$

3. Thermodynamics of a Generalized NUT-Metric

The NUT metric reads

$$dl^2 = \left(1 - \frac{R_g R + 2n^2}{R^2 + n^2}\right)\left(dx^0 + 4n\sin^2\frac{\theta}{2}d\phi\right)^2 - \frac{dR^2}{1 - \frac{R_g R + 2n^2}{R^2 + n^2}} \\ -(R^2 + n^2)(d\theta^2 + \sin^2\theta d\phi^2), \tag{10}$$

where n is a NUT parameter, $R_g = 2G^*M/c^2$ is a gravitational radius, x^0 is a time coordinate; R, θ, ϕ are spatial spherical coordinates. In the case of R^2, $R_g R \gg n^2$ this metric behaves like Schwarzschild metric or Reissner–Nordström one if a term $\frac{{q_e}^2 + q_m^2}{R^2+n^2}$ is added to $\Phi(R) = 1 - \frac{R_g R + 2n^2}{R^2+n^2}$. Here q_e and q_m are electric and magnetic charges, respectively. Since a following approximation $\Phi^{-1}(R)R^2 \approx (R^2 + n^2)$, holds, one can transform the expression (10) as

$$dl^2 = \Phi(R)\left(dx^0 + 4n\sin^2(\theta/2)\, d\phi\right)^2 - \frac{dR^2}{\Phi(R)} - \frac{R^2}{\Phi(R)}(d\theta^2 + \sin^2\theta\, d\phi^2). \tag{11}$$

At R^2, $R_g R < n^2$ one can add a correction $(\Lambda/3)\left(R^2 + 5n^2\right) - \frac{8(\Lambda/3)n^4 - {q_e}^2 - q_m^2}{R^2+n^2}$ to $\Phi(R)$ in the metric (10). Then for negative cosmological constant Λ, $\Lambda < 0$ the NUT-metric is transformed to the form:

$$dl^2 = \tilde{\Phi}(R)\left(dx^0 + 4n\sin^2(\theta/2)\, d\phi\right)^2 - \frac{dR^2}{\tilde{\Phi}(R)} - \frac{R^2}{\tilde{\Phi}(R)}(d\theta^2 + \sin^2\theta\, d\phi^2) \tag{12}$$

with a following lapse function $\tilde{\Phi}(R)$ [26, 58]:

$$\tilde{\Phi}(R) = 1 - \frac{1}{3}|\Lambda|\left(R^2 + 5n^2\right) + \frac{q_e^2 + q_m^2}{n^2 + R^2} \\ -\frac{1}{n^2 + R^2}\left(RR_g - \frac{8|\Lambda|n^4}{3} + 2n^2\right). \tag{13}$$

While $R^2 \ll n^2$, $R \sim R_g$, $R_g^2 \sim -2n^2 + |\Lambda|n^4 + {q_e}^2 + q_m^2$ (or $R^2 \ll \left|n^2\right|$, $R \sim R_g$, $R_g^2 \sim 2\left|n^2\right| + \Lambda n^4 + {q_e}^2 + q_m^2$ for $n^2 < 0$), the substitution $R^2/\tilde{\Phi}(R) \approx R^2/\left(1 - \frac{|\Lambda|}{3}R^2\right) \to R^2$ leads to the metric (12) obtains the features of anti–de Sitter metric ($\tilde{\Phi}(R) \to 1 - |\Lambda|R^2/3$, $\Lambda < 0$).

The expression for the electrical potential may be taken as

$$\varphi_e = q_e\left(\frac{1}{R_+} - \frac{1}{R}\right),$$

where q_e/R_+ defines the asymptotic value of the electric potential measured at infinity. A Hawking temperature T and an entropy S may be found from

$$S = \pi R_+^2, \quad T = \frac{1}{4\pi} \frac{\partial \Phi(R)}{\partial R}\bigg|_{R \to R_+},$$

where outer-horizon radius R_+ is given by the zeroes of the lapse function $\Phi(R)$: $\Phi(R_+) = 0$ (Here and in subsequent the tilde symbol will be omitted). The Hawking temperature T of charged generalized-NUT black hole has the form

$$T = \frac{1}{4\sqrt{\pi S}} \left[\frac{2S}{\pi n^2 + S} \left(-\frac{\Lambda S}{\pi} - \frac{5\Lambda n^2}{3} + 1 \right) - \frac{\Lambda S}{\pi} \right. \\ \left. - \frac{\pi}{\pi n^2 + S} \left(q_m^2 + 8\Lambda n^4 - 2n^2 + \frac{S\varphi_e^2}{\pi} \right) + \frac{5\Lambda n^2}{3} - 1 \right]. \tag{14}$$

A Gibbs free energy specific to the mixed grand-canonical-canonical ensemble involving a varying electrical charge q_e and a fixed magnetic charge q_m is as follows

$$G = M - TS - \varphi_e q_e.$$

For a charged generalized NUT black hole the Gibbs free energy G in terms of the variables φ_e, q_m and S reads

$$G = \frac{1}{4\sqrt{\pi S}} \left[-\frac{S^2}{\pi\left(\pi n^2 + S\right)} \left(\pi \Lambda n^2 - \Lambda S + 2\pi \left(-\frac{5\Lambda n^2}{3} + 1 \right) \right) \right. \\ + S\left(-2\Lambda n^2 - 5\Lambda n^2 + 3\right) + \frac{\pi q_m^2 \left(2\pi n^2 + 3S\right)}{\pi n^2 + S} + 2\pi n^2 \left(-\frac{5\Lambda n^2}{3} \right. \\ \left. \left. + 1 \right) - \frac{\pi\left(3S + 2\pi n^2\right)}{\pi n^2 + S} \left(\frac{8\Lambda n^4}{3} - 2n^2 \right) - \frac{\left(S + 2\pi n^2\right) S \varphi_e^2}{\pi n^2 + S} \right]. \tag{15}$$

In Figs. 1 and 2 the dependencies of Gibbs energy G on the Hawking temperature T and the plot of T with the entropy S can hold cusp points. In cusp points, there is sudden change in a physical solution such as a pitchfork bifurcation, and spontaneous breaking of system symmetry happens [59].

4. An Axially Symmetric Metric of Thermodynamic Phase Space

It was shown in [23, 53, 54] that the thermodynamic metric (9) holds a following explicit form:

$$dl^2 = L(\vec{r}, \dot{\vec{r}}, \dot{\xi}) ds^2 = -\tilde{p}|V| r^5 e^{\frac{2|V|\xi}{r}} \frac{\dot{\xi}}{\dot{r}} d\xi\, ds + \tilde{U}(\xi, r)\dot{\xi}\, ds^2 \\ + m \frac{(\dot{r}^2 + r^2\dot{\phi}^2)}{2\dot{\xi}} ds^2 \tag{16}$$

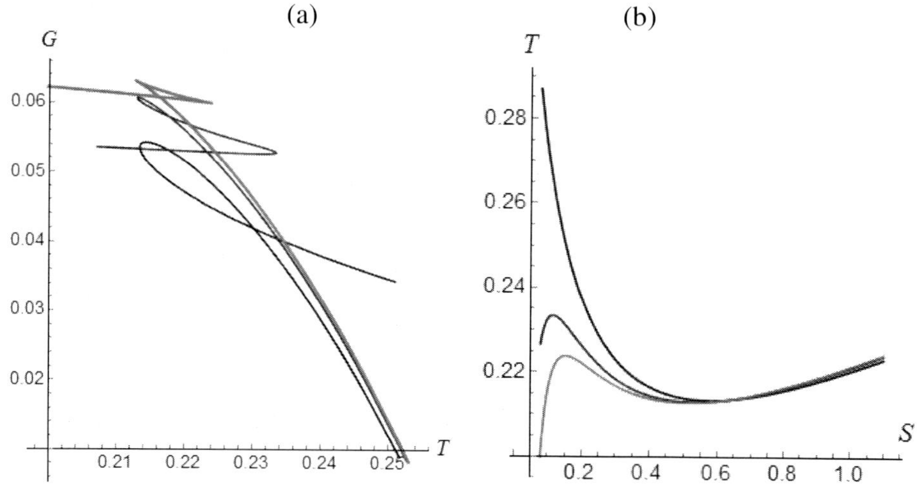

Figure 1. The dependencies of Gibbs energy on the Hawking temperature (a) and the temperature on the entropy (b). The black, blue and magenta curves correspond to $n = 0.1$, $n = 0.05$ and $n = 0.01$, respectively, at $\Lambda = -0.01$, $q_m = 0.09$, $\varphi_e = 0.6$.

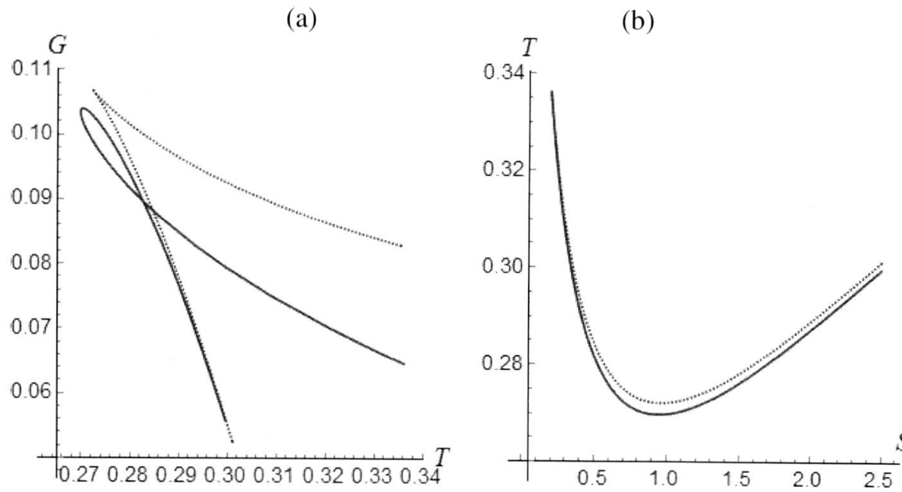

Figure 2. The dependencies of Gibbs energy on the Hawking temperature (a) and the temperature on the entropy (b). The blue and black curves correspond to $n = 0.1$ and $n = 0.01$, respectively, at $\Lambda = -0.1$, $q_m = 0.09$, $\varphi_e = 0.01$.

and describes the first-order 2D-phase transition with relaxation time distributions. Here $L(\vec{r}, \dot{\vec{r}}, \xi, \dot{\xi})$ is the Lagrangian in polar coordinates (r, ϕ), $\dot{r} = \frac{dr}{ds}$ and $\dot{\phi} = \frac{d\phi}{ds}$, the function $\tilde{U}(\xi, r)$ named the reduced potential function is expressed through the electrocapillary-force potential function $U(\xi, r)$ as $\tilde{U}(\xi, r) = U(\xi, r) - \tilde{p}|V|r^5 e^{\frac{2|V|\xi}{r}}/\dot{r} - mc^2$ and can be chosen in the following forms:

$$\begin{aligned}\tilde{U}_1 = \tilde{p}|V|^2\xi \left(e^{\frac{2\xi|V|}{r}} \left(\frac{1}{2}r^4 + \frac{1}{3}r^3\xi|V| + \frac{1}{3}r^2\xi^2|V|^2 + \frac{2}{3}r\xi^3|V|^3 \right) \right. \\ \left. - \frac{4}{3}\xi^4|V|^4 \mathrm{Ei}\left(\frac{2\xi|V|}{r}\right)\right) \frac{\dot{\xi}}{\dot{r}} - \tilde{p}\left(e^{\frac{2\xi|V|}{r}} \left(-\frac{3r^5}{4} + r^4\xi|V| \right.\right. \\ \left. + \frac{3}{4}r^3\xi^2|V|^2 + \frac{5}{6}r^2\xi^3|V|^3 + \frac{11}{6}r\xi^4|V|^4 - \frac{\xi^5|V|^5}{3} \right) \\ \left. - \frac{2}{3}\xi^5|V|^5 \left(6 - \frac{\xi|V|}{r}\right) \mathrm{Ei}\left(\frac{2\xi|V|}{r}\right)\right) - mc^2,\end{aligned} \tag{17}$$

$$\begin{aligned}\tilde{U}_2 = \tilde{p}\left\{ \left[-\frac{4}{3}r^5 + \frac{16}{15}(|V|\xi)r^4 + \frac{1}{30}(|V|\xi)^2 r^3 \right.\right. \\ \left. + \frac{1}{45}(|V|\xi)^3 r^2 + \frac{1}{45}(|V|\xi)^4 r + \frac{2}{45}(|V|\xi)^5 \right] e^{\frac{2|V|\xi}{r}} \\ \left. + \frac{4}{45}\left((|V|\,\xi)^5 - \frac{(|V|\xi)^6}{r} \right) \mathrm{Ei}\left(\frac{2|V|\xi}{r}\right) \right\} - mc^2,\end{aligned} \tag{18}$$

$$\tilde{U}_3 = \tilde{U}_2 - \frac{4}{45}(|V|\,\xi)^5 \,\mathrm{Ei}\left(\frac{2|V|\xi}{r}\right)\tilde{p}; \tag{19}$$

V characterizes the expansion rate, $\tilde{p}$ and m are the model parameters, Ei is the special function *exponential integral*. The thermodynamic metric (16) is implemented on a contact statistical manifold for the first-order 2D-phase transition. In Fig.3 the typical dependencies of potential functions U_1 and U_2 on radius r are shown. As one can see, the approximate function U_2 has not so sharp local minimum as the function U_1.

Let us rewrite the metric (16) in the following form:

$$\begin{aligned}dl^2 = L(\vec{r}, \dot{\vec{r}}, \dot{\xi})ds^2 = -\tilde{p}|V|r^5 e^{\frac{2|V|\xi}{r}} \frac{\dot{\xi}}{\dot{r}} d\xi\, ds + \tilde{U}(\xi, r)\dot{\xi}\, ds^2 \\ + \frac{m}{2\dot{\xi}}\left(dr^2 + r^2 d\phi^2\right),\end{aligned} \tag{20}$$

where $d\vec{r} = \vec{r}_\xi\, d\xi$, $d\phi = \phi_\xi\, d\xi$. Then one can observe that the metric (20) with a pseudo-lapse function $\dot{\xi}$ and a pseudo-time s in the entropy representation is very similar to the NUT-metric (12).

Further fitting the lapse functions Φ with the thermodynamic velocity $\dot{\xi}$ we will demonstrate that there exist first-order phase transitions of the spacetime with the generalized NUT-like metric because a Berwald curvature for the metric (20) diverges and changes sign.

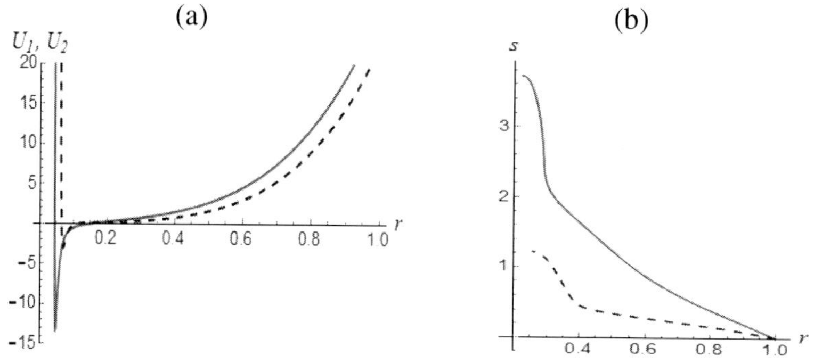

Figure 3. (a) The typical dependencies of potential functions U_1 (solid) and U_2 (dashed) on radius r at following values of parameters: $\dot{r}/\dot{\xi} = 0.027$, $V = 0.224$, $\xi = 2$, $\tilde{p} = 1$. (b) The isotherms $s(r)$ (geodesics) obtained by numerical calculations. Thermodynamic metric function parameters and potential functions used: $\tilde{p} = 1$, $mc^2 = 1$, $m = 10^{-5}$, U_1 (solid); (b) $\tilde{p} = 1$, $m = 1$, $c = 1$, U_2 (dashed).

One can transform the metric (16) to a following auxiliary pseudo-Finsler metric function:

$$F_i^2 = -L(\vec{r}, \vec{\dot{r}}, \dot{\xi})\dot{\xi} \equiv A\frac{\dot{\xi}^3}{\dot{r}} + B_i\dot{\xi}^2 - m\frac{(\dot{r}^2 + r^2\dot{\phi}^2)}{2}, \quad i = 1, 2; \tag{21}$$

where $A = \tilde{p}|V|r^5 e^{\frac{2|V|\xi}{r}}$, $B_i = -\tilde{U}_i$, $i = 1, 2$. The main difference between the cases with the potential functions consists in the fact that all terms in F_1^2 depend on the thermodynamic velocity $\dot{r}$. For F_2^2 the first and third terms depend on $\dot{r}$ only.

Let us use an Euler–Lagrange equation to calculate the thermodynamic velocity $\dot{\xi}(r)$ determining the lapse function $f(R)$ and to approximate it by a set of lapse functions $\Phi(R)$ (13). The Euler–Lagrange equation reads

$$\frac{dy^i}{ds} + 2G^i = 0, \qquad G^i = \frac{1}{4}g^{il}\left\{2\frac{\partial g_{jl}}{\partial x^k} - \frac{\partial g_{jk}}{\partial x^l}\right\}y^j y^k, \tag{22}$$

where $g_{ij}\left(x^k, y^k\right) = \frac{1}{2}\frac{\partial^2 F^2}{\partial y^i \partial y^j}$, $x^j = (\xi(s), r(s), \phi(s))$ and $y^j \equiv (\dot{\xi} = \frac{d\xi}{ds}, \dot{r}(s), \dot{\phi}(s))$, $(r \equiv R)$. The semispray coefficients G^i in the explicit form are given by the expressions

$$G^1 = \frac{A}{4\Delta_2}\left[\frac{1}{2}A_\xi\frac{\dot{\xi}^{6\,4}}{\dot{r}} + B_\xi\frac{\dot{\xi}^5}{\dot{r}^3} + \frac{1}{2}B_r\frac{\dot{\xi}^4}{\dot{r}^2} + \frac{3}{2}Cr\frac{\dot{\xi}^2\dot{\phi}^2}{\dot{r}^2}\right]$$
$$-\frac{C}{8\Delta_2}\left[2A_\xi\frac{\dot{\xi}^3}{\dot{r}} + B_\xi\dot{\xi}^2 + 3A_r\dot{\xi}^2 + 2B_r\dot{\xi}\dot{r}\right],$$
$$G^2 = \frac{3}{4}\frac{A}{\Delta_2}\left[-\frac{1}{2}A_r\frac{\dot{\xi}^4}{\dot{r}^2} + \frac{1}{2}B_\xi\frac{\dot{\xi}^4}{\dot{r}^2} + Cr\frac{\dot{\xi}\dot{\phi}^2}{\dot{r}}\right] \tag{23}$$
$$-\frac{B}{4\Delta_2}\left[A_\xi\frac{\dot{\xi}^4}{\dot{r}^2} + 2A_r\frac{\dot{\xi}^3}{\dot{r}} + B_r\dot{\xi}^2 - Cr\dot{\phi}^2\right],$$
$$G^3 = \frac{1}{r}\dot{r}\dot{\phi},$$

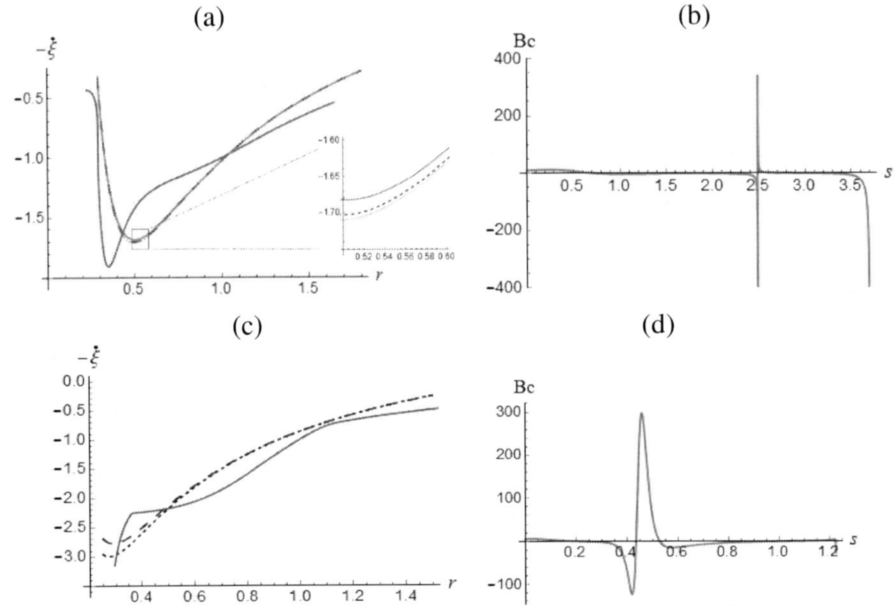

Figure 4. The dependencies $\dot{\xi}$ on r along the geodesics $r(s)$ and its approximations by lapse functions $\Phi(R)$, $R = r$ (13) (a,c). The dependencies of Berwald curvature $B_C(s)$ on evolution parameter (b,d). Thermodynamic metric-function parameters and potential functions used for numerical simulation at $V = 0.224, \tilde{p} = 1$: (a, b) $mc^2 = 1, \quad m = 10^{-5}$, U_1; (c, d) $m = 1, \quad c = 1, U_2$. Approximation parameters: (a) $\Lambda = -0.01, r_g = 2.7$, $q_e^2 + q_m^2 = 0.82, n = 0.1$ (grey solid line), $n = 0.05$ (blue dashed line), $n = 0.01$ (green dotted line); (c) $\Lambda = -0.1, r_g = 2.2, q_e^2 + q_m^2 = 0.55, n = 0.1$ (blue solid line), $n = 0.01$ (black dashed line).

$$\Delta_2 = \frac{3A^2}{4}\left(\frac{\dot{\xi}}{\dot{r}}\right)^4 + AB\left(\frac{\dot{\xi}}{\dot{r}}\right)^3 - \frac{3AC}{2}\frac{\dot{\xi}}{\dot{r}} - \frac{BC}{2},$$

$$A_\xi = \frac{\partial A}{\partial \xi}, \quad A_r = \frac{\partial A}{\partial r}, \quad B_\xi = \frac{\partial B}{\partial \xi}, \quad B_r = \frac{\partial B}{\partial r}.$$

Figs. 4a,c demonstrate the dependencies of $\dot{\xi}(r)$ along the thermodynamic geodesics $r(s)$ obtained by the numerical simulation for potential functions U_1 and U_2. The corresponding approximations by the generalized NUT-metric has been performed (see Fig. 4). As one can see, the dependence of $\dot{\xi}(r)$ for potentials U_i, $i = 1, 2$ are in good agreement with the lapse functions $\Phi(R)$, $R = r$. It should be noted that the lapse functions $\Phi(R)$ are weakly dependent on the NUT-parameter n. For example, the curves obtained for $n = 0.1$, 0.05 and 0.01 practically coincide, as shown in Fig. 4a.

The dependencies of Gibbs energy G on the Hawking temperature T with the fitting parameters obtained from the geodesics determined by the metric function (16) are presented above in Fig. 1. The dependence with two cusp bifurcations and an additional sudden-change point named swallowtail catastrophe (Fig. 1a) for the case of the pseudo-Finsler function F_1^2 is typical for the phase transitions of first order. In next section a Berwald curvature will testify existence of such transitions.

5. Thermodynamic Curvature

Similarly to second order phase transitions, the first order phase transitions can be geometrically interpreted as singularities of a space curvature [42, 60]. The latter implies that the thermodynamic curvature tends to infinity in the points where derivations of a thermodynamic potential diverge. The zero-derivations give a significant arbitrariness in a choice of the metric function.

Let $\rho(x, y)$ be a statistical distribution of nuclei in an ensemble defined by initial conditions $\left(x_0^i, y_0^i\right)$. We consider the generalized continuity equation [61]:

$$\begin{gathered}
\frac{\partial \rho}{\partial s}+\sum_{i=1}^{3}\left[\frac{\delta}{\delta x^i}\left(\rho y^i\right)+\frac{\delta}{\delta y^i}\left(\rho \dot{y}^i\right)\right] \\
=\frac{\partial \rho}{\partial s}+\sum_{i=1}^{3}\left[\frac{\partial}{\partial x^i}\left(\rho y^i\right)-N_i^j \frac{\partial}{\partial y^j}\left(\rho y^i\right)+\frac{\partial}{\partial y^i}\left(\rho \dot{y}^i\right)\right] \\
=\frac{\partial \rho}{\partial s}+\sum_{i=1}^{3}\left[\rho \frac{\partial y^i}{\partial x^i}+y^i \frac{\partial \rho}{\partial x^i}+\rho \frac{\partial \dot{y}^i}{\partial y^i}+\dot{y}^i \frac{\partial \rho}{\partial y^i}\right. \\
\left.-\rho N_i^j \frac{\partial y^i}{\partial y^j}-N_i^j y^i \frac{\partial \rho}{\partial y^j}\right]=\frac{\partial \rho}{\partial s}+\sum_{i=1}^{3}\left[\dot{x}^i \frac{\partial \rho}{\partial x^i}+\dot{y}^i \frac{\partial \rho}{\partial y^i}\right] \\
+\sum_{i=1}^{3}\left[\rho \frac{\partial y^i}{\partial x^i}-3 \rho N_i^i-N_i^j y^i \frac{\partial \rho}{\partial y^j}\right]=0,
\end{gathered} \tag{24}$$

where $N_j^i = \partial G^i/\partial y^j$ is nonlinear connection. Due to the scale invariance of phase transition [57, 62], the distribution function can not depend upon evolution parameter s, therefore one has the identity

$$\frac{d\rho}{ds}=\frac{\partial \rho}{\partial s}+\sum_{i=1}^{3}\left[\dot{x}^i \frac{\partial \rho}{\partial x^i}+\dot{y}^i \frac{\partial \rho}{\partial y^i}\right]=0. \tag{25}$$

Accounting of the equality $\frac{\partial y^i}{\partial x^i} = 0$ and the relation (25) eq. (24) takes the form

$$3N_i^i+N_i^j y^i \frac{\partial \ln \rho}{\partial y^j}=0. \tag{26}$$

The nonlinear connection coefficients N_j^i are homogeneous functions of degree one in y^l, therefore the following relation holds $N_j^i = \frac{\partial N_j^i}{\partial y^k} y^k = N_{jk}^i y^k$. Then the equation (26) can be rewritten as:

$$\left(3N_{jk}^j+N_{ik}^j y^i \frac{\partial\ \ln \rho}{\partial y^j}\right) y^k=0. \tag{27}$$

The relation (27) can be transformed into another form. First, let us define the covariant vector $Y_i = g_{ik}y^k$; its inner product with original vector is given as $Y_i y^i = g_{ik}y^k y^i = F^2$. Then we have

$$\bar{Y}_i \equiv \frac{Y_i}{F^2}, \quad \bar{Y}_i\, y^i = 1. \tag{28}$$

Using the last identity the relation (27) can be written in the form

$$\begin{aligned}&\left(3N^j_{jk}\,\bar{Y}_i\, y^i + N^j_{ik} y^i \frac{\partial\, \ln\rho}{\partial y^j}\right) y^k \\ &= \left(3N^j_{jk}\,\bar{Y}_i\, + N^j_{ik}\frac{\partial\, \ln\rho}{\partial y^j}\right) y^i y^k = 0.\end{aligned} \tag{29}$$

In the used model there exist a geometrical object – the mean Berwald curvature tensor E_{ij} [63]

$$E_{ij} = \frac{1}{2}\frac{\partial^2}{\partial y^i \partial y^j}\sum_k \frac{\partial G^k}{\partial y^k},$$

with the following property:

$$E_{kl}y^k y^l = 0.$$

Let us assume that the following equality takes place:

$$3N^j_{jk}\frac{g_{il}y^l}{F^2} + N^j_{ik}\frac{\partial\, \ln\rho}{\partial y^j} = \alpha E_{ik}, \tag{30}$$

where α is some function depending on x^i, y^i. In what follows we restrict ourselves with the case when α is a numeric parameter. We define the Berwald scalar curvature B_C as a convolution of E-tensor with the metric tensor:

$$B_C = g^{ij}E_{ij}. \tag{31}$$

Then eq. (30) can be transformed to the equation

$$3g^{ik}N^j_{jk}\frac{g_{il}y^l}{F^2} + g^{ik}N^j_{ik}\frac{\partial\, \ln\rho}{\partial y^j} = \alpha B_C. \tag{32}$$

For the Gibbs distribution [62], we have $\rho = e^{\theta(\Omega - H)}$ where Ω is the thermodynamic potential, θ is the inverse temperature, H is the internal energy. Then the equation (32) can be rewritten as:

$$3g^{ik}N^j_{jk}\frac{g_{il}y^l}{F^2} + \theta g^{ik}N^j_{ik}\frac{\partial\Omega}{\partial y^j} - \theta g^{ik}N^j_{ik}\frac{\partial H}{\partial y^j} = \alpha B_C. \tag{33}$$

The expression for the internal energy will be use in the form:

$$H = \frac{m(\dot{r}^2 + r^2\dot{\varphi}^2) + F^2}{\dot{\xi}^2} - mc^2.$$

For subsequent it is convenient in eq. (33) to transfer from differentiation over variables y^j to differentiation on volume. With this in mind let us use the formula $d\tilde{V} = \sigma_F dr\, d\varphi\, dt$ for a volume differential in Finsler space that the differential is written as:

$$W \equiv \frac{d\tilde{V}}{(ds)^3} = \sigma_F y^1 y^2 y^3. \tag{34}$$

The following formulas for derivatives hold:

$$\frac{\partial}{\partial y^i} = \frac{\partial W}{\partial y^i}\frac{\partial}{\partial W} = \sigma_F y^j y^k \frac{\partial}{\partial W}, \qquad i \neq j \neq k, \tag{35}$$

$$\frac{\partial}{\partial W} = \frac{1}{\sigma_F}\left(\frac{1}{y^2 y^3}\frac{\partial}{\partial y^1} + \frac{1}{y^3 y^1}\frac{\partial}{\partial y^2} + \frac{1}{y^1 y^2}\frac{\partial}{\partial y^3}\right). \tag{36}$$

With account of (35–36) eq. (33) is transformed into the following one:

$$\begin{gathered} 3g^{ik}N^j_{jk}\frac{g_{il}y^l}{F^2} + \theta\sigma_F g^{ik}\frac{\partial\Omega}{\partial W}\left(N^1_{ik}y^2y^3 + N^2_{ik}y^1y^3 + N^3_{ik}y^1y^2\right) \\ -\theta g^{ik}N^j_{ik}\frac{\partial H}{\partial y^j} = \alpha B_C, \end{gathered} \tag{37}$$

the last can be resolved in respect to $\partial\Omega/\partial W$:

$$\begin{gathered} \frac{\partial\Omega}{\partial W} = \frac{1}{\theta\sigma_F\left(\beta^1y^2y^3 + \beta^2y^3y^1 + \beta^3y^1y^2\right)} \\ \times\left(\alpha B_C - 3g^{ik}N^j_{jk}\frac{g_{il}y^l}{F^2} + \theta g^{ik}N^j_{ik}\frac{\partial H}{\partial y^j}\right), \end{gathered} \tag{38}$$

$$\beta^l \equiv g^{ik}N^l_{ik}.$$

Let us remind the standard definition of compressibility through the thermodynamic potential Ω:

$$\kappa = -\frac{1}{\tilde{V}}\left(\frac{\partial\tilde{V}}{\partial p}\right)_T = \frac{1}{\tilde{V}}\left(\frac{\partial^2\Omega}{\partial\tilde{V}^2}\right)^{-1}.$$

Then

$$\begin{aligned} \kappa &= -\frac{1}{W}\left(\frac{\partial W}{\partial\Delta p}\right)_T = \frac{1}{W}\left(\frac{\partial^2\Omega}{\partial W^2}\right)^{-1} \\ &= \frac{1}{W}\left(\frac{3}{\theta\sigma_F^2}\left(\frac{1}{y^2y^3}\frac{\partial}{\partial y^1} + \frac{1}{y^3y^1}\frac{\partial}{\partial y^2} + \frac{1}{y^1y^2}\frac{\partial}{\partial y^3}\right)\right. \\ &\quad\times\left.\left[\frac{\alpha B_C - 3g^{ik}N^j_{jk}\,g_{il}y^l/F^2 + \theta g^{ik}N^j_{ik}\partial H/\partial y^j}{\left(\beta^1y^2y^3 + \beta^2y^3y^1 + \beta^3y^1y^2\right)}\right]\right)^{-1}. \end{aligned} \tag{39}$$

It is possible to demonstrate that compressibility κ can approximately be obtained in the following form [54, 64]

$$\kappa = -\frac{1}{2}\frac{\theta\sigma_F}{\alpha B_C}\left(\beta^1y^2y^3 + \beta^2y^3y^1 + \beta^3y^1y^2\right). \tag{40}$$

Phase transitions are revealed by a plateau of the isotherms $s(r)$ presented in Fig. 3b and 5b,e, respectively. As one can see from Figs. 4b,d and 5c, 5f, the Berwald curvature B_C changes sign in the phase transitions. Comparison between Figs. 4b,d shows that the increase of the number of cusp points and the swallowtail point narrow the phase transition sharply but do not change its qualitative character. Because of this we will further analyze the geometrodynamics in the case of U_2. The lapse function $\dot{\xi}$ is nonzero in Fig.5a. It trends to zero in Fig. 5d in the phase-transition region, and, accordingly, one observes an emerging black hole because B_C of the metastable "liquid phase" state is negative. After the phase transition the state of the black hole is stabilized if $\dot{\xi} \to 0$, as B_C becomes positive (see Fig. 5f). If $\dot{\xi}$ does not tend to zero, after the phase transition B_C tends to a negative value (see Fig. 5a,c) and, accordingly, the state of the black hole is unstable. Accordingly, the compressibility (40) diverges and changes sign.

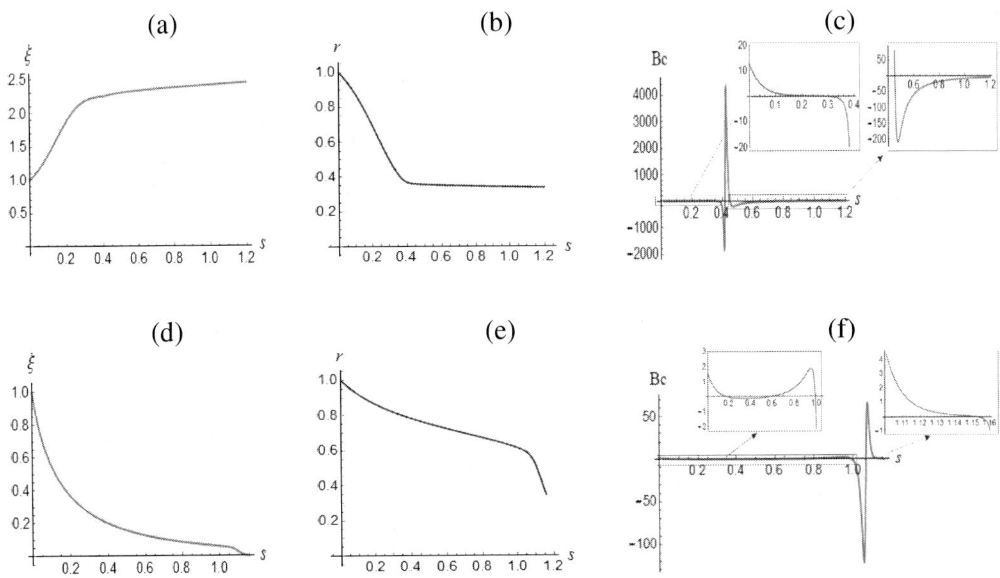

Figure 5. The dependencies $\dot{\xi}(s)$ (a, d), $r(s)$ (b, e), $B_C(s)$ (c, f) along the geodesics for the thermodynamic metric spaces with the potential U_2. Metric function parameters used for numerical simulation: $\tilde{p} = 1, \quad m = 1, \quad c = 1$; $V = 0.05$ (a-c), $V = 10$ (d-f).

Thus, since in the entropy representation the Berwald curvature B_C diverges and changes sign and meanwhile in the free-energy representation the black NUT-hole is metastable, one can propose that it is possible that the true vacuum bubbles are originated at very small values of the NUT-parameter, when the mass-term contribution in the NUT-metric increases. In next section we prove this conjecture using the simplified potential function $\tilde{U}_3$.

6. Scalar Field in the Phase Transition between False and True Vacua

A spacetime metric can be obtained from the metric (16) by the Wick rotations $\xi \to -i\,\tau$ and $s \to = -i\,x^0$:

$$d\tilde{l}^2 = \tilde{p}\,|V|\,r^5 \exp\left(\frac{-i2|V|\tau}{r}\right)\frac{\dot{\xi}}{\dot{r}}d\tau\,dx^0 - \tilde{U}_3(-i\tau, r)\dot{\xi}\,(dx^0)^2 \\ -\frac{m}{2\dot{\xi}}(dr^2 + r^2 d\phi^2), \tag{41}$$

where $dR = \dot{r}dx^0$, $d\phi = \dot{\phi}dx^0$, and $R \equiv r$. Let $\dot{\xi}$ be a known lapse function $f(R)$: $\dot{\xi} = f(R)m/2$. Then, the metric (41) can be considered as a metric of 4D-surface $\theta{=}\pi/2$ in the contact 5D-manifold:

$$d\tilde{l}^2 = \frac{m}{2}f(R)\left[-\tilde{U}_3(-i\tau, R)\left(dx^0\right)^2 + \exp\left(\frac{-i2|V|\tau}{R}\right)\frac{\tilde{p}\,|V|\,R^5}{\dot{r}}d\tau\,dx^0\right] \\ -\frac{\left[dR^2 + R^2\left(d\theta^2 + \sin^2\theta d\phi^2\right)\right]}{f(R)}, \\ \theta = \pi/2\,. \tag{42}$$

Let us choose some another 4D-surface in the contact 5D-manifold so that angle θ takes arbitrary values, while ranges of variables R, ϕ are dynamically limited by the variable τ. This can be realized in the following way. Let us impose a gauge condition on the scalar field $\varphi \equiv e^{\frac{-i|V|\tau}{R}}$:

$$\frac{4\tilde{p}}{45}\frac{(|V|\tau)^6}{R} = 1. \tag{43}$$

We consider the case of small values of rate $|V| \to 0$ when the function Ei $\left(-i\frac{2|V|\tau}{R}\right)$ turns to a constant value $\Re e\ \mathrm{Ei}(i0) = -\infty$, $\Im m\ \mathrm{Ei}(i0) = \pi/2$. Then, while the condition (43) is imposed, the scalar field φ oscillates strongly,

$$\tilde{U}_3(-i\tau, R) = -\mathrm{Ei}\,(i0) \ \text{ at } \ |V| \to 0, \quad \text{and} \quad R = \frac{4\tilde{p}}{45}(|V|\tau)^6. \tag{44}$$

Due to the constraint (43), one has $\dot{r} = \frac{dR}{d\tau}\frac{d\tau}{-i\,dx^0} = \frac{24\tilde{p}}{45}|V|^6\tau^5\frac{d\tau}{-i\,dx^0}$. Then changing $\left(dx^0\right)^2 \to (m\,/2)\left(dx^0\right)^2$ the metric (42) takes the form

$$dl^2_{renorm} = f(R)\left[\Im m\,\mathrm{Ei}\,(i0) - i\exp\left(\frac{-i2|V|\tau}{R}\right)\frac{45}{24}\frac{R^5}{|V|^5\tau^5}\right]\left(dx^0\right)^2 \\ -\frac{\left[dR^2 + R^2\left(d\theta^2 + \sin^2\theta d\phi^2\right)\right]}{f(R)}, \tag{45}$$

where $dl^2_{renorm} = d\tilde{l}^2 - \Re e\,\mathrm{Ei}\,(i0)\,f(r)\left(dx^0\right)^2$.

Now, one has to exclude the 5-th coordinate τ. A phase $\frac{2|V|\tau}{R}$ in exponential term $\exp\left(\frac{-i2|V|\tau}{R}\right)$ of 5-th dimension varies in the range from 0 up to π. Therefore, the phase $\frac{|V|\tau}{R}$ is defined by the angle $\theta/2$:

$$dl^2_{renorm} = f(R)\left\{\Im m\, \mathrm{Ei}\,(i0) - i\left(\cos\frac{\theta}{2} - i\sin\frac{\theta}{2}\right)^2 \frac{45}{24}\frac{R^5}{|V|^5\tau^5}\right\} \times \left(dx^0\right)^2 - \frac{\left[dR^2 + R^2\left(d\theta^2 + \sin^2\theta d\phi^2\right)\right]}{f(R)}. \qquad (46)$$

At $\tau \to \infty$ the expression (46) is an expansion over a small parameter $1/\tau$:

$$dl^2_{renorm} = f(R)\left[d\tilde{x}^0 - i\frac{\left(\cos\frac{\theta}{2} - i\sin\frac{\theta}{2}\right)^2}{48\,\Im m\, \mathrm{Ei}\,(i0)}\frac{45R^5 d\tilde{x}^0}{|V|^5\tau^5}\right]^2 - \frac{\left[dR^2 + R^2\left(d\theta^2 + \sin^2\theta d\phi^2\right)\right]}{f(R)},$$

where $d\tilde{x}^0 = \left(\Im m\, \mathrm{Ei}\right)^{1/2}(i0)\, dx^0$. Then, taking into account that $0 \le \frac{R}{|V|\tau} \le 1$ to impose the constraint of the form $2\pi\left(\frac{R}{|V|\tau}\right)^5 d\tilde{x}^0 = d\phi$ on the angle ϕ, one gets

$$dl^2_{renorm} = f(R)\left[d\tilde{x}^0 - \frac{45i}{24\pi}\left(\cot\frac{\theta}{2} - i\right)^2 \sin^2\frac{\theta}{2}\, d\phi/(2\pi)\right]^2 - \frac{\left[dR^2 + R^2\left(d\theta^2 + \sin^2\theta d\phi^2\right)\right]}{f(R)}. \qquad (47)$$

Comparison between (12) and (47) gives the expression for the NUT-parameter, which becomes an imaginary constant at $\theta \to \pi$, 0:

$$n\sin^2\frac{\theta}{2}\bigg|_{\theta\to\pi,\,0} \equiv -\frac{45i}{192\pi^2}\sin^2\frac{\theta}{2}\left(\cot\frac{\theta}{2} - i\right)^2\bigg|_{\theta\to\pi,\,0} = \pm\frac{45i}{192\pi^2}. \qquad (48)$$

Let us show that the parameter V is the control parameter of the model. To do this, let express the field $\varphi = e^{-i\frac{2|V|\tau}{R}}$ through $\xi(x^0)$, $r(x^0)$. Due to the homogeneity of degree two of functions G^j (23) in velocities $y^j = \left(\frac{dt}{ds}, \frac{dr}{ds}, \frac{d\phi}{ds}\right)$ the changes obtained by the Wick rotation:

$$s \to -ix^0, \quad \frac{dx^j}{ds} \to \frac{dx^j}{d(-ix^0)} = i\frac{dx^j}{dx^0},$$

$$\frac{d^2x^j}{ds^2} \to \frac{d^2x^j}{d(-ix^0)^2} = -\frac{d^2x^j}{dx^{0\,2}}, \quad x^j = (t, r, \phi)$$

give the same system (22) of equations for the geodesics $\xi(x^0)$, $r(x^0)$, $\phi(x^0)$ so that $\tau \to \xi(x^0)$, $R = r(x^0)$. Then one gets

$$\varphi = e^{-i\frac{2|V|\tau}{R}} = e^{-i\frac{2|V|\xi(x^0)}{r(x^0)}},$$

where $\xi(x^0)$ and $r(x^0)$ are the solution of the system of equations (22).

Numerical simulations of geodesics in the configurational space, isotherms $s(r)$, and appropriate fields φ for both cases have been performed. The numerical results are presented in Fig. 6. As one can see in Fig 6c, for very small values of the parameter V the field $\varphi(x_0)$ does not have minima that corresponds to the absence of phase transition on the isotherm in Fig 6a. At increase of V the field $\varphi(x_0)$ is characterized by the existence of two minima corresponding to different vacuum states (Fig. 6f). The geodesic has a typical form for a dynamical system with a limit circle and this can be associated with the presence of isotherm plateau (Figs. 6d–e) [65].

Because the field φ is stationary in the absence of the phase transition (Fig. 6a–c), the NUT-parameter takes on a large imaginary value and the spacetime is similarly to free-mass anti–de Sitter one. The phase transition is characterized by a change of field state from a steady state to strongly oscillating one (Fig. 6d–f). Meanwhile the NUT-parameter (48) can become zero. When the NUT-parameter is close to zero, the mass-term contribution in the generalized NUT metric (12–13) increases so that the metric becomes the metric of curved spacetime during the phase transition. Interpreting of this appearing mass as non-zero vacuum average of the field, we conclude that the transition from the false vacuum to the true one has occurred.

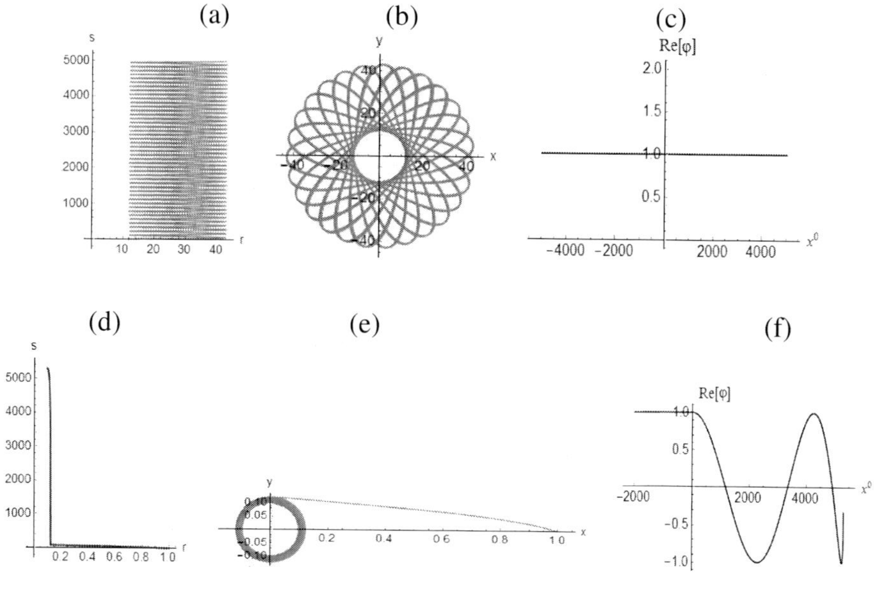

Figure 6. Isotherms $s(r)$ (a, d), geodesics (b, e) in configurational space with the potential $\tilde{U}_3$ and scalar field $\varphi(x^0)$ (c, f) at $V = 10^{-25}$, $\tilde{p} = 10^{-8}$ (a-c) and $V = 7 \cdot 10^{-3}$, $\tilde{p} = 1$ (d-f).

Conclusion

So, phase transitions in the axially-symmetric cosmological generalized NUT-like model occur through the bifurcations of pitchfork type. In the dependence of Gibbs free energy on the Hawking temperature appears caustic of swallowtail type, at the parameters of which the generalized NUT spacetime is in metastable state. The NUT parameter is determined by the scalar field, potential of which takes on a constant value before the phase transition, that corresponds to false vacuum in the cosmological model. The spacetime curves when this field transits from steady state to strongly oscillating state, that reveals transition of false vacuum into true one.

References

[1] Gogberashvili M. (2018) *J. Advances in High Energy Physics* **2018**, 4653202.

[2] Ghosh B. (2016) *Pramana* **87**, 43.

[3] Gurtler M, Ilgenfritz E-M, Schiller A. (1997) *Phys. Rev.* D **56**, 3888.

[4] Csikor F, Fodor Z, Heitger J. (1999) *Phys. Rev. Lett.* **82**, 21.

[5] Witten E. (1984) *Phys. Rev.* D **30**, 272.

[6] Kamionkowski M, Kosowsky A, Turner M S. (1994) *Phys. Rev.* D **49**, 2837

[7] Anderson G W, Hall L J. (1992) *Phys. Rev.* D **45**(8), 2685

[8] Brauner T, Tenkanen T V I, Tranberg A, Vuorinen A, Weir D J. (2017) *JHEP* **1703**, 007.

[9] Linde A D. (1990) *Physics of elementary particles and inflationary cosmology* (Moscow: Science)

[10] Zeldovich Y B, Kobzarev I Y, Okun L B. (1974) *J. Exper. Theor. Phys.* (Zh. Eksp. Teor. Fiz.) **67**, 3.

[11] Kempfer B, Lukash B, Paal G. (1990) *Physics of elementary particles and atomic nucleus* **22**, 63.

[12] Rao V U M, Suryanarayana G, Aditya Y. (2016) *Advances in Astrophysics* **1**, 62.

[13] Dutta S, Jain A, Soni R. (2013) *JHEP* **2013**, 060.

[14] Padmanabhan T. (2003) *Phys. Rept.* **380**, 235.

[15] Banerjee R, Modak S K, Roychowdhury D. (2012) *JHEP* **1210**, 125.

[16] Chaturvedi P, Das A, Sengupta G. (2017) *Eur. Phys. J.* C **77**, 110.

[17] Sahay A, Sarkar T, Sengupta G. (2010) *JHEP* **10**, 118.

[18] Avrami M. (1939) *J. Chem. Phys.* **7**, 1103; (1940) *J. Chem. Phys.* **8**, 212; (1941) *J. Chem. Phys.* **9**, 177.

[19] Jinno R, Leeb S, Seong H, Takimoto M. (2017) CTPU-17-28 KEK-TH-1990; *arXiv:1708.01253v2.*

[20] Megevand A, Ramirez S. (2018) *arXiv:1710.06279v2.*

[21] Weir D J. (2018) HIP-2017-06/TH; *arXiv:1705.01783v2.*

[22] Balazs C, Fowlie A, Mazumdar A, White G. (2017) *Phys. Rev.* D **95**, 043505.

[23] Ivanov I P. (2017) *Prog. Part. Nucl. Phys.* **95**, 160.

[24] Turner M S, Weinberg E J, Widrow L M. (1992) *Phys. Rev.* D **46**, 2384.

[25] Hiramatsu T, Kawasaki M, Saikawa K. (2014) *J. Cosmol. Astropart. Phys.* **2014**, 31.

[26] Vladimirov Y S. *Geometrophysics* (BINOM, Moscow, 2010).

[27] Vilenkin A, Ford L H. (1982) *Phys. Rev.* D **26**, 1231.

[28] Starobinsky A A. (1982) *Phys. Lett.* **117**, 175.

[29] Ken'ichi Saikawa. (2017) *Universe* **3**, 40.

[30] Hawking S W. (1975) *Commun. Math. Phys.* **43**, 199.

[31] Bekenstein J D. (1973) *Phys. Rev.* D **7**, 2333.

[32] Hawking S W, Page Don N. (1983) *Commun. Math. Phys.* **87**, 577.

[33] Cvetic M, Gubser S S. (1999) *JHEP* **07**, 010.

[34] Chaturvedi P, Sengupta G. (2015) *Preprint arXiv:1501.07748v1.*

[35] Simovic F, Mann R B. (2018) *arXiv:1807.11875v1.*

[36] Xu Z, Hou X, Wang J. (2018) *arXiv:1711.04538v2.*

[37] Multamki T, Vilja I .(2006) *Phys. Rev.* D **74**, 064022.

[38] Gutirrez-Pieres A C., Lopez-Monsalvo C S (2013) *Phys. Lett.* B **718**, 1493.

[39] Nashed G G L, Bamba K. (2019) *arXiv:1902.08020v1.*

[40] Sakti M F A R, Suroso A, Zen F P. (2019) *arXiv:1901.09163v1.*

[41] Schaf J. (2018) *J. Modern Physics* **9**, 1883.

[42] Bravetti A, Lopez-Monsalvo C S, Nettel F, Quevedo H. (2014) *J. Geom. Phys.* **81**, 1.

[43] Bravetti A, Lopez-Monsalvo C S, Nettel F. (2015) *Entropy* **17**, 6150.

[44] Hawking S. (1966) *Astrophys. J.* **145**, 544.

[45] Bittencourt E, Salim J, dos Santos G B. (2014) *Int. J. Nonlin. Phenom. in Complex Sys.* **17**, 352.

[46] Balan V, Grushevskaya H V, Krylova N G, Neagu M. (2016) *Int. J. Nonlin. Phenom. in Complex Sys.* **19**, 223.

[47] Quevedo H. (2007) *J. Math. Phys.* 48, 13506.

[48] Quevedo H, Quevedo M N. (2011) *EJTP - Zacatecas Proc.* **II**, 1.

[49] Quevedo H, Sanchez A, Taj S, Vazquez A. (2011) *Gen. Rel. Grav.* 43 1153.

[50] Boyer C P. (2011) *SIGMA* **7**, 058.

[51] Arnold V I, Novikov S P. *Dynamical Systems IV: Symplectic Geometry and Its Applications* (Springer, Berlin, 2001).

[52] Mrugała R. (2000) *Rep. Math. Phys.* **46**, 461.

[53] Krylova N G. (2017) *Proceedings of NAS of Belatus. Ser. Phys.-Math.* **53**(3), 66. (in Russian)

[54] Krylova N G, Grushevskaya H V. (2018) *Proceedings of NAS of Belatus. Ser. Phys.-Math.* **54**(1), 84. (in Russian)

[55] Nemilov S V. *Optical study of materials* (SPbSU ITMO Publishing, Saint-Petersburg, 2009). (in Russian)

[56] Kuni F, Shchekin A K, Grinin A P. (2009) *Physics-Uspekhi (Advances in Physical Sciences)* **171**, 345. (in Russian)

[57] Gitterman M, Halpern V. *Phase Transitions* (World Scientific, London, 2004).

[58] Grushevskaya H V, Krylova N G. (2018) *J. of Phys.* CS **1051**, 012013.

[59] Thompson J M T. *Instabilities and Catastrophes in Science and Engineering* (Wiley, New York, 1982).

[60] Bravetti A, Nettel F. (2014) *Phys. Rev.* D **90**, 044064; arXiv:1208.0399v2.

[61] Voicu N. (2011) *Progress In Electromagnetics Research.* **113**, 83.

[62] Landau L D, Lifshits E M. *Statistical physics* (Science, Moscow, 1976).

[63] Bao D, Chern S S, Shen Z. *An introduction to Riemann-Finsler geometry* (Springer, Berlin, 2000).

[64] Balan V, Grushevskaya H, Krylova N, Neagu M, Oana A. (2013) *Applied Sciences* **15**, 30.

[65] Neagu M., Krylova N G, Grushevskaya H V. (2013) *J. Math. Phys.* **54**, 031508.

In: Future Relativity, Gravitation, Cosmology
Editors: V. V. Dvoeglazov et al.
ISBN: 979-8-88697-455-3

Chapter 13

Fermion with Three Mass Parameters

V. V. Kisel[1,*]***, V. A. Pletyukhov***[2,†]***, E. M. Ovsiyuk***[3,‡]***,***
Ya. A. Voynova[4,§] ***and V. M. Red'kov***[4,¶]
[1] Belarus State University of Informatics and Radio-Electronics
[2] Brest State University, Belarus
[3] Mozyr State Pedagogical University, Belarus
[4] Institute of Physics, National Academy of Sciences of Belarus

Abstract

In the chapter, starting from a 20-component wave function in general Gel'fand-Yaglom approach a new wave equation for spin 1/2 fermion, which is characterized by three mass parameters, is derived. In absence of external field, three involved bispinors obey three separate Dirac-like equations with different masses M_1, M_2, M_3. In presence of external electromagnetic fields or non-Euclidean background with non-vanishing Ricci scalar curvature, the main equation is not split into separated equations, instead a quite definite mixing of three Dirac-like equations arises. It is shown that a generalized equation for Majorana particle with three mass parameters exists as well, such a generalized Majorana equation is not split into three separated equations in curved space-time if the Ricci scalar of the space-time model does not vanish.

PACS 1130, 0230, 0365.

Keywords: Lorentz group, extended sets of representations, generalized wave equation, fermion, mass parameters, electromagnetic field, gravitational field.

1. Introduction: Fermion with Three Mass Parameters

In the context of existence of the similar neutrinos of different masses, we examine a possibility within the theory of relativistic wave equations to describe a spin 1/2 particle with

*Corresponding Author's Email: vasiliy-bspu@mail.ru.
†Corresponding Author's Email: terphys@brsu.brest.by.
‡Corresponding Author's Email: e.ovsiyuk@mail.ru.
§Corresponding Author's Email: voinyuschka@mail.ru.
¶Corresponding Author's Email: redkov@ifanbel.bas-net.by.

three mass parameters. In general, existence of more general wave equations than commonly used ones is well known within the Gel'fand–Yaglom formalism – see references [1] – [58] and also books [59]–[61].

Recently, a model for a spin 1/2 particle with two mass parameters was developed [56], [57]. Properties of that model are as follows. The main wave equation for a 16-component field is presented in spin-tensor form, it is formulated with the use of the 4×4 Dirac matrices. For two main bispinors – they completely determine the initial 16-component wave function – in absence of external fields two separate Dirac-like equations are derived, they differ in masses M_1 and M_2. However, in presence of external electromagnetic field or gravitational one with non-vanishing Ricci scalar, the wave equation for two bispinor does not split into separated equations, instead a quite definite mixing of two Dirac-like equations arises. It was shown that a generalized equation for Majorana particle with two mass parameters exists as well, such a generalized Majorana equation is not trivial as well if the Ricci scalar does not vanish.

In the present chapter, we extend analysis from [56] to a fermion with three mass parameters. The model is based on extended set of irreducible representations of the Lorentz group (adhere notations from [61])

$$T = (0, \frac{1}{2}) \oplus (\frac{1}{2}, 0) \oplus (0, \frac{1}{2})' \oplus (\frac{1}{2}, 0)' \oplus (1, \frac{1}{2}) \oplus (\frac{1}{2}, 1). \tag{1}$$

Omitting rather laborious and technical analysis in the frames of Gel'fand–Yaglom approach, we start with the resulting system of equations in spin-tensor form

$$\begin{gathered}
c_1 \hat{\partial}(\gamma_\mu \Psi_\mu) + \frac{c_3}{\sqrt{6}} \Big(\hat{\partial}(\gamma_\mu \Psi_\mu) - 4(\partial_\mu \Psi_\mu) \Big) + M(\gamma_\mu \Psi_\mu) = 0, \\
c_2 \hat{\partial} \Psi_0 - i \frac{4c_4}{\sqrt{6}} \left(-\frac{1}{4} \hat{\partial}(\gamma_\mu \Psi_\mu) + (\partial_\mu \Psi_\mu) \right) + M \Psi_0 = 0, \\
-\frac{2 f c_3^*}{\sqrt{6}} \left(\partial_\lambda (\gamma_\mu \Psi_\mu) - \frac{1}{4} \gamma_\lambda \hat{\partial}(\gamma_\mu \Psi_\mu) \right) + i \frac{2 g c_4^*}{\sqrt{6}} \left(\partial_\lambda \Psi_0 - \frac{1}{4} \gamma_\lambda \hat{\partial} \Psi_0 \right) \\
+ M \left(\Psi_\lambda - \frac{1}{4} \gamma_\lambda (\gamma_\mu \Psi_\mu) \right) = 0,
\end{gathered} \tag{2}$$

where 20-component wave function consists of a bispinor Ψ_0 and a vector-bispinor Ψ_μ; the system contains a number of numerical parameters: the quantities c_1, c_2 are real; c_3, c_4 are complex; $f, g \in \{\pm 1\}$. Physical sense of these parameters becomes clear below. Dirac matrices are specified in Weyl spinor basis

$$\begin{gathered}
\gamma_\mu = \frac{1}{i} \begin{vmatrix} 0 & \sigma^\mu \\ \sigma^\mu & 0 \end{vmatrix}, \quad \sigma^4 = \begin{vmatrix} i & 0 \\ 0 & i \end{vmatrix}, \\
\sigma^1 = \begin{vmatrix} 0 & 1 \\ 1 & 0 \end{vmatrix}, \sigma^2 = \begin{vmatrix} 0 & -i \\ i & 0 \end{vmatrix}, \sigma^3 = \begin{vmatrix} 1 & 0 \\ 0 & -1 \end{vmatrix}.
\end{gathered} \tag{3}$$

In Minkowski space, the ict-metric is used.

2. Reformulation of the Initial Equations

The system (2) may be presented differently, which is substantial point for further studying.

Let us act on the first equation in (2) by operator $\frac{1}{4}\gamma_\lambda$, it yields

$$\frac{1}{4}c_1\gamma_\lambda\hat{\partial}(\gamma_\mu\psi_\mu) + \frac{c_3}{\sqrt{6}}\left(\frac{1}{4}\gamma_\lambda\hat{\partial}(\gamma_\mu\psi_\mu) - \gamma_\lambda(\partial_\mu\psi_\mu)\right) + \frac{M}{4}\gamma_\lambda(\gamma_\mu\psi_\mu) = 0.$$

Summing this equation with the third one in (2) we obtain

$$-\frac{2fc_3^*}{\sqrt{6}}\left(\partial_\lambda(\gamma_\mu\psi_\mu) - \frac{1}{4}\gamma_\lambda\hat{\partial}(\gamma_\mu\psi_\mu)\right) + i\frac{2gc_4^*}{\sqrt{6}}\left(\partial_\lambda\psi_0 - \frac{1}{4}\gamma_\lambda\hat{\partial}\psi_0\right)$$
$$+\frac{1}{4}c_1\gamma_\lambda\hat{\partial}(\gamma_\mu\psi_\mu) - \frac{c_3}{\sqrt{6}}\left(\gamma_\lambda(\partial_\mu\psi_\mu) - \frac{1}{4}\gamma_\lambda\hat{\partial}(\gamma_\mu\psi_\mu)\right) + M\psi_\lambda = 0.$$

In fact, this equation together with the second equation in (2) makes the system that is equivalent to the initial system (2):

$$c_2\hat{\partial}\psi_0 - i\frac{4c_4}{\sqrt{6}}\left(-\frac{1}{4}\hat{\partial}(\gamma_\mu\psi_\mu) + (\partial_\mu\psi_\mu)\right) + M\psi_0 = 0,$$
$$-\frac{2fc_3^*}{\sqrt{6}}(\partial_\lambda\left(\gamma_\mu\psi_\mu) - \frac{1}{4}\gamma_\lambda\hat{\partial}(\gamma_\mu\psi_\mu)\right) + i\frac{2gc_4^*}{\sqrt{6}}\left(\partial_\lambda\psi_0 - \frac{1}{4}\gamma_\lambda\hat{\partial}\psi_0\right) \tag{4}$$
$$+\frac{1}{4}c_1\gamma_\lambda\hat{\partial}(\gamma_\mu\psi_\mu) - \frac{c_3}{\sqrt{6}}\left(\gamma_\lambda(\partial_\mu\psi_\mu) - \frac{1}{4}\gamma_\lambda\hat{\partial}(\gamma_\mu\psi_\mu)\right) + M\psi_\lambda = 0.$$

To prove this, it suffices to multiply the second equation in (4) by the matrix γ_λ and take into account identity $\gamma_\lambda\gamma_\lambda = 4$. Then we derive

$$c_1\hat{\partial}(\gamma_\mu\psi_\mu) + \frac{c_3}{\sqrt{6}}\left(\hat{\partial}(\gamma_\mu\psi_\mu) - 4(\partial_\mu\psi_\mu)\right) + M(\gamma_\mu\psi_\mu) = 0,$$

which coincides with the first equation in (2).

From the second equation in (4), we can see that the vector-bispinor ψ_μ is expressed through the three bispinors, ψ_0, $(\gamma_\mu\psi_\mu)$, and $(\partial_\mu\psi_\mu)$. Let us re-formulate the whole initial system (2) in such a way that it would contain only three bispinors: $(\gamma_\mu\psi_\mu)$, ψ_0, $(\partial_\mu\psi_\mu)$.

To this end, we act on the second equation in (4) by operator ∂_λ, so obtaining

$$-f\frac{\sqrt{6}}{4}c_3^*\partial_\lambda\partial_\lambda(\gamma_\mu\psi_\mu) + ig\frac{\sqrt{6}}{4}c_4^*\partial_\lambda\partial_\lambda\psi_0 + \frac{1}{4}c_1\partial_\lambda\partial_\lambda(\gamma_\mu\psi_\mu)$$
$$-\frac{c_3}{\sqrt{6}}\hat{\partial}(\partial_\mu\psi_\mu) + \frac{c_3}{4\sqrt{6}}\partial_\lambda\partial_\lambda(\gamma_\mu\psi_\mu) + M(\partial_\mu\psi_\mu) = 0. \tag{5}$$

Now, let us act on the first equation in (4) by operator $\hat{\partial}$, this yields

$$c_1\partial_\lambda\partial_\lambda(\gamma_\mu\psi_\mu) + \frac{c_3}{\sqrt{6}}\partial_\lambda\partial_\lambda(\gamma_\mu\psi_\mu) = \frac{4c_3}{\sqrt{6}}\hat{\partial}(\partial_\mu\psi_\mu) - M\hat{\partial}(\gamma_\mu\psi_\mu),$$

with this in mind eq. (5) can be re-written as follows

$$-f\frac{\sqrt{6}}{4}c_3^*\partial_\lambda\partial_\lambda(\gamma_\mu\psi_\mu) + ig\frac{\sqrt{6}}{4}c_4^*\partial_\lambda\partial_\lambda\psi_0 + M(\partial_\mu\psi_\mu) - \frac{M}{4}\hat{\partial}(\gamma_\mu\psi_\mu) = 0. \tag{6}$$

Now, acting on eq. (5) by operator ∂_λ we get

$$\partial_\lambda \partial_\lambda (\gamma_\mu \psi_\mu) = \frac{1}{c_1 + c_3/\sqrt{6}} \left[\frac{4c_3}{\sqrt{6}} \hat{\partial}(\partial_\mu \psi_\mu) - M \hat{\partial}(\gamma_\mu \psi_\mu) \right]. \tag{7}$$

With eq. (7) in mind, eq. (5) reduces to the form

$$-\frac{f|c_3|^2}{c_1 + c_3/\sqrt{6}} \hat{\partial}(\partial_\mu \psi_\mu) + M \frac{\sqrt{6}}{4} \frac{f c_3^*}{c_1 + c_3/\sqrt{6}}$$
$$+ig \frac{\sqrt{6}}{4} c_4^* \partial_\lambda \partial_\lambda \psi_0 + M(\partial_\mu \psi_\mu) - \frac{M}{4} \hat{\partial}(\gamma_\mu \psi_\mu) = 0. \tag{8}$$

Also, acting on eq. (4) by $\hat{\partial}$ we derive

$$c_2 \partial_\lambda \partial_\lambda \Psi_0 + i \frac{c_4}{\sqrt{6}} \partial_\lambda \partial_\lambda (\gamma_\mu \psi_\mu) = \hat{\partial} \left[i \frac{4}{\sqrt{6}} c_4 (\partial_\mu \psi_\mu) - M \psi_0 \right],$$

which (alow for eq. (7)) can be re-written as

$$c_2 \partial_\lambda \partial_\lambda \psi_0 + i \frac{c_4}{\sqrt{6}} \frac{1}{c_1 + c_3/\sqrt{6}} \left[\frac{4c_3}{\sqrt{6}} \hat{\partial}(\partial_\mu \psi_\mu) - M \hat{\partial}(\gamma_\mu \psi_\mu) \right]$$
$$= \hat{\partial} \left[i \frac{4}{\sqrt{6}} c_4 (\partial_\mu \psi_\mu) - M \psi_0 \right],$$

whence for the term $\partial_\lambda \partial_\lambda \psi_0$ we derive

$$\partial_\lambda \partial_\lambda \psi_0 = \frac{i}{c_2} \left[-\frac{2}{3} \frac{c_3 c_4}{c_1 + \frac{c_3}{\sqrt{6}}} + \frac{4c_4}{\sqrt{6}} \right] \hat{\partial}(\partial_\mu \psi_\mu)$$
$$+i \frac{M}{c_2 \sqrt{6}} \cdot \frac{c_4}{c_1 + \frac{c_3}{\sqrt{6}}} \hat{\partial}(\gamma_\mu \Psi_\mu) - \frac{M}{c_2} \hat{\partial} \Psi_0.$$

Therefore, eq. (8) reads

$$-\frac{f c_2 |c_3|^2 + g c_1 |c_4|^2}{c_2 (c_1 + \frac{c_3}{\sqrt{6}})} \hat{\partial}(\partial_\mu \psi_\mu) - igM \frac{\sqrt{6}}{4} \frac{c_4^*}{c_2} \hat{\partial} \psi_0 + M(\partial_\mu \psi_\mu)$$
$$+\frac{M}{4} \frac{1}{c_2 (c_1 + \frac{c_3}{\sqrt{6}})} \left[\sqrt{6} f c_2 c_3^* - g|c_4|^2 - c_2 (c_1 + \frac{c_3}{\sqrt{6}}) \right] \hat{\partial}(\gamma_\mu \psi_\mu) = 0. \tag{9}$$

Eq. (9) is one of equations we need – it does not contain the vector-bispinor ψ_λ.

To derive another equation of that type it suffices to express the term $(\partial_\mu \psi_\mu)$ from eq. (9) and substitute the result in the firs equation of the system (2). In this way we get

$$\frac{1}{c_2 (c_1 + \frac{c_3}{\sqrt{6}})} \left[c_1 c_2 (c_1 + \frac{c_3}{\sqrt{6}}) + f c_2 |c_3|^2 - \frac{g}{\sqrt{6}} c_3 |c_4|^2 \right] \hat{\partial}(\gamma_\mu \psi_\mu)$$
$$-\frac{4c_3}{M\sqrt{6}} \frac{f c_2 |c_3|^2 + g c_1 |c_4|^2}{c_2 (c_1 + \frac{c_3}{\sqrt{6}})} \hat{\partial}(\partial_\mu \psi_\mu) - ig \frac{c_3 c_4^*}{c_2} \hat{\partial} \psi_0 + M(\gamma_\mu \psi_\mu) = 0. \tag{10}$$

Similarly, expressing the term $(\partial_\mu \psi_\mu)$ from (5) and substituting it into the second equation in (2), we derive

$$\frac{c_2^2 + g|c_4|^2}{c_2}\hat{\partial}\psi_0 - i\frac{4c_4}{M\sqrt{6}}\frac{fc_2|c_3|^2 + gc_1|c_4|^2}{c_2(c_1 + \frac{c_3}{\sqrt{6}})}\hat{\partial}(\partial_\mu \psi_\mu)$$
$$+\frac{i}{\sqrt{6}}\frac{c_4}{c_2(c_1 + \frac{c_3}{\sqrt{6}})}\left[\sqrt{6}fc_2c_3^* - g|c_4|^2\right]\hat{\partial}(\gamma_\mu \psi_\mu) + M\Psi_0 = 0. \tag{11}$$

Thus, the needed form of the main system with respect to bispinors $(\gamma_\mu \psi_\mu)$, ψ_0, $(\partial_\mu \psi_\mu)$ is as follows:

$$\frac{1}{c_2(c_1 + c_3/\sqrt{6})}\left\{c_1c_2(c_1 + c_3/\sqrt{6}) + fc_2|c_3|^2 - \frac{g}{\sqrt{6}}c_3|c_4|^2\right\}\hat{\partial}(\gamma_\mu \psi_\mu)$$
$$-ig\frac{c_3c_4^*}{c_2}\hat{\partial}\psi_0 - \frac{4c_3}{M\sqrt{6}}\cdot\frac{fc_2|c_3|^2 + gc_1|c_4|^2}{c_2(c_1 + c_3/\sqrt{6})}\hat{\partial}(\partial_\mu \psi_\mu) + M(\gamma_\mu \psi_\mu) = 0, \tag{12}$$

$$\frac{i}{\sqrt{6}}\frac{c_4}{c_2(c_1 + c_3/\sqrt{6})}(\sqrt{6}fc_2c_3^* - g|c_4|^2)\hat{\partial}(\gamma_\mu \psi_\mu) + \frac{c_2^2 + g|c_4|^2}{c_2}\hat{\partial}\psi_0$$
$$-i\frac{4c_4}{M\sqrt{6}}\frac{fc_2|c_3|^2 + gc_1|c_4|^2}{c_2(c_1 + c_3/\sqrt{6})}\hat{\partial}(\partial_\mu \psi_\mu) + M\psi_0 = 0, \tag{13}$$

$$+\frac{M}{4}\frac{1}{c_2(c_1 + c_3/\sqrt{6})}\left\{\sqrt{6}fc_2c_3^* - g|c_4|^2 - c_2(c_1 + c_3/\sqrt{6})\right\}\hat{\partial}(\gamma_\mu \psi_\mu)$$
$$-igM\frac{\sqrt{6}}{4}\frac{c_4^*}{c_2}\hat{\partial}\psi_0 - \frac{fc_2|c_3|^2 + gc_1|c_4|^2}{c_2(c_1 + c_3/\sqrt{6})}\hat{\partial}(\partial_\mu \psi_\mu) + M(\partial_\mu \psi_\mu) = 0. \tag{14}$$

This system may be presented in the matrix form

$$K\hat{\partial}\begin{vmatrix} \gamma_\mu \psi_\mu \\ \psi_0 \\ \partial_\mu \psi_\mu \end{vmatrix} = M\begin{vmatrix} \gamma_\mu \psi_\mu \\ \psi_0 \\ \partial_\mu \psi_\mu \end{vmatrix}. \tag{15}$$

3. Characteristic Equation and Possible Values of Masses

Let us find the characteristic equation for the matrix of the first order system (15) with respect to bispinors $\gamma_\mu \psi_\mu$, ψ_0, $\partial_\mu \psi_\mu$; It has the form $\det F = 0$, where the matrix K is (its entries are listed by columns)

$$\left| \begin{array}{l} \frac{1}{c_2(c_1+\frac{c_3}{\sqrt{6}})}\{c_1c_2(c_1 + \frac{c_3}{\sqrt{6}}) + fc_2|c_3|^2 - \frac{g}{\sqrt{6}}c_3|c_4|^2\} - \lambda \\ \frac{i}{\sqrt{6}}\cdot\frac{c_4}{c_2(c_1+\frac{c_3}{\sqrt{6}})}\{\sqrt{6}fc_2c_3^* - g|c_4|^2\} \\ \frac{M}{4}\cdot\frac{1}{c_2(c_1+\frac{c_3}{\sqrt{6}})}\{\sqrt{6}fc_2c_3^* - g|c_4|^2 - c_2(c_1 + \frac{c_3}{\sqrt{6}})\} \end{array} \right.$$

$$\begin{array}{cc} -ig\frac{c_3c_4^*}{c_2} & -\frac{4c_3}{M\sqrt{6}} \cdot \frac{fc_2|c_3|^2+gc_1|c_4|^2}{c_2(c_1+\frac{c_3}{\sqrt{6}})} \\ \frac{c_2^2+g|c_4|^2}{c_2} - \lambda & -i\frac{4c_4}{M\sqrt{6}} \cdot \frac{fc_2|c_3|^2+gc_1|c_4|^2}{c_2(c_1+\frac{c_3}{\sqrt{6}})} \\ -igM\frac{\sqrt{6}}{4} \cdot \frac{c_4^*}{c_2} & -\frac{fc_2|c_3|^2+gc_1|c_4|^2}{c_2(c_1+\frac{c_3}{\sqrt{6}})} - \lambda \end{array} \right| .$$

In this way we get

$$\lambda^3 - \lambda^2(c_1 + c_2) + \lambda(c_1c_2 - f|c_3|^2 \\ -g|c_4|^2) + (fc_2|c_3|^2 + gc_1|c_4|^2) = 0. \tag{16}$$

The matrix of the system (15) can be transformed to a diagonal form

$$K = \left| \begin{array}{ccc} A_1 & B_1 & R_1 \\ A_2 & B_2 & R_2 \\ A_3 & B_3 & R_3 \end{array} \right| \quad \Longrightarrow \quad \left| \begin{array}{ccc} \lambda_1 & 0 & 0 \\ 0 & \lambda_2 & 0 \\ 0 & 0 & \lambda_3 \end{array} \right| , \tag{17}$$

and the system will take the form of three separated equations of the Dirac type but with different masses:

$$M_1 = \frac{M}{\lambda_1}, \quad M_2 = \frac{M}{\lambda_2}, \quad M_3 = \frac{M}{\lambda_3} . \tag{18}$$

That transformation is done in accordance with the following procedure

$$(K - M)\Psi = 0, \quad \Psi' = S\Psi, \quad SKS^{-1} = \hat{K}' = \left| \begin{array}{ccc} \lambda_1 & 0 & 0 \\ 0 & \lambda_2 & 0 \\ 0 & 0 & \lambda_3 \end{array} \right| . \tag{19}$$

The matrix S obeys the following relations

$$S = \left| \begin{array}{ccc} a_1 & a_2 & a_3 \\ b_1 & b_2 & b_3 \\ r_1 & r_2 & r_3 \end{array} \right| ,$$

$$\left| \begin{array}{ccc} a_1 & a_2 & a_3 \\ b_1 & b_2 & b_3 \\ r_1 & r_2 & r_3 \end{array} \right| \left| \begin{array}{ccc} A_1 & B_1 & R_1 \\ A_2 & B_2 & R_2 \\ A_3 & B_3 & R_3 \end{array} \right| = \left| \begin{array}{ccc} \lambda_1 & 0 & 0 \\ 0 & \lambda_2 & 0 \\ 0 & 0 & \lambda_3 \end{array} \right| \left| \begin{array}{ccc} a_1 & a_2 & a_3 \\ b_1 & b_2 & b_3 \\ r_1 & r_2 & r_3 \end{array} \right| ,$$

whence three linear subsystems follow

$$a_1A_1 + a_2A_2 + a_3A_3 = \lambda_1 a_1 , \\ a_1B_1 + a_2B_2 + a_3B_3 = \lambda_1 a_2 , \\ a_1R_1 + a_2R_2 + a_3R_3 = \lambda_1 a_3 ; \tag{20}$$

$$b_1A_1 + b_2A_2 + b_3A_3 = \lambda_2 b_1 , \\ b_1B_1 + b_2B_2 + b_3B_3 = \lambda_2 b_2, \\ b_1R_1 + b_2R_2 + b_3R_3 = \lambda_2 b_3 ; \tag{21}$$

$$r_1 A_1 + r_2 A_2 + r_3 A_3 = \lambda_3 r_1 \,,$$
$$r_1 B_1 + r_2 B_2 + r_3 B_3 = \lambda_3 r_2,$$
$$r_1 R_1 + r_2 R_2 + r_3 R_3 = \lambda_3 r_3. \qquad (22)$$

These systems are in standard way. For instance, we get

$$a_2 = a_1 \left[-igc_3c_4^*(c_1 + \frac{c_3}{\sqrt{6}})\lambda_1 \right]$$

$$\times \left[\lambda_1 c_2(\lambda_1 - c_2) + (\lambda_1 - c_2)[fc_2|c_3|^2 + gc_1|c_4|^2] - \lambda_1(c_1 + \frac{c_3}{\sqrt{6}})g|c_4|^2 \right]^{-1} ,$$

$$a_3 = -a_1 \frac{4}{\sqrt{6}} \frac{1}{M} \left[c_3(\lambda_1 - c_2)[fc_2|c_3|^2 + gc_1|c_4|^2] \right.$$

$$\left. \times \left[\lambda_1 c_2(\lambda_1 - c_2) + (\lambda_1 - c_2)[fc_2|c_3|^2 + gc_1|c_4|^2] - \lambda_1(c_1 + \frac{c_3}{\sqrt{6}})g|c_4|^2 \right]^{-1} \right. .$$

At these a_2, a_3, the first equation in (20) reduces to the above characteristic equation

$$c_2^2(c_1 + \frac{c_3}{\sqrt{6}})^2 \{ \lambda_1^3 - \lambda_1^2(c_1 + c_2) + \lambda_1(c_1c_2 - f|c_3|^2$$

$$-g|c_4|^2) + (fc_2|c_3|^2 + gc_1|c_4|^2 \} = 0 \,,$$

where $\lambda = \lambda_1$.

Similarly we get solutions of the system (21):

$$b_2 = b_1 \left[-igc_3c_4^*(c_1 + \frac{c_3}{\sqrt{6}})\lambda_2 \right]$$

$$\times \left[\lambda_2 c_2(\lambda_2 - c_2) + (\lambda_2 - c_2)[fc_2|c_3|^2 + gc_1|c_4|^2] - \lambda_2(c_1 + \frac{c_3}{\sqrt{6}})g|c_4|^2 \right]^{-1} ,$$

$$b_3 = -b_1 \frac{4}{\sqrt{6}} \frac{1}{M} \left[c_3(\lambda_2 - c_2)[fc_2|c_3|^2 + gc_1|c_4|^2] \right]$$

$$\times \left[\lambda_2 c_2(\lambda_2 - c_2) + (\lambda_2 - c_2)[fc_2|c_3|^2 + gc_1|c_4|^2] - \lambda_2(c_1 + \frac{c_3}{\sqrt{6}})g|c_4|^2 \right]^{-1} ,$$

Similar results are for the system (22):

$$r_2 = r_1 \left[-igc_3c_4^*(c_1 + \frac{c_3}{\sqrt{6}})\lambda_3 \right]$$

$$\times \left[\lambda_3 c_2(\lambda_3 - c_2) + (\lambda_3 - c_2)[fc_2|c_3|^2 + gc_1|c_4|^2] - \lambda_3(c_1 + \frac{c_3}{\sqrt{6}})g|c_4|^2 \right]^{-1} ,$$

$$r_3 = -r_1 \frac{4}{\sqrt{6}} \frac{1}{M} \left[c_3(\lambda_3 - c_2)[fc_2|c_3|^2 + gc_1|c_4|^2] \right]$$

$$\times \left[\lambda_3 c_2(\lambda_3 - c_2) + (\lambda_3 - c_2)[fc_2|c_3|^2 + gc_1|c_4|^2] - \lambda_3(c_1 + \frac{c_3}{\sqrt{6}})g|c_4|^2 \right]^{-1} .$$

4. On Possible Values for Mass Parameters

Characteristic equation may be presented in the short form $\lambda^3 + \lambda^2 a + \lambda b + c = 0$, where (recall that $f, g \in \{-1, +1\}$)

$$a = -(c_1 + c_2),\ b = c_1 c_2 - f|c_3|^2 - g|c_4|^2,\ c = f c_2 |c_3|^2 + g c_1 |c_4|^2. \tag{23}$$

After standard change of the variable we get to a reduced cubic equation

$$y^3 + py + q = 0, \quad y = \lambda - \frac{c_1 + c_3}{3} = \lambda + \frac{a}{3},$$

where

$$p = -\frac{a^3}{3} + b = -\frac{1}{3}(c_1 + c_3)^3 + c_1 c_2 - f|c_3|^2 - g|c_4|^2,$$

$$q = 2(\frac{a}{3})^3 - \frac{ab}{3} + c$$

$$= -\frac{2}{27} c_1 + c_3{}^3 + \frac{c_1 + c_2}{3}\{c_1 c_2 - f|c_3|^2 - g|c_4|^2\} + \{f|c_3|^2 + g|c_4|^2\}.$$

By physical reason, we assume real-valuedness of all three roots of characteristic equation. It is possible if the following inequalities are valid

$$p < 0, \qquad Q < 0, \quad \text{where} \quad Q = (\frac{p}{3})^3 + (\frac{q}{2})^2. \tag{24}$$

Solutions can be presented in trigonometric form:

$$\lambda_1 = y_1 + \frac{c_1 + c_2}{3} = 2\sqrt{-\frac{p}{3}} \cos \frac{\alpha}{3} + \frac{c_1 + c_2}{3},$$

$$\lambda_2 = y_2 + \frac{c_1 + c_2}{3} = -2\sqrt{-\frac{p}{3}} \cos(\frac{\alpha}{3} - \frac{\pi}{3}) + \frac{c_1 + c_2}{3}, \tag{25}$$

$$\lambda_3 = y_3 + \frac{c_1 + c_2}{3} = -2\sqrt{-\frac{p}{3}} \cos(\frac{\alpha}{3} + \frac{\pi}{3}) + \frac{c_1 + c_2}{3}.$$

We may readily derive relations (remember on real-valuedness of the roots):

$$\lambda_1 + \lambda_2 + \lambda_3 = c_1 + c_2, \quad \lambda_1^2 + \lambda_2^2 + \lambda_3^2 = c_1^2 + c_2^2 + 2f|c_3|^2 + 2g|c_4|^2 > 0,$$

$$\lambda_1 \lambda_2 + \lambda_1 \lambda_3 + \lambda_2 \lambda_2 = c_1 c_2 - f|c_3|^2 - g|c_4|^2, \quad \lambda_1 \lambda_2 \lambda_3 = -f c_2 |c_3|^2 - g c_1 |c_4|^2,$$

$$p = \frac{1}{6}\{(\lambda_1 + \lambda_2 + \lambda_3)^2 - 3(\lambda_1^2 + \lambda_2^2 + \lambda_3^2)\} < 0,$$

$$q = \frac{1}{6}\{\frac{5}{9}(\lambda_1 + \lambda_2 + \lambda_3)^2 - (\lambda_1^2 + \lambda_2^2 + \lambda_3^2)\}(\lambda_1 + \lambda_2 + \lambda_3) - \lambda_1 \lambda_2 \lambda_3,$$

$$\cos \frac{\alpha}{3} = \frac{2\lambda_1 - \lambda_2 - \lambda_3}{\sqrt{-2[(\lambda_1 + \lambda_2 + \lambda_3)^2 - 3(\lambda_1^2 + \lambda_2^2 + \lambda_3^2)]}},$$

$$\cos(\frac{\alpha}{3} - \frac{\pi}{3}) = \frac{2\lambda_2 - \lambda_1 - \lambda_3}{\sqrt{-2[(\lambda_1 + \lambda_2 + \lambda_3)^2 - 3(\lambda_1^2 + \lambda_2^2 + \lambda_3^2)]}},$$

$$\cos(\frac{\alpha}{3}+\frac{\pi}{3}) = \frac{2\lambda_3-\lambda_1-\lambda_2}{\sqrt{-2[(\lambda_1+\lambda_2+\lambda_3)^2-3(\lambda_1^2+\lambda_2^2+\lambda_3^2)]}}.$$

In the first place we are interested in positive roots. Evidently, three positive roots may arise only if

$$c_1>0, \quad c_2>0, \quad f=-1, \quad g=-1. \tag{26}$$

It is convenient to use the notations $|c_4|^2=a^2,\ |c_3|^2=b^2$. From the constraints

$$\lambda_1+\lambda_2=c_1+c_2-\lambda_3, \quad \lambda_1\lambda_2=\frac{c_1a^2+c_2b^2}{\lambda_3},$$

follow expressions for λ_1, λ_2 follow

$$\lambda_{1,2}=\frac{c_1+c_2-\lambda_3}{2}\pm\sqrt{(\frac{c_1+c_2-\lambda_3}{2})^2-\frac{c_1a^2+c_2b^2}{\lambda_3}}, \tag{27}$$

under square root we should have positive number. Note that due to inequality

$$\lambda_1+\lambda_2=c_1+c_2-\lambda_3>0,$$

we have inequalities $0<\lambda_3<c_1+c_2$. Therefore the following parametrization is possible:

$$\lambda_3=(c_1+c_2)\cos\alpha, \qquad \alpha\in(0,\frac{\pi}{2}), \quad \Gamma=\frac{c_1a^2+c_2b^2}{(c_1+c_2)^3}. \tag{28}$$

Correspondingly, for the roots we get expressions

$$\lambda_{1,2}=\lambda_3\cdot\frac{1}{\cos\alpha}\left(\sin^2(\alpha/2)\pm\sqrt{\sin^4(\alpha/2)-\Gamma/\cos\alpha}\right), \tag{29}$$

where the term under the square root should be positive: $\cos\alpha\ \sin^4(\alpha/2)>\Gamma$. Further we obtain expressions for three possible mass parameters:

$$M_3=\frac{M}{\lambda_3}=\frac{M}{(c_1+c_2)}\frac{1}{\cos\alpha}=\mu\frac{1}{\cos\alpha},$$
$$M_1=\frac{M}{\lambda_2}=\frac{\mu}{\sin^2(\alpha/2)-\sqrt{\sin^4(\alpha/2)-\Gamma/\cos\alpha}}, \tag{30}$$
$$M_2=\frac{M}{\lambda_2}=\frac{\mu}{\sin^2(\alpha/2)+\sqrt{\sin^4(\alpha/2)-\Gamma/\cos\alpha}}.$$

There exist special cases; for instance, the following one

$$a\to0, b\to0 \implies \Gamma\to0,\ M_3=\frac{\mu}{\cos\alpha}, M_1\to\infty, M_2=\frac{\mu}{1-\cos^2\alpha}, \tag{31}$$

if additionally $\alpha=0$ or $\alpha=\pi/2$ then yet another mass will be infinite.

Let us study inequality $\cos\alpha\sin^4(\alpha/2)>\Gamma$, or

$$f(\alpha)=\cos\alpha\,\frac{(1-\cos^2\alpha)^2}{4}>\Gamma, \quad f(\alpha=0)=0, \qquad f(\alpha=\frac{\pi}{2})=0.$$

In the variable $x = \cos\alpha$ we have

$$f(x) = \frac{1}{4}x(1-x^2)^2 > \Gamma, \quad x \in (0,1),$$

in the interval $x \in (0,1)$ there exists a point of a local maximum α_0:

$$\frac{df}{dx} = \frac{(1-x^2)(1-5x^2)}{4}, \quad \cos\alpha_0 = \frac{1}{\sqrt{5}}, \quad f(\alpha_0) = \frac{4}{25\sqrt{5}}.$$

Thus, we have the following constraint for $f(\alpha)$:

$$\Gamma < \Gamma_0 = \frac{4}{25\sqrt{5}}, \quad f(\alpha) > \Gamma,$$

and there exists some finite interval containing the point α_0.

The value $\Gamma_0 = 4/25\sqrt{5}$ is special, because the interval for α degenerates into a single point: $\alpha = \alpha_0$ and we have situation when two masses are fixed and equal to each other:

$$\cos\alpha_0 = \frac{1}{\sqrt{5}}, \quad M_3 = \frac{\mu}{\cos\alpha_0} = \mu\sqrt{5},$$
$$M_{1,2} = \frac{\mu}{(1-\cos\alpha_0)/2} = \mu\frac{2\sqrt{5}}{\sqrt{5}-1}. \quad (32)$$

5. The Case of a Free Particle

In is convenient to employ the shortening notations:

$$A_1 = \frac{1}{c_2(c_1 + c_3/\sqrt{6})}\left[c_2(c_1^2 + f|c_3|^2) + \frac{c_3}{\sqrt{6}}(c_1c_2 - g|c_4|^2\right],$$

$$A_2 = \frac{i}{\sqrt{6}}\frac{c_4}{c_2(c_1 + c_3/\sqrt{6})}[f\sqrt{6}c_2c_3^* - g|c_4|^2],$$

$$A_3 = \frac{M}{4}\frac{1}{c_2(c_1 + c_3/\sqrt{6})}\left[f\sqrt{6}c_2c_3^* - g|c_4|^2 - c_2(c_1 + c_3/\sqrt{6})\right],$$

$$B_1 = -ig\frac{c_3c_4^*}{c_2}, \qquad B_2 = \frac{c_2^2 + g|c_4|^2}{c_2},$$

$$B_3 = -igM\frac{\sqrt{6}\,c_4^*}{4\;c_2},$$

$$R_1 = -\frac{4}{M}\frac{c_3}{\sqrt{6}}\frac{fc_2|c_3|^2 + gc_1|c_4|^2}{c_2(c_1 + c_3/\sqrt{6})}, \qquad R_2 = -\frac{4i}{M}\frac{c_4}{\sqrt{6}}\frac{fc_2|c_3|^2 + gc_1|c_4|^2}{c_2(c_1 + c_3/\sqrt{6})},$$

$$R_3 = -\frac{fc_2|c_3|^2 + gc_1|c_4|^2}{c_2(c_1 + c_3/\sqrt{6})}.$$

Correspondingly, the equations (12)–(14) read

$$A_1\hat{\partial}\,(\gamma_\mu\Psi_\mu) + B_1\hat{\partial}\,\Psi_0 + R_1\hat{\partial}\,(\partial_\mu\Psi_\mu) + M(\gamma_\mu\Psi_\mu) = 0,$$
$$A_2\hat{\partial}\,(\gamma_\mu\Psi_\mu) + B_2\hat{\partial}\,\Psi_0 + R_2\hat{\partial}\,(\partial_\mu\Psi_\mu) + M\Psi_0 = 0, \qquad (33)$$
$$A_3\hat{\partial}\,(\gamma_\mu\Psi_\mu) + B_3\hat{\partial}\,\Psi_0 + R_3\hat{\partial}\,(\partial_\mu\Psi_\mu) + M(\partial_\mu\Psi_\mu) = 0.$$

As shown above, this system may be transformed into separate equations for new three bispinors. We are to detail explicit form of these separate equations.

To this end, first we sum three equations with coefficients a_1, a_2, a_3, this yields

$$\begin{aligned}(a_1A_1 + a_2A_2 + a_2A_3)\hat{\partial}\,(\gamma_\mu\Psi_\mu)\\ +(a_1B_1 + a_2B_2 + a_3B_3)\hat{\partial}\,\Psi_0\\ +(a_1R_1 + a_2R_2 + a_3R_3)\hat{\partial}\,(\partial_\mu\Psi_\mu)\\ +M\{a_1(\gamma_\mu\Psi_\mu) + a_2\Psi_0 + a_3(\partial_\mu\Psi_\mu)\} = 0,\end{aligned}$$

taking in mind identities

$$a_1A_1 + a_2A_2 + a_3A_3 = \lambda_1 a_1,$$
$$a_1B_1 + a_2B_2 + a_3B_3 = \lambda_1 a_2,$$
$$a_1R_1 + a_2R_2 + a_3R_3 = \lambda_1 a_3,$$

the last equation is re-written as

$$\lambda_1 a_1\hat{\partial}(\gamma_\mu\Psi_\mu) + \lambda_1 a_2\hat{\partial}\Psi_0 + \lambda_1 a_3\hat{\partial}(\partial_\mu\Psi_\mu)$$
$$+M\{a_1(\gamma_\mu\Psi_\mu) + a_2\Psi_0 + a_3(\partial_\mu\Psi_\mu)\} = 0.$$

Thus, for a new bispinor we produce a Dirac-like equation

$$\Phi_1 = a_1(\gamma_\mu\Psi_\mu) + a_2\Psi_0 + a_3(\partial_\mu\Psi_\mu),$$
$$(\hat{\partial} + \frac{M}{\lambda_1})\Phi_1 = 0, \quad M_1 = M/\lambda_1. \qquad (34)$$

Similarly, the root λ_2 enters the system

$$b_1A_1 + b_2A_2 + b_3A_3 = \lambda_2 b_1,$$
$$b_1B_1 + b_2B_2 + b_3B_3 = \lambda_2 b_2,$$
$$b_1R_1 + b_2R_2 + b_3R_3 = \lambda_2 b_3,$$

its solution is known. Turning to equations (33), we sum them with coefficients b_1, b_2, b_3. In this way, for a new bispinor we get a Dirac-like equation

$$\Phi_2 = b_1(\gamma_\mu\Psi_\mu) + b_2\Psi_0 + b_3(\partial_\mu\Psi_\mu),$$
$$(\hat{\partial} + \frac{M}{\lambda_2})\Phi_2 = 0\,, \quad M_2 = M/\lambda_2. \qquad (35)$$

Finally, turning to the system which contains the root λ_3:

$$r_1 A_1 + r_2 A_2 + r_3 A_3 = \lambda_3 r_1,$$
$$r_1 B_1 + r_2 B_2 + r_3 B_3 = \lambda_3 r_2,$$
$$r_1 R_1 + r_2 R_2 + r_3 R_3 = \lambda_3 r_3,$$

and acting as above, we obtain a Dirac-like equation for the third bispinor

$$\Phi_3 = r_1(\gamma_\mu \Psi_\mu) + r_2 \Psi_0 + r_3(\partial_\mu \Psi_\mu),$$
$$(\hat{\partial} + \frac{M}{\lambda_3})\Phi_3 = 0 \,, \quad M_3 = M/\lambda_3. \tag{36}$$

6. Presence of Electromagnetic Fields

Presence of electromagnetic fields is taken into account by modifying the derivative: $D_\mu = \partial_\mu - ieA_\mu(x)$. We proceed with equations (4), but now in presence of external field (let $\hat{D} = \gamma_\mu D_\mu$):

$$c_2 \hat{D}\Psi_0 - \frac{4ic_4}{\sqrt{6}} \left[(D_\mu \Psi_\mu) - \frac{1}{4}\hat{D}(\gamma_\mu \Psi_\mu) \right] + M\Psi_0 = 0, \tag{37}$$

$$-\frac{2fc_3^*}{\sqrt{6}} \left[D_\lambda(\gamma_\mu \Psi_\mu) - \frac{1}{4}\gamma_\lambda \hat{D}(\gamma_\mu \Psi_\mu) \right]$$
$$+\frac{2igc_4^*}{\sqrt{6}} \left[D_\lambda \Psi_0 - \frac{1}{4}\gamma_\lambda \hat{D}\Psi_0 \right] + \frac{c_1}{4}\gamma_\lambda \hat{D}(\gamma_\mu \Psi_\mu)$$
$$-\frac{c_3}{\sqrt{6}} \left[\gamma_\lambda (D_\mu \Psi_\mu) - \frac{1}{4}\gamma_\lambda \hat{D}(\gamma_\mu \Psi_\mu) \right] + M\Psi_\lambda = 0. \tag{38}$$

We are to perform some transformation over this system. First, let us act on eq.(38) by the matrix γ_λ, this results in

$$(c_1 + \frac{c_3}{\sqrt{6}})\hat{D}(\gamma_\mu \Psi_\mu) - \frac{4c_3}{\sqrt{6}}(D_\mu \Psi_\mu) + M(\gamma_\mu \Psi_\mu) = 0. \tag{39}$$

Now, act on eq. (38) by operator D_λ:

$$-\frac{2fc_3^*}{\sqrt{6}} \left[D^2(\gamma_\mu \Psi_\mu) - \frac{1}{4}\hat{D}\hat{D}(\gamma_\mu \Psi_\mu) \right]$$
$$+\frac{2igc_4^*}{\sqrt{6}} \left[D^2 \Psi_0 - \frac{1}{4}\hat{D}\hat{D}\Psi_0 \right] + \frac{c_1}{4}\hat{D}\hat{D}(\gamma_\mu \Psi_\mu)$$
$$-\frac{c_3}{\sqrt{6}} \left[\hat{D}(D_\mu \Psi_\mu) - \frac{1}{4}\hat{D}\hat{D}(\gamma_\mu \Psi_\mu) \right] + M(D_\lambda \Psi_\lambda) = 0, \tag{40}$$

where $D^2 = D_\lambda D_\lambda$. It is readily proved the identity

$$\hat{D}\hat{D} = D_\lambda D_\rho \, [\, \frac{1}{2}(\gamma_\lambda \gamma_\rho - \gamma_\rho \gamma_\lambda) + \frac{1}{2}(\gamma_\lambda \gamma_\rho + \gamma_\rho \gamma_\lambda) \,]$$
$$= [\, D^2 + (D_\lambda D_\rho - D_\rho D_\lambda)\sigma_{\lambda\rho} \,]. \tag{41}$$

Therefore, eq. (40) reads

$$-\frac{2fc_3^*}{\sqrt{6}}\left[\frac{3}{4}D^2(\gamma_\mu\Psi_\mu)-\frac{1}{4}(D_\lambda D_\rho-D_\rho D_\lambda)\sigma_{\lambda\rho}(\gamma_\mu\Psi_\mu)\right]$$
$$+\frac{2igc_4^*}{\sqrt{6}}\left[\frac{3}{4}D^2\Psi_0-\frac{1}{4}(D_\lambda D_\rho-D_\rho D_\lambda)\sigma_{\lambda\rho}\Psi_0\right]$$
$$+\frac{c_1}{4}[(D_\lambda D_\rho-D_\rho D_\lambda)\sigma_{\lambda\rho}+D^2](\gamma_\mu\Psi_\mu)$$
$$-\frac{c_3}{\sqrt{6}}\left[\hat{D}(D_\mu\Psi_\mu)-\frac{1}{4}[(D_\lambda D_\rho-D_\rho D_\lambda)\sigma_{\lambda\rho}+D^2](\gamma_\mu\Psi_\mu)\right]$$
$$+M(D_\lambda\Psi_\lambda)=0. \tag{42}$$

Note the commutator $D_\lambda D_\rho - D_\rho D_\lambda = (-ieF_{\lambda\rho})$. Eq. (42) can be re-written as follows

$$-\frac{2fc_3^*}{\sqrt{6}}\left[\frac{3}{4}D^2\,(\gamma_\mu\Psi_\mu)-\frac{1}{4}(-ieF_{\lambda\rho})\,\sigma_{\lambda\rho}\,(\gamma_\mu\Psi_\mu)\right]$$
$$+\frac{2igc_4^*}{\sqrt{6}}\left[\frac{3}{4}D^2\,\Psi_0-\frac{1}{4}(-ieF_{\lambda\rho})\,\sigma_{\lambda\rho}\,\Psi_0\right]-\frac{c_3}{\sqrt{6}}\hat{D}\,(D_\mu\Psi_\mu)$$
$$+\frac{1}{4}(c_1+\frac{c_3}{\sqrt{6}})\,[\,D^2+(-ieF_{\lambda\rho})\,\sigma_{\lambda\rho}\,]\,(\gamma_\mu\Psi_\mu)+M\,(D_\lambda\Psi_\lambda)=0. \tag{43}$$

Acting on (39) by operator $\hat{D}$, we get

$$(c_1+\frac{c_3}{\sqrt{6}})[\,D^2+(-ieF_{\lambda\rho})\sigma_{\lambda\rho}\,]\,(\gamma_\mu\Psi_\mu)=\frac{4c_3}{\sqrt{6}}\hat{D}(D_\mu\Psi_\mu)-M\hat{D}(\gamma_\mu\Psi_\mu). \tag{44}$$

With this relation in mind, eq. (43) reduces to the form

equation III

$$-\frac{2fc_3^*}{\sqrt{6}}\left[\frac{3}{4}D^2-\frac{1}{4}(-ieF_{\lambda\rho})\,\sigma_{\lambda\rho}\right](\gamma_\mu\Psi_\mu)$$
$$+\frac{2igc_4^*}{\sqrt{6}}\left[\frac{3}{4}D^2-\frac{1}{4}(-ieF_{\lambda\rho})\,\sigma_{\lambda\rho}\right]\Psi_0-\frac{M}{4}\hat{D}(\gamma_\mu\Psi_\mu)+M\,(D_\lambda\Psi_\lambda)=0. \tag{45}$$

Let us write down two other equations:

equation I

$$c_2\hat{D}\Psi_0-\frac{4ic_4}{\sqrt{6}}\left[(D_\mu\Psi_\mu)-\frac{1}{4}\hat{D}(\gamma_\mu\Psi_\mu)\right]+M\Psi_0=0, \tag{46}$$

equation II

$$(c_1+\frac{c_3}{\sqrt{6}})\hat{D}(\gamma_\mu\Psi_\mu)-\frac{4c_3}{\sqrt{6}}(D_\mu\Psi_\mu)+M(\gamma_\mu\Psi_\mu)=0. \tag{47}$$

We can see that equation III contains the 2-nd order operator D^2. The task is to get the system of first order for three bispinors $(\gamma_\mu\Psi_\mu)$, Ψ_0, $(D_\mu\Psi_\mu)$. Therefore we are to exclude operator D^2 from eq. (45).

To this end, first let us re-write eq. (45) differently

$$-\frac{\sqrt{6}fc_3^*}{4}D^2(\gamma_\mu\Psi_\mu)+\frac{fc_3^*}{2\sqrt{6}}(-ieF_{\lambda\rho})\,\sigma_{\lambda\rho}\,(\gamma_\mu\Psi_\mu)$$

$$+\frac{i\sqrt{6}gc_4^*}{4}D^2\Psi_0-\frac{igc_4^*}{2\sqrt{6}}(-ieF_{\lambda\rho})\,\sigma_{\lambda\rho}\Psi_0-\frac{1}{4}M\hat{D}(\gamma_\mu\Psi_\mu)+M(D_\lambda\Psi_\lambda)=0.$$

With the help of eq. (44) in the form

$$D^2(\gamma_\mu\Psi_\mu)=(+ieF_{\lambda\rho})\sigma_{\lambda\rho}(\gamma_\mu\Psi_\mu)$$

$$+\frac{4}{\sqrt{6}}\frac{c_3}{(c_1+c_3/\sqrt{6})}\hat{D}(D_\mu\Psi_\mu)\frac{M}{(c_1+c_3/\sqrt{6})}\hat{D}(\gamma_\mu\Psi_\mu),$$

in eq. (45) we transform the first term

$$-\frac{\sqrt{6}fc_3^*}{4}\frac{4}{\sqrt{6}}\frac{c_3}{(c_1+c_3/\sqrt{6})}\hat{D}(D_\mu\Psi_\mu)+\frac{M}{4}\frac{\sqrt{6}fc_3^*}{(c_1+c_3/\sqrt{6})}\hat{D}(\gamma_\mu\Psi_\mu)$$

$$+\frac{i\sqrt{6}gc_4^*}{4}D^2\Psi_0-\frac{igc_4^*}{2\sqrt{6}}(-ieF_{\lambda\rho})\,\sigma_{\lambda\rho}\Psi_0$$

$$+fc_3^*\,(\,\frac{\sqrt{6}}{4}+\frac{1}{2\sqrt{6}}\,)\,(-ieF_{\lambda\rho})\,\sigma_{\lambda\rho}\,(\gamma_\mu\Psi_\mu)$$

$$-\frac{1}{4}M\hat{D}(\gamma_\mu\Psi_\mu)+M(D_\lambda\Psi_\lambda)=0.$$

So reducing eq. (45) to the form

$$-\frac{f|c_3|^2}{(c_1+c_3/\sqrt{6})}\hat{D}(D_\mu\Psi_\mu)+\frac{M}{4}\,[\,\frac{\sqrt{6}fc_3^*}{(c_1+c_3/\sqrt{6})}-1\,]\,\hat{D}(\gamma_\mu\Psi_\mu)$$

$$+\frac{i\sqrt{6}gc_4^*}{4}D^2\Psi_0-\frac{igc_4^*}{2\sqrt{6}}(-ieF_{\lambda\rho})\sigma_{\lambda\rho}\Psi_0$$

$$+\frac{2}{\sqrt{6}}fc_3^*(-ieF_{\lambda\rho})\sigma_{\lambda\rho}(\gamma_\mu\Psi_\mu)+M(D_\lambda\Psi_\lambda)=0. \tag{48}$$

Now we turn to eq. (46) and act on it by operator $\hat{D}$:

$$c_2\hat{D}\hat{D}\Psi_0-\frac{4ic_4}{\sqrt{6}}\left[\hat{D}(D_\mu\Psi_\mu)-\frac{1}{4}\hat{D}\hat{D}(\gamma_\mu\Psi_\mu)\right]+M\hat{D}\Psi_0=0;$$

whence (allowing for identity $\hat{D}\hat{D}=D^2+(-ieF_{\lambda\rho})\sigma_{\lambda\rho}$) it follows

$$D^2\Psi_0=-(-ieF_{\lambda\rho})\sigma_{\lambda\rho}\Psi_0+\frac{4ic_4}{\sqrt{6}c_2}\hat{D}(D_\mu\Psi_\mu)$$

$$-\frac{ic_4}{\sqrt{6}c_2}D^2(\gamma_\mu\Psi_\mu)-\frac{ic_4}{\sqrt{6}c_2}(-ieF_{\lambda\rho})\sigma_{\lambda\rho}(\gamma_\mu\Psi_\mu)-\frac{M}{c_2}\hat{D}\Psi_0.$$

The term $D^2(\gamma_\mu \Psi_\mu)$ may be expressed as follows

$$D^2(\gamma_\mu \Psi_\mu) = (+ieF_{\lambda\rho})\sigma_{\lambda\rho}(\gamma_\mu \Psi_\mu)$$

$$+\frac{4}{\sqrt{6}}\frac{c_3}{(c_1+c_3/\sqrt{6})}\hat{D}(D_\mu \Psi_\mu) - \frac{M}{(c_1+c_3/\sqrt{6})}\hat{D}(\gamma_\mu \Psi_\mu),$$

therefore the previous relation takes the form

$$D^2\Psi_0 = -\frac{M}{c_2}\hat{D}\Psi_0 - (-ieF_{\lambda\rho})\sigma_{\lambda\rho}\Psi_0$$
$$+\frac{4i}{\sqrt{6}}\frac{c_1 c_4}{c_2(c_1+c_3/\sqrt{6})}\hat{D}(D_\mu \Psi_\mu) + iM\frac{1}{\sqrt{6}}\frac{c_4}{c_2(c_1+c_3/\sqrt{6})}\hat{D}(\gamma_\mu \Psi_\mu). \quad (49)$$

Applying the last relation, we can exclude the quadratic term $D^2\Psi_0$ in eq. (48), so deriving

$$\left[-\frac{f|c_3|^2}{(c_1+c_3/\sqrt{6})} + \frac{4i}{\sqrt{6}}\frac{c_1 c_4}{c_2(c_1+c_3/\sqrt{6})}\right]\hat{D}(D_\mu \Psi_\mu)$$
$$+\left[\frac{M}{4}\frac{\sqrt{6}fc_3^* - (c_1+c_3/\sqrt{6})}{(c_1+c_3/\sqrt{6})} - \frac{M}{4}\frac{gc_4^* c_4}{c_2(c_1+c_3/\sqrt{6})}\right]\hat{D}(\gamma_\mu \Psi_\mu)$$
$$-\frac{i\sqrt{6}gc_4^*}{4}\frac{M}{c_2}\hat{D}\Psi_0 - \left(\frac{i\sqrt{6}gc_4^*}{4} + \frac{igc_4^*}{2\sqrt{6}}\right)(-ieF_{\lambda\rho})\sigma_{\lambda\rho}\Psi_0$$
$$+\frac{2}{\sqrt{6}}fc_3^*\,(-ieF_{\lambda\rho})\sigma_{\lambda\rho}\,(\gamma_\mu \Psi_\mu) + M\,(D_\mu \Psi_\mu) = 0.$$

Thus, equation III reads

$$+\frac{M}{4}\frac{\sqrt{6}fc_2c_3^* - c_2(c_1+c_3/\sqrt{6}) - g|c_4|^2}{c_2(c_1+c_3/\sqrt{6})}\hat{D}(\gamma_\mu \Psi_\mu) - i\frac{\sqrt{6}}{4}M\,\frac{gc_4^*}{c_2}\,\hat{D}\Psi_0$$
$$-\frac{fc_2|c_3|^2 + gc_1|c_4|^2}{c_2(c_1+c_3/\sqrt{6})}\,\hat{D}(D_\mu \Psi_\mu) - i\frac{2}{\sqrt{6}}\,gc_4^*\,(-ieF_{\lambda\rho})\sigma_{\lambda\rho}\,\Psi_0$$
$$+\frac{2}{\sqrt{6}}fc_3^*\,(-ieF_{\lambda\rho})\sigma_{\lambda\rho}\,(\gamma_\mu \Psi_\mu) + M\,(D_\mu \Psi_\mu) = 0, \quad (50)$$

it does not contain the 2-nd order operator D^2.

7. Diagonalizing the System in Electromagnetic Fields

We start with three equations

$$c_2 \hat{D}\Psi_0 - \frac{4ic_4}{\sqrt{6}} \left[(D_\mu \Psi_\mu) - \frac{1}{4} \hat{D}(\gamma_\mu \Psi_\mu) \right] + M\Psi_0 = 0, \tag{51}$$

$$(c_1 + \frac{c_3}{\sqrt{6}}) \hat{D}(\gamma_\mu \Psi_\mu) - \frac{4c_3}{\sqrt{6}} (D_\mu \Psi_\mu) + M(\gamma_\mu \Psi_\mu) = 0, \tag{52}$$

$$+\frac{M}{4} \frac{\sqrt{6} f c_2 c_3^* - c_2(c_1 + c_3/\sqrt{6}) - g|c_4|^2}{c_2(c_1 + c_3/\sqrt{6})} \hat{D}(\gamma_\mu \Psi_\mu) - i\frac{\sqrt{6}}{4} M \frac{gc_4^*}{c_2} \hat{D}\Psi_0$$
$$-\frac{fc_2|c_3|^2 + gc_1|c_4|^2}{c_2(c_1 + c_3/\sqrt{6})} \hat{D}(D_\mu \Psi_\mu) - i\frac{2}{\sqrt{6}} gc_4^* (-ieF_{\lambda\rho})\sigma_{\lambda\rho} \Psi_0$$
$$+\frac{2}{\sqrt{6}} fc_3^* (-ieF_{\lambda\rho})\sigma_{\lambda\rho} (\gamma_\mu \Psi_\mu) + M (D_\mu \Psi_\mu) = 0. \tag{53}$$

The task is to transform the system to a form which is similar to that existing for the free particle, but takes into account the presence of external field. To this end, first let us consider eq. (51):

$$c_2 \hat{D}\Psi_0 + \frac{ic_4}{\sqrt{6}} \hat{D}(\gamma_\mu \Psi_\mu) - \frac{4ic_4}{\sqrt{6}} (D_\mu \Psi_\mu) + M\Psi_0 = 0 .$$

The term $(D_\mu \Psi_\mu)$ is excluded with eq. (53), so we obtain

$$\frac{c_2^2 + g|c_4|^2}{c_2} \hat{D}\Psi_0 + \frac{ic_4}{\sqrt{6}} \frac{\sqrt{6} f c_2 c_3^* - g|c_4|^2}{c_2(c_1 + c_3/\sqrt{6})} \hat{D}(\gamma_\mu \Psi_\mu)$$
$$-\frac{4ic_4}{\sqrt{6}} \frac{1}{M} \frac{fc_2|c_3|^2 + gc_1|c_4|^2}{c_2(c_1 + c_3/\sqrt{6})} + \frac{4}{3} \frac{1}{M} g|c_4|^2 (-ieF_{\lambda\rho})\sigma_{\lambda\rho}\Psi_0$$
$$+\frac{4i}{3} \frac{1}{M} fc_3^* c_4 (-ieF_{\lambda\rho})\sigma_{\lambda\rho} (\gamma_\mu \Psi_\mu) + M\Psi_0 = 0. \tag{54}$$

Similarly, from equation II

$$(c_1 + \frac{c_3}{\sqrt{6}}) \hat{D}(\gamma_\mu \Psi_\mu) - \frac{4c_3}{\sqrt{6}} (D_\mu \Psi_\mu) + M(\gamma_\mu \Psi_\mu) = 0 ,$$

we can exclude the term $(D_\mu \Psi_\mu)$ as well

$$\frac{1}{c_2(c_1 + c_3/\sqrt{6})} \left[c_2(c_1^2 + f|c_3|^2) + \frac{c_3}{\sqrt{6}} (c_1 c_2 - g_4|c^2) \right] \hat{D}(\gamma_\mu \Psi_\mu)$$
$$-\frac{4}{\sqrt{6}} \frac{c_3}{M} \frac{fc_2|c_3|^2 + gc_1|c_4|^2}{c_2(c_1 + c_3/\sqrt{6})} \hat{D}(D_\mu \Psi_\mu) - i\frac{gc_3 c_4^*}{c_2} \hat{D}\Psi_0$$
$$-\frac{4ic_3}{3} \frac{1}{M} gc_4^* (-ieF_{\lambda\rho})\sigma_{\lambda\rho} \Psi_0$$
$$+\frac{4c_3}{3} \frac{1}{M} fc_3^* (-ieF_{\lambda\rho})\sigma_{\lambda\rho} (\gamma_\mu \Psi_\mu) + M(\gamma_\mu \Psi_\mu) = 0. \tag{55}$$

Let us introduce notations:

$$\gamma_\mu \Psi_\mu = \bar{\Phi}_1, \quad \bar{\Phi}_2 = \Psi_0, \quad \bar{\Phi}_3 = D_\mu \Psi_\mu \,, \quad (-ieF_{\lambda\rho})\sigma_{\lambda\rho} = \Sigma \,. \tag{56}$$

Thus, we have the following three equations

$$\frac{1}{c_2(c_1 + c_3/\sqrt{6})} \left[c_2(c_1^2 + f|c_3|^2) + \frac{c_3}{\sqrt{6}}(c_1 c_2 - g_4|c^2) \right] \hat{D}\bar{\Phi}_1$$

$$-i\frac{g c_3 c_4^*}{c_2} \hat{D}\bar{\Phi}_2 - \frac{4}{\sqrt{6}} \frac{c_3}{M} \frac{f c_2 |c_3|^2 + g c_1 |c_4|^2}{c_2(c_1 + c_3/\sqrt{6})} \hat{D}\bar{\Phi}_3$$

$$+ \frac{4c_3}{3} \frac{1}{M} f c_3^* \Sigma \, \bar{\Phi}_1 - \frac{4ic_3}{3} \frac{1}{M} g c_4^* \Sigma \, \bar{\Phi}_2 + M\bar{\Phi}_1 = 0,$$

$$\frac{ic_4}{\sqrt{6}} \frac{\sqrt{6} f c_2 c_3^* - g|c_4|^2}{c_2(c_1 + c_3/\sqrt{6})} \hat{D}\bar{\Phi}_1$$

$$+ \frac{c_2^2 + g|c_4|^2}{c_2} \hat{D}\bar{\Phi}_2 - \frac{4ic_4}{\sqrt{6}} \frac{1}{M} \frac{f c_2 |c_3|^2 + g c_1 |c_4|^2}{c_2(c_1 + c_3/\sqrt{6})} \hat{D}\bar{\Phi}_3$$

$$+ \frac{4i}{3} \frac{1}{M} f c_3^* c_4 \Sigma \, \bar{\Phi}_1 + \frac{4}{3} \frac{1}{M} g|c_4|^2 \Sigma \bar{\Phi}_2 + M\Psi_0 = 0,$$

$$+ \frac{M}{4} \frac{\sqrt{6} f c_2 c_3^* - c_2(c_1 + c_3/\sqrt{6}) - g|c_4|^2}{c_2(c_1 + c_3/\sqrt{6})} \hat{D}\bar{\Phi}_1$$

$$+ i\frac{\sqrt{6}}{4} M \frac{g c_4^*}{c_2} \hat{D}\bar{\Phi}_2 - \frac{f c_2 |c_3|^2 + g c_1 |c_4|^2}{c_2(c_1 + c_3/\sqrt{6})} \hat{D}\bar{\Phi}_3$$

$$+ \frac{2}{\sqrt{6}} f c_3^* \, \Sigma \, \bar{\Phi}_1 - i\frac{2}{\sqrt{6}} \, g c_4^* \, \Sigma \bar{\Phi}_2 + M \, \bar{\Phi}_3 = 0.$$

Having used the previously introduced notations $A_i, B_i, R_i, \; i = 1, 2, 3$, we may rewrite the last system in a shorter form:

$$\begin{aligned}
&A_1 \hat{D} \, \bar{\Phi}_1 + B_1 \hat{D} \, \bar{\Phi}_2 + R_1 \hat{D} \, \bar{\Phi}_3 \\
+M \, \bar{\Phi}_1 + \frac{4}{3} &\frac{f|c_3|^2}{M} \Sigma \, \bar{\Phi}_1 - \frac{4i}{3} \frac{g c_3 c_4^*}{M} \Sigma \, \bar{\Phi}_2 = 0, \\
&A_2 \hat{D} \, \bar{\Phi}_1 + B_2 \hat{D} \, \bar{\Phi}_2 + R_2 \hat{D} \, \bar{\Phi}_3 \\
+M \, \Phi_2 + \frac{4i}{3} &\frac{f c_3^* c_4}{M} \Sigma \, \bar{\Phi}_1 + \frac{4}{3} \frac{g|c_4|^2}{M} \Sigma \, \bar{\Phi}_2 = 0, \\
&A_3 \hat{D} \, \bar{\Phi}_1 + B_3 \hat{D} \, \bar{\Phi}_2 + R_3 \hat{D} \, \bar{\Phi}_3 \\
+M \, \bar{\Phi}_3 + \frac{2}{\sqrt{6}} &f c_3^* \, \Sigma \, \bar{\Phi}_1 - i\frac{2}{\sqrt{6}} \, g c_4^* \Sigma \, \bar{\Phi}_2 = 0.
\end{aligned} \tag{57}$$

Now, we should act in the same manner as for the free case. We multiply equations 1,2,3 by a_1, a_2, a_3 and sum the results:

$$(a_1 A_1 + a_2 A_2 + a_3 A_3)\hat{D}\bar{\Phi}_1 + (a_1 B_1 + a_2 B_2 + a_3 B_3)\hat{D}\bar{\Phi}_2$$

$$+(a_1R_1+a_2R_2+a_3R_3)\hat{D}\bar{\Phi}_3+M(a_1\bar{\Phi}_1+a_2\bar{\Phi}_2+a_3\bar{\Phi}_3)$$
$$+(a_1\frac{4}{3}\frac{f|c_3|^2}{M}+a_2\frac{4i}{3}\frac{fc_3^*c_4}{M}+a_3\frac{2}{\sqrt{6}}fc_3^*)\Sigma\,\bar{\Phi}_1$$
$$+(-a_1\frac{4i}{3}\frac{gc_3c_4^*}{M}+a_2\frac{4}{3}\frac{g|c_4|^2}{M}-a_3\frac{2i}{\sqrt{6}}\,gc_4^*)\Sigma\,\bar{\Phi}_2=0;$$

whence taking in mind three relations

$$a_1A_1+a_2A_2+a_3A_3=\lambda_1a_1,$$
$$a_1B_1+a_2B_2+a_3B_3=\lambda_1a_2,$$
$$a_1R_1+a_2R_2+a_3R_3=\lambda_1a_3,$$

we get

$$\lambda_1\hat{D}(a_1\bar{\Phi}_1+a_2\bar{\Phi}_2+a_3\bar{\Phi}_3)+M(a_1\bar{\Phi}_1+a_2\bar{\Phi}_2a_3\bar{\Phi}_3)$$
$$+(\,a_1\frac{4}{3}\frac{f|c_3|^2}{M}+a_2\frac{4i}{3}\frac{fc_3^*c_4}{M}+a_3\frac{2}{\sqrt{6}}fc_3^*)\Sigma\bar{\Phi}_1$$
$$+(-a_1\frac{4i}{3}\frac{gc_3c_4^*}{M}+a_2\frac{4}{3}\frac{g|c_4|^2}{M}-a_3\frac{2i}{\sqrt{6}}gc_4^*)\Sigma\bar{\Phi}_2=0. \tag{58}$$

Similarly, we obtain yet two equations (in fact, changing a_i on b_i and on r_i)

$$\lambda_1\hat{D}(b_1\bar{\Phi}_1+b_2\bar{\Phi}_2+b_3\bar{\Phi}_3)+M(b_1\bar{\Phi}_1+b_2\bar{\Phi}_2+b_3\bar{\Phi}_3)$$
$$+(b_1\frac{4}{3}\frac{f|c_3|^2}{M}+b_2\frac{4i}{3}\frac{fc_3^*c_4}{M}+b_3\frac{2}{\sqrt{6}}fc_3^*)\Sigma\bar{\Phi}_1$$
$$+(-b_1\frac{4i}{3}\frac{gc_3c_4^*}{M}+b_2\frac{4}{3}\frac{g|c_4|^2}{M}-b_3\frac{2i}{\sqrt{6}}gc_4^*)\Sigma\bar{\Phi}_2=0\,; \tag{59}$$

$$\lambda_1\hat{D}(r_1\bar{\Phi}_1+r_2\bar{\Phi}_2+r_3\bar{\Phi}_3)+M(r_1\bar{\Phi}_1+r_2\bar{\Phi}_2+r_3\bar{\Phi}_3)$$
$$+(r_1\frac{4}{3}\frac{f|c_3|^2}{M}+r_2\frac{4i}{3}\frac{fc_3^*c_4}{M}+r_3\frac{2}{\sqrt{6}}fc_3^*)\Sigma\bar{\Phi}_1$$
$$+(-r_1\frac{4i}{3}\frac{gc_3c_4^*}{M}+r_2\frac{4}{3}\frac{g|c_4|^2}{M}-r_3\frac{2i}{\sqrt{6}}gc_4^*)\Sigma\bar{\Phi}_2=0\,. \tag{60}$$

Introducing three new bispinors

$$\Phi_1=a_1\bar{\Phi}_1+a_2\bar{\Phi}_2+a_3\bar{\Phi}_3,$$
$$\Phi_2=b_1\bar{\Phi}_1+b_2\bar{\Phi}_2+b_3\bar{\Phi}_3,$$
$$\Phi_3=r_1\bar{\Phi}_1+r_2\bar{\Phi}_2+r_3\bar{\Phi}_3, \tag{61}$$

we may present the above system shorter

$$\lambda_1\hat{D}\Phi_1+M\Phi_1$$
$$+(a_1\frac{4}{3}\frac{f|c_3|^2}{M}+a_2\frac{4i}{3}\frac{fc_3^*c_4}{M}+a_3\frac{2}{\sqrt{6}}fc_3^*)\Sigma\,\bar{\Phi}_1$$
$$+(-a_1\frac{4i}{3}\frac{gc_3c_4^*}{M}+a_2\frac{4}{3}\frac{g|c_4|^2}{M}-a_3\frac{2i}{\sqrt{6}}\,gc_4^*)\Sigma\,\bar{\Phi}_2=0, \tag{62}$$

$$\lambda_1 \hat{D}\Phi_2 + M\Phi_2$$
$$+(b_1 \frac{4}{3}\frac{f|c_3|^2}{M} + b_2 \frac{4i}{3}\frac{f c_3^* c_4}{M} + b_3 \frac{2}{\sqrt{6}} f c_3^*) \Sigma\, \bar{\Phi}_1$$
$$+(-b_1 \frac{4i}{3}\frac{g c_3 c_4^*}{M} + b_2 \frac{4}{3}\frac{g|c_4|^2}{M} - b_3 \frac{2i}{\sqrt{6}}\, g c_4^*) \Sigma\, \bar{\Phi}_2 = 0, \tag{63}$$

$$\lambda_1 \hat{D}\Phi_3 + M\Phi_3$$
$$+\left(r_1 \frac{4}{3}\frac{f|c_3|^2}{M} + r_2 \frac{4i}{3}\frac{f c_3^* c_4}{M} + r_3 \frac{2}{\sqrt{6}} f c_3^*\right) \Sigma\, \bar{\Phi}_1$$
$$+\left(-r_1 \frac{4i}{3}\frac{g c_3 c_4^*}{M} + r_2 \frac{4}{3}\frac{g|c_4|^2}{M} - r_3 \frac{2i}{\sqrt{6}}\, g c_4^*\right) \Sigma\, \bar{\Phi}_2 = 0. \tag{64}$$

8. Mixing the Components in the System

Relation between two sets of bispinors $\bar{\Phi}_i$ and Φ_i is determined by the formula

$$\begin{vmatrix} \Phi_1 \\ \Phi_2 \\ \Phi_3 \end{vmatrix} = \begin{vmatrix} a_1 & a_2 & a_3 \\ b_1 & b_2 & b_3 \\ r_1 & r_2 & r_3 \end{vmatrix} \begin{vmatrix} \bar{\Phi}_1 \\ \bar{\Phi}_2 \\ \bar{\Phi}_3 \end{vmatrix}, \qquad \Phi = S\bar{\Phi}. \tag{65}$$

In fact, we need the inverse matrix:

$$\bar{\Phi} = S^{-1}\Phi, \qquad S^{-1} = \frac{1}{\det S} \begin{vmatrix} S_{11} & S_{21} & S_{31} \\ S_{12} & S_{22} & S_{32} \\ S_{13} & S_{22} & S_{33} \end{vmatrix},$$

$$S^{-1} = \frac{1}{\det S} \begin{vmatrix} (b_2 r_3 - b_3 r_2) & -(a_2 r_3 - a_3 r_2) & (a_2 b_3 - a_3 b_2) \\ -(b_1 r_3 - b_3 r_1) & (a_1 r_3 - a_3 r_1) & -(a_1 b_3 - a_3 b_1) \\ (b_1 r_2 - b_2 r_1) & -(a_1 r_2 - a_2 r_1) & (a_1 b_2 - a_2 b_1) \end{vmatrix},$$

$$\det S = r_1(a_2 b_3 - a_3 b_2) - r_2(a_1 b_3 - a_3 b_1) + r_3(a_1 b_2 - a_2 b_1). \tag{66}$$

The elements of the matrix S were given above. Because a_1, b_1, r_1 may be arbitrary, it is convenient to fix them so that the elements of the matrix be most simple (we collect them into three sets)

$$a_1 = \lambda_1\, c_2(\lambda_1 - c_2) + (\lambda_1 - c_2)\, [f c_2 |c_3|^2 + g c_1 |c_4|^2]$$
$$-\lambda_1\, (c_1 + c_3/\sqrt{6})\, g|c_4|^2,$$
$$b_1 = \lambda_2\, c_2(\lambda_2 - c_2) + (\lambda_2 - c_2)\, [f c_2 |c_3|^2 + g c_1 |c_4|^2]$$
$$-\lambda_2\, (c_1 + c_3/\sqrt{6})\, g|c_4|^2,$$
$$r_1 = \lambda_3\, c_2(\lambda_3 - c_2) + (\lambda_3 - c_2)\, [f c_2 |c_3|^2 + g c_1 |c_4|^2]$$
$$-\lambda_3\, (c_1 + c_3/\sqrt{6})\, g|c_4|^2,$$

$$a_2 = -igc_3c_4^*\,(c_1 + c_3/\sqrt{6})\lambda_1, \quad b_2 = -igc_3c_4^*\,(c_1 + c_3/\sqrt{6})\lambda_2,$$
$$r_2 = -igc_3c_4^*\,(c_1 + c_3/\sqrt{6})\lambda_3,$$
$$a_3 = -\frac{4}{\sqrt{6}}\frac{1}{M}\,c_3(\lambda_1 - c_2)\,[fc_2|c_3|^2 + gc_1|c_4|^2],$$
$$b_3 = -\frac{4}{\sqrt{6}}\frac{1}{M}\,c_3(\lambda_2 - c_2)\,[fc_2|c_3|^2 + gc_1|c_4|^2],$$
$$r_3 = -\frac{4}{\sqrt{6}}\frac{1}{M}\,c_3(\lambda_3 - c_2)\,[fc_2|c_3|^2 + gc_1|c_4|^2].$$

With notation

$$K = [fc_2|c_3|^2 + gc_1|c_4|^2], \qquad L = (c_1 + c_3/\sqrt{6}), \tag{67}$$

the formulas read shorter

$$a_1 = \lambda_1\,c_2(\lambda_1 - c_2) + (\lambda_1 - c_2)\,K - \lambda_1\,L\,g|c_4|^2,$$
$$b_1 = \lambda_2\,c_2(\lambda_2 - c_2) + (\lambda_2 - c_2)\,K - \lambda_2\,L\,g|c_4|^2,$$
$$r_1 = \lambda_3\,c_2(\lambda_3 - c_2) + (\lambda_3 - c_2)\,K - \lambda_3\,L\,g|c_4|^2,$$

$$a_2 = -igc_3c_4^*\,L\,\lambda_1\,, \quad b_2 = -igc_3c_4^*\,L\,\lambda_2\,, \quad r_2 = -igc_3c_4^*\,L\,\lambda_3,$$

$$a_3 = -\frac{4}{\sqrt{6}}\frac{1}{M}c_3(\lambda_1 - c_2)K,\; b_3 = -\frac{4}{\sqrt{6}}\frac{1}{M}c_3(\lambda_2 - c_2)K,$$
$$r_3 = -\frac{4}{\sqrt{6}}\frac{1}{M}\,c_3(\lambda_3 - c_2)K.$$

Let us collect results related to the matrix S^{-1} together.

$$\det S = -\frac{2\,i\,\sqrt{6}}{3}\,\frac{{c_2}^2{c_3}^2c_4^*\,(\lambda_1 - \lambda_3)\,(\lambda_1 - \lambda_2)\,(\lambda_2 - \lambda_3)\,KLg}{M}; \tag{68}$$

and the complete inverse matrix S^{-1} (its entries are listed by columns)

$$S^{-1} = \left| \begin{array}{cc} \frac{1}{(\lambda_1-\lambda_3)(\lambda_1-\lambda_2)c_2} & -\frac{1}{(\lambda_2-\lambda_3)(\lambda_1-\lambda_2)c_2} \\ \frac{iLg(|c_4|)^2+i(-\lambda_2+c_2)(-\lambda_3+c_2)}{(\lambda_1-\lambda_3)(\lambda_1-\lambda_2)Lc_4^*\,gc_3\,c_2} & \frac{-iLg(|c_4|)^2-i(-\lambda_1+c_2)(-\lambda_3+c_2)}{(\lambda_2-\lambda_3)(\lambda_1-\lambda_2)Lc_4^*\,gc_3\,c_2} \\ \frac{1}{4}\,\frac{M\sqrt{6}(\lambda_2\,\lambda_3+K)}{c_3\,c_2\,(\lambda_1-\lambda_3)(\lambda_1-\lambda_2)K} & -\frac{1}{4}\,\frac{M\sqrt{6}(\lambda_1\,\lambda_3+K)}{c_3\,c_2\,(\lambda_2-\lambda_3)(\lambda_1-\lambda_2)K} \end{array} \right.$$

$$\left. \begin{array}{c} \frac{1}{(\lambda_2-\lambda_3)(\lambda_1-\lambda_3)c_2} \\ \frac{iLg(|c_4|)^2+i(-\lambda_1+c_2)(-\lambda_2+c_2)}{(\lambda_2-\lambda_3)(\lambda_1-\lambda_3)Lc_4^*\,gc_3\,c_2} \\ \frac{1}{4}\,\frac{M\sqrt{6}(\lambda_1\,\lambda_2+K)}{c_3\,c_2\,(\lambda_2-\lambda_3)(\lambda_1-\lambda_3)K} \end{array} \right| . \tag{69}$$

Let turn to equations (62)–(64) and calculate the term

$$\left\{ a_1 \frac{4}{3}\frac{f|c_3|^2}{M} + a_2 \frac{4i}{3}\frac{f c_3^* c_4}{M} + a_3 \frac{2}{\sqrt{6}} f c_3^* \right\} \Sigma\, \bar{\Phi}_1$$

$$= \lambda_1(\lambda_1 - c_2) \frac{4}{3M} f c_2 |c_3|^2\, \Sigma\, \bar{\Phi}_1,$$

and the term

$$\left\{ -a_1 \frac{4i}{3}\frac{g c_3 c_4^*}{M} + a_2 \frac{4}{3}\frac{g|c_4|^2}{M} - a_3 \frac{2i}{\sqrt{6}}\, g c_4^* \right\} \Sigma\, \bar{\Phi}_2$$

$$= -\lambda_1(\lambda_1 - c_2) \frac{4i}{3}\frac{1}{M}\, g c_2 c_3 c_4^*\, \Sigma\, \bar{\Phi}_2 .$$

Therefore, eq. (62) takes the form

$$\lambda_1 \hat{D} \Phi_1 + M \Phi_1 + \frac{4 c_2 c_3}{3M} \lambda_1(\lambda_1 - c_2)\, \Sigma \left(f c_3^*\, \bar{\Phi}_1 - i g c_4^* \bar{\Phi}_2 \right) = 0\,. \tag{70}$$

Similarly find

$$\left\{ b_1 \frac{4}{3}\frac{f|c_3|^2}{M} + b_2 \frac{4i}{3}\frac{f c_3^* c_4}{M} + b_3 \frac{2}{\sqrt{6}} f c_3^* \right\} \Sigma\, \bar{\Phi}_1$$

$$= \lambda_2(\lambda_2 - c_2) \frac{4}{3M} f c_2 |c_3|^2\, \Sigma\, \bar{\Phi}_1,$$

and

$$\left\{ -b_1 \frac{4i}{3}\frac{g c_3 c_4^*}{M} + b_2 \frac{4}{3}\frac{g|c_4|^2}{M} - b_3 \frac{2i}{\sqrt{6}}\, g c_4^* \right\} \Sigma\, \bar{\Phi}_2$$

$$= -\lambda_2(\lambda_2 - c_2) \frac{4i}{3M} g c_2 c_3 c_4^* \Sigma\, \bar{\Phi}_2 .$$

Therefore, eq. (63) takes the form

$$\lambda_2 \hat{D} \Phi_2 + M \Phi_2 + \frac{4 c_2 c_3}{3M} \lambda_2(\lambda_2 - c_2)\, \Sigma \left(f c_3^*\, \bar{\Phi}_1 - i g c_4^* \bar{\Phi}_2 \right) = 0. \tag{71}$$

The third equation (64) is presented as

$$\lambda_3 \hat{D} \Phi_3 + M \Phi_3 + \frac{4 c_2 c_3}{3M} \lambda_3(\lambda_3 - c_2)\, \Sigma \left(f c_3^*\, \bar{\Phi}_1 - i g c_4^* \bar{\Phi}_2 \right) = 0. \tag{72}$$

Collect three equations together:

$$\begin{gathered}
\lambda_1 \hat{D} \Phi_1(x) + M \Phi_1(x) \\
+ \frac{4 c_2 c_3}{3M} \lambda_1(\lambda_1 - c_2)\, \Sigma(x) \left[f c_3^*\, \bar{\Phi}_1(x) - i g c_1^* \bar{\Phi}_2(x) \right] = 0, \\
\lambda_2 \hat{D} \Phi_2(x) + M \Phi_2(x) \\
+ \frac{4 c_2 c_3}{3M} \lambda_2(\lambda_2 - c_2)\, \Sigma(x) \left[f c_3^*\, \bar{\Phi}_1(x) - i g c_4^* \bar{\Phi}_2(x) \right] = 0, \\
\lambda_3 \hat{D} \Phi_3(x) + M \Phi_3(x) \\
+ \frac{4 c_2 c_3}{3M} \lambda_3(\lambda_3 - c_2)\, \Sigma(x) \left[f c_3^*\, \bar{\Phi}_1(x) - i g c_4^* \bar{\Phi}_2(x) \right] = 0.
\end{gathered} \tag{73}$$

Taking in mind the explicit form of inverse matrix S^{-1}, the functions $\bar{\Phi}_1$ and $\bar{\Phi}_2$ are expressed in terms of Φ_1, Φ_2, Φ_3 by the formulas

$$\bar{\Phi}_1 = \frac{1}{(\lambda_1 - \lambda_3)(\lambda_1 - \lambda_2)\, c_2}\Phi_1 - \frac{1}{(\lambda_2 - \lambda_3)(\lambda_1 - \lambda_2)\, c_2}\Phi_2$$

$$+\frac{1}{(\lambda_2 - \lambda_3)(\lambda_1 - \lambda_3)\, c_2}\Phi_3,$$

and

$$\bar{\Phi}_2 = \frac{iLg\,(|c_4|)^2 + i\,(-\lambda_2 + c_2)\,(-\lambda_3 + c_2)}{(\lambda_1 - \lambda_3)(\lambda_1 - \lambda_2)\, Lc_4^* g c_3\, c_2}\Phi_1$$

$$+\frac{-iLg\,(|c_4|)^2 - i\,(-\lambda_1 + c_2)\,(-\lambda_3 + c_2)}{(\lambda_2 - \lambda_3)(\lambda_1 - \lambda_2)\, Lc_4^* g c_3\, c_2}\Phi_2$$

$$+\frac{iLg\,(|c_4|)^2 + i\,(-\lambda_1 + c_2)\,(-\lambda_2 + c_2)}{(\lambda_2 - \lambda_3)(\lambda_1 - \lambda_3)\, Lc_4^* g c_3\, c_2}\Phi_3,$$

we find explicit form for the term

$$fc_3^*\,\bar{\Phi}_1(x) - igc_4^*\bar{\Phi}_2(x) = L_1\Phi_1(x) + L_2\Phi_2(x) + L_3\Phi_3(x) = \Phi(x)\,, \qquad (74)$$

where L_i turn to be rather symmetrical:

$$L_1 = \frac{gL|c_4|^2 + fL|c_3|^2 + c_2{}^2 - c_2(\lambda_2 + \lambda_3) + \lambda_2\,\lambda_3}{Lc_2\, c_3(\lambda_1 - \lambda_2)(\lambda_1 - \lambda_3)},$$
$$L_2 = \frac{gL|c_4|^2 + fL|c_3|^2 + c_2{}^2 - c_2(\lambda_3 + \lambda_1) + \lambda_3\,\lambda_1}{Lc_2\, c_3(\lambda_2 - \lambda_3)(\lambda_2 - \lambda_1)},$$
$$L_3 = \frac{gL|c_4|^2 + fL|c_3|^2 + c_2{}^2 - c_2(\lambda_1 + \lambda_2) + \lambda_1\,\lambda_2}{Lc_2\, c_3(\lambda_3 - \lambda_1)(\lambda_3 - \lambda_2)}; \qquad (75)$$

recall the notation $L = (c_1 + c_3/\sqrt{6})$; and remember that positive values for three masses appear only when $f = -1$ and $g = -1$. With notations

$$Y_i = \frac{4c_2c_3}{3M}(\lambda_i - c_2), \quad i = 1, 2, 3, \qquad (76)$$

equations (73) may be presented as

$$\hat{D}\Phi_1(x) + M_1\Phi_1(x) + Y_1\,\Sigma(x)\Phi(x) = 0,$$
$$\hat{D}\Phi_2(x) + M_2\Phi_2(x) + Y_2\,\Sigma(x)\Phi(x) = 0,$$
$$\hat{D}\Phi_3(x) + M_3\Phi_3(x) + Y_3\,\Sigma(x)\Phi(x) = 0, \qquad (77)$$

where $\Sigma(x) = -ieF_{ab}(x)\sigma_{ab}$.

9. Extension to General Relativity

In order to follow extension of the model from flat Minkowski space to any Riemannian space-time we should turn back to the system (51)–(53) and make several simple modifications in it.

1. Taking in mind that in Minkowski space the ict-metrics was used, however in Riemannian space we use the metrics $g_{\alpha\beta}(x)$, related to signature $(+,-,-,-)$, we must make the following change:

$$M \Longrightarrow iM. \tag{78}$$

2. Now Dirac matrices in spinor basis are determined by the formulas

$$\gamma^0 = \begin{vmatrix} 0 & I \\ I & 0 \end{vmatrix}, \quad \gamma^i = \begin{vmatrix} 0 & -\sigma^i \\ \sigma^i & 0 \end{vmatrix}. \tag{79}$$

3. Derivatives are modified according to the rules [60]

$$\begin{gathered} D_\alpha(x) = \partial_\alpha + ieA_\alpha(x) \quad \Longrightarrow \\ D_\alpha(x) = \nabla_\alpha + \Gamma_\alpha(x) + ieA_\alpha(x)\,, \quad \hat{D} = \gamma^\alpha(x) D_\alpha(x), \end{gathered} \tag{80}$$

where $\Gamma_\alpha(x)$ is bispinor connection by Tetrode-Weyl-Fock-Ivanenko [60], and $\gamma^\alpha(x) = \gamma^a e^\alpha_{(a)}(x)$.

4. Note important commutation rules [60]:

$$\begin{aligned} &\gamma^\rho(x) D_\beta = D_\beta \gamma^\rho(x)\,, \qquad D_\sigma(x) g_{\alpha\beta}(x) = g_{\alpha\beta}(x) D_\sigma(x), \\ &\hat{D}\hat{D} = D_\alpha D_\beta \Big(\frac{\gamma^\alpha\gamma^\beta + \gamma^\beta\gamma^\alpha}{2} + \frac{\gamma^\alpha\gamma^\beta - \gamma^\beta\gamma^\alpha}{2}\Big) = \Box - \Sigma(x), \\ &D^2 = D^\alpha D_\alpha, \quad \Sigma(x) = (-ieF_{\alpha\beta}\sigma^{\alpha\beta}(x) + \frac{R}{4}), \end{aligned} \tag{81}$$

where $R(x)$ is the Ricci scalar.

Therefore, now we use the modified system of equations (51)–(53):

$$c_2 \hat{D}\Psi_0 - \frac{4ic_4}{\sqrt{6}} \Big[(D^\mu \Psi_\mu) - \frac{1}{4}\hat{D}(\gamma^\mu(x)\Psi_\mu) \Big] + iM\Psi_0 = 0, \tag{82}$$

$$(c_1 + \frac{c_3}{\sqrt{6}})\hat{D}(\gamma^\mu(x)\Psi_\mu) - \frac{4c_3}{\sqrt{6}}(D^\mu\Psi_\mu) + iM(\gamma^\mu(x)\Psi_\mu) = 0, \tag{83}$$

$$\begin{aligned} &+\frac{iM}{4}\frac{\sqrt{6} f c_2 c_3^* - c_2(c_1 + c_3/\sqrt{6}) - g|c_4|^2}{c_2(c_1 + c_3/\sqrt{6})}\ \hat{D}(\gamma^\mu(x)\Psi_\mu) - i\frac{\sqrt{6}}{4} iM\ \frac{g c_4^*}{c_2}\ \hat{D}\Psi_0 \\ &-\frac{f c_2|c_3|^2 + g c_1 |c_4|^2}{c_2(c_1 + c_3/\sqrt{6})}\ \hat{D}(D^\mu\Psi_\mu) - i\frac{2}{\sqrt{6}}\ g c_4^*\ (-ieF_{\lambda\rho}\sigma_{\lambda\rho}(x) + \frac{R}{4})\ \Psi_0 \\ &+\frac{2}{\sqrt{6}} f c_3^*\ (-ieF_{\lambda\rho}\sigma_{\lambda\rho} + \frac{R}{4})\ (\gamma^\mu(x)\Psi_\mu) + iM\ (D^\mu\Psi_\mu) = 0. \end{aligned} \tag{84}$$

Repeating with minor changes the above calculations, we arrive at the following generally covariant system

$$
\begin{aligned}
&i\hat{D}(x)\Phi_1(x) - M_1\Phi_1(x) + iY_1\,\Sigma(x)\Phi(x) = 0,\\
&i\hat{D}(x)\Phi_2(x) - M_2\Phi_2(x) + iY_2\,\Sigma(x)\Phi(x) = 0,\\
&i\hat{D}(x)\Phi_3(x) - M_3\Phi_3(x) + iY_3\,\Sigma(x)\Phi(x) = 0,
\end{aligned} \tag{85}
$$

where

$$
\begin{aligned}
&\hat{D}(x) = \gamma^\alpha(x)D_\alpha, \quad D_\alpha(x) = \nabla_\alpha + \Gamma_\alpha(x) + ieA_\alpha(x),\\
&Y_i = \frac{4c_2c_3}{3iM}(\lambda_i - c_2), \quad \Phi = L_1\Phi_1(x) + L_2\Phi_2(x) + L_3\Phi_3(x),\\
&\Sigma(x) = -ieF_{\lambda\rho}(x)\sigma_{\lambda\rho}(x) + \frac{R(x)}{4}.
\end{aligned} \tag{86}
$$

Because bispinors are scalars with respect to general relativity, in (86) the covariant derivative ∇_α acts as ordinary derivative ∂_α.

One important remark should be done. Evidently, that the system (85)–(86) allows for restriction to Majorana case (that is when $e = 0$), then instead of (85)–(86) we have

$$
\begin{aligned}
&i\gamma^\alpha(x)(\nabla_\alpha + \Gamma_\alpha)\Phi_1(x) - M_1\Phi_1(x) + Y_1' \frac{R(x)}{4}\Phi(x) = 0,\\
&i\gamma^\alpha(x)(\nabla_\alpha + \Gamma_\alpha)\Phi_2(x) - M_2\Phi_2(x) + Y_2' \frac{R(x)}{4}\Phi(x) = 0,\\
&i\gamma^\alpha(x)(\nabla_\alpha + \Gamma_\alpha)\Phi_3(x) - M_3\Phi_3(x) + Y_3' \frac{R(x)}{4}\Phi(x) = 0,
\end{aligned} \tag{87}
$$

where

$$
Y_i' = \frac{4c_2c_3}{3M}(\lambda_i - c_2), \quad \Phi = L_1\Phi_1(x) + L_2\Phi_2(x) + L_3\Phi_3(x). \tag{88}
$$

In any Majorana basis the properties hold

$$
[i\gamma^\alpha(x)]^* = i\gamma^\alpha(x) \quad [\Gamma_\alpha(x)]^* = \Gamma_\alpha(x);
$$

therefore bispinors $\Phi_1(x), \Phi_2(x), \Phi_3(x)$ and $\Phi(x)$ may be real or pure imaginary. This means that a generalized equation for Majorana particle with three mass parameters exists as well.

Conclusion

In the paper, starting from general Gel'fand-Yaglom approach a new 20-component wave equation for spin 1/2 fermion, which is characterized by three mass parameters, is derived. On the base of 20-component wave function, three auxiliary bispinors are introduce, in absence of external field they obey three separate Dirac-like equations with different masses M_1, M_2, M_3. In presence of external fields, electromagnetic field and gravitational non-Euclidean background with non-vanishing Ricci scalar curvature, the main equation is not split into separated equations, instead a quite definite mixing of three Dirac-like equations arises. It is shown that a generalized equation for Majorana particle with three mass parameters exists as well, such a generalized Majorana equation is not split into three separated equations in curved space-time if the Ricci scalar of the space-time model does not vanish.

Acknowledgments

The present work was developed within the framework of the project *BRFFR A18-RA-015*(2018-2020), set within the framework of the agreement between the National Academy of Sciences of Belarus and the Romanian Academy. Also this work was supported by the project *BRFFR F18Y-009* (2018-2020).

References

[1] V. L. Ginzburg, Ya. A. Smorodinsky. On wave equations for particles with variable spin. *Zh. Eksp. Teor. Fiz.* 1943. **13**. 274 (in Russian).

[2] V. L. Ginzburg, To the theory of exited states of elementary particles. *Zh. Eksp. Teor. Fiz.* 1943. **13**. 33–58 (in Russian).

[3] A. S. Davydov. Wave equations of a particle having spin $3/2$ in absence of field. *Zh. Eksp. Teor. Fiz.* 1943. **13**. no 9-10. 313–319 (in Russian).

[4] H. J. Bhabha, Harish-Chandra. On the theory of point particles. *Proc. Roy. Soc. London.* A. 1944. **183**. 134–141.

[5] H. J. Bhabha. Relativistic wave equations for the proton. *Proc. Indian Acad. Sci.* A. 1945. **21**. 241–264.

[6] H. J. Bhabha. Relativistic wave equations for elementary particles. *Rev. Mod. Phys.* 1945. **17**. no 2-3. 200–215.

[7] H. J. Bhabha. The theory of the elementary particles. *Rep. Progr. Phys.* 1946. **10**. 253–271.

[8] Harish-Chandra. On the equations of motion of point particles. *Proc. Roy. Soc. London.* A. 1946. **185**. 269–287.

[9] Harish-Chandra, Relativistic equations for elementary particles. *Proc. Roy. Soc. London.* A. 1948. **192**. 195–218.

[10] I. M. Gel'fand, A. M. Yaglom. General relativistic invariant equations snd infinitely dimensional representation of the Lorentz group. *Zh. Eksp. Teor. Fiz.* 1948. **18**. no 8. 703–733 (in Russian).

[11] I. M. Gel'fand, A.M. Yaglom. Pauli theorem for general relativistic invariant wave equations. *Zh. Eksp. Teor. Fiz.* 1948. **18**. no 12. 1096–1104 (in Russian).

[12] I. M. Gel'fand, A. M. Yaglom. Charge conjugation for general relativistic invariant wave equations. *Zh. Eksp. Teor. Fiz.* 1948. **18**. no 12. 1105–1111.

[13] H. J. Bhabha. Theory of elementary particles-fields. Lectures Delivered at 2nd Summer *Seminar Canadian Math. Congress*: held at University of British Columbia, Aug-1949. 1–103.

[14] F. I. Fedorov. To the problem of solving relativistic wave equations. *Doklady AN USSR*. 1949. **65**. no 6. 813–814 (in Russian).

[15] E. E. Fradkin. To the theory of particles with higher spins. *Zh. Eksp. Teor. Fiz.* 1950. **20**. no 1. 27–38 (in Russian).

[16] F. I. Fedorov. On minimal polynomials of matrices of relativistic wave equations. *Doklady AN USSR*. 1951. **79**. no 5. 787–790 (in Russian).

[17] F. I. Fedorov. *To the theory of a spin 2 particle.* Uchionye Zapiski Beloruaasian State University. ser. Phys.-mat. 1951. no 12. 156–173 (in Russian).

[18] H. J. Bhabha. An equation for a particle with two mass states and positive charge density. *Phil. Mag. 1952. Ser VII.* **43**. 33–47.

[19] F. I. Fedorov. Generalized relativistic wave equations. *Doklady AN USSR*. 1952. **82**. no 1. 37–40 (in Russian).

[20] M. Petras. A contribution of the theory of the Pauli–Fierz's equations a particle with spin $3/2$. *Czech. J. Phys.* 1955. **5**. no 2. 169–170.

[21] M. Petras. A note to Bhabha's equation for a particle with maximum spin $3/2$. *Czech. J. Phys.* 1955. **5**. no 3. 418–419.

[22] V. Ya. Fainberg, To the interactionb theory of the particles of the higher spins with electromagnetic and meson fields. *Trudy FIAN USSR*. 1955. **6**. 269–332 (in Russian).

[23] V. L. Ginzburg. On relativistic wave equations with a mass spectrum. *Acta Phys. Pol.* 1956. **15**. 163–175.

[24] H. Shimazu. A relativistic wave equation for a particle with two mass states of spin 1 and 0. *Progress of Theoretical Physics.* 1956. **16**. no 4. 285–298.

[25] T. Regge. On properties of the particle with spin 2. *Nuovo Cimento.* 1957. **5**. no 2. 325–326.

[26] P. G. Bergmann, A. I. Janis. Subsidiary conditions in covariant theories. *Phys. Rev.* 1958. **111**. no 4. 1191–1200.

[27] H. A. Buchdahl. On the compatibility of relativistic wave equations for particles of higher spin in the presence of a gravitational field. *Nuovo Cim.* 1958. **10**. 96–103.

[28] L. A. Shelepin. Covariant theory of relativictic wave equations. *Nucl. Phys.* 1962. **33**. no 4. 580–593.

[29] A. Z. Capri. First order wave equations for multi-mass fermions. *Nuovo Cim.* B. 1969. **64**. no 1. 151–158.

[30] A. Aurilia, H. Umezawa. Theory of high spin fields. *Phys. Rev.* 1969. **182**. no 5. 1682–1694.

[31] F.I. Fedorov, V.A. Pletyukhov. Wave equations with repeated representations of the Lorentz group. *Proceedings of the National academy of sciences of Belarus. Phys.-Math. series.* 1969. no 6. 81–88 (in Russian).

[32] V. A. Pletyukhov, F. I. Fedorov. The wave equation with repeated representations for spin 0 particle. *Proceedings of the National academy of sciences of Belarus. Phys.-Math. series.* 1970. no 2. 79–85 (in Russian).

[33] F. I. Fedorov, V. A. Pletyukhov. Wave equations with repeated representations of the Lorentz group. Half-integer spin. *Proceedings of the National academy of sciences of Belarus. Phys.-Math. series.* 1970. no 3. 78–83 (in Russian).

[34] V. A. Pletyukhov, F. I. Fedorov. Wave equation with repeated reprentations for a spin 1 particle. *Proceedings of the National academy of sciences of Belarus. Phys.-Math. series.* 1970. no 3. 84–92 (in Russian).

[35] A. Shamaly, A. Z. Capri. First-order wave equations for integral spin. *Nuovo Cim.* B. 1971. **2**. no 2. 235–253.

[36] V. Amar, U. Dozzio. Finite dimensional Gel'fand–Yaglom equations for arbitrary integral spin. *Nuovo Cim.* B. 1972. **9**. 53–63.

[37] A. Z. Capri. Electromagnetic properties of a new spin-$1/2$ field. *Progr. Theor. Phys.* 1972. **48**. no 4. 1364–1374.

[38] A. Shamaly, A.Z. Capri. Unified theories for massive spin 1 fields. *Can. J. Phys.* 1973. **51**. no 14. 1467–1470.

[39] M. A. K. Khalil. Properties of a 20-component spin $1/2$ relativistic wave equation. *Phys. Rev.* D. 1977. **15**. no 6. 1532–1539.

[40] A. S. Wightman. Invariant wave equations: general theory and applications to the external field problem. *Lecture Notes in Physics.* 1978. **73**. 1–101.

[41] M. A. K. Khalil. An equivalence of relativistic field equations. *Nuovo Cimento.* A. 1978. **45**. no 3. 389–404.

[42] L. Garding. Mathematics of invariant wave equations. *Lect. Notes in Physics.* 1978. **73**. 102–164.

[43] J. P. Gazeau. L'equation de Dirac avec masse et spin arbitrares: une construction simple et naturelle [The Dirac equation with arbitrary mass and spin: a simple and natural construction]. *J. Phys, G. Nucl. Phys.* 1980. **6**. no 12. 1459–1475.

[44] W. Cox. Higher-rank representations for zero-spin filds theories. *J. Phys.* A. 1982. **15**. 627–635.

[45] W. Cox. First-order formulation of massive spin-2 field theories. *J. Phys.* A. 1982. **15**. 253–268.

[46] P. M. Mathews, B. Vijayalakshmi, M.Sivakuma. On the admissible Lorentz group representations in unique-mass, unique-spin relativistic wave equations. *Phys.* A. 1982. **15**. no 11. 1579–1582.

[47] P. M. Mathews, B. Vijayalakshmi. On inequivalent classes unique-mass-spin relativistic wave equations involving repeated irreducible representations with arbitrary multiplicities. *J. Math. Phys.* 1984. **25**. no 4. 1080–1087.

[48] W. Cox. On the Lagrangian and Hamiltonian constraint algorithms for the Rarita-Schwinger field coupled to an external electromagnetic field. *J. Phys.* A. 1989. **22**. no 10. 1599–1608.

[49] S. Deser, A. Waldron. Inconsistencies of massive charged gravitating higher spins.*Nucl. Phys.* B. 2002. **631**. 369–387.

[50] V. Simulik. Relativistic wave equations of arbitrary spin in quantum mechanics and field theory, example spin $S = 2$. *J. of Phys. Conference Series*, 2017. **804**. no 1. paper 012040.

[51] E. M. Ovsiyuk, V. V. Kisel, Y. A. Voynova, O. V. Veko, V. M. Red'kov Spin 1/2 particle with anomalous magnetic moment in a uniform magnetic field, exact solutions *Nonlinear Phenomena in Complex Systems.* 2016. **19**. no 2. 153–165.

[52] V. Kisel, Ya. Voynova, E. Ovsiyuk, V. Balan, V. Red'kov. Spin 1 Particle with Anomalous Magnetic Moment in the External Uniform Magnetic Field. *Nonlinear Phenomena in Complex Systems.* 2017. **20**. no 1. 21–39.

[53] E. M. Ovsiyuk, Ya. A. Voynova, V. V. Kisel, V. Balan, V. M. Red'kov Spin 1 Particle with Anomalous Magnetic Moment in the External Uniform Electric Field Chapter in: *Quaternions: Theory and Applications.* Editor: Sandra Griffin. – Nova Science Publishers, Inc. USA, 2017. – P. 47–84.

[54] E. M. Ovsiyuk, Ya. A. Voynova, V. V. Kisel, V. Balan, V. M. Red'kov Spin 1 Particle with Anomalous Magnetic Moment in the External Uniform Electric Field Chapter in: *Quaternions: Theory and Applications.* Editor: Sandra Griffin. – Nova Science Publishers, Inc. USA, 2017. – P. 47–84.

[55] V. V. Kisel, E. M. Ovsiyuk, Ya. A. Voynova, V. M. Red'kov. Quantum mechanics of spin 1 particle with quadrupole moment in external uniform magetic field. *Problems of Physics, Mathemativs, and Thechnics.* 2017. no 3(32). P. 18–27.

[56] V. V. Kisel, V. A. Pletyukhov, V. V. Gilewsky, E. M. Ovsiyuk, O. V. Veko, V. M. Red'kov. Spin 1/2 particle with two mass states. interaction with external fields. *Nonlinear Phenomena in Complex Systems.* 2017. **20**. 4. 404–423.

[57] E. M. Ovsiyuk, O. V. Veko, Ya. A. Voynova, V. V. Kisel, V. Balan, V. M. Red'kov. Spin 1/2 particle with two masses in magnetic field. *Applied Sciences.* 2018. **20**. 148–166.

[58] E. M. Ovsiyuk, O. V. Veko, Ya. A. Voynova, V. M. Red'kov, V. V. Kisel. N. V. Samsonenko. Spin 1/2 particle with two masses in external magnetic field. *J. Mech. Cont. and Math. Sci.* Special Issue. 2019. no 1. P. 651–660

[59] I. M. Gel'fand, R. A. Minlos, Z. Ya. Shapiro. *Representations of the rotation and Lorentz groups and their applications.* Translated from the Russian edition (1958) by G. Cummins and T. Boddington. Pergamon, London; Macmillan, New York, 1963.

[60] V. M. Red'kov. *Fields in Riemannian space and the Lorentz group.* Publishing House: Belarusian Science, Minsk, 2009.

[61] V. V. Kisel, E. M. Ovsiyuk, V. Balan, O. V. Veko, V. M. Red'kov. Elementary particles with internal structure in external field. *Vol. I. General theory.* – Nova Science Publishers, Inc. USA, 2018; *Vol. II. Physical problems.* – Nova Science Publishers, Inc. USA, 2018.

In: Future Relativity, Gravitation, Cosmology
Editors: V. V. Dvoeglazov et al.
ISBN: 979-8-88697-455-3

Chapter 14

FRADKIN EQUATION FOR A SPIN 3/2 PARTICLE IN ELECTROMAGNETIC AND GRAVITATIONAL FIELDS

A. V. Ivashkevich[1,*]***, E. M. Ovsiyuk***[2,†]***, V. V. Kisel***[3,‡] ***and V. M. Red'kov***[1,§]
[1]Institute of Physics, National Academy of Sciences of Belarus
[2]Mozyr State Pedagogical University, Belarus
[3]Belarus State University of Informatics and Radio-Electronics, Belarus

Abstract

Many year ago, a special generalized equation for a spin 3/2 particle, different from Pauli–Fierz and Rarita–Schwinger model, was proposed by Fradkin. To the present time it is not clear which additional structure underlies this extended wave equation. We investigate this model systematically, applying the general Gel'fand–Yaglom formalism. Having used a standard set of requirements: relativistic invariance, single nonzero mass and single spin S=3/2, P-symmetry, existence of Lagrangian formulation for the model, we derive a set of spinor equations. The 20-component wave function consists of bispinor and vector-bispinor. It is shown that in the free case the Fradkin equation reduces to minimal Pauli–Fierz and Rarita–Schwinger equation for a bisbinor. In presence of external electromagnetic fields, the minimal form the Fradkin equation for a bispinor function contains an additional interaction term governed by electromagnetic tensor $F_{\alpha\beta}$. Finally we take into account the external curved space-time background, in generally covariant case the Fradkin equation contains additional gravitational interaction through the Ricci tensor $R_{\alpha\beta}$. If the electric charge of the particle is zero, The Fradkin model remains correct and describes a neutral spin 3/2 particle of Majorana type interacting additionally with geometrical background by means of the Ricci tensor.

PACS 1130, 0230, 0365.

Keywords: Lorentz group, extended set of representations, generalized wave equation, fermion, intrinsic electromagnetic structure, external fields, curved space-time.

*Corresponding Author's Email:ivashkevich.alina@yandex.by.
†Corresponding Author's Email: e.ovsiyuk@mail.ru.
‡Corresponding Author's Email: vasiliy-bspu@mail.ru.
§Corresponding Author's Email: redkov@ifanbel.bas-net.by.

1. Introduction

The general theory of relativistic equations for elementary particles had started with the investigations by Dirac [1], Majorana [2], Fierz and Pauli [3, 4]. In the present time, in addition to simplest and commonly used equations [1]–[20] for particles of spins $S = 0, \frac{1}{2}, 1, \frac{3}{2}, 2$ many other and more complicated equations have been proposed [5]–[45]: Ginzburg [5, 21], Davydov [6], Bhabha and Harish-Chandra [7]–[10], [15], Gel'fand and Yaglom [11], Fradkin [12], Fedorov [13, 14, 16, 42], Regge [17], Petras [18, 19], Fainberg [20], Ginzburg [21], Shimazu [22], Gel'fand, Minlos, and Shapiro [23], Shelepin [24, 25], Capri [26, 27, 33], Fedorov and Pletyukhov [28]–[31], Shamaly [32, 34], Khalil [35, 37], Wightman [36], Garding [38], Cox [39, 40, 44], Mathews,Vijayalakshmi and Sivakuma [41, 43], Deser and Waldron [45].

These generalized models are based on the use of extended sets of irreducible representations of the Lorentz group. There are known equations with several mass parameters ($S = 0, \frac{1}{2}, 1$), with several spin parameters, with additional intrinsic electromagnetic characteristics: anomalous magnetic moment ($S = \frac{1}{2}, 1, 2$), quadruple moment ($S = 1$), polarisability ($S = 0, 1$), Darwin-Cox structure ($S = 0$), and other. Such additional intrinsic characteristics physically manifest themselves in presence of external fields, for instance electromagnetic and gravitational.

In particular, many years ago, a special generalized equation for a spin 3/2 particle, different from Pauli–Fierz and Rarita–Schwinger model, was proposed by Fradkin [12]. To the present time it is not clear which additional structure underlies this extended wave equation. We investigate this model systematically, applying the general Gel'fand–Yaglom formalism [11]. Also see more recent publications concerning this field in [46]–[54], some results from them will be used in the present work.

2. Gel'fand–Yaglom Basis

In the frames of general theory of the first-order relativistic wave equations [2, 42, 52, 53]

$$(\Gamma_\mu \partial_\mu + M)\Phi = 0 \tag{1}$$

with the use of extended sets of Lorentz group representations, we will examine in detail an approach proposed by Fradkin [12] do describe a particle with spin 3/2. In this model, a 20-component wave function Ψ of the particle transforms as the following set of irreducible representations of the proper Lorentz group (enumerated by numbers 1, ..., 6):

$$\begin{array}{c} (1/2, 0) \sim 1, \quad (0, 1/2) \sim 2, \quad (1, 1/2) \sim 5, \\ (1/2, 1) \sim 6, \quad (1/2, 0)' \sim 3, \quad (0, 1/2)' \sim 4\,. \end{array} \tag{2}$$

The corresponding linking scheme is

$$\begin{array}{ccccccc} & & 4 & - & 3 & & \\ & & | & & | & & \\ 5 & - & 1 & - & 2 & - & 6 \end{array}$$

| |

3 – – – – – – – 4.

In accordance with the general theory, first we are do determine the matrix Γ_4 in (modified) Gel'fand–Yaglom basis (for more detail see in [52], [53]):

$$\begin{array}{lllll}
\epsilon^{1}_{\frac12,\frac12},\epsilon^{1}_{\frac12,-\frac12}, & \epsilon^{2}_{\frac12,\frac12},\epsilon^{2}_{\frac12,-\frac12}; & \epsilon^{3}_{\frac12,\frac12},\epsilon^{3}_{\frac12,-\frac12}, & \epsilon^{4}_{\frac12,\frac12},\epsilon^{4}_{\frac12,-\frac12}; & \epsilon^{6}_{\frac12,\frac12},\epsilon^{6}_{\frac12,-\frac12}, \\
\epsilon^{5}_{\frac12,\frac12},\epsilon^{5}_{\frac12,-\frac12}, & \epsilon^{6}_{\frac32,\frac32},\epsilon^{6}_{\frac32,-\frac32}, & \epsilon^{5}_{\frac32,\frac32},\epsilon^{5}_{\frac32,-\frac32}; & \epsilon^{6}_{\frac32,\frac12},\epsilon^{6}_{\frac32,-\frac12}, & \epsilon^{5}_{\frac32,\frac12},\epsilon^{5}_{\frac32,-\frac12}\,.
\end{array} \tag{3}$$

Spin-blocks $C^{(\frac12)}$ and $C^{(\frac32)}$ of the matrix Γ_4 may have the structure

$$C^{(\frac12)} = \left|\begin{array}{cccccccccccc}
0 & 0 & c_{12}^{(\frac12)} & 0 & 0 & 0 & c_{14}^{(\frac12)} & 0 & 0 & 0 & c_{15}^{(\frac12)} & 0 \\
0 & 0 & 0 & c_{12}^{(\frac12)} & 0 & 0 & 0 & c_{14}^{(\frac12)} & 0 & 0 & 0 & c_{15}^{(\frac12)} \\
c_{21}^{(\frac12)} & 0 & 0 & 0 & c_{23}^{(\frac12)} & 0 & 0 & 0 & c_{26}^{(\frac12)} & 0 & 0 & 0 \\
0 & c_{21}^{(\frac12)} & 0 & 0 & 0 & c_{23}^{(\frac12)} & 0 & 0 & 0 & c_{26}^{(\frac12)} & 0 & 0 \\
0 & 0 & c_{32}^{(\frac12)} & 0 & 0 & 0 & c_{34}^{(\frac12)} & 0 & 0 & 0 & c_{35}^{(\frac12)} & 0 \\
0 & 0 & 0 & c_{32}^{(\frac12)} & 0 & 0 & 0 & c_{34}^{(\frac12)} & 0 & 0 & 0 & c_{35}^{(\frac12)} \\
c_{41}^{(\frac12)} & 0 & 0 & 0 & c_{43}^{(\frac12)} & 0 & 0 & 0 & c_{46}^{(\frac12)} & 0 & 0 & 0 \\
0 & c_{41}^{(\frac12)} & 0 & 0 & 0 & c_{43}^{(\frac12)} & 0 & 0 & 0 & c_{46}^{(\frac12)} & 0 & 0 \\
0 & 0 & c_{62}^{(\frac12)} & 0 & 0 & 0 & c_{64}^{(\frac12)} & 0 & 0 & 0 & c_{65}^{(\frac12)} & 0 \\
0 & 0 & 0 & c_{62}^{(\frac12)} & 0 & 0 & 0 & c_{64}^{(\frac12)} & 0 & 0 & 0 & c_{65}^{(\frac12)} \\
c_{51}^{(\frac12)} & 0 & 0 & 0 & c_{53}^{(\frac12)} & 0 & 0 & 0 & c_{56}^{(\frac12)} & 0 & 0 & 0 \\
0 & c_{51}^{(\frac12)} & 0 & 0 & 0 & c_{53}^{(\frac12)} & 0 & 0 & 0 & c_{56}^{(\frac12)} & 0 & 0
\end{array}\right|,$$

$$C^{(\frac32)} = \left|\begin{array}{cccccccc}
0 & 0 & c_{65}^{(\frac32)} & 0 & 0 & 0 & 0 & 0 \\
0 & 0 & 0 & c_{65}^{(\frac32)} & 0 & 0 & 0 & 0 \\
c_{56}^{(\frac32)} & 0 & 0 & 0 & 0 & 0 & 0 & 0 \\
0 & c_{56}^{(\frac32)} & 0 & 0 & 0 & 0 & 0 & 0 \\
0 & 0 & 0 & 0 & 0 & 0 & c_{65}^{(\frac32)} & 0 \\
0 & 0 & 0 & 0 & 0 & 0 & 0 & c_{65}^{(\frac32)} \\
0 & 0 & 0 & 0 & c_{56}^{(\frac32)} & 0 & 0 & 0 \\
0 & 0 & 0 & 0 & 0 & c_{56}^{(\frac32)} & 0 & 0
\end{array}\right|.$$

Due to P-invariance of the model, we have the following restrictions on the numerical coefficients c_{ij}:

$$c_{56}^{(\frac32)} = c_{65}^{(\frac32)}, \quad c_{56}^{(\frac12)} = c_{65}^{(\frac12)}, \quad c_{12}^{(\frac12)} = c_{21}^{(\frac12)}, \quad c_{14}^{(\frac12)} = c_{23}^{(\frac12)}, \quad c_{32}^{(\frac12)} = c_{41}^{(\frac12)},$$
$$c_{34}^{(\frac12)} = c_{43}^{(\frac12)}, \quad c_{15}^{(\frac12)} = c_{26}^{(\frac12)}, \quad c_{35}^{(\frac12)} = c_{46}^{(\frac12)}, \quad c_{62}^{(\frac12)} = c_{51}^{(\frac12)}, \quad c_{64}^{(\frac12)} = c_{53}^{(\frac12)};$$

therefore spin-blocks may be presented shorter

$$C^{(\frac12)} = \left|\begin{array}{ccc}
c_{12}^{(\frac12)} & c_{14}^{(\frac12)} & c_{15}^{(\frac12)} \\
c_{32}^{(\frac12)} & c_{34}^{(\frac12)} & c_{35}^{(\frac12)} \\
c_{62}^{(\frac12)} & c_{64}^{(\frac12)} & c_{65}^{(\frac12)}
\end{array}\right| \otimes \gamma_4\,, \qquad C^{(\frac32)} = c_{65}^{(\frac32)} I_2 \otimes \gamma_4\,. \tag{4}$$

According to general theory, possible structure of the coefficients $c^{(s)}_{\tau\tau'}$ may be as follows

$$l_+ = l + l', \qquad l_- = l - l',$$

$$\tau \sim (l,l'), \tau' \sim (l-1/2, l'-1/2)\,, \quad c^{(s)}_{\tau\tau'} = c_{\tau\tau'}\sqrt{(s-l_+)(s+l_++1)}\,,$$
$$\tau \sim (l,l'), \tau' \sim (l+1/2, l'+1/2)\,, \quad c^{(s)}_{\tau\tau'} = c_{\tau\tau'}\sqrt{(s-l_+-1)(s+l_++2)}\,,$$
$$\tau \sim (l,l'), \tau' \sim (l-1/2, l'+1/2)\,, \quad c^{(s)}_{\tau\tau'} = c_{\tau\tau'}\sqrt{(s+l_-)(s-l_-+1)}\,,$$
$$\tau \sim (l,l'), \tau' \sim (l+1/2, l'-1/2)\,, \quad c^{(s)}_{\tau\tau'} = c_{\tau\tau'}\sqrt{(s-l_-)(s+l_-+1)}\,,$$

with this in mind we simplify spin-blocks to the form

$$C^{(\frac{1}{2})} = \left| \begin{array}{ccc} c_{12} & c_{14} & i\sqrt{3}c_{15} \\ c_{32} & c_{34} & i\sqrt{3}c_{35} \\ i\sqrt{3}c_{62} & i\sqrt{3}c_{64} & c_{65} \end{array} \right| \otimes \gamma_4\,, \quad C^{(\frac{3}{2})} = 2c_{65} I_2 \otimes \gamma_4\,. \tag{5}$$

Due to uniqueness of spin and mass of a described particle, we have restriction $c_{65} = 1/2$, which simplifies the spin-blocks yet more:

$$C^{(\frac{1}{2})} = \left| \begin{array}{ccc} c_{12} & c_{14} & i\sqrt{3}c_{15} \\ c_{32} & c_{34} & i\sqrt{3}c_{35} \\ i\sqrt{3}c_{62} & i\sqrt{3}c_{64} & 1/2 \end{array} \right| \otimes \gamma_4\,, \quad C^{(\frac{3}{2})} = I_2 \otimes \gamma_4\,. \tag{6}$$

Minimal polynomial of the matrix $\beta^{(\frac{1}{2})}$,

$$\beta^{(\frac{1}{2})} = \left| \begin{array}{ccc} c_{12} & c_{14} & i\sqrt{3}c_{15} \\ c_{32} & c_{34} & i\sqrt{3}c_{35} \\ i\sqrt{3}c_{62} & i\sqrt{3}c_{64} & 1/2 \end{array} \right|, \tag{7}$$

may be of two different forms:

$$[\beta^{(\frac{1}{2})}]^2 = 0\,, \quad \text{or} \quad [\beta^{(\frac{1}{2})}]^3 = 0\,. \tag{8}$$

The first variant leads to a more simple and well-known Pauli–Fierz model.

By this reason, further we will examine in detail the second variant. After performing needed calculation from second equation in (8) we derive restrictions for coefficients c_{kn}:

$$\begin{gathered} c_{12} + c_{34} = -1/2\,, c_{12}c_{34} - c_{32}c_{14} + 3(c_{15}c_{62} + c_{35}c_{64}) = 1/4\,, \\ (c_{12}c_{34} - c_{32}c_{14}) \\ +6(c_{62}c_{34}c_{15} + c_{64}c_{35}c_{12} - c_{32}c_{64}c_{15} - c_{14}c_{35}c_{62}) = 0. \end{gathered} \tag{9}$$

Now we are to take into account consequences from existence of a matrix η of the invariant bilinear form (this ensures possibility of Lagrangian description of the model). Possible structure of two blocks of that matrix is

$$\eta^{(\frac{1}{2})} = \left| \begin{array}{ccc} a & 0 & 0 \\ 0 & b & 0 \\ 0 & 0 & -1 \end{array} \right| \otimes \gamma_4, \quad \eta^{(\frac{3}{2})} = I_2 \otimes \gamma_4. \tag{10}$$

Allowing for standard properties of the bilinear for matrix η we find the following restrictions on coefficients: c_{12}, c_{34} must be real-valued, and remaining coefficients must obey the constraints

$$c_{62} = ac^*_{15}\,, \quad c_{64} = bc^*_{35}\,, \quad c_{32} = abc^*_{14} \quad (a, b = \pm 1)\,. \tag{11}$$

Thus, we arrive at a more specified form for spin-block

$$\beta^{(\frac{1}{2})} = \left| \begin{array}{ccc} c_{12} & c_{14} & i\sqrt{3}c_{15} \\ abc^*_{14} & c_{34} & i\sqrt{3}c_{35} \\ i\sqrt{3}ac^*_{15} & i\sqrt{3}bc^*_{35} & 1/2 \end{array} \right|, \tag{12}$$

involved in (11) coefficients must obey three constraints:

$$\begin{gathered} c_{12} + c_{34} = -1/2\,, \\ c_{12}c_{34} - abc_{14}c^*_{14} + 3(ac_{15}c^*_{15} + bc_{35}c^*_{35}) = 1/4\,, \\ (c_{12}c_{34} - abc_{14}c^*_{14}) + 6(ac^*_{15}c_{34}c_{15} + bc^*_{35}c_{35}c_{12} \\ -ac^*_{14}c^*_{35}c_{15} - ac_{14}c_{35}c^*_{15}) = 0\,. \end{gathered} \tag{13}$$

3. Spin-Vector Representation of the Wave Equations

Spinorial structure of the 20-component wave function Ψ, its representations in modified GY-basis and in canonical basis are given by the formulas

$$\Psi_{spin.} = \left| \begin{array}{c} \Psi^{\dot{1}} \\ \Psi^{\dot{2}} \\ \Psi_1 \\ \Psi_2 \\ \Psi'^{\dot{1}} \\ \Psi'^{\dot{2}} \\ \Psi'_1 \\ \Psi'_2 \\ \\ \Psi^{\dot{1}}_{(11)} \\ \Psi^{\dot{1}}_{(12)} \\ \Psi^{\dot{1}}_{(22)} \\ \Psi^{\dot{2}}_{(11)} \\ \Psi^{\dot{2}}_{(12)} \\ \Psi^{\dot{2}}_{(22)} \\ \\ \Psi_1^{(\dot{1}\dot{1})} \\ \Psi_1^{(\dot{1}\dot{2})} \\ \Psi_1^{(\dot{2}\dot{2})} \\ \Psi_2^{(\dot{1}\dot{1})} \\ \Psi_2^{(\dot{1}\dot{2})} \\ \Psi_2^{(\dot{2}\dot{2})} \end{array} \right|, \quad \Psi_{G.-Y.m.} = \left| \begin{array}{c} \Psi^{(1/2,0)}_{1/2,1/2} \\ \Psi^{(1/2,0)}_{1/2,-1/2} \\ \Psi^{(0,1/2)}_{1/2,1/2} \\ \Psi^{(0,1/2)}_{1/2,-1/2} \\ \Psi'^{(1/2,0)}_{1/2,1/2} \\ \Psi'^{(1/2,0)}_{1/2,-1/2} \\ \Psi'^{(0,1/2)}_{1/2,1/2} \\ \Psi'^{(0,1/2)}_{1/2,-1/2} \\ \Psi^{(1/2,1)}_{1/2,1/2} \\ \Psi^{(1/2,1)}_{1/2,-1/2} \\ \Psi^{(1,1/2)}_{1/2,1/2} \\ \Psi^{(1,1/2)}_{1/2,-1/2} \\ \Psi^{(1/2,1)}_{3/2,3/2} \\ \Psi^{(1/2,1)}_{3/2,-3/2} \\ \Psi^{(1,1/2)}_{3/2,3/2} \\ \Psi^{(1,1/2)}_{3/2,-3/2} \\ \Psi^{(1/2,1)}_{3/2,1/2} \\ \Psi^{(1/2,1)}_{3/2,-1/2} \\ \Psi^{(1,1/2)}_{3/2,1/2} \\ \Psi^{(1,1/2)}_{3/2,-1/2} \end{array} \right|, \quad \Psi_{can.} = \left| \begin{array}{c} \Psi^{(1/2,0)}_{(1/2,0)} \\ \Psi^{(1/2,0)}_{(-1/2,0)} \\ \Psi^{(0,1/2)}_{(0,1/2)} \\ \Psi^{(0,1/2)}_{(0,-1/2)} \\ \Psi'^{(1/2,0)}_{(1/2,0)} \\ \Psi'^{(1/2,0)}_{(-1/2,0)} \\ \Psi'^{(0,1/2)}_{(0,1/2)} \\ \Psi'^{(0,1/2)}_{(0,-1/2)} \\ \Psi^{(1/2,1)}_{(1/2,1)} \\ \Psi^{(1/2,1)}_{(1/2,0)} \\ \Psi^{(1/2,1)}_{(1/2,-1)} \\ \Psi^{(1/2,1)}_{(-1/2,1)} \\ \Psi^{(1/2,1)}_{(-1/2,0)} \\ \Psi^{(1/2,1)}_{(-1/2,-1)} \\ \Psi^{(1,1/2)}_{(1,1/2)} \\ \Psi^{(1,1/2)}_{(0,1/2)} \\ \Psi^{(1,1/2)}_{(-1,1/2)} \\ \Psi^{(1,1/2)}_{(1,-1/2)} \\ \Psi^{(1,1/2)}_{(0,-1/2)} \\ \Psi^{(1,1/2)}_{(-1,-1/2)} \end{array} \right|.$$

After transforming the matrix Γ_4 from modified Gel'fand–Yaglom basis to spinorial one, we find (omitting technical details) relation which determines action of the matrix $\Gamma_4^{spin.}$ on the wave function $\Psi_{spin.}$:

$$\Gamma_4^{spin.}\Psi_{spin.} =$$

$$\left| \begin{array}{c} c_{12}\frac{1}{i}(\sigma^4)^{\dot{a}b}\psi_b + c_{14}\frac{1}{i}(\sigma^4)^{\dot{a}b}\psi'_b - \sqrt{2}c_{15}(\sigma^4)^b_{\dot{c}}\psi_b^{(\dot{a}\dot{c})} \\ c_{12}\frac{1}{i}(\sigma^4)_{a\dot{b}}\psi^{\dot{b}} + c_{14}\frac{1}{i}(\sigma^4)_{a\dot{b}}\psi'^{\dot{b}} - \sqrt{2}c_{15}(\sigma^4)^c_{\dot{b}}\psi^{\dot{b}}_{(ac)} \\ c_{32}\frac{1}{i}(\sigma^4)^{\dot{a}b}\psi_b + c_{34}\frac{1}{i}(\sigma^4)^{\dot{a}b}\psi'_b - \sqrt{2}c_{35}(\sigma^4)^b_{\dot{c}}\psi_b^{(\dot{a}\dot{c})} \\ c_{32}\frac{1}{i}(\sigma^4)_{a\dot{b}}\psi^{\dot{b}} + c_{34}\frac{1}{i}(\sigma^4)_{a\dot{b}}\psi'^{\dot{b}} - \sqrt{2}c_{35}(\sigma^4)^c_{\dot{b}}\psi^{\dot{b}}_{(ac)} \\ \frac{c_{62}}{\sqrt{2}}((\sigma^4)^{\dot{c}}_a\psi_b + (\sigma^4)^{\dot{c}}_b\psi_a) + \frac{c_{64}}{\sqrt{2}}((\sigma^4)^{\dot{c}}_a\psi'_b + (\sigma^4)^{\dot{c}}_b\psi'_a) + \frac{1}{2}((\sigma^4)_{\dot{n}a}\psi_b^{(\dot{n}\dot{c})} + (\sigma^4)_{\dot{n}b}\psi_a^{(\dot{n}\dot{c})}) \\ \frac{c_{62}}{\sqrt{2}}((\sigma^4)^{\dot{a}}_c\psi^{\dot{b}} + (\sigma^4)^{\dot{b}}_c\psi^{\dot{a}}) + \frac{c_{64}}{\sqrt{2}}((\sigma^4)^{\dot{a}}_c\psi'^{\dot{b}} + (\sigma^4)^{\dot{b}}_c\psi'^{\dot{a}}) + \frac{1}{2}((\sigma^4)^{n\dot{a}}\psi^{\dot{b}}_{(nc)} + (\sigma^4)^{n\dot{b}}\psi^{\dot{a}}_{(nc)}) \end{array} \right| , \tag{14}$$

where

$$(\sigma^4)_{\dot{a}b} = \left| \begin{array}{cc} i & 0 \\ 0 & i \end{array} \right| .$$

Similarly, there are defined remaining three matrices $\Gamma_{1,2,3}^{spin.}$, to get them it suffices to make formal change $\sigma^4 \mapsto \sigma^\mu$:

$$(\sigma^1)_{\dot{a}b} = \left| \begin{array}{cc} 0 & 1 \\ 1 & 0 \end{array} \right| , \quad (\sigma^2)_{\dot{a}b} = \left| \begin{array}{cc} 0 & -i \\ i & 0 \end{array} \right| , \quad (\sigma^3)_{\dot{a}b} = \left| \begin{array}{cc} 1 & 0 \\ 0 & -1 \end{array} \right| . \tag{15}$$

It is convenient to employ shortening notations for coefficients:

$$\lambda_1 = c_{12}\,, \quad \lambda_2 = c_{14}\,, \quad \lambda_3 = -i\sqrt{2}c_{15}\,, \quad \lambda_4 = c_{32}\,,$$
$$\lambda_5 = c_{34}\,, \quad \lambda_6 = -i\sqrt{2}c_{35}\,, \quad \lambda_7 = i\sqrt{2}c_{62}\,, \quad \lambda_8 = i\sqrt{2}c_{64}\,, \tag{16}$$

then equations in spinor form read

$$\lambda_1\partial^{\dot{a}b}\psi_b + \lambda_2\partial^{\dot{a}b}\psi'_b + \lambda_3\partial^b_{\dot{c}}\psi_b^{(\dot{a}\dot{c})} + M\psi^{\dot{a}} = 0\,, \tag{17}$$

$$\lambda_1\partial_{a\dot{b}}\psi^{\dot{b}} + \lambda_2\partial_{a\dot{b}}\psi'^{\dot{b}} + \lambda_3\partial^c_{\dot{b}}\psi^{\dot{b}}_{(ac)} + M\psi_a = 0\,, \tag{18}$$

$$\lambda_4\partial^{\dot{a}b}\psi_b + \lambda_5\partial^{\dot{a}b}\psi'_b + \lambda_6\partial^b_{\dot{c}}\psi_b^{(\dot{a}\dot{c})} + M\psi'^{\dot{a}} = 0\,, \tag{19}$$

$$\lambda_4\partial_{a\dot{b}}\psi^{\dot{b}} + \lambda_5\partial_{a\dot{b}}\psi'^{\dot{b}} + \lambda_6\partial^c_{\dot{b}}\psi^{\dot{b}}_{(ac)} + M\psi'_a = 0\,, \tag{20}$$

$$\frac{\lambda_7}{2}(\partial^{\dot{c}}_a\psi_b + \partial^{\dot{c}}_b\psi_a) + \frac{\lambda_8}{2}(\partial^{\dot{c}}_a\psi'_b + \partial^{\dot{c}}_b\psi'_a)$$
$$+\frac{1}{2}(\partial_{\dot{n}a}\psi_b^{(\dot{n}\dot{c})} + \partial_{\dot{n}b}\psi_a^{(\dot{n}\dot{c})}) + M\psi^{\dot{c}}_{(ab)} = 0\,, \tag{21}$$

$$\frac{\lambda_7}{2}(\partial^{\dot{a}}_c\psi^{\dot{b}} + \partial^{\dot{b}}_c\psi^{\dot{a}}) + \frac{\lambda_8}{2}(\partial^{\dot{a}}_c\psi'^{\dot{b}} + \partial^{\dot{b}}_c\psi'^{\dot{a}})$$
$$+\frac{1}{2}(\partial^{n\dot{a}}\psi^{\dot{b}}_{(nc)} + \partial^{n\dot{b}}\psi^{\dot{a}}_{(nc)}) + M\psi_c^{(\dot{a}\dot{b})} = 0\,, \tag{22}$$

where $\partial_{a\dot{b}} = \frac{1}{i}\partial_\mu \sigma^\mu_{a\dot{b}}$. Numerical coefficients λ_i obey the restrictions

$$\begin{gathered}\lambda_1 + \lambda_5 = -\frac{1}{2}, \quad \lambda_1\lambda_5 - \lambda_2\lambda_4 + \frac{3}{2}(\lambda_3\lambda_7 + \lambda_6\lambda_8) = \frac{1}{4},\\ \lambda_1\lambda_5 - \lambda_2\lambda_4 = 3[(\lambda_2\lambda_6 - \lambda_3\lambda_5)\lambda_7 + (\lambda_3\lambda_4 - \lambda_1\lambda_6)\lambda_8],\\ \lambda_1^* = \lambda_1, \ \lambda_5^* = \lambda_5, \ \lambda_7^* = c_1\lambda_3, \ \lambda_8^* = c_2\lambda_6, \ \lambda_4 = c_1 c_2 \lambda_2^*,\\ c_1 = \pm 1, \quad c_2 = \pm 1.\end{gathered} \tag{23}$$

In fact, here we have four variants

$$\begin{array}{llll}(+,+) \ , & \lambda_7^* = +\lambda_3, & \lambda_8^* = +\lambda_6, & \lambda_4 = +\lambda_2^*;\\ (-,-) \ , & \lambda_7^* = -\lambda_3, & \lambda_8^* = -\lambda_6, & \lambda_4 = +\lambda_2^*;\\ (+,-) \ , & \lambda_7^* = +\lambda_3, & \lambda_8^* = -\lambda_6, & \lambda_4 = -\lambda_2^*;\\ (-,+) \ , & \lambda_7^* = -\lambda_3, & \lambda_8^* = +\lambda_6, & \lambda_4 = -\lambda_2^*.\end{array} \tag{24}$$

Spinors entering equations (17)–(22) may be presented as spin-vectors, according to the formulas

$$\begin{gathered}\psi_c^{(\dot{a}\dot{b})} = \frac{1}{2}(\sigma_c^{\mu\dot{a}}\psi_\mu^{\dot{b}} + \sigma_c^{\mu\dot{b}}\psi_\mu^{\dot{a}}), \quad \psi^{\dot{c}}_{(ab)} = \frac{1}{2}(\sigma_a^{\mu\dot{c}}\psi_{\mu b} + \sigma_b^{\mu\dot{c}}\psi_{\mu a}),\\ \psi_a = \sigma^\mu_{a\dot{b}}\psi_\mu^{\dot{b}}, \quad \psi^{\dot{a}} = \sigma^{\mu\dot{a}b}\psi_{\mu b}, \quad \psi'_a = \psi_{0a}, \quad \psi'^{\dot{a}} = \psi_0^{\dot{a}}.\end{gathered} \tag{25}$$

Let us consider Eq. (17), with (25) in mind it reads

$$\lambda_1\partial^{\dot{a}b}\sigma^\mu_{b\dot{c}}\psi_\mu^{\dot{c}} + \lambda_2\partial^{\dot{a}b}\psi_{0b} + \frac{\lambda_3}{2}\partial^b_{\dot{c}}(\sigma_b^{\mu\dot{a}}\psi_\mu^{\dot{c}} + \sigma_b^{\mu\dot{c}}\psi_\mu^{\dot{a}}) + M\sigma^{\mu\dot{a}b}\psi_{\mu b} = 0,$$

or

$$\lambda_1\partial^{\dot{a}b}\sigma^\mu_{b\dot{c}}\psi_\mu^{\dot{c}} + \lambda_2\partial^{\dot{a}b}\psi_{0b} + \frac{\lambda_3}{2}\{-\sigma^{\mu\dot{a}b}\partial_{b\dot{c}}\psi_\mu^{\dot{c}} + \frac{2}{i}\partial_\mu\psi_\mu^{\dot{a}}\} + M\sigma^{\mu\dot{a}b}\psi_{\mu b} = 0. \tag{26}$$

Similarly, Eq. (18) gives

$$\lambda_1\partial_{a\dot{b}}\sigma^{\mu\dot{b}c}\psi_{\mu c} + \lambda_2\partial_{a\dot{b}}\psi_0^{\dot{b}} + \frac{\lambda_3}{2}\{-\sigma^\mu_{a\dot{b}}\partial^{\dot{b}c}\psi_{\mu c} + \frac{2}{i}\partial_\mu\psi_{\mu a}\} + M\sigma^\mu_{a\dot{b}}\psi_\mu^{\dot{b}} = 0. \tag{27}$$

With notations

$$\psi_\mu = \begin{vmatrix}\psi_\mu^{\dot{a}}\\ \psi_{\mu b}\end{vmatrix}, \quad \psi_0 = \begin{vmatrix}\psi_0^{\dot{a}}\\ \psi_{0b}\end{vmatrix}, \quad \gamma_\mu = \frac{1}{i}\begin{vmatrix}0 & \sigma^{\mu\dot{a}b}\\ \sigma^\mu_{a\dot{b}} & 0\end{vmatrix}, \tag{28}$$

equations (26) and (27) may be jointed into one:

$$\lambda_1\hat{\partial}(\gamma_\mu\psi_\mu) - i\lambda_2\hat{\partial}\psi_0 + \frac{\lambda_3}{2}\ [\hat{\partial}(\gamma_\mu\psi_\mu) - 4(\partial_\mu\psi_\mu)] + M(\gamma_\mu\psi_\mu) = 0, \tag{29}$$

this equation contains bispinor ψ_0 and vector-bispinor ψ_μ. Acting in similar manner, from equations (19) and (20) we derive

$$\lambda_4\partial^{\dot{a}b}\sigma^\mu_{b\dot{c}}\psi_\mu^{\dot{c}} + \lambda_5\partial^{\dot{a}b}\psi_{0b} + \frac{\lambda_6}{2}\{-\sigma^{\mu\dot{a}b}\partial_{b\dot{c}}\psi_\mu^{\dot{c}} + \frac{2}{i}\partial_\mu\psi_\mu^{\dot{a}}\} + M\psi_0^{\dot{a}} = 0, \tag{30}$$

$$\lambda_4\partial_{a\dot{b}}\sigma^{\mu\dot{b}c}\psi_{\mu c} + \lambda_5\partial_{a\dot{b}}\psi_0^{\dot{b}} + \frac{\lambda_6}{2}\{-\sigma^\mu_{a\dot{b}}\partial^{\dot{b}c}\psi_{\mu c} + \frac{2}{i}\partial_\mu\psi_{\mu a}\} + M\psi_{0a} = 0, \tag{31}$$

after joining them into one equation we get

$$i\lambda_4\hat{\partial}(\gamma_\mu\psi_\mu) + \lambda_5\hat{\partial}\psi_0 - 2i\lambda_6[(\partial_\mu\psi_\mu) - \frac{1}{4}\hat{\partial}(\gamma_\mu\psi_\mu)] + M\psi_0 = 0\,. \tag{32}$$

Now, we consider Eq. (22), it yields

$$\frac{\lambda_7}{2}\,(\,\partial_c^{\dot a}\sigma^{\mu\dot b k}\psi_{\mu k} + \partial_c^{\dot b}\sigma^{\mu\dot a k}\psi_{\mu k}\,) + \frac{\lambda_8}{2}\,(\,\partial_c^{\dot a}\psi_0^{\dot b} + \partial_c^{\dot b}\psi_0^{\dot a}\,)$$

$$+\frac{1}{4}\,(\,\partial^{n\dot a}(\sigma_n^{\mu\dot b}\psi_{\mu c} + \sigma_c^{\mu\dot b}\psi_{\mu n}) + \partial^{n\dot b}(\sigma_n^{\mu\dot a}\psi_{\mu c} + \sigma_c^{\mu\dot a}\psi_{\mu n})\,]$$

$$+\frac{M}{2}(\sigma_c^{\mu\dot a}\psi_\mu^{\dot b} + \sigma_c^{\mu\dot b}\psi_\mu^{\dot a}) = 0\,.$$

After multiplying it by $\sigma_{\dot a}^{\lambda c}$, it gives

$$\frac{\lambda_7}{2}\,(\,\frac{2}{i}\partial_\lambda\sigma^{\mu\dot b k}\psi_{\mu k} - \partial^{\dot b c}\sigma_{c\dot a}^{\lambda}\sigma^{\mu\dot a k}\psi_{\mu k}\,) + \frac{\lambda_8}{2}\,(\,\frac{2}{i}\partial_\lambda\psi_0^{\dot b} - \partial^{\dot b c}\sigma_{c\dot a}^{\lambda}\psi_0^{\dot a}\,)$$

$$+\frac{1}{4}\,(\,\sigma^{\mu\dot b n}\partial_{n\dot a}\sigma^{\lambda\dot a c}\psi_{\mu c} - \sigma^{\mu\dot b c}\sigma_{c\dot a}^{\lambda}\partial^{\dot a n}\psi_{\mu n} - \partial^{\dot b n}\sigma^{\mu n\dot a}\sigma^{\lambda\dot a c}\psi_{\mu c} + 2\partial^{\dot b n}\psi_{\lambda n}\,)$$

$$+\frac{M}{2}\,(\,2\psi_\lambda^{\dot b} - \sigma^{\mu\dot b c}\sigma_{c\dot a}^{\lambda}\psi_\mu^{\dot a}\,) = 0\,. \tag{33}$$

Similarly from Eq. (21) we derive

$$\frac{\lambda_7}{2}\,(\,\partial_a^{\dot c}\sigma^{\mu}_{b\dot k}\psi_\mu^{\dot k} + \partial_b^{\dot c}\sigma^{\mu}_{a\dot k}\psi_\mu^{\dot k}\,) + \frac{\lambda_8}{2}\,(\,\partial_a^{\dot c}\psi_{0b} + \partial_b^{\dot c}\psi_{0a}\,)$$

$$+\frac{1}{4}\,[\,\partial_{\dot n a}(\sigma_b^{\mu\dot n}\psi_\mu^{\dot c} + \sigma_b^{\mu\dot c}\psi_\mu^{\dot n}) + \partial_{\dot n b}(\sigma_a^{\mu\dot n}\psi_\mu^{\dot c} + \sigma_a^{\mu\dot c}\psi_\mu^{\dot n})\,]$$

$$+\frac{M}{2}\,(\sigma_a^{\mu\dot c}\psi_{\mu b} + \sigma_b^{\mu\dot c}\psi_{\mu a}\,) = 0\,,$$

$$\frac{\lambda_7}{2}\,(\frac{2}{i}\partial_\lambda\sigma^{\mu}_{b\dot k}\psi_\mu^{\dot k} - \partial_{b\dot c}\sigma^{\lambda\dot c a}\sigma^{\mu}_{a\dot k}\psi_\mu^{\dot k}) + \frac{\lambda_8}{2}(\frac{2}{i}\partial_\lambda\psi_{0b} - \partial_{b\dot c}\sigma^{\lambda\dot c a}\psi_{0a})$$

$$+\frac{1}{4}\,(\,\sigma^{\mu}_{b\dot n}\partial^{\dot n a}\sigma^{\lambda}_{a\dot c}\psi_\mu^{\dot c} - \sigma^{\mu}_{b\dot c}\sigma^{\lambda\dot c a}\partial_{a\dot n}\psi_\mu^{\dot n} - \partial_{b\dot n}\sigma^{\mu\dot n a}\sigma^{\lambda}_{a\dot c}\psi_\mu^{\dot c} + 2\partial_{b\dot n}\psi_\lambda^{\dot n}\,)$$

$$+\frac{M}{2}\,(2\psi_{\mu b} - \sigma^{\mu}_{b\dot c}\sigma^{\lambda\dot c a}\psi_{\mu a}) = 0\,. \tag{34}$$

Joining equations (33) and (34) into one, we find third equations for vector-bispinor

$$\frac{\lambda_7}{2}[2\partial_\lambda(\gamma_\mu\psi_\mu) + \hat{\partial}\gamma_\lambda(\gamma_\mu\psi_\mu)] + \frac{\lambda_8}{2}(\frac{2}{i}\partial_\lambda\psi_0 - i\hat{\partial}\gamma_\lambda\psi_0)$$

$$+\frac{1}{4}(-\gamma_\mu\hat{\partial}\gamma_\lambda\psi_\mu + \gamma_\mu\gamma_\lambda\hat{\partial}\psi_\mu + \hat{\partial}\gamma_\mu\gamma_\lambda\psi_\mu + 2\hat{\partial}\psi_\lambda)$$

$$+\frac{M}{2}(2\psi_\lambda + \gamma_\mu\gamma_\lambda\psi_\mu) = 0\,,$$

or

$$\lambda_7\{\partial_\lambda(\gamma_\mu\psi_\mu) - \frac{1}{4}\gamma_\lambda\hat{\partial}(\gamma_\mu\psi_\mu)\} - i\lambda_8\{\partial_\lambda\psi_0 - \frac{1}{4}\gamma_\lambda\hat{\partial}\psi_0\}$$
$$+\frac{1}{8}\{3\gamma_\lambda\hat{\partial}(\gamma_\mu\psi_\mu) - 4\partial_\lambda(\gamma_\mu\psi_\mu) + 8\hat{\partial}\psi_\lambda - 4\gamma_\lambda(\partial_\mu\psi_\mu)\}$$
$$+M\{\psi_\lambda - \frac{1}{4}\gamma_\lambda(\gamma_\mu\psi_\mu)\} = 0\,. \tag{35}$$

Thus, the complete system of spin-vector equations for Fradkin model is

$$\lambda_1\hat{\partial}(\gamma_\mu\psi_\mu) - i\lambda_2\hat{\partial}\psi_0 + \frac{\lambda_3}{2}\{\hat{\partial}(\gamma_\mu\psi_\mu) - 4(\partial_\mu\psi_\mu)\} + M(\gamma_\mu\psi_\mu) = 0\,, \tag{36}$$

$$i\lambda_4\hat{\partial}(\gamma_\mu\psi_\mu) + \lambda_5\hat{\partial}\psi_0 - 2i\lambda_6\{(\partial_\mu\psi_\mu) - \frac{1}{4}\hat{\partial}(\gamma_\mu\psi_\mu)\} + M\psi_0 = 0\,, \tag{37}$$

$$\lambda_7\{\partial_\lambda(\gamma_\mu\psi_\mu) - \frac{1}{4}\gamma_\lambda\hat{\partial}(\gamma_\mu\psi_\mu)\} - i\lambda_8\{\partial_\lambda\psi_0 - \frac{1}{4}\gamma_\lambda\hat{\partial}\psi_0\}$$
$$+\frac{1}{8}\{3\gamma_\lambda\hat{\partial}(\gamma_\mu\psi_\mu) - 4\partial_\lambda(\gamma_\mu\psi_\mu) + 8\hat{\partial}\psi_\lambda - 4\gamma_\lambda(\partial_\mu\psi_\mu)\}$$
$$+M\{\psi_\lambda - \frac{1}{4}\gamma_\lambda(\gamma_\mu\psi_\mu)\} = 0\,. \tag{38}$$

It should be noted that the convolution of the last equation with γ_λ gives four identities $0 = 0$, Indeed, we get

$$\lambda_7\ \{\hat{\partial}(\gamma_\mu\psi_\mu) - \hat{\partial}(\gamma_\mu\psi_\mu)\ \} - i\{\hat{\partial}\psi_0 - \hat{\partial}\psi_0\}$$
$$+\frac{1}{8}\{12\hat{\partial}(\gamma_\mu\psi_\mu) - 4\hat{\partial}(\gamma_\mu\psi_\mu) + 8\gamma_\lambda(\gamma_\mu\partial_\mu)\psi_\lambda - 16(\partial_\mu\psi_\mu)\}$$
$$+M\{\gamma_\lambda\psi_\lambda - \gamma_\mu\psi_\mu\} = 0,$$

whence it follows

$$\hat{\partial}(\gamma_\mu\psi_\mu) + (-\gamma_\mu\gamma_\lambda + 2\delta_{\mu\lambda})\partial_\mu\psi_\lambda - 2(\partial_\mu\psi_\mu) \equiv 0\,.$$

This means that in fact in the system (38) we have instead 24 equations only 20 independent ones. This correlates with the total number of components in bispinor and vector-bispinor.

The system (36)–(38) may be transformed to other (but equivalent) form, when one equation contains the term $M\psi_\lambda$ instead of combination $M\{\psi_\lambda - \frac{1}{4}\gamma_\lambda(\gamma_\mu\psi_\mu)\}$ – see (38). Indeed, let us multiply Eq. (36) by γ_λ, this results in

$$\frac{\lambda_1}{4}\gamma_\lambda\hat{\partial}(\gamma_\mu\psi_\mu) - \frac{i}{4}\lambda_2\gamma_\lambda\hat{\partial}\psi_0 - \frac{\lambda_3}{2}\{-\frac{1}{4}\gamma_\lambda\hat{\partial}(\gamma_\mu\psi_\mu) + \gamma_\lambda(\partial_\mu\psi_\mu)\}$$
$$+\frac{M}{4}\gamma_\lambda(\gamma_\mu\psi_\mu) = 0\,. \tag{39}$$

In turn, after convoluting the last equation with γ_λ we again obtain Eq. (36). All the more, summing Eq. (38) and Eq. (39), we obtain

$$(\lambda_7 - \frac{1}{2})\partial_\lambda(\gamma_\mu\psi_\mu) + \frac{1}{4}(\lambda_1 + \frac{\lambda_3}{2} - \lambda_7 + \frac{3}{2})\gamma_\lambda\hat{\partial}(\gamma_\mu\psi_\mu) - i\lambda_8\partial_\lambda\psi_0$$
$$+\frac{i}{4}(\lambda_8 - \lambda_2)\gamma_\lambda\hat{\partial}\psi_0 + \hat{\partial}\psi_\lambda - \frac{1}{2}(1 + \lambda_3)\gamma_\lambda(\partial_\mu\psi_\mu) + M\psi_\lambda = 0. \tag{40}$$

Eq. (40) contains both Eq. (36) and (38). Indeed, convoluting Eq.(40) with γ_λ leads us to Eq. (36), and therefore we can derive Eq. (39) and Eq. (38).

Thus, the complete system of equation for Fradkin particle may be presented as a set of two equations

$$i\lambda_4 \hat{\partial}(\gamma_\mu \psi_\mu) + \lambda_5 \hat{\partial}\psi_0 - 2i\lambda_6 \{(\partial_\mu \psi_\mu) - \frac{1}{4}\hat{\partial}(\gamma_\mu \psi_\mu)\} + M\psi_0 = 0\,, \tag{41}$$

$$(\lambda_7 - \frac{1}{2})\partial_\lambda(\gamma_\mu \psi_\mu) + \frac{1}{4}(\lambda_1 + \frac{\lambda_3}{2} - \lambda_7 + \frac{3}{2})\gamma_\lambda \hat{\partial}(\gamma_\mu \psi_\mu) - i\lambda_8 \partial_\lambda \psi_0$$
$$+\frac{i}{4}(\lambda_8 - \lambda_2)\gamma_\lambda \hat{\partial}\psi_0 + \hat{\partial}\psi_\lambda - \frac{1}{2}(1 + \lambda_3)\gamma_\lambda(\partial_\mu \psi_\mu) + M\psi_\lambda = 0\,. \tag{42}$$

4. Relation between Fradkin and Pauli–Fierz Models

Now we consider relationships between Fradkin and Pauli–Fierz models. We start with Eqs. (36)–(38). Let us multiply Eq. (36) by λ_7, and Eq. (37) – by $(-i\lambda_8)$. After summing two results, we get

$$(\lambda_1\lambda_7 + \lambda_4\lambda_8)\hat{\partial}(\gamma_\mu \psi_\mu) - i(\lambda_2\lambda_7 + \lambda_5\lambda_8)\hat{\partial}\psi_0$$
$$-2(\lambda_3\lambda_7 + \lambda_6\lambda_8)\,[\,(\partial_\mu \psi_\mu) - \frac{1}{4}\hat{\partial}(\gamma_\mu \psi_\mu)\,]$$
$$+M\,[\,\lambda_7(\gamma_\mu \psi_\mu) - i\lambda_8 \psi_0\,] = 0\,,$$

whence taking in mind identities (see (23))

$$\lambda_1\lambda_7 + \lambda_4\lambda_8 = (-\frac{1}{2} - \lambda_5)\lambda_7 + \lambda_4\lambda_8 = -\frac{\lambda_7}{2} + (\lambda_4\lambda_8 - \lambda_5\lambda_7)\,,$$
$$\lambda_2\lambda_7 + \lambda_5\lambda_8 = (-\frac{1}{2} - \lambda_1)\lambda_8 + \lambda_2\lambda_7 = -\frac{\lambda_8}{2} + (\lambda_2\lambda_7 - \lambda_1\lambda_8)\,,$$

we derive

$$[-\frac{\lambda_7}{2} + (\lambda_4\lambda_8 - \lambda_5\lambda_7)]\hat{\partial}(\gamma_\mu \psi_\mu) - i[-\frac{\lambda_8}{2} + (\lambda_2\lambda_7 - \lambda_1\lambda_8)]\hat{\partial}\psi_0$$
$$-2(\lambda_3\lambda_7 + \lambda_6\lambda_8)\,[\,(\partial_\mu \psi_\mu) - \frac{1}{4}\hat{\partial}(\gamma_\mu \psi_\mu)\,] + M\,[\,\lambda_7(\gamma_\mu \psi_\mu) - i\lambda_8 \psi_0\,] = 0\,.$$

The last after re-grouping the terms takes on the form

$$(-\frac{1}{2}\hat{\partial} + M)\,[\,\lambda_7(\gamma_\mu \psi_\mu) - i\lambda_8 \psi_0\,]$$
$$+\hat{\partial}\,[\,(\lambda_4\lambda_8 - \lambda_5\lambda_7)(\gamma_\mu \psi_\mu) - i(\lambda_2\lambda_7 - \lambda_1\lambda_8)\psi_0\,]$$
$$-2(\lambda_3\lambda_7 + \lambda_6\lambda_8)\,[\,(\partial_\mu \psi_\mu) - \frac{1}{4}\hat{\partial}(\gamma_\mu \psi_\mu)\,] = 0\,. \tag{43}$$

Similarly, consider Eqs. (36)–(37); multiplying the first by $(\lambda_4\lambda_8 - \lambda_5\lambda_7)$, and the second by $[-i(\lambda_2\lambda_7 - \lambda_1\lambda_8)]$, after summing two results we obtain

$$M\,[\,(\lambda_4\lambda_8 - \lambda_5\lambda_7)(\gamma_\mu \psi_\mu) - i(\lambda_2\lambda_7 - \lambda_1\lambda_8)\psi_0\,]$$

$$-2[\,\lambda_3(\lambda_4\lambda_8-\lambda_5\lambda_7)+\lambda_6(\lambda_2\lambda_7-\lambda_1\lambda_8)\,]\,[\,(\partial_\mu\psi_\mu)-\frac{1}{4}\hat{\partial}(\gamma_\mu\psi_\mu)\,]$$
$$+[\,\lambda_1(\lambda_4\lambda_8-\lambda_5\lambda_7)+\lambda_4(\lambda_2\lambda_7-\lambda_1\lambda_8)\,]\,\hat{\partial}(\gamma_\mu\psi_\mu)$$
$$-i[\,\lambda_2(\lambda_4\lambda_8-\lambda_5\lambda_7)+\lambda_5(\lambda_2\lambda_7-\lambda_1\lambda_8)\,]\,\hat{\partial}\psi_0=0\,.$$

Allowing for identities (see (23))

$$\lambda_3(\lambda_4\lambda_8-\lambda_5\lambda_7)+\lambda_6(\lambda_2\lambda_7-\lambda_1\lambda_8)=(1/3)(\lambda_1\lambda_5-\lambda_2\lambda_4),$$
$$\lambda_1(\lambda_4\lambda_8-\lambda_5\lambda_7)+\lambda_4(\lambda_2\lambda_7-\lambda_1\lambda_8)=-\lambda_7(\lambda_1\lambda_5-\lambda_4\lambda_2),$$
$$\lambda_2(\lambda_4\lambda_8-\lambda_5\lambda_7)+\lambda_5(\lambda_2\lambda_7-\lambda_1\lambda_8)=-\lambda_8(\lambda_1\lambda_5-\lambda_2\lambda_4),$$

we produce

$$M\,[\,(\lambda_4\lambda_8-\lambda_5\lambda_7)(\gamma_\mu\psi_\mu)-i(\lambda_2\lambda_7-\lambda_1\lambda_8)\psi_0\,]$$
$$-\frac{2}{3}(\lambda_1\lambda_5-\lambda_2\lambda_4)\,[\,(\partial_\mu\psi_\mu)-\frac{1}{4}\hat{\partial}(\gamma_\mu\psi_\mu)\,]$$
$$-\lambda_7(\lambda_1\lambda_5-\lambda_4\lambda_2)\,\hat{\partial}(\gamma_\mu\psi_\mu)+i\lambda_8(\lambda_1\lambda_5-\lambda_2\lambda_4)\,\hat{\partial}\psi_0=0\,,$$

or

$$M\,[\,(\lambda_4\lambda_8-\lambda_5\lambda_7)(\gamma_\mu\psi_\mu)-i(\lambda_2\lambda_7-\lambda_1\lambda_8)\psi_0\,]$$
$$=(\lambda_1\lambda_5-\lambda_2\lambda_4)\,\left\{\frac{2}{3}\,[\,(\partial_\mu\psi_\mu)-\frac{1}{4}\hat{\partial}(\gamma_\mu\psi_\mu)\,]+\hat{\partial}\,[\,\lambda_7(\gamma_\mu\psi_\mu)-i\lambda_8\psi_0\,]\right\}. \tag{44}$$

Now, with the help of expression for the term (see Eq. (44))

$$[\,(\lambda_4\lambda_8-\lambda_5\lambda_7)(\gamma_\mu\psi_\mu)-i(\lambda_2\lambda_7-\lambda_1\lambda_8)\psi_0\,],$$

we transform the second line in (43):

$$(-\frac{1}{2}\hat{\partial}+M)\,[\,\lambda_7(\gamma_\mu\psi_\mu)-i\lambda_8\psi_0\,]$$
$$+\frac{1}{M}(\lambda_1\lambda_5-\lambda_2\lambda_4)\,\hat{\partial}\left\{\frac{2}{3}\,[\,(\partial_\mu\psi_\mu)-\frac{1}{4}\hat{\partial}(\gamma_\mu\psi_\mu)\,]+\hat{\partial}\,[\,\lambda_7(\gamma_\mu\psi_\mu)-i\lambda_8\psi_0\,]\right\}$$
$$-2(\lambda_3\lambda_7+\lambda_6\lambda_8)\,[\,(\partial_\mu\psi_\mu)-\frac{1}{4}\hat{\partial}(\gamma_\mu\psi_\mu)\,]=0\,. \tag{45}$$

In turn, the last line in (45) may be transformed with the help of the identity

$$\lambda_3\lambda_7+\lambda_6\lambda_8=\frac{1}{6}-\frac{2}{3}(\lambda_1\lambda_5-\lambda_2\lambda_4);$$

in this way we derive

$$(-\frac{1}{2}\hat{\partial}+M)\,[\lambda_7(\gamma_\mu\psi_\mu)-i\lambda_8\psi_0\,]-\frac{1}{3}\,[\,(\partial_\mu\psi_\mu)-\frac{1}{4}\hat{\partial}(\gamma_\mu\psi_\mu)\,]$$
$$+\frac{1}{M}(\lambda_1\lambda_5-\lambda_2\lambda_4)\,\hat{\partial}\left\{\frac{2}{3}\,[\,(\partial_\mu\psi_\mu)-\frac{1}{4}\hat{\partial}(\gamma_\mu\psi_\mu)\,]+\hat{\partial}\,[\lambda_7(\gamma_\mu\psi_\mu)-i\lambda_8\psi_0\,]\right\}$$

$$+\frac{4}{3}(\lambda_1\lambda_5-\lambda_2\lambda_4)\;[\;(\partial_\mu\psi_\mu)-\frac{1}{4}\hat{\partial}(\gamma_\mu\psi_\mu)\;]=0\,. \tag{46}$$

Now, let us act by operator ∂_λ on Eq. (38), this results in

$$\lambda_7\;[\;\partial_\lambda\partial_\lambda(\gamma_\mu\psi_\mu)-\frac{1}{4}\hat{\partial}\hat{\partial}(\gamma_\mu\psi_\mu)\;]-i\lambda_8\;[\;\partial_\lambda\partial_\lambda\psi_0-\frac{1}{4}\hat{\partial}\hat{\partial}\psi_0\;]$$

$$+\frac{1}{8}\;[\;3\hat{\partial}\hat{\partial}(\gamma_\mu\psi_\mu)-4\partial_\lambda\partial_\lambda(\gamma_\mu\psi_\mu)+8\hat{\partial}(\partial_\lambda\psi_\lambda)-4\hat{\partial}(\gamma_\mu\psi_\mu)\;]$$

$$+M[\partial_\lambda\psi_\lambda-\frac{1}{4}\hat{\partial}(\gamma_\mu\psi_\mu)]=0\,,$$

or after simplifying

$$[\;(\partial_\mu\psi_\mu)-\frac{1}{4}\hat{\partial}(\gamma_\mu\psi_\mu)\;]$$

$$=\frac{1}{M}\left\{-\frac{3}{4}\partial_\mu\partial_\mu[\lambda_7(\gamma_\mu\psi_\mu)-i\lambda_8\psi_0]-\frac{1}{2}\;[\;\hat{\partial}(\partial_\mu\psi_\mu)-\frac{1}{4}\partial_\mu\partial_\mu(\gamma_\mu\psi_\mu)\;]\right\}. \tag{47}$$

Taking in mind (47), from (46) we derive

$$\left(-\frac{1}{2}\hat{\partial}+M\right)[\lambda_7(\gamma_\mu\psi_\mu)-i\lambda_8\psi_0\;]-\frac{1}{3}\left((\partial_\mu\psi_\mu)-\frac{1}{4}\hat{\partial}(\gamma_\mu\psi_\mu)\right)$$

$$+\frac{1}{M}(\lambda_1\lambda_5-\lambda_2\lambda_4)\;\hat{\partial}\left\{\frac{2}{3}\;[\;(\partial_\mu\psi_\mu)-\frac{1}{4}\hat{\partial}(\gamma_\mu\psi_\mu)\;]+\hat{\partial}\;[\lambda_7(\gamma_\mu\psi_\mu)-i\lambda_8\psi_0\;]\right\}$$

$$+\frac{4}{3}(\lambda_1\lambda_5-\lambda_2\lambda_4)\frac{1}{M}\left\{-\frac{3}{4}\partial_\mu\partial_\mu[\lambda_7(\gamma_\mu\psi_\mu)-i\lambda_8\psi_0]\right.$$

$$\left.-\frac{1}{2}\;[\;\hat{\partial}(\partial_\mu\psi_\mu)-\frac{1}{4}\partial_\mu\partial_\mu(\gamma_\mu\psi_\mu)\;]\right\}=0\,. \tag{48}$$

The terms, proportional to $(\lambda_1\lambda_5-\lambda_2\lambda_4)$ cancel each other, so we arrive at

$$\left(-\frac{1}{2}\hat{\partial}+M\right)[\;\lambda_7(\gamma_\mu\psi_\mu)-i\lambda_8\psi_0\;]-\frac{1}{3}\left((\partial_\mu\psi_\mu)-\frac{1}{4}\hat{\partial}(\gamma_\mu\psi_\mu)\right)=0\,. \tag{49}$$

It is convenient to write down here the above equation (38) as well

$$\lambda_7\left(\partial_\lambda(\gamma_\mu\psi_\mu)-\frac{1}{4}\gamma_\lambda\hat{\partial}(\gamma_\mu\psi_\mu)\right)-i\lambda_8\left(\partial_\lambda\psi_0-\frac{1}{4}\gamma_\lambda\hat{\partial}\psi_0\right)$$

$$+\frac{1}{8}\left(3\gamma_\lambda\hat{\partial}(\gamma_\mu\psi_\mu)-4\partial_\lambda(\gamma_\mu\psi_\mu)+8\hat{\partial}\psi_\lambda-4\gamma_\lambda(\partial_\mu\psi_\mu)\right)$$

$$+M\left(\psi_\lambda-\frac{1}{4}\gamma_\lambda(\gamma_\mu\psi_\mu)\right)=0\,. \tag{50}$$

In total, two equations, (49) and (50) are consequences of initial Fradkin system (36)–(38). Equations (49) and (50) may be reduced to the form when instead of two combinations

$$[\;\lambda_7(\gamma_\mu\psi_\mu)-i\lambda_8\psi_0],\quad[\psi_\lambda-\frac{1}{4}\gamma_\lambda(\gamma_\mu\psi_\mu)\,], \tag{51}$$

only one (and other) vector-bispinor enters the equations. To this end, we introduce a new vector-bispinor

$$\Phi_\lambda = [\,\psi_\lambda - \frac{1}{4}\gamma_\lambda(\gamma_\mu\psi_\mu)\,] + \frac{1}{4}\gamma_\lambda[\,\lambda_7(\gamma_\mu\psi_\mu) - i\lambda_8\psi_0\,]\,. \tag{52}$$

Identities are readily checked

$$\gamma_\mu\Phi_\mu = \lambda_7(\gamma_\mu\psi_\mu) - i\lambda_8\psi_0\,, \quad \Phi_\lambda - \frac{1}{4}\gamma_\lambda(\gamma_\mu\Phi_\mu) = \psi_\lambda - \frac{1}{4}\gamma_\lambda(\gamma_\mu\psi_\mu)\,. \tag{53}$$

Correspondingly, Eq. (49) may be re-written differently

$$-\frac{1}{2}\hat{\partial}(\gamma_\mu\Phi_\mu) - \frac{1}{3}\partial_\lambda\left(\Phi_\lambda - \frac{1}{4}\gamma_\lambda(\gamma_\mu\Phi_\mu)\right) + M(\gamma_\mu\Phi_\mu) = 0\,, \tag{54}$$

and Eq. (50) is re-written as follows

$$\left(\partial_\lambda - \frac{1}{4}\gamma_\lambda\hat{\partial}\right)(\gamma_\mu\Phi_\mu) + M\left(\Phi_\lambda - \frac{1}{4}\gamma_\lambda(\gamma_\mu\Phi_\mu)\right)$$
$$+\frac{1}{8}\left(8\hat{\partial}\psi_\lambda + 3\gamma_\lambda\hat{\partial}(\gamma_\mu\psi_\mu) - 4\partial_\lambda(\gamma_\mu\psi_\mu) - 4\gamma_\lambda(\partial_\mu\psi_\mu)\right) = 0\,. \tag{55}$$

Due to identity

$$8\hat{\partial}\Phi_\lambda + 3\gamma_\lambda\hat{\partial}(\gamma_\mu\Phi_\mu) - 4\partial_\lambda(\gamma_\mu\Phi_\mu) - 4\gamma_\lambda(\partial_\mu\Phi_\mu)$$
$$= 8\hat{\partial}\psi_\lambda + 3\gamma_\lambda\hat{\partial}(\gamma_\mu\psi_\mu) - 4\partial_\lambda(\gamma_\mu\psi_\mu) - 4\gamma_\lambda(\partial_\mu\psi_\mu)\,,$$

we may change Eq. (55) to the form

$$\left(\partial_\lambda - \frac{1}{4}\gamma_\lambda\hat{\partial}\right)(\gamma_\mu\Phi_\mu) + M\left(\Phi_\lambda - \frac{1}{4}\gamma_\lambda(\gamma_\mu\Phi_\mu)\right)$$
$$+\frac{1}{8}\left(8\hat{\partial}\Phi_\lambda + 3\gamma_\lambda\hat{\partial}(\gamma_\mu\Phi_\mu) - 4\partial_\lambda(\gamma_\mu\Phi_\mu) - 4\gamma_\lambda(\partial_\mu\Phi_\mu)\right) = 0\,. \tag{56}$$

We can see that equations (54) and (56) contain only one vector-bispinor Φ_λ.

Finally, equations (54) and (56) may be jointed into one equation with respect to the vector-bispinor Φ_λ. To this end, let us multiply Eq. (54) by $\frac{1}{4}\gamma_\lambda$, this results in

$$-\frac{1}{8}\gamma_\lambda\hat{\partial}(\gamma_\mu\Phi_\mu) - \frac{1}{12}\gamma_\lambda\left((\partial_\mu\Phi_\mu) - \frac{1}{4}\hat{\partial}(\gamma_\mu\Phi_\mu)\right) + \frac{M}{4}\gamma_\lambda(\gamma_\mu\Phi_\mu) = 0\,. \tag{57}$$

From (57), it follows Eq. (54), to prove this it suffices to multiply Eq. (57) by γ_ν and take into account the properties of Dirac matrices. Finally, summing equations (56) and (57), we arrive at the needed result

$$\frac{1}{8}\left\{8\hat{\partial}\Phi_\lambda + \frac{1}{6}\gamma_\lambda\hat{\partial}(\gamma_\mu\Phi_\mu) + 4\partial_\lambda(\gamma_\mu\Phi_\mu) - \frac{14}{3}\gamma_\lambda(\partial_\mu\Phi_\mu)\right\} + M\Phi_\lambda = 0\,. \tag{58}$$

It coincides with minimal equation by Pauli and Fierz for a spin 3/2 particle.

It is easily verified that Eq. (58) is equivalent to the system of (54) and (56). To this end, let us multiply Eq. (58) by γ_ν, and make convolution over free vector indices, so producing

$$-\frac{5}{12}\hat{\partial}(\gamma_\mu \Phi_\mu) - \frac{1}{3}(\partial_\mu \Phi_\mu) + M(\gamma_\mu \Phi_\mu) = 0\,,$$

which coincides with Eq. (54), and therefore leads to Eq. (57). Correspondingly, subtracting Eq. (58) from (57) we obtain Eq. (56).

Additionally, let us derive relationships between two bispinors $\gamma_\mu \psi_\mu,\ \psi_0$ and the new function Φ_μ. To this end, from Eq. (47) we have

$$\hat{\partial}\,[\,\lambda_7(\gamma_\mu \psi_\mu) - i\lambda_8 \psi_0\,] + \frac{2}{3}[\,(\partial_\mu \psi_\mu) - \frac{1}{4}\hat{\partial}(\gamma_\mu \psi_\mu)\,] = 2M[\,\lambda_7(\gamma_\mu \psi_\mu) - i\lambda_8 \psi_0\,]\,.$$

Taking this into account in (43), we derive

$$(\lambda_4\lambda_8 - \lambda_5\lambda_7)(\gamma_\mu \psi_\mu) - i(\lambda_2\lambda_7 - \lambda_1\lambda_8)\psi_0$$
$$= 2(\lambda_1\lambda_5 - \lambda_2\lambda_4)[\lambda_7(\gamma_\mu \psi_\mu) - i\lambda_8 \psi_0]\,,$$

or (see (52))

$$(\lambda_4\lambda_8 - \lambda_5\lambda_7)(\gamma_\mu \psi_\mu) - i(\lambda_2\lambda_7 - \lambda_1\lambda_8)\psi_0 = 2(\lambda_1\lambda_5 - \lambda_2\lambda_4)(\gamma_\mu \Phi_\mu)\,.$$

The last equation together with (52) permits determine components ψ_0 and $\gamma_\mu \psi_\mu$ through Φ_μ:

$$\psi_0 = \frac{i}{B}\{\,(\lambda_4\lambda_8 - \lambda_5\lambda_7) - 2\lambda_7(\lambda_1\lambda_5 - \lambda_2\lambda_4)\,\}\,(\gamma_\mu \Phi_\mu)\,, \tag{59}$$

$$\gamma_\mu \psi_\mu = -\frac{1}{B}\{\,(\lambda_2\lambda_7 - \lambda_1\lambda_8) - 2\lambda_8(\lambda_1\lambda_5 - \lambda_2\lambda_4)\,\}\,(\gamma_\mu \Phi_\mu)\,, \tag{60}$$

where

$$B = \lambda_8(\lambda_4\lambda_8 - \lambda_5\lambda_7) - \lambda_7(\lambda_2\lambda_7 - \lambda_1\lambda_8)\,.$$

Therefore, in absence of external fields, extended and minimal equations for spin 3/2 particle are equivalent.

Minimal equation (58) may be presented in the matrix form (1)

$$(\Gamma_\mu \partial_\mu + M)\Phi = 0\,,$$

when the matrices Γ_μ are given by the formula

$$\Gamma_\mu = \gamma_\mu \otimes e^{\nu,\nu} + \frac{1}{48}\gamma_\rho\gamma_\mu\gamma_\sigma \otimes e^{\rho,\sigma} + \frac{1}{2}\gamma_\rho \otimes e^{\mu,\rho} - \frac{7}{12}\gamma_\rho \otimes e^{\rho,\mu}\,. \tag{61}$$

In this representation, expression for Γ_μ contains triple product of the Dirac matrices. The structure of main matrices may be simplified. To this end, let us introduce new wave function $\Phi' = R\Phi$, correspondingly the Γ_μ transforms by the rule $\Gamma'_\mu = R\Gamma_\mu R^{-1}$. Let the used transformation is as shown below

$$R = I \otimes I - \frac{4\sqrt{3}+7}{4}\gamma_\rho\gamma_\lambda \otimes e^{\rho,\lambda},$$
$$R^{-1} = I \otimes I - \frac{\sqrt{3}+2}{8\sqrt{3}}\gamma_\alpha\gamma_\beta \otimes e^{\alpha,\beta}. \tag{62}$$

In this way, we derive a new form of equation for vector-bispinor Ψ'_λ (in so-called Petras basis) when the matrices Γ'_μ are

$$\Gamma'_\mu = \gamma_\mu \otimes e^{\nu,\nu} + \frac{1}{\sqrt{3}}(\gamma_\rho \otimes e^{\rho,\mu} - \gamma_\rho \otimes e^{\mu,\rho})\,. \tag{63}$$

In explicit form, the minimal Pauli–Fierz equation for spin 3/2 particle in Petras basis reads simpler

$$\gamma_\mu \partial_\mu \Phi_\lambda + \frac{1}{\sqrt{3}}\left[\, \gamma_\lambda(\partial_\mu \Phi_\mu) - \partial_\lambda(\gamma_\mu \Phi_\mu)\,\right] + M\Phi_\lambda = 0\,. \tag{64}$$

5. External Electromagnetic Field

Now we are to take into account the presence of external electromagnetic fields, adhering the same way of work as in a free particle case. We start with equations (36)–(38), but with generalized derivative $D_\mu = \partial_\mu - ieA_\mu$ (A_μ stands for an electromagnetic field):

$$\lambda_1 \hat{D}(\gamma_\mu \psi_\mu) - \lambda_2 \hat{D}\psi_0 - 2\lambda_3\left((D_\mu \psi_\mu) - \frac{1}{4}\hat{D}(\gamma_\mu \psi_\mu)\right) + M(\gamma_\mu \psi_\mu) = 0\,, \tag{65}$$

$$i\lambda_4 \hat{D}(\gamma_\mu \psi_\mu) + \lambda_5 \hat{D}\psi_0 - 2i\lambda_6\left((D_\mu \psi_\mu) - \frac{1}{4}\hat{D}(\gamma_\mu \psi_\mu)\right) + M\psi_0 = 0\,, \tag{66}$$

$$\begin{aligned} &\left(D_\lambda - \frac{1}{4}\gamma_\lambda \hat{D}\right)[\lambda_7(\gamma_\mu \psi_\mu) - i\lambda_8 \psi_0] \\ &+\frac{1}{8}\left[8\hat{D}\psi_\lambda + 3\gamma_\lambda \hat{D}(\gamma_\mu \psi_\mu) - 4D_\lambda(\gamma_\mu \psi_\mu) - 4\gamma_\lambda(D_\mu \psi_\mu)\right] \\ &+M\left(\psi_\lambda - \frac{1}{4}\gamma_\lambda(\gamma_\mu \psi_\mu)\right) = 0\,. \end{aligned} \tag{67}$$

Let us multiply Eq. (65) by λ_7, and Eq. (66) – by $(-i\lambda_8)$, and sum the results. Allowing for restrictions (23) for λ_i we get

$$\begin{aligned} &\left(-\frac{1}{2}\hat{D} + M\right)\;[\lambda_7(\gamma_\mu \psi_\mu) - i\lambda_8 \psi_0] \\ &+\hat{D}\,[\,(\lambda_4\lambda_8 - \lambda_5\lambda_7)(\gamma_\mu \psi_\mu) - i(\lambda_2\lambda_7 - \lambda_1\lambda_8)\psi_0\,] \\ &-\frac{2}{3}\left(\frac{1}{2} - 2(\lambda_1\lambda_5 - \lambda_2\lambda_4)\right)\left((D_\mu \psi_\mu) - \frac{1}{4}\hat{D}(\gamma_\mu \psi_\mu)\right) = 0\,. \end{aligned} \tag{68}$$

Similarly, we multiply Eq. (65) by $(\lambda_4\lambda_8 - \lambda_5\lambda_7)$, and Eq. (66) – by $[-i(\lambda_4\lambda_8 - \lambda_5\lambda_7)]$, after summing results we obtain

$$\begin{aligned} &M\;[\,(\lambda_4\lambda_8 - \lambda_5\lambda_7)(\gamma_\mu \psi_\mu) - i(\lambda_2\lambda_7 - \lambda_1\lambda_8)\psi_0\,] \\ &= \frac{2}{3}(\lambda_1\lambda_5 - \lambda_2\lambda_4)\left((D_\mu \psi_\mu) - \frac{1}{4}\hat{D}\gamma_\mu \psi_\mu)\right) \\ &\quad +(\lambda_1\lambda_5 - \lambda_2\lambda_4)\hat{D}\,[\,\lambda_7(\gamma_\mu \psi_\mu) - i\lambda_8 \psi_0\,]\,. \end{aligned} \tag{69}$$

Expressing from (69) bispinor combination

$$(\lambda_4\lambda_8 - \lambda_5\lambda_7)(\gamma_\mu\psi_\mu) - i(\lambda_2\lambda_7 - \lambda_1\lambda_8)\psi_0$$

and substituting this into (68), we get

$$\Big(-\frac{1}{2}\hat{D} + M\Big)\ [\,\lambda_7(\gamma_\mu\psi_\mu) - i\lambda_8\psi_0\,] - \frac{1}{3}\Big((D_\mu\psi_\mu) - \frac{1}{4}\hat{D}(\gamma_\mu\psi_\mu)\Big)$$
$$+\frac{1}{M}(\lambda_1\lambda_5 - \lambda_2\lambda_4)\Big\{\hat{D}\hat{D}[\lambda_7(\gamma_\mu\psi_\mu) - i\lambda_8\psi_0] + \frac{2}{3}\hat{D}\Big((D_\mu\psi_\mu) - \frac{1}{4}\hat{D}(\gamma_\mu\psi_\mu)\Big)$$
$$+\frac{4}{3}\,M\Big((D_\mu\psi_\mu) - \frac{1}{4}\hat{D}(\gamma_\mu\psi_\mu)\Big)\Big\} = 0\,. \tag{70}$$

Now, let us act on Eq. (67) by operator D_λ, so producing

$$M\Big((D_\mu\psi_\mu) - \frac{1}{4}\hat{D}(\gamma_\mu\psi_\mu)\Big)$$
$$= -\frac{1}{8}\Big(8D_\mu\hat{D}\psi_\mu + 3\hat{D}\hat{D}(\gamma_\mu\psi_\mu) - 4D_\nu D_\nu(\gamma_\mu\psi_\mu) - 4\hat{D}(D_\mu\psi_\mu)\Big)$$
$$-\Big(D_\mu D_\mu - \frac{1}{4}\hat{D}\hat{D}\Big)\ [\,\lambda_7(\gamma_\mu\psi_\mu) - i\lambda_8\psi_0\,]\,. \tag{71}$$

Expressing from (71) the term $M[(D_\mu\psi_\mu) - \frac{1}{4}\hat{D}(\gamma_\mu\psi_\mu)]$ and substituting this into Eq. (70), we get

$$\Big(-\frac{1}{2}\hat{D} + M\Big)[\,\lambda_7(\gamma_\mu\psi_\mu) - i\lambda_8\psi_0\,] - \frac{1}{3}\Big((D_\mu\psi_\mu) - \frac{1}{4}\hat{D}(\gamma_\mu\psi_\mu)\Big)$$

$$+\frac{1}{M}(\lambda_1\lambda_5 - \lambda_2\lambda_4)\Big\{\hat{D}\hat{D}[\lambda_7(\gamma_\mu\psi_\mu) - i\lambda_8\psi_0] + \frac{2}{3}\hat{D}\Big((D_\mu\psi_\mu) - \frac{1}{4}\hat{D}(\gamma_\mu\psi_\mu)\Big)$$
$$-\frac{4}{3}(D_\nu D_\nu - \frac{1}{4}\hat{D}\hat{D})[\lambda_7(\gamma_\mu\psi_\mu) - i\lambda_8\psi_0]$$

$$-\frac{1}{6}\Big[8D_\mu\hat{D}\psi_\mu + 3\hat{D}\hat{D}(\gamma_\mu\psi_\mu) - 4D_\nu D_\nu(\gamma_\mu\psi_\mu) - 4\hat{D}(D_\mu\psi_\mu)\Big]\Big\} = 0\,. \tag{72}$$

Taking in mind operator identities

$$\hat{D}\hat{D} = D_\mu D_\nu\gamma_\mu\gamma_\nu = D_\mu D_\nu\Big(\frac{1}{2}(\gamma_\mu\gamma_\nu + \gamma_\nu\gamma_\mu) + \frac{1}{2}(\gamma_\mu\gamma_\nu - \gamma_\nu\gamma_\mu)\Big)$$
$$= D_\mu D_\mu + \frac{1}{2}D_\mu D_\nu(\gamma_\mu\gamma_\nu - \gamma_\nu\gamma_\mu) = D_\mu D_\mu - \frac{ie}{4}F_{[\mu\nu]}(\gamma_\mu\gamma_\nu - \gamma_\nu\gamma_\mu)\,,$$

we derive the formula

$$\hat{D}\hat{D}[\lambda_7(\gamma_\mu\psi_\mu) - i\lambda_8\psi_0] - \frac{4}{3}\Big(D_\nu D_\nu - \frac{1}{4}\hat{D}\hat{D}][\lambda_7(\gamma_\mu\psi_\mu) - i\lambda_8\psi_0\Big)$$
$$= -\frac{ie}{3}F_{[\mu\nu]}(\gamma_\mu\gamma_\nu - \gamma_\nu\gamma_\mu)\ [\lambda_7(\gamma_\rho\psi_\rho) - i\lambda_8\psi_0]\,.$$

Correspondingly, Eq. (72) takes the form

$$\left(-\frac{1}{2}\hat{D}+M\right)[\,\lambda_7(\gamma_\mu\psi_\mu)-i\lambda_8\psi_0\,]$$

$$-\frac{1}{3}\left[(D_\mu\psi_\mu)-\frac{1}{4}\hat{D}(\gamma_\mu\psi_\mu)\right]+\frac{1}{M}(\lambda_1\lambda_5-\lambda_2\lambda_4)$$

$$\times\left\{-\frac{ie}{3}F_{[\mu\nu]}(\gamma_\mu\gamma_\nu-\gamma_\nu\gamma_\mu)[\lambda_7(\gamma_\rho\psi_\rho)-i\lambda_8\psi_0]+\frac{2}{3}\hat{D}[(D_\mu\psi_\mu)-\frac{1}{4}\hat{D}(\gamma_\mu\psi_\mu)]\right.$$

$$\left.-\frac{1}{6}\left[8D_\mu\hat{D}\psi_\mu+3\hat{D}\hat{D}(\gamma_\mu\psi_\mu)-4D_\nu D_\nu(\gamma_\mu\psi_\mu)-4\hat{D}(D_\mu\psi_\mu)\right]\right\}=0. \tag{73}$$

Using identity

$$D_\mu\hat{D}=\hat{D}D_\mu+\gamma_\nu(-ieF_{[\mu\nu]})\,,$$

we may simplify one term in (73):

$$\frac{2}{3}\hat{D}[(D_\mu\psi_\mu)-\frac{1}{4}\hat{D}(\gamma_\mu\psi_\mu)]$$

$$-\frac{1}{6}\left(8D_\mu\hat{D}\psi_\mu+3\hat{D}\hat{D}(\gamma_\mu\psi_\mu)-4D_\nu D_\nu(\gamma_\mu\psi_\mu)-4\hat{D}(D_\mu\psi_\mu)\right)$$

$$=+\frac{ie}{6}F_{[\mu\nu]}(\gamma_\mu\gamma_\nu-\gamma_\nu\gamma_\mu)(\gamma_\lambda\psi_\lambda)+\frac{4}{3}\gamma_\nu(-ieF_{[\mu\nu]})\psi_\mu\,.$$

The last relation permits to transform Eq. (73) into the following one

$$\left(-\frac{1}{2}\hat{D}+M\right)[\,\lambda_7(\gamma_\mu\psi_\mu)-i\lambda_8\psi_0\,]-\frac{1}{3}\left((D_\mu\psi_\mu)-\frac{1}{4}\hat{D}(\gamma_\mu\psi_\mu)\right)$$

$$+\frac{1}{M}(\lambda_1\lambda_5-\lambda_2\lambda_4)\left\{-\frac{ie}{3}F_{[\mu\nu]}(\gamma_\mu\gamma_\nu-\gamma_\nu\gamma_\mu)[\lambda_7(\gamma_\rho\psi_\rho)-i\lambda_8\psi_0]\right.$$

$$\left.+\frac{ie}{6}F_{[\mu\nu]}(\gamma_\mu\gamma_\nu-\gamma_\nu\gamma_\mu)(\gamma_\lambda\psi_\lambda)+\frac{4ie}{3}F_{[\mu\nu]}\gamma_\nu\psi_\mu\right\}=0\,. \tag{74}$$

Now we are to perform manipulations with Eq. (74). First, let us re-write it differently

$$\left(-\frac{1}{2}\hat{D}+M\right)[\,\lambda_7(\gamma_\mu\psi_\mu)-i\lambda_8\psi_0\,]-\frac{1}{3}\left((D_\mu\psi_\mu)-\frac{1}{4}\hat{D}(\gamma_\mu\psi_\mu)\right)$$

$$+\frac{1}{M}(\lambda_1\lambda_5-\lambda_2\lambda_4)\left\{\frac{2}{3}ieF_{[\mu\nu]}\gamma_\nu\gamma_\mu\,[\,\lambda_7(\gamma_\rho\psi_\rho)-i\lambda_8\psi_0\,]\right.$$

$$\left.-\frac{1}{3}ieF_{[\mu\nu]}\gamma_\nu\gamma_\mu(\gamma_\rho\psi_\rho)+\frac{4}{3}ieF_{[\mu\nu]}\gamma_\nu\psi_\mu\right\}=0\,.$$

In the last term we may separate the multiplier

$$\left(-\frac{1}{2}\hat{D}+M\right)(\lambda_7(\gamma_\mu\psi_\mu)-i\lambda_8\psi_0)-\frac{1}{3}\left((D_\mu\psi_\mu)-\frac{1}{4}\hat{D}(\gamma_\mu\psi_\mu)\right)$$

$$+\frac{1}{M}(\lambda_1\lambda_5 - \lambda_2\lambda_4)\frac{2}{3}ieF_{[\mu\nu]}\gamma_\nu$$

$$\times\left\{\gamma_\mu[\lambda_7(\gamma_\rho\psi_\rho) - i\lambda_8\psi_0] + 2[\psi_\mu - \frac{1}{4}\gamma_\mu(\gamma_\rho\psi_\rho)]\right\} = 0\,. \tag{75}$$

By analogy with the free particle case, let us introduce the new vector-bispinor Φ_λ (see (52), (53)):

$$\Phi_\lambda = [\psi_\lambda - \frac{1}{4}\gamma_\lambda(\gamma_\mu\psi_\mu)] + \frac{1}{4}\gamma_\lambda[\lambda_7(\gamma_\mu\psi_\mu) - i\lambda_8\psi_0]\,, \tag{76}$$

$$\lambda_7(\gamma_\mu\psi_\mu) - i\lambda_8\psi_0 = \gamma_\mu\Phi_\mu\,, \quad \psi_\lambda - \frac{1}{4}\gamma_\lambda(\gamma_\mu\psi_\mu) = \Phi_\lambda - \frac{1}{4}\gamma_\lambda(\gamma_\mu\Phi_\mu)\,. \tag{77}$$

So, Eq. (75) takes the form

$$\left(-\frac{1}{2}\hat{D} + M\right)(\gamma_\rho\Phi_\rho) - \frac{1}{3}\left((D_\mu\Phi_\mu) - \frac{1}{4}\hat{D}(\gamma_\mu\Phi_\mu)\right)$$

$$+\frac{1}{M}(\lambda_1\lambda_5 - \lambda_2\lambda_4)\frac{4}{3}ieF_{[\mu\nu]}\gamma_\nu\left(\Phi_\mu + \frac{1}{4}\gamma_\mu(\gamma_\rho\Phi_\rho)\right) = 0\,, \tag{78}$$

or differently

$$\frac{1}{4}\gamma_\lambda\left(-\frac{1}{2}\hat{D} + M\right)(\gamma_\rho\Phi_\rho) - \frac{1}{12}\gamma_\lambda\left((D_\mu\Phi_\mu) - \frac{1}{4}\hat{D}(\gamma_\mu\Phi_\mu)\right)$$

$$+\frac{(\lambda_1\lambda_5 - \lambda_2\lambda_4)}{3M}ieF_{[\mu\nu]}\gamma_\lambda\gamma_\nu\left(\Phi_\mu + \frac{1}{4}\gamma_\mu(\gamma_\rho\Phi_\rho)\right) = 0\,. \tag{79}$$

In terms of Φ_λ, Eq. (67) reads

$$(\,D_\lambda - \frac{1}{4}\gamma_\lambda\hat{D}\,)(\gamma_\rho\Phi_\rho) + M\,[\,\Phi_\lambda - \frac{1}{4}\gamma_\lambda(\gamma_\mu\Phi_\mu)\,]$$

$$+\frac{1}{8}\Big[8\hat{D}\psi_\lambda + 3\gamma_\lambda\hat{D}(\gamma_\mu\psi_\mu) - 4D_\lambda(\gamma_\mu\psi_\mu) - 4\gamma_\lambda(D_\mu\psi_\mu)\Big] = 0\,. \tag{80}$$

Taking in mind identity

$$8\hat{D}\Phi_\lambda + 3\gamma_\lambda\hat{D}(\gamma_\mu\Phi_\mu) - 4D_\lambda(\gamma_\mu\Phi_\mu) - 4\gamma_\lambda(D_\mu\Phi_\mu)$$
$$= 8\hat{D}\psi_\lambda + 3\gamma_\lambda\hat{D}(\gamma_\mu\psi_\mu) - 4D_\lambda(\gamma_\mu\psi_\mu) - 4\gamma_\lambda(D_\mu\psi_\mu)\,,$$

the previous equation may be re-written differently

$$(\,D_\lambda - \frac{1}{4}\gamma_\lambda\hat{D}\,)(\gamma_\rho\Phi_\rho) + M\,[\,\Phi_\lambda - \frac{1}{4}\gamma_\lambda(\gamma_\mu\Phi_\mu)]$$

$$+\frac{1}{8}\Big[8\hat{D}\Phi_\lambda + 3\gamma_\lambda\hat{D}(\gamma_\mu\Phi_\mu) - 4D_\lambda(\gamma_\mu\Phi_\mu) - 4\gamma_\lambda(D_\mu\Phi_\mu)\Big] = 0\,. \tag{81}$$

After summing equations (79) and (81), we arrive at the needed equation for vector-bispinor Φ_μ:

$$\frac{1}{8}\left(8\hat{D}\Phi_\lambda + \frac{1}{6}\gamma_\lambda\hat{D}\gamma_\rho\Phi_\rho + 4D_\lambda\gamma_\rho\Phi_\rho - \frac{14}{3}\gamma_\lambda D_\rho\Phi_\rho\right)M\Phi_\lambda$$

$$-\frac{(\lambda_1\lambda_5 - \lambda_2\lambda_4)}{3M}\gamma_\lambda\gamma_\nu\,ieF_{[\nu\mu]}\left(\Phi_\mu + \frac{1}{4}\gamma_\mu\gamma_\rho\Phi_\rho\right) = 0\,. \tag{82}$$

This is the Fradkin equation for spin 3/2 particle in presence of external electromagnetic field. In comparison with the Pauli-Fierz equation, it contains an additional interaction term. Eq. (82) refers to a particle with intrinsic electromagnetic characteristics governed by the factor $(\lambda_1\lambda_5 - \lambda_2\lambda_4)/3M$.

The last equation (82) may be presented in the matrix form

$$(D_\mu\Gamma_\mu + Q + M)\,\Phi = 0\,, \tag{83}$$

where the matrices Γ_μ are given by the formula (58)

$$\Gamma_\mu = \gamma_\mu \otimes e^{\nu,\nu} + \frac{1}{48}\gamma_\rho\gamma_\mu\gamma_\sigma \otimes e^{\rho,\sigma} + \frac{1}{2}\gamma_\rho \otimes e^{\mu,\rho} - \frac{7}{12}\gamma_\rho \otimes e^{\rho,\mu}\,, \tag{84}$$

and the additional interaction term is presented as follows

$$Q = \frac{ie}{3M}(\lambda_1\lambda_5 - \lambda_2\lambda_4)\left(F_{[\mu\nu]}\gamma_\rho\gamma_\nu\right) \otimes e^{\alpha,\alpha}\left\{I \otimes e^{\rho,\mu} + \frac{1}{4}\gamma_\mu\gamma_\eta \otimes e^{\rho,\eta}\right\}. \tag{85}$$

In Petras basis, the last equation becomes simpler

$$\{D_\mu\Gamma'_\mu + Q' + M\}\Phi' = 0\,, \tag{86}$$

where

$$\Gamma'_\mu = R\Gamma_\mu R^{-1} = \gamma_\mu \otimes e^{\nu,\nu} + \frac{1}{\sqrt{3}}(\gamma_\rho \otimes e^{\rho,\mu} - \gamma_\rho \otimes e^{\mu,\rho})\,, \tag{87}$$

and

$$Q' = -\frac{ie}{M}(1 + \frac{2}{\sqrt{3}})(\lambda_1\lambda_5 - \lambda_2\lambda_4)F_{[\mu\nu]}$$
$$\times(\gamma_\rho\gamma_\nu \otimes e^{\alpha,\alpha})\left(I \otimes e^{\rho,\mu} - \frac{1}{2\sqrt{3}}\gamma_\mu\gamma_\eta \otimes e^{\rho,\eta}\right). \tag{88}$$

Explicit form of the wave equation in Petras basis is

$$D_\mu\gamma_\mu\Phi'_\lambda + \frac{1}{\sqrt{3}}(D_\mu\gamma_\lambda\Phi'_\mu - D_\lambda\gamma_\mu\Phi'_\mu)$$
$$+\frac{2ie}{M}(1 + \frac{2}{\sqrt{3}})(\lambda_1\lambda_5 - \lambda_2\lambda_4)$$
$$\times\left(\gamma_\lambda F_{[\nu\mu]}\gamma_\nu\Psi_\mu - \frac{1}{2\sqrt{3}}\gamma_\lambda F_{[\nu\mu]}\gamma_\nu\gamma_\mu\gamma_\eta\Phi'_\eta\right) + M\Phi'_\lambda = 0\,. \tag{89}$$

6. The Wave Equation in Riemannian Space-Time

To extend the Fradkin model to the case of Riemannian space-time, we are to make several formal changes. Instead of ict-metrics in Minkowski space, we will use the metric tensor $g_{\alpha\beta}(x)$ with the signature $(+,-,-,-)$. Correspondingly, we are to employ other Dirac matrices (for definiteness assuming the Weyl basis)

$$\gamma^0 = \begin{vmatrix} 0 & I \\ I & 0 \end{vmatrix}, \quad \gamma^i = \begin{vmatrix} 0 & -\sigma^i \\ \sigma^i & 0 \end{vmatrix}, \quad M \Longrightarrow iM\,. \tag{90}$$

Derivatives are changed to more general form [51]:

$$D_\alpha(x) = \nabla_\alpha + \Gamma_\alpha(x) + ieA_\alpha(x)\,, \quad \hat{D} = \gamma^\alpha(x) D_\alpha(x)\,, \tag{91}$$

where $\Gamma_\alpha(x)$ stands for bispinor connection [51], and local Dirac matrices [51] are determined through tetrad: $\gamma^\alpha(x) = \gamma^a e^\alpha_{(a)}(x)$. We should recall important commutation rules and identities [51]:

$$\hat{D}(x) = \gamma^\rho(x) D_\beta = D_\beta \gamma^\rho(x)\,, \qquad D_\sigma(x) g_{\alpha\beta}(x) = g_{\alpha\beta}(x) D_\sigma(x)\,,$$

$$\sigma^{\nu\rho}(x) = \frac{\gamma^\nu(x)\gamma^\rho(x) - \gamma^\rho(x)\gamma^\nu(x)}{4}\,,$$

$$D_\alpha D_\beta - D_\beta D_\alpha = ieF_{\alpha\beta} + \frac{1}{2}\sigma^{\nu\rho}(x) R_{\nu\rho\alpha\beta}(x)\,,$$

$$D_\mu \hat{D} - \hat{D} D_\mu = (ieF_{\mu\nu} + \Sigma_{\mu\nu}), \quad \Sigma_{\mu\nu} = \frac{1}{2}\sigma^{\rho\sigma}(x) R_{\rho\sigma\mu\nu}(x)\,,$$

$$\hat{D}\hat{D} = D_\alpha D_\beta \Big(g^{\alpha\beta}(x) + \frac{\gamma^\alpha\gamma^\beta - \gamma^\beta\gamma^\alpha}{2}\Big) = D^2 + \Sigma(x)\,, \quad D^2 = D^\alpha D_\alpha\,,$$

$$\Sigma(x) = \left[ieF_{\alpha\beta} + \frac{1}{2}\sigma^{\nu\rho}(x) R_{\nu\rho\alpha\beta}(x)\right] \sigma^{\alpha\beta}(x) = ieF_{\alpha\beta}\sigma^{\alpha\beta}(x) + \frac{R(x)}{4}\,,$$

where Riemann curvature tensor and Ricci scalar are used. There exist some peculiarities in defining the covariant matrix $\gamma^5(x)$:

$$\gamma^5(x) = \frac{i}{4!}\epsilon_{\alpha\beta\rho\sigma}(x)\; \gamma^\alpha(x)\gamma^\beta(x)\gamma^\rho(x)\gamma^\sigma(x)\,,$$

$$\epsilon^{\alpha\beta\rho\sigma}(x) = \epsilon^{abcd}\; e^\alpha_{(a)}(x)\; e^\beta_{(b)}(x)\; e^\rho_{(c)}(x)\; e^\sigma_{(d)}(x)\,, \quad \epsilon_{0123} = -1\,. \tag{92}$$

Covariant Levi-Civita symbol $\epsilon^{\alpha\beta\rho\sigma}(x)$ transforms by the rule

$$\epsilon'^{\alpha\beta\rho\sigma}(x) = -\det[L_{ai}(x)]\; \epsilon^{\alpha\beta\rho\sigma}(x)\,. \tag{93}$$

In particular, with respect to the tetrad P-reflection it behaves as a pseudo-scalar [51]:

$$\epsilon^{(p)\alpha\beta\rho\sigma}(x) = (-1)\; \epsilon^{\alpha\beta\rho\sigma}(x)\,. \tag{94}$$

We readily derive the following identity

$$\gamma^5(x) = \frac{i}{4!}\epsilon^{abcd} e_{\alpha(a)} e_{\beta(b)} e_{\rho(c)} e_{\sigma(d)}\; e^\alpha_{(m)}\gamma^m e^\beta_{(n)}\gamma^n e^\rho_{(k)}\gamma^k e^\sigma_{(l)}\gamma^l)$$

$$= \frac{i}{4!}\epsilon_{mnkl}\gamma^m\gamma^n\gamma^k\gamma^l = \gamma^5\,. \tag{95}$$

All calculationw with Eq. (74) are repeated with only minor changes. Instead of (74), we will obtain

$$\Big(-\frac{1}{2}\hat{D} + iM\Big)\,[\,\lambda_7(\gamma^\mu\psi_\mu) - i\lambda_8\psi_0\,] - \frac{1}{3}\Big((D^\mu\psi_\mu) - \frac{1}{4}\hat{D}(\gamma^\mu\psi_\mu)\Big)$$

$$+\frac{1}{iM}(\lambda_1\lambda_5 - \lambda_2\lambda_4)\Big\{-\frac{1}{3}(ieF_{\mu\nu} + \Sigma_{\mu\nu})(\gamma^\mu\gamma^\nu - \gamma^\nu\gamma^\mu)[\lambda_7(\gamma^\rho\psi_\rho) - i\lambda_8\psi_0]$$

$$+\frac{1}{6}(ieF_{\mu\nu} + \Sigma_{\mu\nu})(\gamma^\mu\gamma^\nu - \gamma^\nu\gamma^\mu)(\gamma^\lambda\psi_\lambda) + \frac{4}{3}(ieF_{\mu\nu} + \Sigma_{\mu\nu})\gamma^\nu\psi^\mu\Big\} = 0\,. \tag{96}$$

Instead of (75), we get a modified equation

$$\left(-\frac{1}{2}\hat{D}+iM\right)(\lambda_7(\gamma^\mu\psi_\mu)-i\lambda_8\psi_0)-\frac{1}{3}\left((D^\mu\psi_\mu)-\frac{1}{4}\hat{D}(\gamma^\mu\psi_\mu)\right)$$

$$+\frac{1}{iM}(\lambda_1\lambda_5-\lambda_2\lambda_4)\frac{2}{3}(ieF_{\mu\nu}+\Sigma_{\mu\nu})$$

$$\times\gamma^\nu\left\{\gamma^\mu\left[\lambda_7(\gamma^\rho\psi_\rho)-i\lambda_8\psi_0\right]+2\left[\psi^\mu-\frac{1}{4}\gamma^\mu(\gamma^\rho\psi_\rho)\right]\right\}=0\,. \tag{97}$$

Further, we transform this equation to the new wave function

$$\Phi_\lambda=[\psi_\lambda-\frac{1}{4}\gamma_\lambda(\gamma^\mu\psi_\mu)]+\frac{1}{4}\gamma_\lambda[\lambda_7(\gamma^\mu\psi_\mu)-i\lambda_8\psi_0]\,,$$

$$\lambda_7(\gamma^\mu\psi_\mu)-i\lambda_8\psi_0=\gamma^\mu\Phi_\mu\,,\quad \psi_\lambda-\frac{1}{4}\gamma_\lambda(\gamma^\mu\psi_\mu)=\Phi_\lambda-\frac{1}{4}\gamma_\lambda(\gamma^\mu\Phi_\mu)\,.$$

So we derive

$$\frac{i}{8}\left(8\hat{D}\Phi^\alpha+\frac{1}{6}\gamma^\alpha(x)\hat{D}\gamma^\rho(x)\Phi_\rho+4D^\alpha\gamma^\rho(x)\Phi_\rho-\frac{14}{3}\gamma^\alpha(x)D^\rho\Phi_\rho\right)$$

$$-M\Phi^\alpha-\frac{(\lambda_1\lambda_5-\lambda_2\lambda_4)}{3M}$$

$$\times\gamma^\alpha(x)\gamma^\nu(x)\left[ieF_{\nu\mu}(x)+\Sigma_{\nu\mu}(x)\right]\left(\Phi^\mu+\frac{1}{4}\gamma^\mu(x)\gamma^\rho(x)\Phi_\rho\right)=0\,, \tag{98}$$

recall that

$$\Sigma_{\nu\mu}(x)=\frac{1}{2}\sigma^{\rho\sigma}(x)R_{\rho\sigma\nu\mu}(x)\,,$$

where $R_{\rho\sigma\mu\nu}(x)$ is the Ricci tensor. Taking into account the symmetry properties of the Riemann tensor, and the multiplication rule for three Dirac matrices, we can transform the term in Eq. (98) as follow

$$\gamma^\nu(ieF_{\nu\mu}+\Sigma_{\nu\mu})=\gamma^\nu ieF_{\nu\mu}+\frac{1}{4}\gamma^\nu\gamma^\alpha\gamma^\beta R_{\alpha\beta\nu\mu}$$

$$=\gamma^\nu ieF_{\nu\mu}+\frac{1}{4}(\gamma^\nu g^{\alpha\beta}-\gamma^\alpha g^{\nu\beta}+\gamma^\beta g^{\nu\alpha}+i\gamma^5\epsilon^{\nu\alpha\beta\rho}\gamma_\rho)R_{\alpha\beta\nu\mu}\,,$$

whence we derive

$$\gamma^\nu(ieF_{\nu\mu}+\Sigma_{\nu\mu})=\gamma^\nu(ieF_{\nu\mu}-\frac{1}{2}R_{\nu\mu})\,, \tag{99}$$

$R_{\nu\mu}$ is the Ricci tensor. Therefore, Eq. (98) may be presented differently

$$i\left(\hat{D}\Phi^\alpha+\frac{1}{48}\gamma^\alpha\hat{D}\gamma^\rho\Phi_\rho+\frac{1}{2}D^\alpha\gamma^\rho\Phi_\rho-\frac{7}{12}\gamma^\alpha D^\rho\Phi_\rho\right)$$

$$-\frac{(\lambda_1\lambda_5-\lambda_2\lambda_4)}{3M}\gamma^\alpha(ieF_{\nu\mu}-\frac{1}{2}R_{\nu\mu})\,\gamma^\nu(\,\Phi^\mu+\frac{1}{4}\gamma^\mu\gamma^\rho\Phi_\rho\,)-M\Phi^\alpha=0. \tag{100}$$

This is the Fradkin equation fora spin 3/2 particle in the curved space-time background.

If the electric charge e of the particle is zero, Eq. (100) becomes simpler

$$i\left(\hat{D}\Phi^{\alpha} + \frac{1}{48}\gamma^{\alpha}\hat{D}\gamma^{\rho}\Phi_{\rho} + \frac{1}{2}D^{\alpha}\gamma^{\rho}\Phi_{\rho} - \frac{7}{12}\gamma^{\alpha}D^{\rho}\Phi_{\rho}\right)$$
$$+\frac{(\lambda_1\lambda_5 - \lambda_2\lambda_4)}{3M}\,\frac{1}{2}R_{\nu\mu}\,\gamma^{\alpha}\gamma^{\nu}(\,\Phi^{\mu} + \frac{1}{4}\gamma^{\mu}\gamma^{\rho}\Phi_{\rho}\,) - M\Phi^{\alpha} = 0\,, \qquad (101)$$

where $D_{\rho} = \nabla_{\rho} + \Gamma_{\rho}(x)$. It should be noted that in any of Majorana bases, the following properties of the Dirac matrices with respect to complex conjugation are valid

$$\gamma^{\alpha})* = -\gamma^{\alpha}, \quad (\Gamma_{\alpha})^{*} = +\Gamma_{\alpha},$$
$$(D_{\alpha})^{*} = +D_{\alpha}, \quad \hat{D}^{*} = -\hat{D}, \quad [\sigma^{\alpha\beta}]^{*} = +\sigma^{\alpha\beta}\,. \qquad (102)$$

Therefore, Eq. (101) does not mix real and imaginary parts of the vector-bispinior Φ_{α}. This means that the theory is valid for neural particles of spin 3/2 as well.

Conclusion

In this chapter we have developed in detail the relativistic theory for a spin 3/2 particle proposed by Fradkin¡ we have taken into account the presence of external electromagnetic fields and the pseudo-Riemanina space time background. At vanishing additional Fradkin parameter Λ this equation reduces to the common Pauli–Fierz theory.

To clarify the physical meaning of the additional parameter Λ, in other paper [54], we have developed the nonrelativistic approximation in Fradkin theory, at this we had into account the presence of only external electromagnetic fields. With the use of the technique of projective operators, we decomposed the wave function into big and small constituents, and then derive a generalized nonrelativistic equation for a 4-component wave function. It is shown that when preserving only the terms of the first order in Fradkin parameter Λ there arises the ordinary nonrelativistic equation for Pauli - Fierz theory without any additional interaction terms with electromagnetic fields. When preserving the terms of the second order in parameter Λ, we obtain nonrelativistic equation with additional interaction, however only with magnetic field. This interaction is quadratic in magnetic field components and is governed by six 4-dimensional matrices. So the Fradkin theory may be understood as relevant to a particle whether with magnetic magnetic quadruple moment.

Acknowledgments

The present work was developed within the framework of the project *BRFFR A18-RA-015 (2018-2020)*, set within the framework of the agreement between the National Academy of Sciences of Belarus and the Romanian Academy. Also this work was supported by the project *BRFFR F18Y-009 (2018-2020).*

References

[1] P. A. M. Dirac. Relativistic wave equations. *Proc. R. Soc. London.* A. 1936. **155**. 447–459.

[2] E. Majorana. Teoria simmetrica dell electrone e dell positrone. *Nuovo Cim.* 1937. **14**. 171–186.

[3] M. Fierz. Uber die relativistische theorie KraftefreierTeilchen mit beliebigem Spin. *Helv. Phys. Acta.* 1939. **12**. 3–37.

[4] W. Pauli, M. Fierz. Uber relativistische Feldleichungen von Teilchen mit beliebigem Spin im elektromagnetishen Feld. *Helv. Phys. Acta.* 1939. **12**. 297–300.

[5] V. L. Ginzburg, Ya. A. Smorodinsky. On wave equations for particles with variable spin. *Zh. Eksp. Teor. Fiz.* 1943. **13**. 274 (in Russian).

[6] A.S. Davydov. Wave equations of a particle having spin $3/2$ in absence of field. *Zh. Eksp. Teor. Fiz.* 1943. **13**. no 9-10. 313–319 (in Russian).

[7] H. J. Bhabha, Harish-Chandra. On the theory of point particles. *Proc. Roy. Soc. London.* A. 1944. **183**. 134–141.

[8] H. J. Bhabha. Relativistic wave equations for the proton. *Proc. Indian Acad. Sci.* A. 1945. **21**. 241–264.

[9] H. J. Bhabha. Relativistic wave equations for elementary particles. *Rev. Mod. Phys.* 1945. **17**. no 2-3. 200–215.

[10] H. J. Bhabha. The theory of the elementary particles. *Rep. Progr. Phys.* 1946. **10**. 253–271.

[11] I. M. Gel'fand, A. M. Yaglom. General relativistic invariant equations snd infinitely dimensional representation of the Lorentz group. *Zh. Eksp. Teor. Fiz.* 1848. **18**. no 8. 703–733 (in Russian).

[12] E. E. Fradkin. To the theory of particles with higher spins. *Zh. Eksp. Teor. Fiz.* 1950. **20**. no. 1. 27–38 (in Russian).

[13] F. I. Fedorov. On minimal polynomials of matrices of relativistic wave equations. *Doklady AN USSR.* 1951. **79**. no 5. 787–790 (in Russian).

[14] F. I. Fedorov. *To the theory of a spin 2 particle.* Uchionye Zapiski Beloruaasian State University. ser. Phys.-mat. 1951. no 12. 156–173 (in Russian).

[15] H. J. Bhabha. An equation for a particle with two mass states and positive charge density. *Phil. Mag.* 1952. **43**. P. 33–47.

[16] F. I. Fedorov. Generalized relativistic wave equations. *Doklady AN USSR.* 1952. **82**. no 1. 37–40 (in Russian).

[17] T. Regge. On properties of the particle with spin 2. *Nuovo Cimento*. 1957. **5**. no 2. 325–326.

[18] M. Petras. A contribution of the theory of the Pauli-Fierz's equations a particle with spin 3/2. *Czech. J. Phys.* 1955. **5**. 169–170.

[19] M. Petras. A note to Bhabha's equation for a particle with maximum spin 3/2. *Czech. J. Phys.* 1955. **5**. 418–419.

[20] V. Ya. Fainberg. To the interaction theory of the particles of the higher spins with electromagnetic and meson fields. *Trudy FIAN USSR*. 1955. **6**. 269–332 (in Russian).

[21] V. L. Ginzburg. On relativistic wave equations with a mass spectrum. *Acta Phys. Pol.* 1956. **15**. P. 163–175.

[22] H. Shimazu. A relativistic wave equation for a particle with two mass states of spin 1 and 0. *Progress of Theoretical Physics*. 1956. **16**. no 4. 285–298.

[23] I. M. Gel'fand, R. A. Minlos, Z. Ya. Shapiro. *Representations of the rotation and Lorentz groups and their applications*. Translated from the Russian edition (1958) by G. Cummins and T. Boddington. Pergamon. London. Macmillan. New York. 1963.

[24] L. A. Shelepin. Covariant theory of relativictic wave equations. *Nucl. Phys.* 1962. **33**. no 4. 580–593.

[25] L. A. Shelepin. Covariant theory of relativictic wave equations for paricle of arbitrary spins. Trudy FIAN USSR. 1964. **30**. 253–321 (in Russian).

[26] A. Z. Capri. Non uniqueness of the spin 1/2 equation. *Phys. Rev.* 1969. **178**. 1811–1815.

[27] A. Z. Capri. First order wave equations for multimass fermions. *Nuovo Cim. B*. 1969. **64**. 151–158.

[28] F. I. Fedorov, V.A. Pletyukhov. Wave equations with repeated representations of the Lorentz group. *Proceedings of the National academy of sciences of Belarus. Phys.-Math. series*. 1969. no 6. 81–88 (in Russian).

[29] V. A. Pletyukhov, F. I. Fedorov. The wave equation with repeated representations for spin 0 particle. Proceedings of the National academy of sciences of Belarus. *Phys.-Math. series*. 1970. no 2. 79–85 (in Russian).

[30] F. I. Fedorov, V. A. Pletyukhov. Wave equations with repeated representations of the Lorentz group. Half-integer spin. *Proceedings of the National academy of sciences of Belarus. Phys.-Math. series*. 1970. no 3. 78–83 (in Russian).

[31] V. A. Pletyukhov, F. I. Fedorov. Wave equation with repeated reprentations for a spin 1 particle. *Proceedings of the National academy of sciences of Belarus. Phys.-Math. series*. 1970. no 3. 84–92 (in Russian).

[32] A. Shamaly, A. Z. Capri. First-order wave equations for integral spin. *Nuovo Cim.* B. 1971. **2**. no 2. 235–253.

[33] A. Z. Capri. Electromagnetic properties of a new spin-1/2 field. *Progr. Theor. Phys.* 1972. **48**. 1364–1374.

[34] A. Shamaly, A. Z. Capri. Unified theories for massive spin 1 fields. *Can. J. Phys.* 1973. **51**. no 14. 1467–1470.

[35] M. A. K. Khalil. Properties of a 20-component spin $1/2$ relativistic wave equation. *Phys. Rev.* D. 1977. **15**. no 6. 1532–1539.

[36] A. S. Wightman. Invariant wave equations: general theory and applications to the external field problem. *Lecture Notes in Physics*. 1978. **73**. 1–101.

[37] M. A. K. Khalil. An equivalence of relativistic field equations. *Nuovo Cimento*. A. 1978. **45**. no 3. 389–404.

[38] L. Garding, L. Mathematics of invariant wave equations. *Lect. Notes in Physics*. 1978. **73**. 102–164.

[39] W. Cox. Higher-rank representations for zero-spin filds theories. *J. Phys.* A. 1982. **15**. 627–635.

[40] W. Cox. First-order formulation of massive spin-2 field theories. *J. Phys.* A. 1982. **15**. 253–268.

[41] P. M. Mathews, B. Vijayalakshmi, M. Sivakuma. On the admissible Lorentz group representations in unique-mass, unique-spin relativistic wave equations. *Phys.* A. 1982. **15**. no 11. 1579–1582.

[42] F. I. Fedorov. *The Lorentz group*. Moscow. Nauka. 1979.

[43] P. M. Mathews, B. Vijayalakshmi. On inequivalent classes unique-mass-spin relativistic wave equations involving repeated irreducible representations with arbitrary multiplicities. *J. Math. Phys.* 1984. **25**. no 4. 1080–1087.

[44] W. Cox. On the Lagrangian and Hamiltonian constraint algorithms for the Rarita-Schwinger field coupled to an external electromagnetic field. *J. Phys.* A. 1989. **22**. no 10. 1599–1608.

[45] S. Deser, A. Waldron. Inconsistencies of massive charged gravitating higher spins. *Nucl. Phys.* B. 2002. **631**. 369–387.

[46] E. M. Ovsiyuk, V. V. Kisel, Y. A. Voynova, O. V. Veko, V. M. Red'kov. Spin 1/2 particle with anomalous magnetic moment in a uniform magnetic field, exact solutions. *Nonlinear Phenomena in Complex Systems*. 2016. **19**. no 2. 153–165.

[47] V. Kisel, Ya. Voynova, E. Ovsiyuk, V. Balan, V. Red'kov. Spin 1 Particle with Anomalous Magnetic Moment in the External Uniform Magnetic Field. *Nonlinear Phenomena in Complex Systems*. 2017. **20**. no 1. 21–39.

[48] V. V. Kisel, E. M. Ovsiyuk, Ya. A. Voynova, V. M. Red'kov. Quantum mechanics of spin 1 particle with quadrupole moment in external uniform magetic field. *Problems of Physics, Mathematics, and Thechnics.* 2017. **32** no 3. 18–27.

[49] V. V. Kisel, V. A. Pletyukhov, V. V. Gilewsky, E. M. Ovsiyuk, O. V. Veko, V. M. Red'kov. Spin 1/2 particle with two mass states, interaction with external fields. *Nonlinear Phenomena in Complex Systems.* 2017. **20**. no 4. 404–423.

[50] E. M. Ovsiyuk, O. V. Veko, Ya. A. Voynova, V. V. Kisel, V. Balan, V. M. Red'kov. Spin 1/2 particle with two masses in magnetic field. *Applied Sciences.* 2018. **20**. 148–166.

[51] V. M. Red'kov. *Fields in Riemannian space and the Lorentz group.* Publishing House: Belarusian Science, Minsk, 2009.

[52] V. A. Pletjukhov, V. M. Red'kov, V. I. Strazhev. Relativistic wave equations and intrinsic degrees of freedom. *Belarusian Science.* Minsk. 2015.

[53] V. V. Kisel, E. M. Ovsiyuk, O. V. Veko, Ya. A. Voynova, V. Balan, V. M. Red'kov. Elementary particles with internal structure in external fields. *I. General theory, II. Physical problems.* New York: Nova Science Publishers Inc. 2018.

[54] A. V. Ivashkevich, O. A. Vasiluyk, V. V. Kisel, V. M. Red'kov. Spin 3/2 particle: the Fradkin theory, non-relativistic approximation. *Nonlinear Dynamics and Applications.* 2021. **27**. 138–175.

In: Future Relativity, Gravitation, Cosmology
Editors: V. V. Dvoeglazov et al.

ISBN: 979-8-88697-455-3
2023 Nova Science Publishers, Inc.

Chapter 15

BLACK HOLE HORIZONS AS PATTERNLESS BINARY MESSAGES AND MARKERS OF DIMENSIONALITY

Szymon Łukaszyk*
Łukaszyk Patent Attorneys, Katowice, Poland

ABSTRACT

This study aims to reconcile quantum theory with the universality of the speed of light in vacuum and its implications on relativity through an information-theoretic approach. We introduce the concepts of a holographic sphere and variational potential. Entropy variation expressed in terms of the information capacity of this sphere results in the concept of binary potential in units of negative, squared speed of light in vacuum. Accordingly, the event horizon is a fundamental holographic sphere in thermodynamic equilibrium with only one exterior side: a noncompressible binary message that maximizes Shannon entropy. Therefore, the Jordan–Brouwer separation theorem and generalized Stokes theorem do not hold for black holes. We introduce the concept of inertial potential and demonstrate its equivalence to the variational potential, which ensures that any inertial acceleration represents a nonequilibrium thermodynamic condition. We introduce the concept of the complementary time period and relate it with the classical time period through integral powers of the imaginary unit to formulate the notions of unobservable velocity and acceleration, which are perpendicular and tangential to the holographic sphere, respectively, and bounded with the observable velocity and acceleration based on Pythagorean relations. We further discuss certain dynamics scenarios between the two masses. The concept of black hole informationless emission is introduced as a complement to informationless Bekenstein absorption and extended to arbitrary wavelengths. Black hole quantum statistics with degeneracy interpreted as the number of Planck areas on the event horizon are discussed. The study concludes that holographic screens and equipotential surfaces are spherical equivalents, and every observer is a sphere in nonequilibrium thermodynamic condition. Lastly, we propose a solution to the black hole information paradox.

* Corresponding Author's Email: szymon@patent.pl.

Keywords: entropic gravity, black hole information paradox, Shannon entropy, Landauer's principle, Axis of Evil (cosmology), black hole quantum statistics, exotic $\mathbb{R}^4$, imaginary time, no-hiding theorem

1. Introduction

An uncharged, nonrotating (Schwarzschild) black hole (BH) can be observed as a 2-sphere; however, one cannot state anything regarding its interior [1], which is equivalent to the fact that it does not have any interior. However, mathematical 2-spheres possess interiors. In general, any $(n-1)$-dimensional topological sphere in $\mathbb{R}^n$ forms a common boundary of two connected components: bounded interior and unbounded exterior, as asserted by the Jordan–Brouwer separation theorem. However, this theorem does not hold for BHs thereby disapproving the correctness of our intuitive concept of the physical space represented as $\mathbb{R}^3$ (or $\mathbb{R}^4$). Additionally, the generalized Stokes theorem does not hold for BHs. A differential form over a BH event horizon does not equal the integral of its exterior derivative over the BH interior, considering there is no such thing as a BH interior.

A BH elevates John Archibald Wheeler's "*it from bit*" conclusion, which states that there is no such thing as space or time or spacetime continuum [2] beyond the microscopic level. This causes us to research nature as a vertex-labeled graph (graph of nature) with certain intrinsic properties reflecting the 2nd law of thermodynamics. The primordial Big Bang singularity (the first point or vertex) expanded this graph of nature into dimensionalities and not into a four-dimensional spacetime [3].

The combinatorial proof of the *H*-theorem derived by Ludwig Boltzmann in 1877 [4] introduced energy quantization, which resulted in the development of quantum theory [5] and created the measurement problem that demands an interpretation. To date, no consensus has been achieved in the scientific community regarding the universal validity of any particular interpretation of the measurement problem [6].

The author is convinced that life is an explanation of the measurement problem performed *hic et nunc* by any particular observer. The measurement yields rational (as rational numbers) and classical (as bits, not qubits) information that is twofold, considering it relates both to the *without* [exterior] and *within* [interior] of things. Quoting Pierre Teilhard de Chardin, "In the eyes of the physicist, nothing exists legitimately, at least up to now, except the *without* of things. The same intellectual attitude is still permissible in the bacteriologist, whose cultures (apart from some substantial difficulties) are treated as laboratory reagents. But it is already more difficult in the realm of plants. It tends to become a gamble in the case of a biologist studying the behavior of insects or coelenterates. It seems merely futile with regard to the vertebrates. Finally, it breaks down completely with man, in whom the existence of a *within* can no longer be evaded, because it is the object of a direct intuition and the substance of all knowledge" [7].

Albert Einstein believed [8] that "the experience of the *now* means something special for man, something essentially different from the past and the future, but that this important difference does not and cannot occur within physics. That this experience cannot be grasped by science seemed to him a matter of painful but inevitable resignation. (…) there is something essential about the *now* which is just outside of the realm of science." The author strongly disagrees with this statement: the 3-dimensional universe emerges out of an uncountably infinite-dimensional graph of nature during observer-dependent measurement. The observer-independent measurements simply do not exist [9, 10, 11]. A local time associated with each observer in special relativity is reversible. Nonetheless, we can now investigate time in other conceptual contexts [12], especially as an emergent, entropic quantity[1].

Presumably, dimension n is a complex number [13], wherein biological evolution is possible only in dimension $n = 3 + 0i$ with 3 spatial dimensions and imaginary time owing to the exotic $\mathbb{R}^4$ property of such space [3], which dimensionality was exploited by biological evolution from among uncountably infinite other dimensionalities. Particularly, the only real imaginary number is $0i = 0$. This zero is the *nunc*, that is the moment that passes for every living biological cell, and it cannot be placed outside the realm of science [8, 12]. Therefore, Lorentz transformations denote the rotations of a 4-ball of a fixed radius R, expressed as

$$x^2 + y^2 + z^2 + (ict)^2 = R^2 . \tag{1}$$

Rather than aiming to propose new physics, this research aims to seek a new meaning to the existing, experimentally verified physical theories, especially to the theory of relativity, which is grounded on the objective, unobservable, extra-instantaneous existence of 3D physical reality (*the universe*), and hence, is not complete.

This paper is structured as follows. Sections 2 and 3 present the discretization in terms of Planck units and the discrete BHs, respectively. Section 4 introduces the concept of a BH informationless emission as complementary to informationless Bekenstein absorption and extends it to arbitrary wavelengths. Subsequently, the author's perspective on the entropic gravity derivation disclosed in [14] along with a simplified model of the spherical universe with two masses is presented in Section 5. Thereafter, the concept of the variational potential and the entropy variation sphere is proposed in Section 6 to display the equivalence of known entropic gravity equations. Section 7 presents the concept of the binary potential to exemplify that an event horizon is a binary message that maximizes the Shannon entropy. The concept of inertial potential is introduced in Section 8. The concept of complementary time period related to *classical* time period by integral powers of the imaginary unit is introduced in Section 9, which includes the kinematics of the two masses on a holographic sphere are, and the significant radii of a constant mass. Section 10 summarizes the Bose–Einstein, Maxwell–Boltzmann, and Fermi–Dirac statistics with degeneracy interpreted as the number of Planck areas on a BH event horizon. A solution to BH information paradox is proposed in Section 11. Section 12 concludes the findings of this study.

The author is of the opinion that the notion of charge and other already unified quantities should emerge from topological and entropic considerations, and hence, charged BHs were deliberately neglected from the considerations of this study.

Baroque derivations were used to support certain controversial claims that largely deviate from the mainstream understanding of physics.

Natural units were also deliberately neglected from the considerations of this study to avoid loss of clarity. The meaning of all seven constants of nature used in this study: c (speed of light in vacuum), e (the base of the natural logarithm), G (the gravitational constant), h (the Planck

---

[1] "Einstein imagined himself riding a photon. But (...) we are too heavy to ride photons or electrons. We cannot possibly replace such airy beings, identify ourselves with them, and describe what they would think, if they were able to think, and what they would experience, if they were able to feel anything" ([12] p. 219).

constant), i (the imaginary unit), k_B (the Boltzmann constant), and π (the ratio of a circle's circumference to its diameter), should not be treated carelessly.

A two-dimensional Boolean lattice (a holographic screen), evolving with time, deduced from unitarity, entropy, and counting arguments has been proposed in [15] to explain the failure of developing completely consistent mathematical models of quantum BHs. This research aims to pursue this idea further in an attempt to reconcile quantum theory with the universality of the speed of light in vacuum and its implications on relativity, in an information-theoretic approach.

2. Discretizations

Discretization in terms of Planck units was adopted in certain segments of this study. These "natural units of measure" introduced by Max Planck in 1899 "are independent of special bodies or substances, thereby necessarily retaining their meaning for all times and for all civilizations, including extraterrestrial and non-human ones" [16].

Where deemed appropriate, uppercase letters denote dimensional quantities and lowercase letters denote Planck units or their multipliers. Masses M were considered as real multiplicities m of the Planck mass m_P without loss of generality (w.l.o.g.), expressed as

$$M \doteq m m_P \quad m \in \mathbb{R}. \tag{2}$$

Additionally, lengths were considered w.l.o.g. as the real multiplicities of the Planck length ℓ_P. Particularly, sphericality was considered w.l.o.g. in terms of the diameter d or radius r multipliers

$$D \doteq d\ell_P \quad R \doteq r\ell_P \quad d, r \in \mathbb{R}. \tag{3}$$

However, the wavelengths were considered as

$$\lambda \doteq l\ell_P \quad l \in \mathbb{R} \setminus \{(-1,1)\}, \tag{4}$$

considering the wavelength denotes the distance over which the shape of the wave repeats, and this distance must be at least equal to the Planck length. No physical models describe smaller lengths, and hence the open set (−1, 1) is forbidden.

Based on Equations (2) and (4), the relation between wavelength l and mass multiplier m can be derived in terms of the Compton wavelength, expressed as

$$\lambda_M \doteq l\ell_P = \frac{h}{c m m_P} \quad \Leftrightarrow \quad lm = 2\pi. \tag{5}$$

On constraining to $l \geq 1$, we get

$$m \leq 2\pi, \tag{6}$$

which can be called the *threshold of distinguishability*, considering it indicates that the Compton wavelengths of masses $M \geq 2\pi m_P$ (1.37E-7 kg) are smaller than the Planck length, which is physically impossible. Such masses no longer exhibit the wave–particle duality and would not interfere with each other in the double-slit experiment. In principle, they are neither bosonic nor fermionic (nor anyonic) and are distinguishable in all instances. Evidently, this does not preclude the distinguishability of lighter masses.

The possible meaning of negative wavelengths and diameters is discussed in Sections 4, 8, 10, and in the Appendices of this paper. The complex wavelength multipliers l expressed in Equation (4) are briefly discussed in Sections 4 and 10. Overall, detailed investigations of these notions are beyond the scope of this study. We note in passing that complex geodesic paths emerge in the presence of BH singularities [17] and when studying entropic dynamics on curved statistical manifolds [18].

3. Discrete Black Holes

On expressing the mass and diameter of the BH w.l.o.g. using Equations (2) and (3), respectively, the Schwarzschild diameter can be expressed as

$$D_{BH} = \frac{4Gm_{BH}}{c^2}\sqrt{\frac{\hbar c}{G}} = 4m_{BH}\ell_P, \tag{7}$$

yielding a simple relationship between the BH mass and the diameter or radius multipliers, given as

$$m_{BH} = \frac{d_{BH}}{4} = \frac{r_{BH}}{2}. \tag{8}$$

Therefore, the BH Compton wavelength (5) can be expressed as

$$l_{BH} = \frac{2\pi}{m_{BH}} = \frac{8\pi}{d_{BH}} \quad \left(\lambda\!\!\!^{-}_{BH} = \frac{2\ell_P^2}{R_{BH}} = \frac{4\ell_P^2}{D_{BH}}\right), \tag{9}$$

and constraining to $l \geq 1$, the BH *threshold of distinguishability* (6) can be rewritten in terms of a BH diameter or radius multipliers as

$$d_{BH} \leq 8\pi, \quad \text{or} \quad r_{BH} \leq 4\pi \tag{10}$$

Therefore, only the BHs with diameter (radius) multipliers below this threshold are adequately small to have Compton wavelengths (5) greater than the Planck length. As we shall see in Section 4, BH threshold of distinguishability (10) is equal to a threshold of BH collapsibility (22).

Equation (3) was used to determine the information capacity of an event horizon as $N_{BH} = \pi D_{BH}^2/\ell_P^2 = \pi d_{BH}^2$. Moreover, the BH surface gravity (82) was combined into the Hawking radiation formula to derive the Hawking radiation temperature T_{BH} as a function of N_{BH} or diameter multiplier d_{BH}

$$T_{BH} = \frac{1}{2\pi d_{BH}} T_P, \tag{11}$$

as shown in Figure 1, where T_P denotes the Planck temperature (cf. Appendix 1).

Given that no physical models define temperatures greater than the Planck temperature, the highest physically meaningful Big Bang temperature corresponds to a BH with $d_{BH} = 1/(2\pi)$, $N_{BH} = 1/(4\pi)$, and $\lfloor N_{BH} \rfloor = 0$. Furthermore, the natural diameter multiplier d_{BH} can be related to the 2π-cycle: BH temperature (11) decreases by a factor of 2π with every integer increment of d. At $d_{BH} = 1$, BH is at a temperature $T_{BH} = T_P/2\pi$, which is a reduced Planck temperature, similar to the reduced Planck constant or reduced wavelength. The BH Hagedorn temperature of approximately 1.7E12 K — *the boiling point of ordinary matter* — is achieved by using the BH with $d_{BH} \approx$ 1.33E19, equivalent to D_{BH} = 2.14E-16 m, which is approximately ¼ of a proton charge radius (8.41E-16 m).

Table 1. Examples of discrete BHs

$d_{BH} = D_{BH}/\ell_P$	$N_{BH} = \pi d_{BH}^2$	$\lfloor N_{BH} \rfloor$	M_{BH}/m_P	T_{BH}/T_P	BH type/comments
$1/(2\pi)$	0.0796	0	0.0398	1	The Big Bang, the Planck temperature
$1/\pi$	0.3183	0	0.0796	0.5	Half of the Planck temperature
$1/\sqrt{\pi}$	1	1	0.1410	0.2821	min 1-bit, MB stat. 1st singularity
$\sqrt{(\ln(4)/\pi)}$	2ln(2)	1	0.1661	0.2396	Landauer 1st bound ($M_{BH}c^2 = T_{BH}k_B\ln(2)$)
$\sqrt{(2/\pi)}$	2	2	0.1995	0.1995	min 2-bit, FD stat. 1st singularity
$2\sqrt{(\ln(2)/\pi)}$	4ln(2)	2	0.2349	0.1694	Landauer 2nd bound (BH entropy)
1	π	3	0.25	0.1592	π-bit, surface gravity equals Planck acceleration
$\sqrt{(4/\pi)}$	4	4	0.2821	0.1410	min 4-bit, one unit of entropy
$\sqrt{2}$	2π	6	0.3536	0.1125	imaginary Planck time at the BH horizon
2	4π	12	0.5	0.0796	12-bit, HUP yields $R_{BH} = \ell_P$
4	16π	50	1	0.0398	$M_{BH} = m_P$
2π	$4\pi^3$	124	1.5708	0.0253	Margolus–Levitin theorem threshold
$4\sqrt{\pi}$	$16\pi^2$	157	1.7725	0.0224	Threshold of informationless stability
8π	$64\pi^3$	1984	2π	$1/(16\pi^2)$	$M_{BH} = 2\pi m_P$, threshold of distinguishability
1.58E45	7.88E90	7.88E90	3.9606E44	1.0071E-46	Sagittarius A*

Table 1 summarizes certain examples of discrete BHs and displays the information capacity, number of bits, mass, and temperature of each hole as a function of its diameter multiplier d_{BH}.

Any vertex v on the surface of an n-ball is related to a certain vertex p, which is the center of this n-ball by the same parameter called radius R. Formally, for a set of vertices V with a generalized distance function $\mu \geq 0$ (cf. Appendix 5), the surface of the n-ball is defined as

$$B(p) = \{v \in V \mid \mu(v, p) = R\}. \tag{12}$$

However, this definition does not hold for BHs in any dimension n. As BHs do not have interiors, no vertex p of the graph of nature can be the BH center. Therefore, to define a BH as a centerless $(n-1)$-sphere, the pairs of vertices on its surface must be related to each other through the same diameter D_{BH}. Therefore, the event horizon of a BH can be defined as

$$B_{BH}(p \in \varnothing) = \left\{ \begin{array}{l} \forall v_i \in V \ \exists! v_j \in V \cap \forall v_k \in V \ \exists! v_l \in V \mid \\ \mu(v_i, v_j) = \mu(v_k, v_l) = D_{BH} \cap \\ \mu(v_i, v_k) = \mu(v_j, v_l) \end{array} \right\}. \tag{13}$$

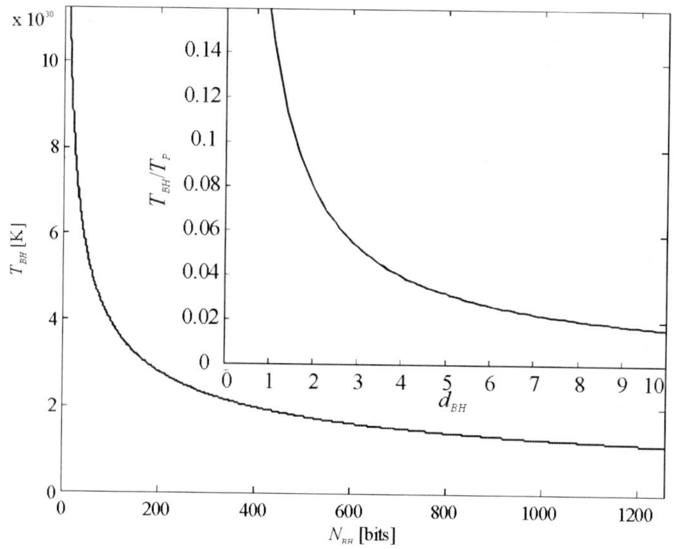

Figure 1. Black hole temperature T_{BH} as a function of its information capacity N_{BH} and diameter multiplier d_{BH}.

Each vertex of the event horizon is associated with precisely one vertex based on the relationship of the diameter D_{BH}, and each pair of vertices non-associated with the relationship of the diameter D_{BH} is associated with each other based on the distance relationship equal to that of their D_{BH} partners. This definition requires at least four vertices defining two diameters, as shown in Figure 2(b) (the lines in the figure do not indicate line segments passing across a BH interior). Therefore, the term Schwarzschild *radius* is a misnomer.

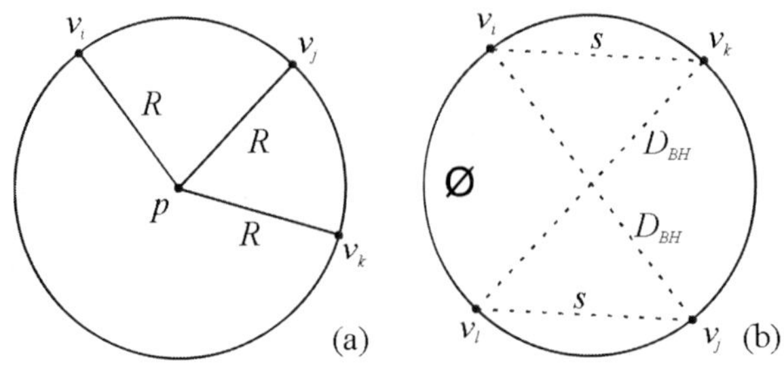

Figure 2. Surface of (a) n-ball, (b) BH.

However, the definition (13) is inappropriate for small BHs. Two micro-BHs in $\mathbb{R}^3$ with triangulated event horizons are shown in Figure 3, wherein π-bit BH ($R_{BH} = \ell_P/2$) comprises four vertices and 4π-bit BH ($R_{BH} = \ell_P$) comprises eleven vertices with the fractional part (52) containing six spherical triangles around the equator. As shown, only the latter event horizon defines only one, precise diameter relation between its two poles 1 and 11. Notably, neither a 1-bit nor a 2-bit BH allows the triangulation.

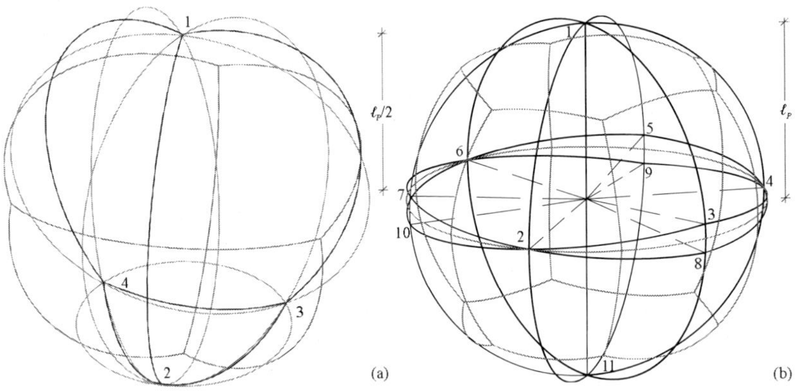

Figure 3. Axonometric views of micro-BHs with Voronoi tessellation (pink). (a) Unipolar π-bit BH and. (b) Bipolar 4π-bit BH.

The Euclidean space $\mathbb{R}^n$, researched as a simplicial n-manifold (i.e., triangulated for $n = 2$), inherits a natural topology from $\mathbb{R}^n$. Moreover, such an approach disentangles the topological (metric-independent) and geometric (metric-dependent) content of the modelled quantities [19, 20]. Therefore, Planck areas on holographic spheres and BHs horizons must be spherical triangles. This is compatible with the Causal Dynamical Triangulations approach, which does not assume a pre-existence of a dimensional space, but focuses on the evolution of the spacetime as such.

BHs are by no means stable entities, as discussed in the subsequent section.

4. Black Hole Diameter Fluctuations

BH energy increases if a photon of energy $E = hc/\lambda = Mc^2$ is incident on a BH event horizon. If we intend to lose all photon information, it should have a sufficiently long wavelength to ensure that the point of collision with the horizon is uncertain [21, 22 (p. 160)]. Therefore, its wavelength must be greater than or equal to the BH radius (26). This thought experiment was famously devised by Jacob Bekenstein to derive a threshold for BH entropy (44). If we assume the condition of equality ($\lambda = R_{BH}$), the Compton mass of the photon is $M = h/(cR_{BH})$, and hence, the new BH radius would be

$$R_{BH}^{A} = R_{BH} + \delta R = \frac{2GM_{BH}}{c^2} + \frac{2G}{c^2}\frac{h}{cR_{BH}}, \tag{14}$$

the new BH area would be

$$A_{BH}^{A} = A_{BH} + \delta A = 4\pi\left(R_{BH} + \delta R\right)^2, \tag{15}$$

and the new BH information capacity would be

$$N_{BH}^{A} = 64\pi^3 \frac{\ell_p^2}{R_{BH}^2} + 32\pi^2 + 4\pi\frac{R_{BH}^2}{\ell_p^2}. \tag{16}$$

Equation (16) attains the minimum for $R_{BH} = 2\ell_P\sqrt{\pi}$, as shown in Figure 4. The BH with this radius (and mass equal to $m_P\sqrt{\pi}$) will increase its information capacity fourfold from $N_{BH} = (4\pi)^2$ to $N_{BH}^A = (8\pi)^2$ after absorbing the photon with mass $M = m_P\sqrt{\pi}$. Furthermore, transmission of another photon with a wavelength corresponding to the new BH radius will increase its information capacity. For radii distinct from $2\ell_P\sqrt{\pi}$, the same BH information capacity can be obtained for two distinct initial radii, and consequently, for two different photon energies.

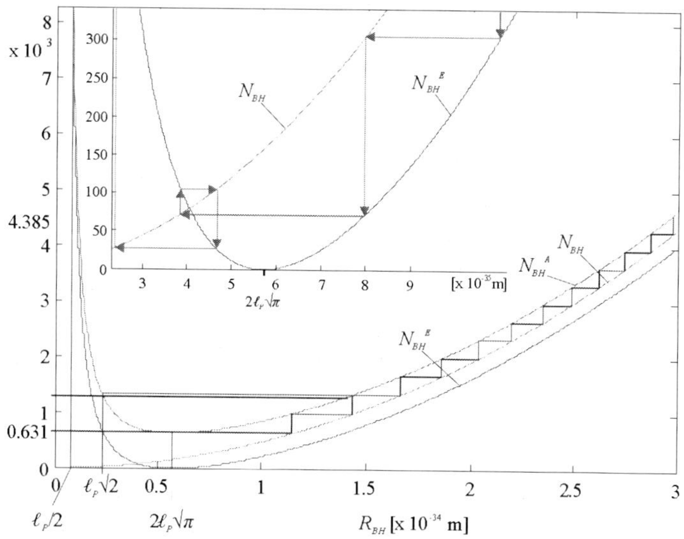

Figure 4. Black hole information capacity: N_{BH}^A (red) - after informationless absorption, N_{BH}^E (blue) - after informationless emission, N_{BH} (green) - initial information capacity.

The informationless BH absorption, depicted by Equation (16), can be reversed to produce the informationless BH emission: we do not intend to know the point of emission on the event horizon. Therefore, by subtracting δR in Equations (14)–(16) we get

$$N_{BH}^{E} = 64\pi^3 \frac{\ell_P^2}{R_{BH}^2} - 32\pi^2 + 4\pi \frac{R_{BH}^2}{\ell_p^2}, \tag{17}$$

which defines the BH information capacity after emitting a photon with wavelength corresponding to its radius. The emission was explained by Hawking [23] and contributes to collapsibility of micro-BHs.

The emission N_{BH}^{E} (17) fluctuates around $R_{BH} = 2\ell_P\sqrt{\pi}$, shown in the inset of Figure 4, and increases for $R_{BH} < 2\ell_P\sqrt{\pi}$, despite the emission of the wavelength. These fluctuations should be considered along with the absorption (16). Upon considering both these processes together, the BH mass along with its diameter fluctuates with time.

The informationless absorption (16) and emission (17) can be extended (cf. Appendix 3) to arbitrary wavelengths l (in general, shedding the informationless property) as

$$N_{BH}^{A/E}(d,l) = 64\pi^3 \frac{1}{l^2} \pm 16\pi^2 \frac{d}{l} + \pi d^2, \tag{18}$$

which can be described more compactly ($N_{BH} = \pi d_{BH}^2$) as a recurrence relation

$$d_{k+1}^{A/E} = \sqrt{64\pi^2 \frac{1}{l^2} \pm 16\pi \frac{d_k}{l} + d_k^2}, \tag{19}$$

where d_{k+1} is a new diameter of the BH (2-sphere) with an initial diameter d_k after absorption (A, "+") or emission (E, "–") of a mass with the Compton wavelength l, as shown in Figure 5.

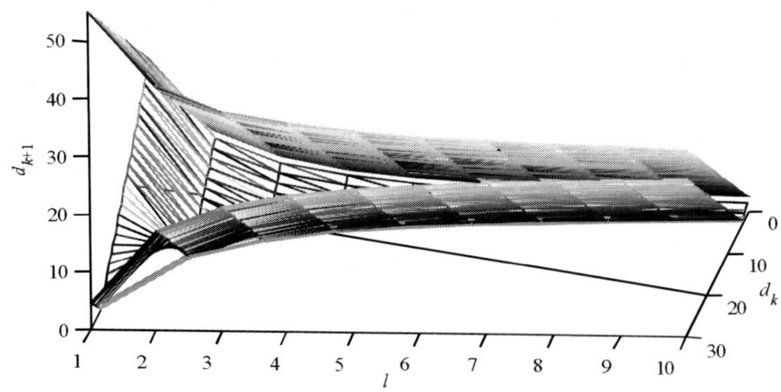

Figure 5. New 2-sphere BH diameter d_{k+1} having initial diameter d_k after absorption or emission of a wavelength l. The sinusoidal characteristic of collapse is due to l drawn as a natural number. For the informationless case, shown in Figure 4, $d = 2l$.

This derivation can be extended to other dimensions *n,* thereby yielding the following recurrence relation for a BH, $(n-1)$-sphere, diameter after absorption (A, "+") or emission (E, "–") of a wavelength l

$$\left(d_{k+1}^{A/E}\right)^{n-1} = \left(d_k \pm \frac{8\pi}{l}\right)^{n-1}. \tag{20}$$

The RHS of Equation (20) shows an initial state, whereas the LHS shows the final state of a BH after absorption/emission of a wavelength l.

Following the limits of one-sided thermodynamic equilibrium, the BHs can certainly be studied in $(n + 0i)$-dimensions with $n \neq 3$, which renders them as markers of dimensionality. Table 2 summarizes the information capacities as the functions of the diameter multiplier d for BHs in a range of integer dimensions [24]. We shall return to it in Section 7. Particularly [24] (cf. also Appendix 15),

$$N_{BH}(n)N_{BH}(2-n) = 4\sin(\pi n/2). \tag{21}$$

There are at least three distinct thresholds of a BH collapse after the emission of wavelength l, which can be postulated in Equation (20):

1. $d_{k+1}^E = 0$ total collapse;
2. $d_{k+1}^E = 1/(2\pi)$ Planck temperature collapse; and
3. $d_{k+1}^E = 1/\sqrt{\pi}$ one-bit collapse.

Table 2. Black hole information capacity N_{BH} in integer dimensions $n = -7, -5, \ldots, 9$

n	$N_{BH}(n)$	n	$N_{BH}(n)$
−7	$3360/(\pi^4 d^8)$	3	πd^2
−5	$-240/(\pi^3 d^6)$	4	$\pi^2 d^3/4$
−3	$24/(\pi^2 d^4)$	5	$\pi^2 d^4/6$
−1	$-4/(\pi d^2)$	6	$\pi^3 d^5/32$
0	0	7	$\pi^3 d^6/60$
1	2	8	$\pi^4 d^7/384$
2	πd	9	$\pi^4 d^8/840$

The total BH collapse occurs after the emission of wavelength $l = 8\pi/d_k$, i.e., after the emission of its entire Compton mass (9). For Sagittarius A*, $l \leq 1.59\text{e-}44 << 1$, so Sagittarius A* will never collapse.

If we set $l \geq 1$ in (20), we obtain

$$d_k \leq 8\pi \approx 25.132, \tag{22}$$

for $d_{k+1}{}^E = 0$,

$$d_k \leq 8\pi + \frac{1}{2\pi} \approx 25.292, \tag{23}$$

for $d_{k+1}{}^E = 1/2\pi$, and

$$d_k \leq 8\pi + \frac{1}{\sqrt{\pi}} \approx 25.697, \tag{24}$$

for $d_{k+1}{}^E = 1/\sqrt{\pi}$, which are the thresholds for the maximum diameter multipliers of collapsible BHs. The lowest threshold $d_k = 8\pi$ corresponds to a BH with mass 2π times greater than the Planck mass. This threshold has been already derived as Equation (10) by considering the BH Compton wavelength and shall be further delivered in Equation (89) by considering the variational potential (78) on the event horizon (88).

The BH diameter does not vary ($d_{k+1}{}^E = d_k$) after the emission of wavelengths listed in Table 3. Equation (20) is symmetrical w.r.t. $n = 1$, i.e., the wavelengths solving $d_{k+1}{}^E = d_k$ are the same for n and $2 - n$, which is a reflection relation around 2. According to prior research [25], an ordinary space of $(m - 2)$-dimension is equivalent to the m-dimensional superspace with an anticommuting coordinate, and a 4 space with only one anticommuting coordinate (time) is equivalent to an ordinary space with negative dimensions of -2.

Table 3. Emission wavelength l maintaining constant BH diameter in integer dimensions $n = -7, -6, \ldots, 9$

n	Emitted l maintaining constant BH diameter
−7, 9	$l_1 = \frac{4\pi}{d_k}, l_{2,3,4,5} = \frac{4\pi}{d_k}\left(1 \pm i\sqrt{3 \pm 2\sqrt{2}}\right), l_{6,7} = \frac{4\pi}{d_k}(1 \pm i)$
−6, 8	complicated and complex solution
−5, 7	$l_1 = \frac{4\pi}{d_k}, l_{2,3} = \frac{4\pi}{d_k}\left(1 \pm \sqrt{3}i\right), l_{4,5} = \frac{4\pi}{d_k}\left(1 \pm \frac{\sqrt{3}}{3}i\right)$
−4, 6	$l_{1,2,3,4} = \frac{4\pi}{d_k}\left(1 \pm \frac{\sqrt{5 \pm 2\sqrt{5}}}{\sqrt{5}}i\right)$
−3, 5	$l_1 = \frac{4\pi}{d_k}, l_{2,3} = \frac{4\pi}{d_k}(1 \pm i)$
−2, 4	$l_{1,2} = \frac{4\pi}{d_k}\left(1 \pm \frac{\sqrt{3}}{3}i\right)$
−1, 3	$l = 4\pi/d_k$
0, 2	contradiction ($1 \neq 0$)
1	identity ($1 = 1$)

The analytical formula $l(m)$ for constant diameter emission in any dimension, if one exists, remains to be researched. The odd dimensionalities admit the real wavelength $l = 4\pi/d_k$ corresponding to precisely half the BH Compton wavelength (9). In even dimensionalities, all constant diameter wavelengths l are complex, thereby indicating the existence of a link with vanishing real volumes and surfaces of m-balls, in even, negative dimensions (owing to the vanishing rational factor g_m, not because of the vanishing diameter) [24].

5. Entropic Work of Gravity

Entropic gravity defines that both inertia and gravity are phenomena emergent on a holographic screen, whereas the classical concepts such as position, velocity, acceleration, mass, and force[2] are far from obvious [26]. Quoting Erik Verlinde, "when a particle has an entropic reason [δS] to be on one side [displaced by δR] of the membrane [holographic screen] and the membrane carries a temperature [T], it [the particle] will experience an effective force [F] equal to [$\delta W =$] $F\delta R = T\delta S$" [26, Equation 3.3].

Based on this entropic work formula, the Second law of Newton [26, Equation 3.5] can be recovered by combining the Unruh temperature T [26, Equation 3.4] with the postulated variation of entropy δS near the screen [26, Equation 3.2], linearized with respect to the reduced Compton wavelength $\lambda\!\!\!^{-} = \hbar/Mc$ by introducing inertial mass M to this equation. Newton's law of gravity [26, Equation 3.9] can be recovered by assuming a sphere partitioned into Planck areas ℓ_P^2. Instead of the Unruh effect, the temperature is derived from the equipartition theorem [26, Equation 3.7] for a degree of freedom, multiplied by the number of these areas. Particularly, the mass–energy equivalence $E = M_2c^2$ [26, Equation 3.8] introduces the 2nd gravitational mass M_2 to the equation. Entropic, emergent gravity [26] has been experimentally confirmed to date [27].

The same entropic work equation can be derived [14] using an entropy formula [14, Equation 1] that reduces to the Bekenstein–Hawking entropy at the event horizon. As such, the temperature [14, Equation 2] is obtained using the Hawking formula and expressed as the gradient of the gravitational potential of the 2nd gravitational mass M_2, which further reduces to the BH temperature at the event horizon. This derivation is certainly compelling to those agreeing with Verlinde's perspective concerning the entropic origin of gravity and inertia and provides new insights into this matter. Let us delve more closely into this derivation.

As shown in Figure 6, we considered a model of a closed 3-dimensional spherical, universe $\mathbb{R}^3$ with an edge of $\partial\mathbb{R}^3$. The concept of an edge $\partial\mathbb{R}^3$ is certainly a strong abuse of notation and does not indicate that the Euclidean space $\mathbb{R}^3$ is *closed*. However, this concept can be introduced in Equation (36), for instance, to prove that the surface integral vanishes if the edge $\partial\mathbb{R}^3$ is moved to infinity. Regardless, we cannot study Euclidean space $\mathbb{R}^3$ without a universe from which to study it. Based on our current knowledge, the radius of the observable universe is approximately 46 billion light years.

At its center, our toy universe contains a Schwarzschild BH with (passive) mass M_{BH}. Its presence is much more formally unsettling than that of the edge $\partial\mathbb{R}^3$, considering it does not contain an interior.

---

2 Time is not included in this enumeration [26, p. 2].

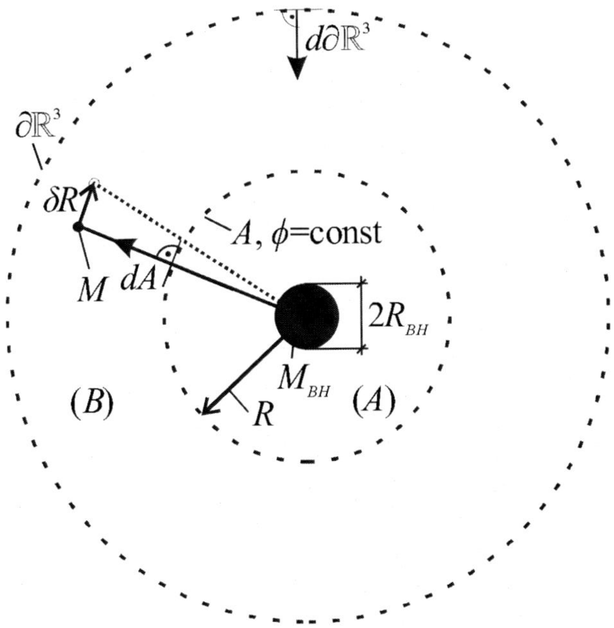

Figure 6. Simplified model of the universe used in this study.

Passive mass M_{BH} generates a conservative and irrotational gravitational field that can be expressed in terms of a scalar, gravitational potential ϕ_g as

$$\phi_g = -\frac{GM_{BH}}{R}, \tag{25}$$

for $R \geq R_{BH}$, where

$$R_{BH} = \frac{2GM_{BH}}{c^2}, \tag{26}$$

represents the radius of the BH event horizon.

The minimum potential ϕ_g occurs at the event horizon and amounts to

$$\phi_{BH} = -\frac{GM_{BH}}{R_{BH}} = -GM_{BH}\frac{c^2}{2GM_{BH}} = -\frac{1}{2}c^2. \tag{27}$$

An equipotential surface A surrounds the M_{BH} in a volume (A) for $R \geq R_{BH}$. Thus, the single-mass M_{BH} system is described by Laplace's equation:

$$\nabla^2\phi_g = 0. \tag{28}$$

Let us now consider a test mass M located in the background of M_{BH} in volume (B), such that $(A) + (B) = \mathbb{R}^3$ and $B = A \cup \partial\mathbb{R}^3$. Owing to this simplified model of a spherical universe including two masses, we do not encounter the three-body problem, and yet this model can adequately demonstrate all the current findings of this study. Moreover, owing to its spherical symmetry, R is the only significant dimension. The system of these two masses M_{BH} and M can be described using Poisson's equation for gravity, given as

$$\nabla^2 \phi_g = 4\pi G\rho, \tag{29}$$

where ρ is the mass density in the volume (B) comprising M. Poisson's Equation (29) is the inhomogeneous version of Laplace's Equation (28) and includes the latter, if $\rho = 0$. It is satisfied on any equipotential surface A and integrating over (A) using the divergence theorem yields

$$4\pi G \iiint_{(A)} \rho dV_A = \iiint_{(A)} \nabla^2 \phi_g dV_A = \oiint_A \nabla_R \phi_g dA, \tag{30}$$

where ∇_R is the potential gradient (25), which is the derivative in the direction of the normal vector on the surface A pointing outside the volume (A).

Newton's law of gravity (48) is a general physical law derived from empirical observations. In this case, the passive mass M_{BH} is located at the center of the simplified model of the spherical universe, which implies that the spherically symmetric gravitational potential $\phi_g(R)$ (25) and mass density $\rho(R)$ satisfy Laplace's equation (28). The 2nd mass M and all the subsequent test masses introduced to this model would induce mass density inhomogeneity and attain the same equilibrium in a *free fall* process induced by $\nabla_R\phi_g$, unless a work is done to disturb it.

Such a disturbing force F — applied to wiggle the test mass M location by a certain δR in a time period δt (doing work δW) — equals to the corresponding variation of the potential energy in volume (B), expressed as

$$\delta W = F \circ \delta R = \delta U = -\iiint_{(B)} \phi_g \delta\rho dV_B = -\frac{1}{4\pi G} \iiint_{(B)} \phi_g \nabla^2 \delta\varphi dV_B, \tag{31}$$

where Equation (29) is used to express the variation in density $\delta\rho$ in the volume (B) containing, the now active, mass M with the corresponding variation $\delta\varphi$ of the potential. Let us refer to δR as the disturbing radius of the active mass M.

Accordingly, integrating Equation (31) over $(B) = \mathbb{R}^3 - (A)$, we get

$$-4\pi G \delta U = \iiint_{\mathbb{R}^3} \phi_g \nabla^2 \delta\varphi dV - \iiint_{(A)} \phi_g \nabla^2 \delta\varphi dV_A, \tag{32}$$

and introducing the 2nd and 3rd terms that negate each other, we get

$$-4\pi G\delta U = \iiint\limits_{\mathbb{R}^3} \phi_g \nabla^2 \delta\varphi dV - \iiint\limits_{(A)} \delta\varphi \nabla^2 \phi_g dV_A + \iiint\limits_{(A)} \delta\varphi \nabla^2 \phi_g dV_A - \iiint\limits_{(A)} \phi_g \nabla^2 \delta\varphi dV_A \,. \tag{33}$$

Thereafter, Green's second identity was used to transform the 3rd and 4th terms into surface integrals, as

$$-4\pi G\delta U = \iiint\limits_{\mathbb{R}^3} \phi_g \nabla^2 \delta\varphi dV - \iiint\limits_{(A)} \delta\varphi \nabla^2 \phi_g dV_A + \oiint\limits_{A} \left(\delta\varphi \nabla_R \phi_g - \phi_g \nabla_R \delta\varphi\right) dA \,. \tag{34}$$

Moreover, the 4th term was integrated over the equipotential surface A to displace the potential ϕ_g before the integral, and the divergence theorem (30) was applied to retrace to the volume integration over (A). However, the 4th term vanished upon employing Poisson's equation (29), considering $\delta\varphi$ contained no sources inside (A). Subsequently, the 2nd term with the volume integration over (A) was further re-expressed as an integration over $(A) = \mathbb{R}^3 - (B)$

$$-4\pi G\delta U = \iiint\limits_{\mathbb{R}^3} \phi_g \nabla^2 \delta\varphi dV - \iiint\limits_{\mathbb{R}^3} \delta\varphi \nabla^2 \phi_g dV + \iiint\limits_{(B)} \delta\varphi \nabla^2 \phi_g dV_B + \oiint\limits_{A} \delta\varphi \nabla_R \phi_g dA \,. \tag{35}$$

Similarly, the 3rd term vanished considering ϕ_g included no sources inside (B). Therefore, using Green's second identity to transform the volume integral over $\mathbb{R}^3$ into the surface integral over $\partial\mathbb{R}^3$, we get

$$-4\pi G\delta U = \oiint\limits_{\partial\mathbb{R}^3} \left(\phi_g \nabla_R \delta\varphi - \delta\varphi \nabla_R \phi_g\right) d\partial\mathbb{R}^3 + \oiint\limits_{A} \delta\varphi \nabla_R \phi_g dA \,, \tag{36}$$

where the 1st surface integral over $\partial\mathbb{R}^3$ vanished as well. Therefore,

$$F \circ \delta R = \delta U = -\frac{1}{4\pi G} \oiint\limits_{A} \nabla_R \phi_g \delta\varphi dA \,. \tag{37}$$

This innovative procedure conveniently transformed the volume integration Equation (31) over (B) defining the universe $\mathbb{R}^3$ with a hole (A) into a surface integration Equation (37) over the equipotential surface A describing only this hole, which expresses the BH M_{BH} event horizon for a limiting case of $R = R_{BH}$.

Both the uniform inertial, that is flat space–time acceleration (a) and the gravitational acceleration (g) can be related to temperature T according to Unruh blackbody-radiation equation

$$T_a = \frac{\hbar}{2\pi c k_B} a \,, \tag{38}$$

or the analogous Hawking blackbody-radiation equation

$$T_g = \frac{\hbar}{2\pi c k_B} g \,. \tag{39}$$

Thus, the temperature can be expressed in terms of the gravitational potential gradient (25) as

$$T_g = \frac{\hbar}{2\pi c k_B} \nabla_R \phi_g \,. \tag{40}$$

To eventually derive Verlinde's entropic gravity, the variation of entropy on A [14, Equation (1)[3]] should be expressed using another function

$$\delta S = -\frac{c k_B}{2G\hbar} \delta\varphi A \,, \tag{41}$$

which expresses the entropy variation as a function of the variation of the potential $\delta\varphi$ induced upon relocating the active mass M in volume (B) by δR, which corresponds to the variation in density $\delta\rho$ in the volume (B), according to Equation (31)

$$\delta dS = -\frac{c k_B}{2G\hbar} \delta\varphi dA \,, \tag{42}$$

on the surface element dA of the surface A.

Thereafter, $\nabla_R \phi_g$ from Equation (40) and $\delta\varphi$ from Equation (42) were substituted into Equation (37), which demonstrates that this is equal to the entropic work

$$F \circ \delta R = -\frac{1}{4\pi G} \oiint_A \left(\frac{2\pi c k_B}{\hbar} T_g \right) \left(-\frac{2G\hbar}{c k_B} \delta dS \right) = \oiint_A T_g \delta dS = T_g \oiint_A \delta dS = T_g \cdot \delta S \,, \tag{43}$$

where T_g can be displaced before the integral, due to integrating over an equipotential surface A. Therefore, work δW and force F can be deemed entropic. The author conjectures that all forces are entropic. Thus, work is both the scalar product of force and variation of location and the regular product of temperature and variation of entropy. However, this does not preclude additional variations of force and temperature.

As discussed in Section 7, the concept of binary potential Equation (54) transforms the entropy variation Equation (41) on the equipotential surface A into a distribution of bits Equation (55) on the holographic screen A, thereby rendering the latter as more general than Equation (41). The author of [14] reported that the "number N of 'bits' on the screen made no appearance here" (i.e., in the derivation of Equation (43)), and apparently disregarded the vitality of a "bit" in relation to the BH information paradox. The entropy formula (41) certainly does not relate to electrodynamics [14], as it reduces to BH entropy [21] (using (27)) as

---

3 The additive constant in this equation can represent the fractional part $\{N_A\}$ of an equipotential surface information capacity.

$$S_{BH} = -\frac{ck_B}{2G\hbar}\phi_{BH}A_{BH} = \frac{k_B c}{2G\hbar}\frac{c^2}{2}A_{BH} = \frac{1}{4}k_B N_{BH}, \tag{44}$$

which is a thermodynamical term. Notably, the entropy of any BH is equal to that of the Schwarzschild hole with the same area (by Penrose process).

Equation (44) can be extended to Equation (41) beyond the event horizon for an arbitrary equipotential A and $\delta\varphi$. However, for radii smaller than the Schwarzschild radius (26), it is meaningless and hence disproves assumption **B** of [14] that $\mathbb{R}^3$ is made up by the union of nonintersecting holographic screens. Screens *passing across* BHs cannot be defined, and therefore, they create voids in such a *holographized* $\mathbb{R}^3$. Thus, if we relied only on the intuitive concept of *space*, we could use neither the divergence theorem nor Green's second identity in the derivation Equation (43). However, they are valid, and we have used them correctly. Only our intuition was misleading.

Neither Verlinde's (46) nor Hossenfelder's (41) entropy equations should be considered as an entropic explanation of any physically observed inverse-square law. Both these equations are related, which will be discussed in the subsequent section.

6. Entropy Variation Sphere and Variational Potential

We compare Hossenfelder's entropy variation equation (41) with Verlinde's formula for the variation in entropy in the vicinity of a holographic screen, expressed as [26, Equation 3.1]

$$\delta S = 2\pi k_B \quad \text{when} \quad \delta R = \frac{\hbar}{Mc}, \tag{45}$$

or [26, Equation 3.2]

$$\delta S = 2\pi k_B \frac{Mc}{\hbar}\delta R = 2\pi k_B \frac{\delta R}{\lambda\!\!\!^{-}_M}, \tag{46}$$

in the case of a test mass M approaching the screen. The factor of 2π included in δS is apparently related to the *threshold of distinguishability* (6).

Equation (46) enables the recovery of the second law of Newton, considering a from the Unruh temperature (38)

$$F \circ \delta R = T_a \cdot \delta S = \frac{\hbar}{2\pi ck_B} a \cdot 2\pi k_B \frac{Mc}{\hbar}\delta R = Ma \cdot \delta R \Leftrightarrow F = Ma = M\frac{\delta R}{\delta t^2}, \tag{47}$$

and Newton's law of gravity, considering Hawking temperature (40) in terms of the gravitational potential (25) of the passive mass M_{BH}

$$F \circ \delta R = T_g \cdot \delta S = \frac{\hbar}{2\pi c k_B}\left(\frac{GM_{BH}}{R^2}\right) \cdot 2\pi k_B \frac{Mc}{\hbar}\delta R = \frac{GM_{BH}M}{R^2} \cdot \delta R \Leftrightarrow F = \frac{GM_{BH}M}{R^2}. \tag{48}$$

Alternatively, the original Verlinde's procedure can be used with the Hawking temperature (40) induced by the equipartition theorem for a single degree of freedom and mass–energy equivalence, as described at the beginning of this Section. Furthermore, Verlinde's Equation (46) directly introduces the Unruh temperature T_a (38) as

$$Ma \circ \delta R = T_a \cdot \delta S = T_a \cdot 2\pi k_B \frac{Mc}{\hbar}\delta R \Leftrightarrow a = \frac{2\pi c k_B}{\hbar} T_a. \tag{49}$$

However, none of these laws can be recovered from Equation (41), whose starting point is the gravitational potential (25) analyzed in the context of Poisson's equation for gravity (29). Therefore, the current approach investigated the recovery of these laws.

Let us define the potential variation on the equipotential surface A, induced by wiggling the active mass M location by δR, as gravitational potential (25) (with "–")

$$\delta\varphi \doteq -\frac{GM}{\delta R} = -\frac{G}{\delta R}\frac{\hbar}{c\lambda_M}. \tag{50}$$

This ensures that the equipotential surface A is a sphere of the disturbing radius δR. We shall further refer to this potential as the "variational potential" and the corresponding sphere as the "entropy variation sphere." By substituting Equation (50) into Equation (41), we recovered Verlinde's entropic equation (46) as

$$\delta S = -\frac{c k_B}{2G\hbar}\left(-\frac{GM}{\delta R}\right) 4\pi\delta R^2 = 2\pi k_B \frac{Mc}{\hbar}\delta R, \tag{51}$$

and hence, we recovered Newton's also the second law and law of universal gravitation.

As expressed in Equation (50), the hypothesis concerning the variational potential $\delta\varphi$ is a consequence of the entropic work derivation presented in [14]. At the onset, A is assumed to surround BH M_{BH} in $\mathbb{R}^3$, as shown in Figure 6, and is not related to δR. However, after completing the derivation, A solely represents a 2-sphere, a set of points situated at a fixed distance from a BH event horizon; each distance pointing to some virtual central point located within the BH with no interior. Equation (41) describes the potential variation on the equipotential surface, whereas Equation (46) was motivated by Bekenstein's original thought experiment [21], which resulted in the BH entropy formula (44). As such, they describe the equipotential sphere from its opposite sides, which is reflected by the inversion of their signs. The sides described in Equations (41) and (46) represent the side of the passive mass M_{BH} and active mass M, respectively, which can be observed by deriving Bekenstein's entropy (44) based on Verlinde's equation (46) (cf. Appendix 6). They are related (51) through the entropy variation sphere, which introduces the sign reversal. In such a radial configuration, a quantity is positive if it is consistent with the radius; else, it is negative. In particular, M_{BH}, M, R, and δR

are parameters of this holographic 2-sphere. Moreover, R denotes a time-independent version of δR, as discussed in the subsequent section.

7. Binary Potential and Binary Messages

Entropy is a measure of information (even if it is missing information [28] or a measure of surprise [29]) and information is measured in bits. According to the holographic principle, the bounded interior and unbounded exterior are separated by a two-dimensional boundary, and by definition, each kth bit is physically represented on this boundary (the holographic sphere A) by a unit Planck area [15, 21, 26]. For an area A, the boundary exhibits the information capacity of $N_A = A/\ell_P^2$ bits. However, this is an imprecise conclusion, considering $N_A \in \mathbb{R}$ and the number of bits is a natural number. Moreover, for a spherical surface, $A = 4\pi R^2$ N_A is a transcendental number: a quotient of the product of a transcendental number (π) with a rational number ($4R^2$) and a rational number (ℓ_P^2).

Therefore, the number of bits provided by the boundary is $\lfloor N_A \rfloor \in \mathbb{N}_0$, where "$\lfloor\rfloor$" is the floor function, that yields the greatest natural number less than N_A. Thus, the fractional part

$$0 < \{N_A\} < 1, \tag{52}$$

can be considered as a signature of the holographic sphere A. Perhaps the continuum hypothesis ensures a unique $\{N_A\}$ for any given holographic sphere, regardless of the simultaneous existence of the same number of bits N_A on an infinitely countable number of other spheres.

Thus, the entropy variation (41) can be expressed as the information capacity of the holographic sphere A

$$\delta S = -\frac{ck_B}{2G\hbar}\delta\varphi N_A \ell_P^2 = -\frac{1}{2}k_B N_A \frac{\delta\varphi}{c^2}. \tag{53}$$

The 2nd term of (53) corresponds to ΔS [26, Equation 3.12], where the Newton potential φ keeps track of the depletion of the entropy per bit on the holographic sphere A.

This equation can be further simplified by postulating the binary variational potential (or simply, binary potential) in the Planck time t_P over the k-th Planck area ℓ_P^2, defined as

$$\delta\varphi_k \doteq -\{0,1\}\ell_P^2/t_P^2 = -\{0,1\}c^2. \tag{54}$$

This relationship between the speed of light and a Planck area is remarkable in the context of the recent discovery that the Planck length and the Planck time can be obtained from a Newtonian force-spring knowing only the value of the speed of light and with no knowledge of the gravitational constant G or the Planck constant $\hbar$ [30]. More specifically, the variational potential requires the notion of time, assumed as imaginary, i.e., $(ic)^2 = -c^2$. This unit also stems from the 2-dimensionality of the Planck area. Such a definition of the binary potential must be true if a bit of information is the only property of ℓ_P^2 on a holographic screen A. Therefore, the equipotential surface A and holographic screen A are the same 2-spheres.

By substituting Equation (54) into Equation (53), we obtain a binary variation of entropy on a holographic sphere A, given as

$$\delta S = -\frac{1}{2} k_B N_A \frac{\delta\varphi}{c^2} = -\frac{1}{2} k_B \sum_{k=1}^{\lfloor N_A \rfloor} \frac{\delta\varphi_k}{c^2} = \frac{1}{2} k_B N_1, \tag{55}$$

where the summation covers all the Planck areas of A, and $N_1 \in \mathbb{N}$ is the number of Planck areas with binary potential $\delta\varphi_k = -c^2$, further called "active Planck areas."

We note in passing that only $n = 2$ provides the first possibility of assigning a bit to a circular segment of the Planck length. This would mean the lack of black holes in lower integer dimensions. In such a 2-dimensional case the binary potential is conjectured to be $\delta\varphi_k(2) = \{0,1\}ic$, so that its square yields $\delta\varphi_k(3) = -\{0,1\}c^2$ (54) (on a Planck area spherical triangle). Following this rule $\delta\varphi_k(4) = -\{0,1\}ic^3$ (on a Planck volume spherical tetrahedron), $\delta\varphi_k(5) = \{0,1\}c^4$, $\delta\varphi_k(6) = \{0,1\}ic^5$, etc.; higher dimensional binary potentials are related to each other by integral powers of the imaginary unit.

On comparing Bekenstein–Hawking entropy (44) with the binary variation of entropy (55), we get

$$S_{BH} = \delta S \Leftrightarrow \frac{1}{4} k_B N_{BH} = \frac{1}{2} k_B N_1 \Leftrightarrow N_1 = \frac{1}{2} N_{BH}, \tag{56}$$

which indicates that the temporary distribution of Planck areas on an event horizon can maximize the Shannon entropy. Therefore, the event horizon is an algorithmically random or patternless [31] binary message comprising a balanced number of Planck areas with binary potential (54) equal to $-c^2$ and 0.

An additional trivial observation is that the average potential of all the Planck areas on the event horizon is equal to the event horizon potential at a vertex, i.e.,

$$\frac{\sum_{k=1}^{N_1}\left(-c^2\right) + \sum_{k=1}^{\lfloor N_{BH} \rfloor - N_1} 0}{N_{BH}} = \frac{-N_1 c^2}{N_{BH}} = \frac{-\frac{1}{2} N_{BH} c^2}{N_{BH}} = -\frac{1}{2} c^2. \tag{57}$$

The two binary strings

$$X_1 = 1\,1\,1\,1\,1\,1\,1\,1\,1\,1\,1\,1\,1\,1\,1\,1\,1\,1\,1\,1, \tag{58}$$

$$X_2 = 1\,0\,1\,1\,0\,1\,0\,0\,1\,0\,1\,0\,0\,0\,1\,0\,1\,1\,0\,1, \tag{59}$$

assumed to be generated by tosses of a fair coin (random variable outcomes) must be considered equally random. However, based on the perspective of Kolmogorov complexity, the second string is more random than the first one considering it is a patternless binary sequence [31]. If strings (58) and (59) were messages transmitted bit by bit over a certain transmission channel, the probabilities for zero and one occurrences in the second message (59) would be equal ($p_0 =$

$p_1 = 1/2$). Therefore the second message would be a noncompressible one, which maximizes the Shannon entropy, as

$$H(X_2) = -\frac{1}{2}\ln\left(\frac{1}{2}\right) - \frac{1}{2}\ln\left(\frac{1}{2}\right) = \ln(2), \tag{60}$$

expressed in nats, whereas the Shannon entropy of the message (58) is zero, as it could be compressed to just one bit. The maximal Shannon entropy of the second message (59) would also minimize the amount of energy required to erase a single bit of information on the event horizon, which is expressed using the Landauer erasure limit [32]

$$E_1 = T_{BH} k_B H(X_2) = T_{BH} k_B \ln(2). \tag{61}$$

On comparing the end terms of Equations (53) and (55), the variational potential can be expressed as

$$\delta\varphi = -\frac{N_1}{N_A} c^2, \tag{62}$$

which is an important relation that also (cf. Table 5) shows that the event horizon is a patternless binary message.

Considering the Hawking BH radiation is a diameter-dependent black-body radiation (27) bearing no information regarding BH interior conditions, it is a patternless message containing the same number of zeros and $-c^2$ terms ($-c^2/2$ on average), as a binary one-time pad secret key. Holographic screens other than event horizons provide more diversified (and less surprising) distributions of zeros and ones with lower Shannon entropies (further discussed in Section 9). However, an event horizon is a patternless binary message (59) and the quantum information of the in-fallen matter cannot be hidden in correlations between the semiclassical Hawking radiation and the *internal states* of the BH, as asserted by the no-hiding theorem [1].

The entropic work on a BH event horizon using the Hawking temperature (40) and Bekenstein–Hawking entropy (44) can be expressed as

$$T_g \cdot \delta S = \left(\frac{\hbar}{2\pi c k_B}\frac{GM_{BH}}{R_{BH}^2}\right)\left(\frac{k_B N_{BH}}{4}\right) = \frac{M_{BH}c^2}{2} = \frac{d_{BH}E_P}{8}, \tag{63}$$

which differs from the BH mass–energy equivalence

$$E_{BH} = M_{BH}c^2 = \frac{1}{4} d_{BH} E_P, \tag{64}$$

where E_P is the Planck energy, and results in two distinct Landauer bounds on a minimum BH information capacity (cf. Appendix 7): $N_{BH} = 2\ln(2)$ (mass–energy equivalence, 1 bit) and N_{BH}

= 4ln(2) (BH entropic work, 2 bits). To assign one unit of entropy on a BH horizon at least four bits (Planck areas) are required.

Heisenberg's Uncertainty Principle (HUP) is commonly expressed as

$$\delta P \delta R \geq \hbar/2, \tag{65}$$

where δP is momentum and δR is the position uncertainty. In an alternative form, the above equation can be expressed as

$$\delta E \delta t \geq \hbar/2, \tag{66}$$

where δE is energy and δt is the time uncertainty. The energy uncertainty δE of one bit can be expressed by the temperature uncertainty using the equipartition theorem ($\delta E = k_B \delta T/2$). Substituting this with the time uncertainty discretized by the Planck time ($\delta t = t\, t_P$, $t \in \mathbb{R}$) into Equation (66) we get the temperature uncertainty

$$\delta T \geq \frac{1}{t} T_P, \tag{67}$$

which is equal to the Planck temperature if $t = 1$. Thus, if the time uncertainty δt of measurement of one bit equals the Planck time, then the temperature uncertainty δT for this measurement equals the Planck temperature.

The nontrivial BH microstate degeneracy starts at 3 vertices [33] of the Voronoi triangulation corresponding to the Delaunay triangulated π-bit BH with 4 vertices (Figure 3(a)). This is interesting in the context of the equipartition theorem for an atom in a monatomic ideal gas, that — assuming that the atom exhibits π degrees (3 bits+) of freedom, instead of three — can be defined as

$$E = \frac{\pi}{2} k_B T. \tag{68}$$

This form enables the recovery of the exact (not approximate) Unruh (38) and Hawking (39) temperature equations with the derivations presented in [34] if energy uncertainty δE is interpreted in terms of temperature uncertainty δT. Notably, equipartition theorem is rigorously proven only for a single degree of freedom (one bit), and the energy (68) corresponds to the π-bit BH, as revealed by substituting BH mass–energy equivalence (64) along with (11) into (68).

Based on the Planck–Einstein (or Compton wavelength) relationship

$$\delta E = h \delta \nu = \frac{2\pi \hbar c}{\delta \lambda}, \tag{69}$$

where the uncertainty of energy δE is expressed as the uncertainty of wavelength $\delta \lambda$ the HUP (66) is expressed as

$$\delta\lambda \leq 4\pi c\delta t \quad \left(l \leq 4\pi t\right). \tag{70}$$

Alternatively, the mass–energy equivalence $\delta E = \delta Mc^2$ can be used to replace energy uncertainty with mass uncertainty δM, and set $c = \delta R/\delta t$ [34] to express HUP (66) as

$$\delta M \delta R \geq \hbar/2c. \tag{71}$$

On substituting $\delta M = 2\pi\hbar/(c\delta\lambda)$ (Compton wavelength of δM) into (71), we get

$$\delta\lambda \leq 4\pi\delta R \quad \left(l \leq 2\pi d\right), \tag{72}$$

which also supports the finding that for $\delta\lambda \geq \ell_P$ ($l \geq 1$) (4), $d_{BH} = 1/(2\pi)$ sets a maximum BH Planck temperature T_P (11). Moreover, by substituting $\delta M = \delta Rc^2/2G$ (Schwarzschild mass corresponding to δR) into (71), we can recover the Planck length [34] as the minimum BH radius, expressed as

$$\delta R^2 \geq \ell_P^2, \quad c^2\delta t^2 \geq \ell_P^2, \tag{73}$$

and hence, the Planck time t_P can be considered as the minimum time period δt on a BH horizon for one bit.

In this section, we introduced the concept of the disturbing time period δt and equated it with time uncertainty. Although we implicitly equated the position uncertainty with the disturbing δR (31) or BH radius R_{BH} (26) (as shall be discussed later, they are additive inverses of each other), it is irrelevant in the current context because the uncertainties in HUP are always nonnegative. These aspects are further discussed in Section 9.

Micro-BHs are inherently unstable and prone to collapses [23]. The HUP, BH geometry, and information theory further deliver arguments in support of this claim. In particular, the BH temperature limit (11) is $d_{BH} = 1/(2\pi)$, whereas $d_{BH} = 1/\sqrt{\pi}$ provides one Planck area (one bit). At least four vertices — the fundamental requirement for defining the event horizon (13) — are given for $d_{BH} = 1$. Moreover, the imaginary time period is exhibited by BH with $d_{BH} = \sqrt{2}$ (103). Twelve Planck areas (with only one precise diameter) correspond to the BH with $d_{BH} = 2$, which is also recovered from the HUP (73). Compared to 7.88E90 Planck areas (Sagittarius A*), the maximization of Kolmogorov complexity of a BH binary message (59) comprising 20 Planck areas is much more challenging.

Generally, the radius of a BH event horizon R_{BH} is derived as a singularity of the Schwarzschild metric, which yields an exact solution to the Einstein field equations without the cosmological constant Λ, thereby describing the gravitational field outside a spherical, uncharged, and nonrotating mass. Note that the cosmological constant introduces concepts like dark matter, dark energy, or dark fluid which are redundant in the framework of Verlinde's entropic gravity [26]. More specifically, the Schwarzschild metric has two singularities: at $R = 0$ ($R = 0$ is always singular in polyspherical coordinates) and at $R = R_{BH}$.

However, R_{BH} is also a solution to the escape velocity threshold, as follows

$$v_E^2 = \frac{2GM}{R} \le c^2, \ \frac{GM}{R_{BH}} = \frac{1}{2}c^2. \tag{74}$$

Therefore, by equating the variational potential (50) with the binary potential (54), we can determine the relationship between the mass M and radius δR. This can be conducted for $\delta\varphi_k = -c^2$ (it is undefined for $\delta\varphi_k = 0$) resulting in

$$\delta R_1 = \frac{GM}{c^2}, \tag{75}$$

which represents half the Schwarzschild radius δR_{BH} of mass M and equals the Planck length if M is equal to the Planck mass. Therefore, we recovered the orbital velocity threshold w.r.t. δR_1, as follows

$$v_O^2 = \frac{GM}{\delta R_1} \le c^2, \tag{76}$$

which expresses the second (or rather, the first) significant velocity related to mass M. This is meaningless in the context of the Schwarzschild metric, as it simply does not encompass orbital velocity (although it encompasses the photon-sphere radius $R_P = 3GM/c^2$). Orbital velocity is included in the Kerr metric, which is another exact solution to the Einstein field equations without the cosmological constant and acts as a generalization of the Schwarzschild metric to the gravitational field prevailing outside a spherical and uncharged mass, rotating with an angular momentum J. In contrast, Birkhoff's theorem asserts that a spherically symmetric solution to Einstein's equations in the vacuum must necessarily be static, and the exterior solution must be described by Schwarzschild metric in the absence of the cosmological constant [35].

Interestingly, for constant M, a hypothetical 2-sphere of radius δR_1 (75) comprises the same number of bits as given by Bekenstein-Hawking entropy (44) (using (2))

$$\begin{aligned} N_{BH} &= 4\pi \frac{4G^2M^2}{c^4}\frac{c^3}{\hbar G} = 16\pi m^2 \\ N_{\delta R_1} &= 4\pi \frac{G^2M^2}{c^4}\frac{c^3}{\hbar G} = 4\pi m^2 = \frac{1}{4}N_{BH} \end{aligned} \tag{77}$$

Furthermore, by equating Verlinde's entropic formula (46) (2nd term) with (53) (2nd term), we can express the variational potential (50) on the entropy variation sphere as a function of the information capacity N_A of the sphere and the reduced Compton wavelength of mass M, given as

$$\delta\varphi = -\frac{4\pi c^2}{N_A}\frac{\delta R}{\lambda\!\!\!^{-}_M}. \tag{78}$$

Thus, combining (78) with the variational potential (50) (2nd term), we get

$$N_A = \frac{4\pi\delta R^2}{\ell_P^2}, \quad \text{and since,} \quad N_A = \frac{4\pi R^2}{\ell_P^2}, \tag{79}$$

we conclude that both δR and R provide the same information capacity N_A on the spherical screen A. However, δR is related to the notion of the time period δt (31), whereas R is not. Thus, we conclude that $R^2 = \delta R^2$. Additionally, δR is an additive inverse of R, as discussed in the subsequent section.

8. Inertial Potential

Here, we introduce the concept of inertial potential of the active mass M, such that

$$\phi_a = \frac{GM}{R}, \tag{80}$$

is an additive inverse of this mass gravitational potential, and is considered an equivalent to the latter. Indeed, the gravitational field and inertia are physically equivalent (equivalence principle), wherein the inertia originates through interactions between all masses in the universe (Mach's principle) on which principles the theory of relativity is based [37]. Potential ϕ relates the notions of mass and space, similar to mass density ρ, Schwarzschild radius (26), or the uncertainty principle. Normal modulation of gravitational potential, caused by rotating bodies is wrongly interpreted as a gravitational wave, understood as a carrier of gravity [36].

Based on the definition of binary potential (54), we observed that radius R emerges as irrelevant in the inertial potential (80). By introducing the inertial potential gradient (83) into the Unruh temperature (38) and substituting it along with the binary variation of entropy (55) and the radius of the holographic sphere $R^2 = N_A\ell_P^2/4\pi$ in Equation (43) we obtain the entropic work (using (62) in the last term), given as

$$\begin{aligned} F \circ \delta R = T_a \cdot \delta S = \frac{\hbar}{2\pi c k_B}\left(-\frac{GM}{R^2}\right)\cdot\left(\frac{1}{2}k_B N_1\right) = \\ = -\frac{\hbar G}{4\pi c} M \frac{4\pi}{N_A \ell_P^2} N_1 = -\frac{N_1}{N_A} M c^2 = M\delta\varphi \end{aligned}, \tag{81}$$

whose modulus attains equilibrium (56) at the event horizon. Therefore, mass M represents the proportionality constant, similar to that in Isaac Newton's second law of motion, and the actual radius R of our toy universe, portrayed in Figure 6, is irrelevant to our considerations.

The supermassive BH Sagittarius A* has an estimated mass $M_{BH} \approx$ 8.62E36 kg, yielding diameter $D_{BH} \approx$ 2.56E10 m. The BH surface gravity can be expressed as[4]

[4] $g = 1/d$, where g is Planck acceleration multiplier.

$$g_{BH} = \frac{GM_{BH}}{R_{BH}^2} = \frac{c^2}{D_{BH}}, \tag{82}$$

which is independent of the gravitational constant G.

This is an extremely large value, even for supermassive BHs. For the BHs listed in Table 1, g_{BH} displays an order of magnitude of the Planck acceleration $a_P = \sqrt{(c^7/\hbar G)} = 5.56\text{E}51$ m/s^2, attained by the π-bit BH. For Sagittarius A*, this is still ~3.51E6 m/s^2. A BH with g_{BH} equal to the Earth's surface gravity (9.98 m/s^2) would require a diameter of D_{BH} = 9E15 m, which is slightly less than one light year.

Such large black holes are not observed in nature.

Thus, any inertial acceleration of an object heavier than the threshold of distinguishability (6) can be considered to be lower than a BH surface gravity. The inertial acceleration a denotes the gradient of inertial potential (80), such that

$$a = \nabla_R \phi_a = -\frac{GM}{R^2} < \frac{c^2}{2R_{BH}}, \tag{83}$$

thereby indicating that the inertial potential of such a mass M at $R_{BH} = R$ (80) must be greater than the BH potential of $-c^2/2$

$$\phi_a = \frac{GM}{R} > -\frac{1}{2}c^2. \tag{84}$$

This is an important finding, since it signifies that the inertial potential of such a mass always describes a non-equilibrium situation, where the equilibrium is defined by the BH thermodynamic equilibrium (56) and Laplace's (28) or Poisson's (29) equation.

Similarly, if we presume that the variational potential (50) should be higher than the BH potential

$$\delta\varphi \doteq -\frac{GM}{\delta R} > -\frac{1}{2}c^2, \tag{85}$$

since in this case mass M is away from the equilibrium (56), we obtain

$$\delta R > \frac{2GM}{c^2}, \tag{86}$$

which indicates that the disturbing radius δR, wiggling mass M against Poisson's equation (29) must be larger than the Schwarzschild radius (26) of mass M. If M is a BH, it cannot be wiggled, as this would invalidate the BH equilibrium (56) condition and Bekenstein's threshold (44) on a BH entropy. The disturbing radius δR should be larger than the Planck length (73).

Based on Equations (79), (81), (84), and (85), we can conclude that the variation potential (50) represents the gravitational potential corresponding to the inertial potential (80), i.e.,

$$\delta R \geq \ell_P, \quad \delta R^2 = R^2, \quad \delta R = -R, \quad \delta\varphi = \phi_a, \tag{87}$$

as shown in Figure 7.

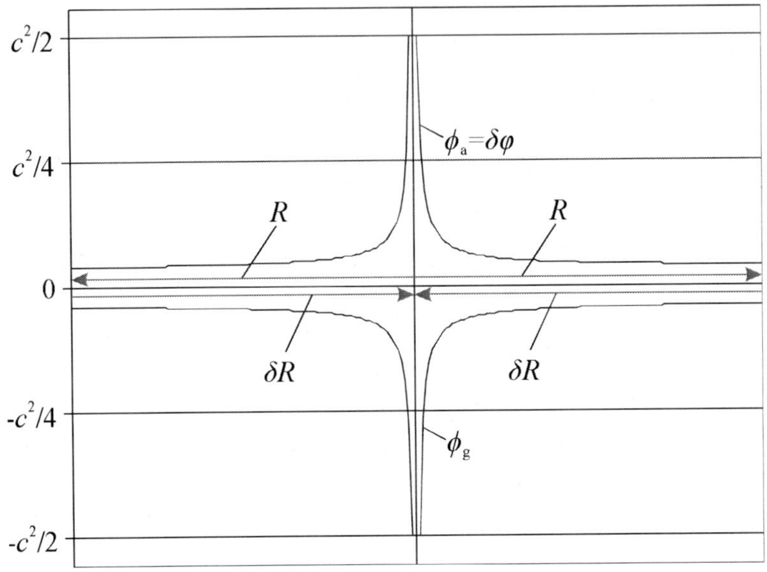

Figure 7. Gravitational, inertial, and variational potentials (not to scale).

The variational potential (78) induced by wiggling the location of the test mass M must be equal to the BH potential of $-c^2/2$, assuming that the test mass M could be wiggled on a BH horizon. Thus, according to Equation (87),

$$\delta\varphi = -\frac{4\pi c^2}{N_A}\frac{\delta R}{\lambdabar_M} \stackrel{BH}{=} \frac{4\pi c^2}{N_{BH}}\frac{R_{BH}}{\lambdabar_M} \doteq -\frac{c^2}{2}. \tag{88}$$

In this case, a simple calculation (cf. Appendix 8) revealed that the Compton wavelength of the active mass M in (88) is the additive inverse of the BH Compton wavelength (9), corresponding to the BH mass required to be emitted to ensure its total collapse (20), given as

$$\lambdabar_M \stackrel{BH}{=} -\frac{2\ell_P^2}{R_{BH}} = -\lambdabar_{BH}. \tag{89}$$

This is not surprising, considering M exhibits an ultimately attained equilibrium (56), and hence, becomes a portion of BH. However, this is negative because M was introduced to M_{BH} in a direction opposite to its radius.

In a subsequent section, we shall investigate the meaning of time period δt and the distance δL tangential to the holographic sphere A.

9. Imaginary Time and Kinematics on Holographic Spheres

Let us postulate the velocity v_R as normal to a holographic sphere, and thus, unobservable. Evidently, only one axis is normal to this sphere at each Planck area, whereas the number of tangential axes is infinite. Let us further postulate that this unobservable velocity v_R is bound with the classical observable velocity

$$v = \frac{\delta L}{\delta t}, \tag{90}$$

where δL denotes a certain length, orthogonal to the disturbing radius δR (and by equation (87) orthogonal to this sphere radius R) based on the Pythagorean relation

$$v^2 + v_R^2 = c^2. \tag{91}$$

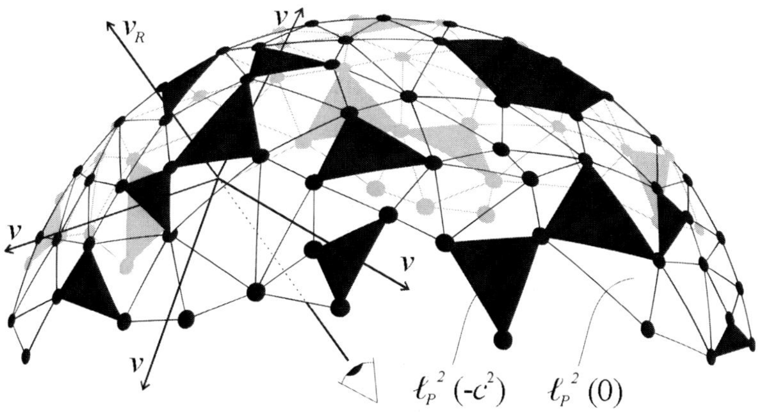

Figure 8. Simplified illustration of visual perception of movement on a holographic patterned sphere (not to scale).

We assumed that δL is observable as moving on the screen owing to the varying Planck areas with their binary potentials (54), as schematically shown in Figure 8. Contrary to interior-less BHs, each observer is bounded within $(2 + i)$-dimensional sphere A, with two spatial dimensions encoding its information capacity N_A with $\lfloor N_A \rfloor$ Planck area spherical triangles and imaginary time, whereas $0i = 0$ denotes the present moment of perception (observation), wherein an active Planck area can become inactive, and vice versa. However, we do not investigate the precise mechanism of this process. It is known that the photoisomerization of a photon in a cone cell of the eye cascades signal transduction.

Based on this postulate (91), the Lorentz factor can be derived as

$$\gamma = \left(1 - \frac{v^2}{c^2}\right)^{-\frac{1}{2}} = \frac{c}{v_R}, \tag{92}$$

the Lorentz contraction (squared) becomes ($0 \le \delta L^2 \le \delta L_0^2$)

$$\delta L^2 = \delta L_0^2 \left(1 - \frac{v^2}{c^2}\right) = \delta L_0^2 \frac{v_R^2}{c^2}, \tag{93}$$

and time dilation (squared) becomes ($0 \le \delta t_0^2 \le \delta t^2$)

$$\delta t^2 = \delta t_0^2 \Big/ \left(1 - \frac{v^2}{c^2}\right) = \delta t_0^2 \frac{c^2}{v_R^2}. \tag{94}$$

Overall, the simple forms of these equations endorse the validity of the postulate (91). According to Equations (38)-(40), the observable acceleration

$$a = \frac{\delta R}{\delta t^2}, \tag{95}$$

is perpendicular to the holographic sphere. Additionally, the application of the disturbing radius δR (31) and disturbing time period δt in observable acceleration a is justified by our previous conclusion (85) that any inertial acceleration introduces a nonequilibrium condition.

Let us then postulate that this observable acceleration a is, similarly to velocity (91), bound with a certain unobservable orthogonal acceleration a_T, which is tangential to the holographic sphere at a given Planck area, based on the similar Pythagorean relation

$$a_T^2 + a^2 = a_P^2 = \frac{c^2}{t_P^2} = \frac{\ell_P^2}{t_P^4}. \tag{96}$$

Moreover, a (95) can be expressed using Unruh temperature (38) as

$$a = \frac{2\pi c k_B}{\hbar} T_a = 2\pi a_P \frac{T_a}{T_P}. \tag{97}$$

Substituting Equation (97) it into Equation (96) we get

$$a_T = a_P \sqrt{1 - 4\pi^2 \tau^2}, \tag{98}$$

where

$$\tau = \frac{T_a}{T_P} = \frac{1}{2\pi} \frac{a}{a_P} \le 1 \quad \tau \in \mathbb{R}, \tag{99}$$

denotes the ratio of the Unruh temperature or acceleration to the Planck temperature or acceleration. If $\tau = 1/(2\pi)$, $a_T = 0$ and the only radial acceleration a is equal to the Planck acceleration, a_P exists on the holographic sphere, which represents the Big Bang condition. If $\tau = 0$ (absolute zero), then $a_T = a_P$ and the radial acceleration a vanishes in (96).

At a BH horizon with $\tau_{BH} = 1/(2\pi d)$ (11), (98) reduces into

$$a_T \overset{BH}{=} a_P \sqrt{1 - \frac{1}{d_{BH}^2}}, \tag{100}$$

which is imaginary if $d_{BH}^2 < 1$. The real values of a_T are first produced with the π-bit BH ($d_{\mathrm{BH}} = 1$) with surface gravity (82) equaling to the Planck acceleration, which also validates the proposed postulate (96).

Moreover, the Unruh temperature (38) can be expressed as

$$T_a = \frac{\hbar}{2\pi c k_B} \frac{\delta R}{\delta t^2} = \frac{r_\delta}{2\pi t^2} T_P, \tag{101}$$

where $\delta R = r_\delta \ell_P$, $\delta t = t t_P$, $r_\delta \in \mathbb{R}$, and $t \in \mathbb{R}$ denote the Planck length and time multipliers in δR and δt. At a BH horizon, this temperature must be equal to the BH temperature (11), which results in the following with (87) (thus, the sign is reversed; $r_\delta = -r_{BH}$):

$$\frac{r_\delta}{2\pi t^2} \overset{BH}{=} \frac{1}{2\pi d_{BH}}, \quad r_\delta\left(-2r_\delta\right) \overset{BH}{=} t^2, \quad \frac{\delta R^2}{\delta t^2} \overset{BH}{=} -\frac{1}{2} c^2. \tag{102}$$

This is equal to the BH potential (27). Since δR is real, we conclude that at a BH horizon time period δt is imaginary

$$\frac{\delta R}{\delta t} \overset{BH}{=} \frac{i}{\sqrt{2}} c, \quad \delta t \overset{BH}{=} \frac{i D_{BH}}{c\sqrt{2}}, \tag{103}$$

and that the unobservable velocity v_R is also imaginary and can be expressed as real

$$v_R = \frac{\delta R}{\delta t_R} = \frac{\delta R}{i \delta t}, \tag{104}$$

upon the introduction of the complementary time period δt_R related to the *classical* time period δt based on integral powers of the imaginaru unit i

$$\delta t_R = i\delta t, \ \delta t_R^2 = -\delta t^2, \ \delta t_R^3 = -i\delta t^3, \ \delta t_R^4 = \delta t^4. \tag{105}$$

Thus, at the BH horizon, Equations (103) and (104) can be combined to derive

$$v_R^2 = \frac{\delta R^2}{\delta t_R^2} \overset{BH}{=} \frac{1}{2} c^2 , \tag{106}$$

which represents the inverse of a BH potential (27), and using (91) provides the observable velocity of

$$v^2 = \frac{\delta L^2}{\delta t^2} \overset{BH}{=} \frac{1}{2} c^2 . \tag{107}$$

Substituting $\delta t = it_P$, the shortest theoretically measurable time period, into (103) yields $d = \sqrt{2}$, which is the minimum BH diameter multiplier allowing for the notion of (imaginary) time.

Furthermore, Equations (102) and (103) yield the acceleration as an additive inverse of the BH surface gravity (82):

$$a = \frac{\delta R}{\delta t^2} \overset{BH}{=} -\frac{c^2}{2\delta R} = \frac{ic}{\sqrt{2}\delta t} , \tag{108}$$

which is congruent with (87).

Based on Equation (104), time dilation (94) can be expressed as

$$\frac{\delta R^2}{\delta t_0^2} = -c^2 , \tag{109}$$

which becomes independent both on the time period δt measured in a moving inertial frame of reference and the velocity v of this frame. The velocity bound (91) and Lorentz contraction (93) results in (cf. Appendix 12)

$$\frac{\delta L_0^2}{\delta L^2} + \frac{\delta L^2}{\delta R^2} = 1 , \tag{110}$$

which is neither time period nor velocity dependent.

Let us now postulate that the unobservable acceleration can presume the form of

$$a_T = \frac{\delta L}{\delta t_R^2} . \tag{111}$$

The current conclusions concerning the velocities (90) and (104) and accelerations (95) and (111) are implied by Mach's principle: observable velocity provides information on a holographic sphere, whereas observable acceleration acts on the holographic sphere and is originated in an interaction between all masses in the universe. To illustrate this principle, we quote Steven Weinberg:

"There is a simple experiment that anyone can perform on a starry night, to clarify the issues raised by Mach's principle. First, stand still, and let your arms hang loose at your sides. Observe the stars are more or less unmoving, and that your arms hang more or less straight down. Then pirouette. The stars will seem to rotate around the zenith, and at the same time your arms will be drawn upward" [38].

The questions that arise are, why do one's arms hang loose when the stars are still, and why they would be drawn upward when the stars rotate? The most important factor for our definitions of the observable velocity (90), acting tangential to the holographic sphere, and the observable acceleration (95), acting perpendicularly to it, is that the *stars* made to whirl around the person standing in the center of a planetarium would not draw the person's arms upward.

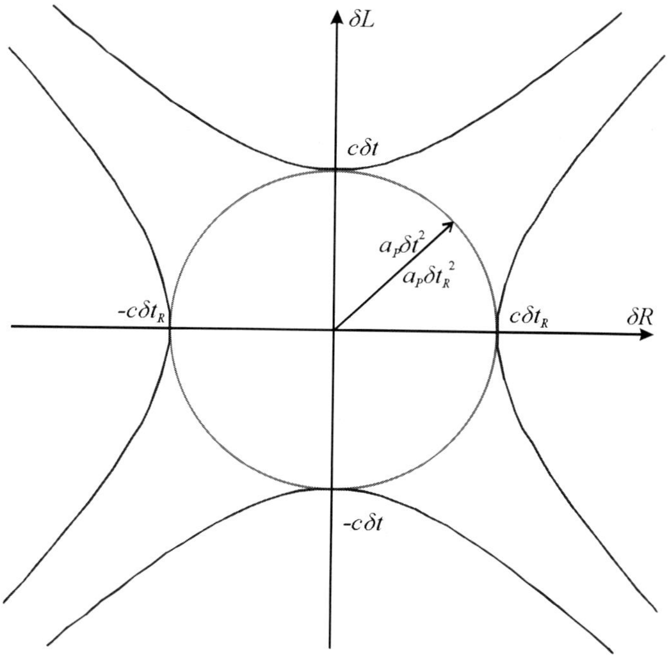

Figure 9. Pythagorean (red) and hyperbolic (blue) relations between δL perpendicular to the disturbing radius δR.

The bounds of velocity (91) and acceleration (96) can now be expressed in terms of δt, using (105) as

$$\frac{\delta L^2}{\delta t^2}+\frac{\delta R^2}{\delta t_R^2}=\frac{\delta L^2}{\delta t^2}-\frac{\delta R^2}{\delta t^2}=c^2, \tag{112}$$

$$\frac{\delta L^2}{\delta t_R^4}+\frac{\delta R^2}{\delta t^4}=\frac{\delta L^2}{\delta t^4}+\frac{\delta R^2}{\delta t^4}=a_P^2=\frac{c^2}{t_P^2}. \tag{113}$$

Equation (112) was divided by δt^2 and both these equations were added to arrive at the relationship:

$$\delta L^2 = \frac{1}{2}\delta t^2\left(a_P^2\delta t^2 + c^2\right), \tag{114}$$

bounding δt with δL. Equations (91) and (96) can be expressed in terms of δt_R, obtaining the bound for δR as

$$\delta R^2 = \frac{1}{2}\delta t^2\left(a_P^2\delta t^2 - c^2\right), \tag{115}$$

which considered in combination yields

$$\delta L^2 + \delta R^2 = a_P^2\delta t^4, \tag{116}$$

corresponding to (96) with substituted (105). Discretized by the Planck length and time for natural multipliers r, t, a, this represents the OEIS A349078 sequence.

As observed, δL and δR are bound by the Pythagorean (circle) relation (116) and by the hyperbolic relation:

$$\delta L^2 - \delta R^2 = c^2\delta t^2, \tag{117}$$

as shown in Figure 9.

Dividing Equations (114) and (115) with $\delta t^2 \neq 0$, we get

$$\frac{\delta L^2}{\delta t^2} = \frac{1}{2}c^2\left(t^2 + 1\right) = v^2 \leq c^2, \tag{118}$$

and

$$\frac{\delta R^2}{\delta t^2} = \frac{1}{2}c^2\left(t^2 - 1\right) = -v_R^2 \leq c^2. \tag{119}$$

Considering, no velocity exceeds the speed of light c, the bounds are set on inequalities (118) and (119), as listed in Table 4. Based on the BH bounds (106) and (107), we can conclude that

$$0 \leq \frac{\delta L^2}{\delta t^2} \leq \frac{1}{2}c^2,\ 0 \leq \frac{\delta R^2}{\delta t^2} \leq \frac{1}{2}c^2. \tag{120}$$

Table 4. Characteristic Planck time multipliers

	0	$c^2/2$	c^2
$\delta L^2/\delta t^2$	$t^2 = -1$	$t^2 = 0$	$t^2 = 1$
$\delta R^2/\delta t^2$	$t^2 = 1$	$t^2 = 2$	$t^2 = 3$

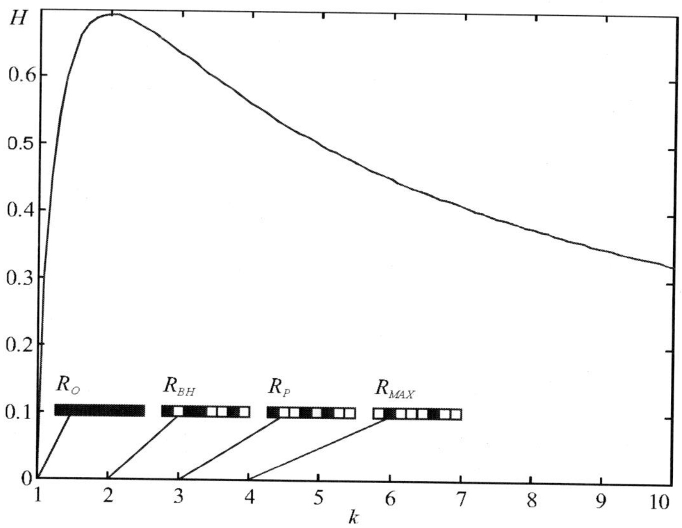

Figure 10. Shannon entropy (in nats) of holographic spheres, as a function of GM/c^2 multiplier k; M = const.

On substituting Equations (118) and (119) into the orbiting condition $v_O^2 \leq v^2 \leq v_E^2$, with (76) and (74) we get

$$\frac{R_{BH}}{R} \leq t^2 + 1 \leq \frac{2R_{BH}}{R}, \tag{121}$$

and

$$\frac{R_{BH}}{R} \leq t^2 - 1 \leq \frac{2R_{BH}}{R}. \tag{122}$$

These are interesting results. Our toy universe, shown in Figure 6, contains a black hole in its center. But it is not necessary for the above relations to be valid. The Schwarzschild radius of the Sun (2E30 kg), for example, amounts to R_{BH} = 2953.34 m.

At BH horizon $R = R_{BH}$, and hence, $0 \leq t^2 \leq 1$ and $2 \leq t^2 \leq 3$ in Equations (121) and (122), which corresponds to the $c^2/2$ to c^2 range in Table 4. Furthermore, we observed that the range 0 to $c^2/2$ in Table 4 is achieved by $0 \leq t^2+1 \leq 1$ ($-1 \leq t^2 \leq 0$) and $0 \leq t^2-1 \leq 1$ ($1 \leq t^2 \leq 2$) with a gap $0 \leq t^2 \leq 1$. Comparing this with Equations (121) and (122), we can conclude that the maximum significant BH radius equals $2R_{BH}$. Notably, there are no celestial objects other than BHs[5] having a potential equal to or lower than $|-c^2/4|$. For a constant $M_{BH} = R_{BH}c^2/2G$ and $R_{MAX} = 2R_{BH}$ the Bekenstein bound [21] yields twice BH entropy (44) (M_{BH} is bounded by R_{BH} but R varies)

$$S \leq \frac{2\pi k_B c R_{MAX} M_{BH}}{\hbar} = 2\pi k_B r_{MAX} m_{BH} = \frac{2\pi k_B c 2R_{BH}}{\hbar}\frac{R_{BH}c^2}{2G} = \frac{1}{2}k_B N_{BH} = k_B N_1, \tag{123}$$

[5] Even white dwarfs (such as Sirius; M = 4.1E30 kg, R = 1.19E9 m → ϕ_g = −2.3E11 m²/s²) and neutron stars (such as PSR J0740+6620; M = 4.14E30 kg, R = 1.5E4 m → ϕ_g = −1.83E16 m²/s²) provide potential lower (up to modulus) than BHs, with the latter being close to $-c^2/4$ = -2.25E16 m²/s².

which also according to Equation (56) equals the number of active Planck areas on a BH horizon. We can express the radius R_k of a holographic sphere of a constant mass $M = mm_P$, $m \in \mathbb{R}$ as a function of GM/c^2 multiplier $k \in \mathbb{R}$, as

$$R_k = k\frac{GM}{c^2} = km\ell_P, \tag{124}$$

which corresponds to the Schwarzschild radius (26) of mass M for $k = 2$. This yields potential $\phi_g(R_k)$ (25) and the number of the active Planck areas (62) at the holographic sphere of radius R_k equal to $N_1 = N_A/k$. Therefore, there are fewer active Planck areas *outside* the BH horizon of mass M than $N_1 = N_A/2$.

Substituting G from ℓ_P^2, (124) we get

$$Mc\frac{\ell_P^2}{R_k} = \frac{\hbar}{k}, \tag{125}$$

which for $k = 2$ corresponds to HUP bound (65). Parameters of the holographic spheres of a constant mass M are listed in Table 5, while Figure 10 shows their binary Shannon entropy (60) (cf. Appendix 14)

$$H(k) = \ln(k) - \frac{k-1}{k}\ln(k-1). \tag{126}$$

Substituting δR^2 from time dilation (109) into Equation (119), we get

$$t^4 - t^2 - 2t_0^2 = 0, \tag{127}$$

where the proper time is $\delta t_0 = it_0t_P$ and $t_0 \in \mathbb{R}$ represents Planck time multiplier. Subsequent t_0^2 from a set of triangular numbers (0, 1, 3, 6,...; OEIS A000217) substituted into $t^2 = (1\pm\sqrt{(1+8t_0^2)})/2$ (127) yield subsequent nonpositive $t^2 = 0, -1, -2, -3,\ldots$ and positive $t^2 = 1, 2, 3,\ldots$ squares of Planck time multipliers.

An outstanding property of Boolean $\{0, 1\}^N$ address space is that the mean Hamming distance between any address a_m and all the other addresses (including a_m) is $N/2$, whereas the variation is $N/4$ [39]. If an $\{N\}$-cube [3] was inscribed in the closed N-ball most of its vertices would lay at or near the equator, whereas the unit of the active Planck area ($-c^2$) suggests a particular orientation of BH address space, with vertex 000… representing one pole and vertex 111… [$-c^2$] representing the other pole. This indicates a link between diameter only dependent Planck BH blackbody spectral radiance (130), (131) and the corresponding patternless BH binary message (59). Interestingly, the fraction of available BH horizon arrangements comprising $N_1 = N_{BH}/2$ active Planck areas with binary potential (54) equal to $-c^2$ decreases from 50% (2-bit BH) to zero with N_{BH} approaching infinity. Using Stirling's approximation for large N_{BH} it amounts

$$\binom{N_{BH}}{N_1}\Big/2^{N_{BH}} \approx \sqrt{\frac{2}{\pi N_{BH}}} = \frac{\sqrt{2}}{\pi d_{BH}}. \tag{128}$$

A relation between a bit of a binary message (58) and a wavelength multiplier l requires further research. It is certainly easier to maintain patternlessness of a longer binary message, which is likely to be responsible for a flatter spectral radiance of larger BHs, such as Sagittarius A* (cf. Appendix 13). Neutron stars and white dwarfs supported against collapse owing to Pauli exclusion principle, also emit blackbody radiation. Therefore, they generate patternless binary messages. Perhaps, just one spherical side of BHs hints their anyonic nature, whereas neutron stars and white dwarfs exhibit fermionic natures.

These considerations are not surprising. We observed and modeled a BH as a 2-sphere, from which electromagnetic radiation cannot escape. However, in the context of the graph of nature, the BH is a patternless binary message. Both δt and δt_R are imaginary and become real ($0i = 0$) only at the moment of interaction/observation of the Planck area ℓ_P^2 at the holographic sphere.

Certain dynamics scenarios can be contemplated in the simplified model of a spherical universe with two masses (Figure 6), as shown in Figure 11. The scenarios involving more than two bodies are not discussed, as they are irrelevant. In any given moment of perception, only two bodies interact with each other through Planck area ℓ_P^2 in the Planck time t_P, and exchange the information in units of 0 or $-c^2$ according to the second law of Newton (47) and Newton's law of gravity (48).

The mass M in a state of a timeless free fall on M_{BH} is shown in Figure 11(a), where $\delta R = 0$. Therefore, no entropic work occurs, and the system is entirely described by Poisson's equation (29) with the inhomogeneity of the mass density included in the $4\pi G\rho$ term. This equation is independent of time or the variations of other quantities (position, velocity) with time. In particular, $\delta\varphi$ is infinite (50) and $A = 0$ for vanishing δR; therefore, the variation of entropy δS is zero (41), such as that in the limiting case of the binary Shannon entropy (for $p_0 = 0$ or $p_1 = 0$), where $0\ln(0)$ is defined as 0.

The entropic work is shown in Figure 11(b–d). The notion of time emerges from the instantaneous variations of entropy (41) induced by the instantaneous variations of δR in time δt (31).

As shown in Figure 11(b), the mass M moves along a spiral trajectory towards M_{BH}. Obviously, the geometry of this trajectory is not described in Figure 11(b). While the variations of entropy δS (41) is defined and nonvanishing ($\delta R \neq 0$ and $\delta R/\delta t \neq 0$), the velocity of M is below the orbital velocity v_O defined by M_{BH} and R ($M << M_{BH}$). Consequently, a holographic 2-sphere A emerges both as the gravitational potential (25) sphere and entropy variation sphere (50). Subsequently, as time progresses, the system aims toward the equilibrium (56), $N_1(t) \to N_{BH}(t)/2$.

As shown in Figure 11(c), the velocity of M is greater than the orbital velocity v_O and less than the escape velocity v_E. Therefore, M orbits surrounding M_{BH}. If the entropic work is being done, $\delta R \neq 0$ and $\delta R/\delta \mathrm{t} \neq 0$, which implies fluctuations in the instantaneous velocity of an orbiting body ranging between the orbital and escape velocities. In this case, $N_1(t) < N_A(t)/2$ and the structure of M and M_{BH} is dissipative [12].

Table 5. Size (R, D, d), potential (ϕ_g), information capacity (N_A), no. and probability of occurrence of active Planck areas (N_1) within all Planck areas (N_A), Bekenstein bound (S/k_B), and Shannon entropy (H) of holographic spheres for a constant mass $M = m\, m_P$, $m \in \mathbb{R}$, $k \in \mathbb{R}$

R	$R_{MIN} = 0$	$R_O = m\ell_P$	$R_{BH} = 2m\ell_P$	$R_P = 3m\ell_P$	$R_{MAX} = 4m\ell_P$	$R_k = km\ell_P$
D	$D_{MIN} = 0$	$D_O = 2m\ell_P$	$D_{BH} = 4m\ell_P$	$D_P = 6m\ell_P$	$D_{MAX} = 8m\ell_P$	$D_k = 2km\ell_P$
$\phi_g = -GM/R$	$-\infty$	$-c^2$	$-c^2/2$	$-c^2/3$	$-c^2/4$	$-c^2/k$
$d = D/\ell_P$	0	$2m$	$4m$	$6m$	$8m$	$2km$
$N_A = \pi d^2$	0	$4\pi m^2$	$16\pi m^2 = N_{BH}$	$36\pi m^2$	$64\pi m^2$	$4\pi k^2 m^2$
$N_1 = -\phi_g N_A/c^2$	0	$4\pi m^2 = N_{BH}/4$	$8\pi m^2 = N_{BH}/2$	$12\pi m^2 = 3N_{BH}/4$	$16\pi m^2 = N_{BH}$	$4\pi k m^2$
$p_1 = N_1/N_A$	undef.	1	1/2	1/3	1/4	$1/k$
S/k_B (rel. to M)	0	$N_{BH}/8 = N_1/4$	$N_{BH}/4 = N_1/2$	$3N_{BH}/8 = 3N_1/4$	$N_{BH}/2 = N_1$	$kN_{BH}/8 = 2\pi k m^2$
$H(N_1/N_A)$	undef.	0	$\ln(2)$	$\ln(3) - 2\ln(2)/3$	$\ln(4) - 3\ln(3)/4$	$\ln(k) - (k-1)\ln(k-1)/k$

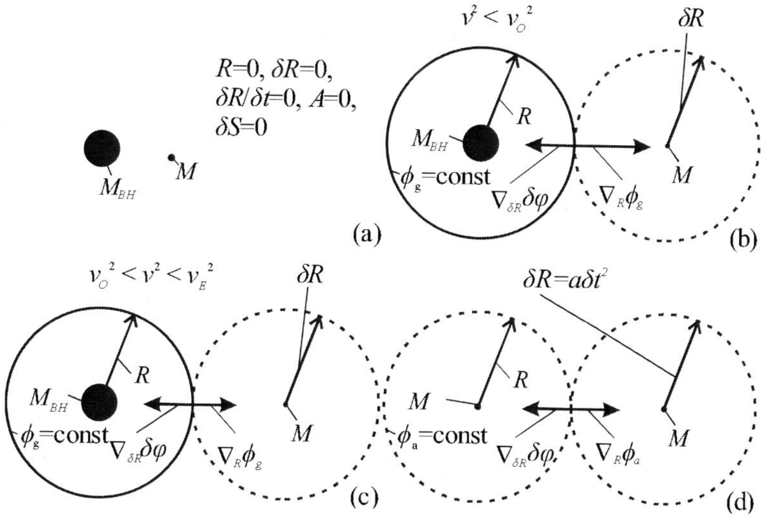

Figure 11. Entropic work of gravity and inertia. (a) Free fall (δR = 0, $\delta R/\delta t$ = 0, A = 0), (b) spiral trajectory fall $v < v_O$, (c) stable orbit $v_O < v < v_E$, and (d) inertial acceleration (not to scale). Gradient sphere (left); entropy variation sphere (right).

As shown in Figure 11(d), M accelerates with an acceleration of $a = \delta R/\delta t^2$ in the absence of background mass. The gradient of the inertial potential (83) provides a conceptual radius R, which defines an emergent instantaneous holographic sphere A, whereas δR provides the entropy variation sphere (50). Eventually, the conceptual R is eliminated in (81), thereby leaving only the conceptual 2nd M. The gradient of the inertial potential (83) on A is normal to both sides of A as the area of the holographic sphere A increases, as the inertial acceleration a increases and decreases as the gravitational acceleration g decreases. Furthermore, we observed that the acceleration a of M produces the conceptual R along with the corresponding variations $\delta\varphi_k$ of $\lfloor N_A \rfloor$ bits on a holographic sphere A. Each distribution of these bits varies from the BH equilibrium (56). Thus, in this case, $N_1(t) < N_A(t)/2$ and the mass M alone is a dissipative structure. The dissipativity of this mass, e.g., a rocket, accelerating by itself, can be attributed to the fact that the inertial acceleration can be a result of only two factors: either the rocket accelerates due to gravity or it is accelerated by an engine. In the former case, the gravity of a stellar object can pull the rocket toward its surface (Figure 11(b)), and the rocket can either escape away or orbit around it (Figure 11(c), dissipative case). In the latter case (Figure 11(d)), the dissipativity is provided only by the engine of the rocket. We neglected the situation of the law of conservation of momentum (interaction of rocket with another moving object).

Therefore, the cases depicted in Figure 11(b–d) are identical in nature. The left sphere displayed in Figure 11 is a gradient sphere, that defines the inertial or gravitational potential along with the information capacity N_A, whereas the right sphere denotes the entropy variation sphere defining δR and N_1. R is an artificial spatial coordinate that enables the introduction of the notion of temperature generating potentials (25) and (80). Similar to the Rayleigh–Bénard convection, the entropic work must involve a heat transfer (38) and (39). Although the equations of motion in the abstract Newtonian dynamics are mathematically reversible in case the Planck time multiplier t is replaced by $-t$, it is not an observed behavior of the macroscopic objects modelled by these equations. Therefore, although both gravity and inertia appear to be reversible, their origin is entropic.

Any geometric considerations pertaining to the structure of $\mathbb{R}^3$ outside the holographic screen are unnecessary. The ancient Greek notion of geometry brings forth an unnecessary burden and axiomatization: *geo-* suggests unobservable extra-instantaneous existence, whereas *-metric* indicates the identity of indiscernibles, which is an ontological principle and the first axiom of the metric. This principle is invalid owing to the Ugly duckling theorem [40, 41] that asserts that any two distinct objects are equally similar. Additionally, the first axiom of the metric is also invalid in the quantum domain [42].

Geometric considerations are even more unnecessary if one explores the graph of nature with certain intrinsic properties that reflect the 2nd law of thermodynamics, wherein the perceived dimensionality of this graph is induced [3] by the peculiar property of the Euclidean space $\mathbb{R}^4$, known as exotic $\mathbb{R}^4$. Particularly, the equipotential/holographic spheres or event horizons are defined by the sets of vertices in the graph of nature with the same potential (during instantaneous observation), which attains minimum for a BH (27). Based on a geometric perspective, an event horizon is a limiting or fundamental screen with only one spherical geometric, exterior side. We used the term "holographic sphere" instead of the term "equipotential sphere," considering the latter —which suggests permanent equality of potential — contradicts the notion of a binary message.

The anomaly observed in the cosmic microwave background (CMB) radiation aligned with the plane of the Solar System, dubbed the "Axis of Evil" [sic], is considered to be an unsolved problem in physics. A report on the quest for proposed solutions deserves a separate article. They include astrophysical foregrounds (such as Solar system dust [43] or contributions from the Kuiper Belt objects [44]), artifacts of faulty data analysis, instrumental systematics, and theoretical/cosmological arguments [45]. However, the proposed models require severe tuning to reproduce the features of the CMB [45, 46, 47].

The author is convinced that this anomaly simply contradicts our intuitive concept of the physical space and could be explained within the framework of the variational (50), binary (54), and inertial (80) potentials, and the velocity (91) and acceleration (96) relations. Although this matter requires further research, this anomaly may be due to the fact that Ampère's right-hand grip rule induces some temperature gradient, which is substantially perpendicular to the ecliptic of the Solar System, and skewness owing to the direction of the Sun's motion.

10. Black Holes Quantum Statistics

Planck's law for blackbody spectral radiance expressed in terms of wavelength λ is

$$B_\lambda(\lambda, T) = \frac{2hc^2}{\lambda^5} \frac{1}{\exp\left(hc/\lambda k_B T\right) - 1}. \tag{129}$$

For a BH, upon substituting $\lambda := l\ell_P$ and BH temperature (38), the above equation can be expressed as

$$B_\lambda(l,d) = \frac{2hc^2}{l^5\ell_P^5}\frac{1}{\exp(4\pi^2 d/l)-1}, \tag{130}$$

for wavelength or

$$B_\nu(l,d) = \frac{2hc}{l^3\ell_P^3}\frac{1}{\exp(4\pi^2 d/l)-1}, \tag{131}$$

for frequency (cf. Appendix 13).

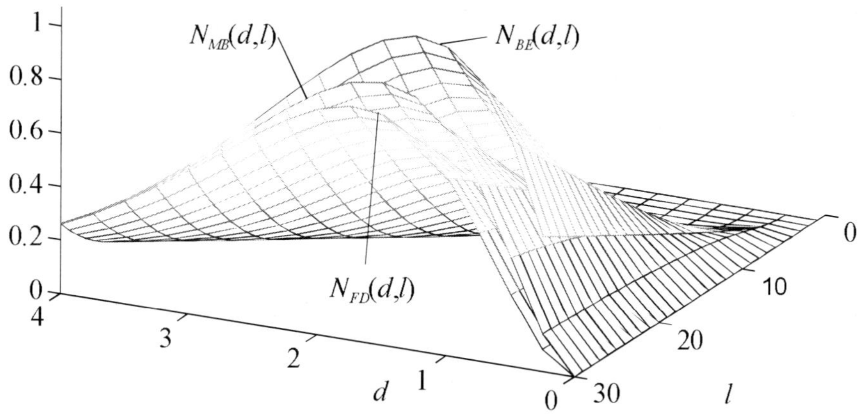

Figure 12. Bose-Einstein, Maxwell–Boltzmann, and Fermi-Dirac 2-sphere BH statistics.

If a BH Planck area emits (absorbs) a photon (boson), as it alters its state from active ($-c^2$) to inactive (0) (or vice versa), and an electron (fermion) energy level is decreased (increased) in this interaction, the event horizon can be described using Bose–Einstein (BE) and Fermi–Dirac (FD) statistics with the degeneracy interpreted as the number of Planck areas πd^2 on the event horizon. No Planck area is distinct and each one is equally capable of emitting/absorbing energy corresponding to the wavelength multiplier l. There are certainly no dedicated microwave or X-ray areas on the horizon. Thus, these statistics along with the Maxwell–Boltzmann (MB) statistics can be described as

$$N_{BE}(d,l) = \frac{\pi d^2}{e^{4\pi^2 d/l}-1}, \tag{132}$$

$$N_{MB}(d,l) = \frac{\pi d^2}{e^{4\pi^2 d/l}}, \text{ and} \tag{133}$$

$$N_{FD}(d,l) = \frac{\pi d^2}{e^{4\pi^2 d/l}+1}, \tag{134}$$

as shown in Figure 12, where *N*(*d*, *l*) represents the average number of bosons, classical *particles*, and fermions, respectively, with wavelength *l*.

To emit/absorb $N > 0$ photons or correspondingly increase/decrease $N > 0$ electron energy levels on average, $N(d, l)$ must be greater or equal to N and the statistics (132)–(134) yield the following wavelength thresholds

$$l_{BE}(d,N) \geq \frac{4\pi^2 d}{\ln(\pi d^2 + N) - \ln(N)}, \tag{135}$$

$$l_{MB}(d,N) \geq \frac{4\pi^2 d}{\ln(\pi d^2) - \ln(N)}, \tag{136}$$

$$l_{FD}(d,N) \geq \frac{4\pi^2 d}{\ln(\pi d^2 - N) - \ln(N)}, \tag{137}$$

shown in Figure 13 for $N = 1$.

Each threshold poses singularity, given as

$$d_{BE}^{\sin} = 0, \quad d_{MB}^{\sin} = \sqrt{\frac{N}{\pi}}, \quad d_{FD}^{\sin} = \sqrt{\frac{2N}{\pi}}, \tag{138}$$

and the minimum, given as

$$d_{BE}^{\min} = \sqrt{-N\left(\frac{W_0(-2e^{-2}) + 2}{\pi W_0(-2e^{-2})}\right)}, \quad d_{MB}^{\min} = e\sqrt{\frac{N}{\pi}}, \quad d_{FD}^{\min} = \sqrt{N\frac{W_0(2e^{-2}) + 2}{\pi W_0(2e^{-2})}}, \tag{139}$$

where $W_0(x)$ is the Lambert *W* function (omega function). Table 6 summarizes the minima along with the corresponding minimum wavelengths.

Perhaps, the negative wavelengths for positive diameters can be interpreted in terms of emission/absorption, wherein the positive and negative wavelengths represent the BH absorption and emission, respectively. Consequently, the BE threshold (135) with no negative wavelengths illustrates that bosons (photons) cannot escape from a BH. However, a relation between a bit (Planck area) of a binary message (58) and a wavelength multiplier *l* remains to be researched.

For $0 < d \leq \sqrt{(N/\pi)}$, the wavelengths *l* obtained from FD bound (137) are complex, as shown in Figure 14. Below this threshold, corresponding to the diameter of a 1-bit BH for $N = 1$, BH emits complex wavelengths. After exceeding this diameter, they become *real,* and hence, observable. Presumably, only the negative and real wavelengths in the FD threshold (137) within the range $\sqrt{(N/\pi)} < d_{BH} \leq \sqrt{(2N/\pi)}$ represent the Hawking radiation of fermions. Interestingly, the real component of the wavelength below $d_{BH} = 1/(2\pi)$ Planck temperature threshold vanishes.

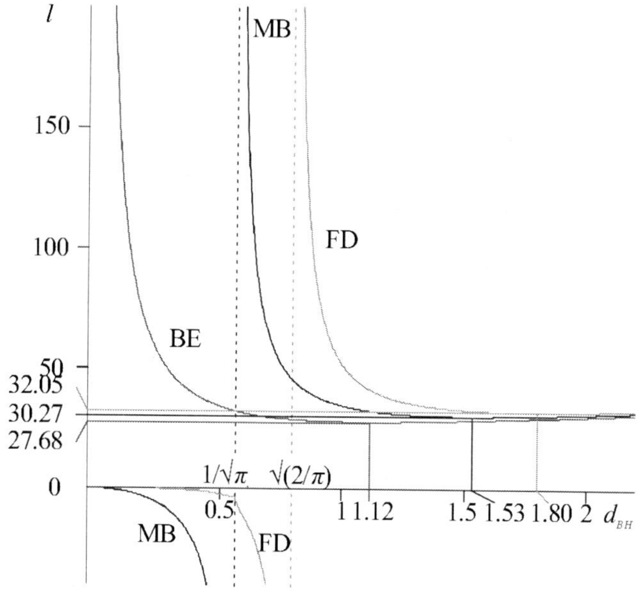

Figure 13. Wavelength thresholds for BE (red), MB (blue), and FD (green) statistics for the 2-sphere BH as a function of its diameter multiplier d_{BH} with $N = 1$.

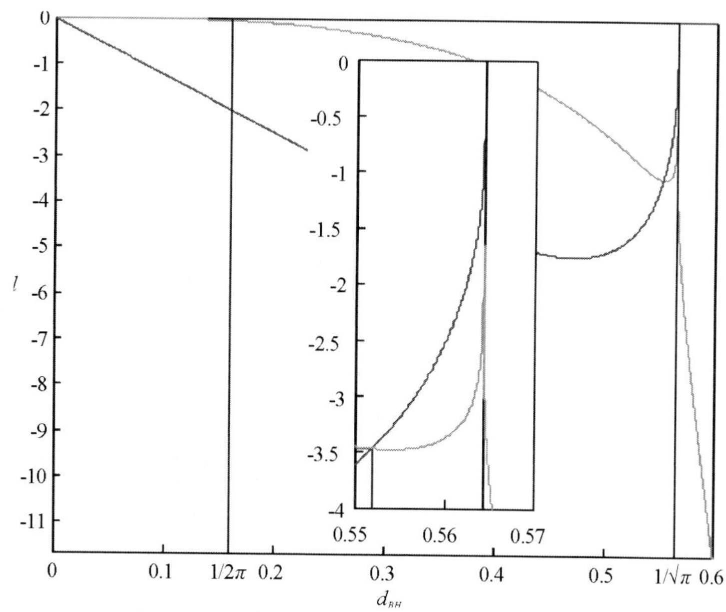

Figure 14. Real (green) and imaginary (red) component of wavelength threshold for FD statistics as a function of d_{BH} in the vicinity of $1/\sqrt{\pi}$ with $N = 1$.

As described below, a more detailed description of a BH interaction with the environment certainly requires further research, while considering the imaginary set of Planck units [48]. The degenerated, patternless BE BH blackbody radiation is characterized by zero Gibbs free energy, whereas the departure from degeneracy is responsible for the unique spectral characteristics of stars [49].

Table 6. BE, MB, and FD thresholds for $N = 1, 2, \ldots, 6$

N	$d^{min}{}_{BE}$	$l^{min}{}_{BE}$	$d^{min}{}_{MB}$	$l^{min}{}_{MB}$	$d^{min}{}_{FD}$	$l^{min}{}_{FD}$
1	1.1173	27.678	1.5336	30.273	1.8007	32.054
2	1.5800	39.142	2.1689	42.812	2.5465	45.332
3	1.9352	47.939	2.6563	52.434	3.1188	55.520
4	2.2345	55.355	3.0673	60.545	3.6013	64.109
5	2.4982	61.889	3.4293	67.692	4.0264	71.676
6	2.7367	67.796	3.7566	74.152	4.4107	78.517

11. Black Hole Information Paradox Solution

To illustrate the black hole information paradox that also applies to neutron stars and white dwarfs emitting patternless blackbody radiation (129), we consider two quantum, two-state *particles* (qubits) A and B entangled in one out of four maximally entangled Bell states

$$|\Phi^{\pm}\rangle = \frac{1}{\sqrt{2}}\left(|0\rangle_A|0\rangle_B \pm |1\rangle_A|1\rangle_B\right), \tag{140}$$

or

$$|\Psi^{\pm}\rangle = \frac{1}{\sqrt{2}}\left(|0\rangle_A|1\rangle_B \pm |1\rangle_A|0\rangle_B\right), \tag{141}$$

the coherence of which is maintained. If we throw one of these *particles* (say B) into a BH event horizon, we lose all the information contained in *particle B*, as asserted by the no-hiding theorem [1]. However, this information is entangled with the information support of the *particle A* through surjective isometry of the state (140) or (141). Does that mean that qubit A can no longer be measured? Or, if it can be measured, yielding - say - $|0\rangle_A$ in a basis $\{|0\rangle, |1\rangle\}$, does that mean that the *particle B* has been earlier *measured* by the BH event horizon also as $|0\rangle_B$ (in the case of $|\Phi^{\pm}\rangle$) or $|1\rangle_B$ (in the case of $|\Psi^{\pm}\rangle$)?

Certainly not. Even if a BH horizon is a quantum system, it cannot stand in the role of an observer [50, 51]. Patternless BH horizon contradicts the entanglement between A and B defined by Equations (140) or (141). Any state of the orthonormal basis of the four Bell states in a Hilbert space $\mathbb{C}^4$ is a unique pattern among a uncountably infinite number of other possible quantum states. Therefore, BH would destroy coherence of the entangled state [52] transforming it into a separable state

$$|\varphi\rangle = \frac{1}{2}\left(|0\rangle_A|0\rangle_B \pm |0\rangle_A|1\rangle_B \pm |1\rangle_A|0\rangle_B \pm |1\rangle_A|1\rangle_B\right), \tag{142}$$

from which the measurement of A will yield an arbitrary, equally probable result. The principle of locality is invalid in the quantum domain [53]. Much of the essence of quantum theory already makes itself known in the case of just two nonorthogonal states [54].

These limits do not hold for observation of *particle B* on a holographic patterned sphere, as shown in Figure 8. If *B* is a photon, observed visually through an eye of a living organism, its photoisomerization induced signal transduction cascade will provide a definite result preserving entanglement. In other words, the information obtained from the measured qubit *B* will penetrate the interior of the observer through a Planck area ℓ_P^2 on a holographic sphere, as one bit of classical information. This cannot happen for an interior-less BH shown in Figure 2(b).

Considering a BH horizon as a quantum system, each BH patternless message (59) has exactly one orthogonal counterpart (e.g., 0101 is orthogonal to 1010). Thus, the number of orthogonal swaps that BH provides is

$$\binom{N_{BH}}{\frac{N_{BH}}{2}}\Big/2 = \frac{N_{BH}!}{2\left[\left(\frac{N_{BH}}{2}\right)!\right]^2} \approx \frac{2^{N_{BH}}}{\sqrt{2\pi N_{BH}}} = \frac{2^{\pi d_{BH}^2}}{\sqrt{2\pi} d_{BH}}, \tag{143}$$

which approaches infinity with N_{BH} approaching infinity, and where Stirling's approximation extends this formula to odd N_{BH}. The Margolus–Levitin theorem and Equation (64) set a bound on a minimum time period required for the BH to transfer from one state to another, orthogonal one, given (with (64)) as

$$\delta t_{ML} \geq \frac{\pi\hbar}{2E_{BH}} = \frac{2\pi}{d_{BH}} t_P, \tag{144}$$

which is longer than the Planck time only for $d_{BH} < 2\pi$, providing another argument for micro BHs instability. Perhaps the Margolus–Levitin bound is not applicable for larger BHs. For 1-bit BH the Margolus–Levitin bound is

$$\delta t_{ML} \overset{d_{BH}=\sqrt{\frac{1}{\pi}}}{\geq} 2\sqrt{\pi^3} t_P \approx 11.137 t_P, \tag{145}$$

which is the minimum time, 1-bit BH needs to swap its Planck area from 0 to 1 $[-c^2]$ or vice versa. A relation between (143) and (144) requires further research; certainly, the more orthogonal swaps available, the shorter the swap time.

CONCLUSION

This study aimed to define BHs (13) as interior-less entities emitting degenerate, patternless binary messages (56), whereas $D_{BH} = \ell_P/(2\pi)$ represented the Big Bang conditions. Micro-BHs are inherently unstable and prone to collapses [23], and hence, this study delivered further arguments in support of this claim. BH threshold of distinguishability (10) was found to be a threshold of BH collapsibility (22). Temperature uncertainty for one degree of freedom is equal to the Planck temperature if time uncertainty equals the Planck time (67), whereas π degrees of freedom enable the recovery (68) of exact Unruh (38) and Hawking (39) temperature equations. The variational potential (50) can be expressed as the ratio of the number of the active Planck areas to the information capacity of the holographic sphere, whereas the entropic work is the product of the test mass and the variational potential (81). Postulated Pythagorean relation (96) between an observable acceleration a and an unobservable orthogonal acceleration a_T derives (100) the smallest BH having surface gravity (82) equaling to the Planck acceleration. Along with the postulated Pythagorean relation (91) between an observable velocity v and an unobservable orthogonal velocity v_R simplifies the Lorentz contraction (93), (110) and time dilation (94), (109) formulas which become independent of the observer's velocity. Both relations (91) and (96) introduce significant BH radius equal to $4GM/c^2$, that is twice its Schwarzschild radius ($2GM/c^2$), and exceeding the BH photon sphere radius ($3GM/c^2$) by GM/c^2. Shannon entropy (126) of holographic spheres, as a function of GM/c^2 multiplier is derived, showing that there are fewer active Planck areas N_1 *outside* the BH horizon than $N_1 = N_A/2$, where N_A is the information capacity of the holographic sphere. Certain dynamics scenarios between the passive mass M_{BH} and the active mass M are discussed. Bose–Einstein (BE), Maxwell–Boltzmann (MB), and Fermi–Dirac (FD) statistics (132)–(134) are derived with the degeneracy interpreted as the number of Planck areas πd^2 on the event horizon. The solution to BH information paradox is proposed (142) and discussed in the context of the Margolus–Levitin theorem.

A relation between a bit of a binary message (59) and wavelength multiplier l of a blackbody radiation requires further research and may shed new light on the observed "Axis of Evil" anomaly. An in-depth study of BH interaction with the environment is required.

Perhaps the most striking conclusion of this study is that every dissipative structure is a sphere in nonequilibrium thermodynamic condition, provided with an interior. Interiors of living cells were exploited by biological evolution, probably beginning with coacervates. An open question is why would a hurricane need an interior?

ACKNOWLEDGMENTS

I truly thank my wife and her mother for their support. I would like to thank Mirek for the insightful discussions and encouragement, especially for noting that a bit is inadequate to be meaningfully assigned to the notion of the potential and that BHs invalidate the Jordan–Brouwer separation theorem, in addition to assisting while solving the sign riddle. I would like to thank my godson Wawrzyniec for meticulously spotting typographical errors and missing commas, and for his report on the "Axis of Evil" state of art. I also thank Piotr for the helpful discussions and for finding certain logical loopholes in the paper, along with noting that the effect of Axis of Evil could be due to Ampère's right-hand grip rule.

REFERENCES

[1] Braunstein S., & Pati A. (2007). Quantum Information Cannot Be Completely Hidden in Correlations: Implications for the Black-Hole Information Paradox. *Phys. Rev. Lett.* 98, 080502. https://doi.org/10.1103/PhysRevLett.98.080502.

[2] Wheeler J. (1989). *Proceedings of the 3rd International Symposium Foundations of Quantum Mechanics in the Light of New Technology. Phys. Soc.* Japan, 434. ISBN4890270035.

[3] Łukaszyk S. (2020). Four Cubes. arXiv:2007.03782 [math.GM]. https://doi.org/10.48550/arXiv.2007. 03782.

[4] Boltzmann L. (1877). *Über die Beziehung zwischen dem zweiten Hauptsatze des mechanischen Wärmetheorie und der Wahrscheinlichkeitsrechnung, respective den Sätzen über das Wärmegleichgewicht* [*On the relationship between the second main theorem of mechanical heat theory and the probability calculation with respect to the results about the heat equilibrium*]. Kaiserlichen Akademie der Wissenschaften, mathematich-naturwissen Cl. LXXVI, Abt II, 373-435. https://doi.org/10.1017/CBO9781139381437.011.

[5] Kuhn T. (1978). *Black-Body Theory and the Quantum Discontinuity*, 1894-1912. University of Chicago Press. ISBN0226458008.

[6] Schlosshauer, M., Kofler, J., & Zeilinger, A. (2013). A Snapshot of Foundational Attitudes Toward Quantum Mechanics. *Studies in History and Philosophy of Science Part B: Studies in History and Philosophy of Modern Physics*, 44(3), 222–230. https://doi.org/10.1016/j.shpsb.2013.04.004.

[7] Teilhard de Chardin P. (1955). *Le Phénomène Humain* [*The Phenomenon of Man*]. Éditions du Seuil. ISBN006090495X.

[8] Carnap R. (1963). *The Philosophy of Rudolf Carnap*. Open Court Publishing. ISBN0812691539.

[9] Brukner Č. (2018). A No-Go Theorem for Observer-Independent Facts. *Entropy* 2018, 20(5), 350. https://doi.org/10.3390/e20050350.

[10] Proietti M., Pickston A., Graffitti F., Barrow P., Kundys D., Branciard C., Ringbauer M., & Fedrizzi A. (2019). Experimental Rejection of Observer Independence in the Quantum World. *Science Advances*, Vol. 5, Issue 9. https://doi.org/10.1126/sciadv.aaw9832.

[11] Bong K-W, Utreras-Alarcón A., Ghafari F., Liang Y-C., Tischler N., Cavalcanti E., Pryde G., & Wiseman H. (2019). Testing the Reality of Wigner's Friend's Experience. *Proceedings vol. 11200, AOS Australian Conference on Optical Fibre Technology (ACOFT) and Australian Conference on Optics, Lasers, and Spectroscopy (ACOLS).* https://doi.org/10.1117/12.2540002.

[12] Prigogine I., & Stengers I. (1984). *Order out of Chaos: Man's New Dialogue with Nature*. Bantam Books. ISBN0553340824.

[13] Manin Y. (2006). The Notion of Dimension in Geometry and Algebra. *Bulletin of the American Mathematical Society*. 43 (2): 139–161. https://doi.org/10.1090/S0273-0979-06-01081-0.

[14] Hossenfelder S. (2010). Comments on and Verlinde's paper "On the Origin of Gravity and the Laws of Newton". arXiv:1003.1015 [gr-qc]. https://doi.org/10.48550/arXiv.1003.1015.

[15] Hooft G. (1993). *Dimensional Reduction in Quantum Gravity.* arXiv:gr-qc/9310026. https://doi.org/ 10.48550/arXiv.gr-qc/9310026.

[16] Planck M. (1900). Über Irreversible Strahlungsvorgänge [About Irreversible Radiation Processes]. *Ann. Phys.*, 306: 69-122. https://doi.org/10.1002/andp.19003060105.

[17] Fidkowski Ł., Hubeny V., Kleban M., & Shenker S. (2004). The Black Hole Singularity in AdS/CFT. *Journal of High Energy Physics*, Vol. 2004. https://doi.org/10.1088/1126-6708/2004/02/014.

[18] Gassner S., & Cafaro C. (2019). Information Geometric Complexity of Entropic Motion on Curved Statistical Manifolds under Different Metrizations of Probability Spaces. *International Journal of Geometric Methods in Modern Physics* Vol. 16, No. 06. https://doi.org/10.1142/S0219887819500828.

[19] Hirani, A. (2003). *Discrete Exterior Calculus.* Dissertation (Ph.D.), California Institute of Technology. https://doi.org/10.7907/ZHY8-V329.

[20] Desbrun, M., Kanso, E., & Tong, Y. (2008). Discrete Differential Forms for Computational Modeling. In: Bobenko, A.I., Sullivan, J. M., Schröder, P., Ziegler, G.M. (eds.) *Discrete Differential Geometry.* Oberwolfach Seminars, Vol. 38. Birkhäuser Basel. https://doi.org/10.1007/978-3-7643-8621-4_16.

[21] Bekenstein J. (1973). Black Holes and Entropy. *Phys. Rev.* D 7, 2333. https://doi.org/ 10.1103/PhysRevD. 7.2333.

[22] Susskind L. (2008). *Black Hole War: My Battle with Stephen Hawking to Make the World Safe for Quantum Mechanics.* Little, Brown and Company. ISBN 9780316016407.

[23] Hawking S. (1974). Black Hole Explosions? *Nature* Vol. 248, 30–31. https://doi.org/ 10.1038/248030a0.

[24] Łukaszyk S. (2022). Novel Recurrence Relations for Volumes and Surfaces of n-Balls, Regular n-Simplices, and n-Orthoplices in Real Dimensions. *Mathematics* 2022, 10(13). https://doi.org/10.3390/ math10132212.

[25] Parisi G. & Sourlas N. (1979). Random Magnetic Fields, Supersymmetry, and Negative Dimensions. *Phys. Rev. Lett.* 43, 744. https://doi.org/10.1103/PhysRevLett.43.744.

[26] Verlinde E. (2010). On the Origin of Gravity and the Laws of Newton. *Journal of High Energy Physics* Vol. 2011, No. 29. https://doi.org/10.1007/JHEP04(2011)029.

[27] Brouwer M., Visser M., Dvornik A., Hoekstra H., Kuijken K., Valentijn E., Bilicki M., Blake C., Brough S., Buddelmeijer H., Erben T., Heymans C., Hildebrandt H., Holwerda B., Hopkins A., Klaes D., Liske J., Loveday J., McFarland J., Nakajima R., Sifón C., & Taylor E. (2017). First test of Verlinde's theory of emergent gravity using weak gravitational lensing measurements. *Monthly Notices of the Royal Astronomical Society*, Vol. 466, Issue 3, 2547–2559. https://doi.org/10.1093/mnras/stw3192.

[28] Ben-Naim A. (2008). *Farewell to Entropy, Statistical Thermodynamics Based on Information.* World Scientific. ISBN9812707077.

[29] Vedral V. (2010). *Decoding Reality: The Universe as Quantum Information.* Oxford University Press. ISBN0199695741.

[30] Haug E. (2020). Finding the Planck Length Multiplied by the Speed of Light Without any Knowledge of G, c, or ħ, using a Newton force spring. *Journal of Physics Communications*, Vol. 4, No 7. https://doi.org/10.1088/2399-6528/ab9dd7.

[31] Chaitin G. (1966). On the Length of Programs for Computing Finite Binary Sequences. *Journal of the ACM* Vol. 13 Issue 4, 547–569. https://doi.org/10.1145/321356.321363.

[32] Landauer R. (1961). Irreversibility and Heat Generation in the Computing Process. *IBM Journal of Research and Development*, Vol. 5, Issue: 3. https://doi.org/10.1147/rd.53.0183.

[33] Davidson A. (2019). From Planck Area to Graph Theory: Topologically Distinct Black Hole Microstates. *Phys. Rev.* D 100, 081502(R). https://doi.org/10.1103/PhysRevD.100.081502.

[34] Scardigli F. (1995). Some Heuristic Semiclassical Derivations of the Planck Length, the Hawking Effect and the Unruh Effect. *Il Nuovo Cimento* B Vol. 110, 1029–1034. https://doi.org/10.1007/BF02726152.

[35] Goutéraux B. (2010). *Black-Hole Solutions to Einstein's Equations in the Presence of Matter and Modifications of Gravitation in Extra Dimensions*. PhD thesis, Univ. Paris-Sud. arXiv:1011.4941 [hep-th]. https://doi.org/10.48550/arXiv.1011.4941.

[36] Szostek R., Góralski P., & Szostek K. (2019). Gravitational Waves in Newton's Gravitation and Criticism of Gravitational Waves Resulting from the General Theory of Relativity (LIGO). *Bulletin of the Karaganda University. Physics series*, No. 4 (96), 39-56, ISSN 2518-7198. https://doi.org/10.20944/ preprints201908.0050.v1.

[37] Weinstein G. (2015). *Einstein's Uniformly Rotating Disk and the Hole Argument*. arXiv:1504.03989 [physics.hist-ph]. https://doi.org/10.48550/arXiv.1504.03989.

[38] Weinberg S. (1972). *Gravitation and Cosmology: Principles and Applications of the General Theory of Relativity*. Wiley. ISBN0471925675.

[39] Kanerva P. (1988). *Sparse Distributed Memory*. MIT Press. ISBN0262514699.

[40] Watanabe S. (1969). *Knowing and Guessing: A Quantitative Study of Inference and Information*. Wiley. ISBN0471921300.

[41] Watanabe S. (1986). Epistemological Relativity - Logico-Linguistic Source of Relativity. *Annals of the Japan Association for the Philosophy of Science*, 7(1), 1–14. https://doi.org/10.4288/jafpos1956.7.1.

[42] Łukaszyk S. (2004). New Concept of Probability Metric and its Applications in Approximation of Scattered Data Sets. *Computational Mechanics* Vol. 33, 299–304. https://doi.org/10.1007/s00466-003-0532-2.

[43] Dikarev V., Preuß O., Solanki S., Krüger H.& Krivov A. (2008). Understanding the WMAP Results: Low-Order Multipoles and Dust in the Vicinity of the Solar System. *Earth, Moon, and Planets* Vol. 102, 555–561. https://doi.org/10.1007/s11038-007-9172-4.

[44] Hansen M., Kim J., Frejsel A., Ramazanov S., Naselsky P., Zhao W., & Burigana C. (2012). Can Residuals of the Solar System Foreground Explain Low Multipole Anomalies of the CMB? *Journal of Cosmology and Astroparticle Physics*, Vol. 2012. https://doi.org/10.1088/1475-7516/2012/10/059.

[45] Copi C., Huterer D., Schwarz D., & Starkman G. (2010). Large-Angle Anomalies in the CMB, *Advances in Astronomy*, Vol. 2010, Article ID 847541. https://doi.org/10.1155/2010/847541.

[46] Kamionkowski M., & Kosowsky A. (1999). The Cosmic Microwave Background and Particle Physics. *Annual Review of Nuclear and Particle Science*, Vol. 49:77-123. https://doi.org/10.1146/annurev.nucl. 49.1.77.

[47] Land K. & Magueijo J. (2005). Examination of Evidence for a Preferred Axis in the Cosmic Radiation Anisotropy. *Phys. Rev. Lett.* 95, 071301. https://doi.org/10.1103/PhysRevLett.95.071301.

[48] Łukaszyk S. (2020). *A Short Note about Graphene and the Fine Structure Constant.* https://doi.org/10.13140/RG.2.2.21718.88641.
[49] Kothari D., & Singh B. (1941). Bose-Einstein Statistics and Degeneracy. Proceedings of the Royal Society of London. *Series A, Mathematical and Physical Sciences* Vol. 178, No. 973. https://doi.org/10.1098/ rspa.1941.0049.
[50] Brukner Č. (2021). *Qubits are not Observers – a No-Go Theorem.* arXiv:2107.03513 [quant-ph]. https://doi.org/10.48550/arXiv.2107.03513.
[51] Pienaar J. (2021). A quintet of Quandaries: Five No-Go Theorems for Relational Quantum Mechanics. *Foundations of Physics* Vol. 51, Article number: 97. https://doi.org/10.1007/s10701-021-00500-6.
[52] Alsing P., & Milburn G. (2003). Teleportation with a Uniformly Accelerated Partner. *Phys. Rev. Lett.* 91, 180404. https://doi.org/10.1103/PhysRevLett.91.180404.
[53] Bell J. (1964). On the Einstein Podolsky Rosen Paradox. *Physics Physique Fizika* 1, 195. https://doi.org/10.1103/PhysicsPhysiqueFizika.1.195.
[54] Fuchs C. (1998). Just Two Nonorthogonal Quantum States. In: Kumar, P., D'Ariano, G.M., Hirota, O. (eds.) *Quantum Communication, Computing, and Measurement* 2. Springer, Boston, MA. https://doi.org/ 10.1007/0-306-47097-7_2.
[55] Axler S., Bourdon P., & Ramey W. (2001). *Harmonic Function Theory. Graduate Texts in Mathematics*, Springer. https://doi.org/10.1007/978-1-4757-8137-3.
[56] Gantumur T. (2010). *Harmonic functions.* McGill University, Math 140. https://www. coursehero.com/ file/114461026/harmonicpdf.

Appendix

1. The alternative form of BH (Hawking) temperature (39) is (with (82)) given as

$$\begin{aligned} T_{BH} &= \frac{\hbar}{2\pi c k_B} g_{BH} = \frac{\hbar}{2\pi c k_B} \frac{c^2}{D_{BH}} = \frac{\hbar}{2\pi c k_B} \frac{c^2}{d_{BH}\ell_P} = \\ &= \frac{\hbar c}{2\pi d_{BH} k_B} \sqrt{\frac{c^3}{\hbar G}} = \frac{1}{2\pi d_{BH}} \sqrt{\frac{\hbar c^5}{G k_B^2}} = \frac{T_P}{2\pi d_{BH}} \end{aligned} . \tag{146}$$

2. The alternative form of Planck–Einstein (Compton) relation is given as

$$E = h\nu = h\frac{c}{\lambda} = \frac{hc}{l\ell_P} = \frac{2\pi\hbar c}{l}\sqrt{\frac{c^3}{\hbar G}} = \frac{2\pi}{l}\frac{k_B}{k_B}\sqrt{\frac{\hbar^2 c^2 c^3}{\hbar G}} = \frac{2\pi}{l} k_B T_P . \tag{147}$$

3. BH diameter fluctuations

After the absorption/emission of a wavelength l, the 2-sphere BH diameter defined by its mass increases(+)/decreases(–) owing to the Compton mass $M = h/(c\lambda)$ of this wavelength. Therefore,

$$D_{BH}^{A/E} = D_{BH} \pm \delta D = d\ell_P \pm \frac{4G}{c^2}\frac{2\pi\hbar}{cl\ell_P}$$
$$= d\ell_P \pm \frac{8\pi\hbar G}{lc^3\ell_P} = d\ell_P \pm \frac{8\pi\ell_P}{l} = \left(d \pm 8\pi/l\right)\ell_P \quad (148)$$

Accordingly, the BH area increases/decreases, such that

$$A_{BH}^{A/E} = \pi\left(d^{A/E}\right)^2 \ell_P^2 = \pi\left(d\ell_P \pm \frac{8\pi G\hbar}{c^3 l\ell_P}\right)^2$$
$$\pi\left(d^{A/E}\right)^2 \ell_P^2 = \pi\left(d\ell_P \pm \frac{8\pi}{l}\ell_P\right)^2 = \pi\ell_P^2\left(d \pm \frac{8\pi}{l}\right)^2 \quad (149)$$

If $l = d/2$ [21, 22] then

$$d_{k+1}^{A/E} = \pm\sqrt{256\pi^2\frac{1}{d_k^2} \pm 32\pi + d_k^2}\,. \quad (150)$$

4. BH diameter (20) after absorption or emission of a wavelength l in dimensions $n = 2, 3,\dots,5$.

n	$(d_{k+1}^{A/E})^{n-1}$
2	$d_k \pm 8\pi/l$
3	$d_k^2 \pm 16\pi d_k/l + 64\pi^2/l^2$
4	$d_k^3 \pm 24\pi d_k^2/l + 192\pi^2 d_k/l^2 \pm 512\pi^3/l^3$
5	$d_k^4 \pm 32\pi d_k^3/l + 384\pi^2 d_k^2/l^2 \pm 2048\pi^3 d_k/l^3 + 4096\pi^4/l^4$

5. Negative and imaginary unit lengths.

There are four possibilities for the volume of unit n-cubic element a^n in n-dimensional space in dependence on what we take as the unit length $a = \{1, -1, i, -i\}$:

$$a = 1 \quad a^n = 1 \quad n \in \mathbb{C}, \quad (151)$$

$$a = -1 \quad a^n = i^{2n} = \begin{cases} 1 & n \text{ is even} \\ -1 & n \text{ is odd} \\ \mathbb{C} & n \in \mathbb{R}\setminus\mathbb{Z} \cup n \in \mathbb{C} \end{cases}, \quad (152)$$

$$a = i \quad a^n = \begin{cases} 1 & n = \pm 0, 4, 8, \ldots \\ -1 & n = \pm 2, 6, 10, \ldots \\ i & n = \ldots, -7, -3, 1, 5, 9, \ldots \\ -i & n = \ldots, -9, -5, -1, 3, 7, \ldots \\ \mathbb{C} & n \in \mathbb{R} \setminus \mathbb{Z} \cup n \in \mathbb{C} \end{cases}, \tag{153}$$

$$a = -i \quad a^n = \begin{cases} 1 & n = \pm 0, 4, 8, \ldots \\ -1 & n = \pm 2, 6, 10, \ldots \\ i & n = \ldots, -9, -5, -1, 3, 7, \ldots \\ -i & n = \ldots, -7, -3, 1, 5, 9, \ldots \\ \mathbb{C} & n \in \mathbb{R} \setminus \mathbb{Z} \cup n \in \mathbb{C} \end{cases}. \tag{154}$$

The unit length a is a scaling factor of a relationship between the vertices of the graph of nature. Generally, it can be real (positive or negative) or imaginary (positive or negative), and this list may not be exhaustive. Any vertex beyond the countably infinite number of vertices can exhibit any relationship with other vertices. However, only a real and positive unit length (151) yields a real and positive unit n-cubic element a^n for any complex dimension n.

6. Bekenstein-Hawking entropy (44) from Verlinde Equation (46)

$$\begin{aligned} \delta S &= 2\pi k_B \frac{Mc}{\hbar} \delta R \overset{BH}{=} -2\pi k_B \frac{M_{BH} c}{\hbar} R_{BH} = -2\pi k_B \frac{c}{\hbar} \frac{R_{BH} c^2}{2G} R_{BH} = \\ &= -\frac{\pi k_B R_{BH}^2}{\ell_P^2} = -\frac{1}{4} k_B N_{BH} \end{aligned}, \tag{155}$$

where Equation (87) was substituted, so the sign was reversed.

7. BH temperature vs. BH energy/work

Comparing dimensionless BH temperature with dimensionless BH energy yields

$$\frac{T_{BH}}{T_P} = \frac{1}{2\pi d_{BH}}, \frac{E_{BH}}{E_P} = \frac{d_{BH}}{4} \Rightarrow d_{BH} = \pm\sqrt{\frac{2}{\pi}}, \tag{156}$$

that is d_{BH} of a minimum 2-bit BH and the 1st singularity (138) of the FD statistics.

Comparing dimensionless BH temperature with BH entropic work (63) yields

$$\frac{T_{BH}}{T_P} = \frac{1}{2\pi d_{BH}}, \frac{W_{BH}}{E_P} = \frac{d_{BH}}{8} \Rightarrow d_{BH} = \pm\sqrt{\frac{4}{\pi}}, \tag{157}$$

that is d_{BH} of a minimum 4-bit BH, one unit of BH entropy (44), and the 2nd singularity (138) of the FD statistics.

Comparing BH mass–energy equivalence (64) with the Landauer limit yields

$$\begin{aligned}
&M_{BH}c^2 = T_{BH}k_B \ln(2) \\
&\frac{d_{BH}\ell_P c^2}{4G}c^2 = \frac{T_P}{2\pi d_{BH}}k_B \ln(2) \\
&\frac{d_{BH}c^4}{4G}\sqrt{\frac{\hbar G}{c^3}} = \sqrt{\frac{\hbar c^5}{Gk_B^2}}\frac{k_B \ln(2)}{2\pi d_{BH}} \\
&\frac{d_{BH}}{2} = \frac{\ln(2)}{\pi d_{BH}} \quad d_{BH} = \sqrt{2}\sqrt{\frac{\ln(2)}{\pi}} \approx 0.6643 \\
&N_{BH} = \pi d_{BH}^2 = 2\ln(2) \approx 1.3863
\end{aligned} \quad . \tag{158}$$

Comparing BH entropic work (63) with the Landauer limit, or equivalently, the BH entropy with the Landauer limit entropy yields

$$\begin{aligned}
&M_{BH}c^2 = T_{BH}k_B \ln(2) \\
&\frac{d_{BH}\ell_P c^2}{4G}c^2 = \frac{T_P}{2\pi d_{BH}}k_B \ln(2) \\
&\frac{d_{BH}c^4}{4G}\sqrt{\frac{\hbar G}{c^3}} = \sqrt{\frac{\hbar c^5}{Gk_B^2}}\frac{k_B \ln(2)}{2\pi d_{BH}} \\
&\frac{d_{BH}}{2} = \frac{\ln(2)}{\pi d_{BH}} \quad d_{BH} = \sqrt{2}\sqrt{\frac{\ln(2)}{\pi}} \approx 0.6643 \\
&N_{BH} = \pi d_{BH}^2 = 2\ln(2) \approx 1.3863
\end{aligned} \quad . \tag{159}$$

8. Variational potential on a BH horizon

$$\frac{4\pi c^2}{N_{BH}}\frac{R_{BH}}{\bar{\lambda}_M} = -\frac{c^2}{2}, \tag{160}$$

$$\frac{4\pi c^2}{N_{BH}}\frac{R_{BH}}{\bar{\lambda}_M} = \frac{4\pi c^2}{4\pi R_{BH}^2/\ell_P^2}\frac{R_{BH}}{\bar{\lambda}_M} = -\frac{c^2}{2}, \tag{161}$$

$$\bar{\lambda}_M = -\frac{2\ell_P^2}{R_{BH}} = -\bar{\lambda}_{BH}\,. \tag{162}$$

Furthermore,

$$N_{BH} = \frac{4\pi R_{BH}^2}{\ell_P^2} \quad \Leftrightarrow \quad R_{BH} = \pm\sqrt{\frac{N_{BH}}{4\pi}}\ell_P, \tag{163}$$

so

$$ƛ_{BH} = \pm 2\ell_P\sqrt{\frac{4\pi}{N_{BH}}} \quad \Leftrightarrow \quad ƛ_M = \mp 2\ell_P\sqrt{\frac{4\pi}{N_{BH}}}. \tag{164}$$

9. Planck units' derivation

Although this derivation is probably established, it has not been determined by the author. We begin with the measured speed of light c and define it to be a quotient of a length unit (ℓ_P) and a time unit (t_P): $c = \ell_P/t_P$. Thereafter, we define a unit of acceleration (a_P) as the time unit required to achieve the speed of light: $a_P = c/t_P$. Subsequently, we assume an inertial force equal to gravitational force between two such mass units (m_P) distanced by the length unit

$$m_P a_P = G\frac{m_P m_P}{\ell_P^2}. \tag{165}$$

Ultimately, a Compton wavelength or Planck–Einstein relation (they are equivalents) for the mass unit is required, which is expressed as

$$m_P = \frac{h}{c\lambda} \quad \Leftrightarrow \quad m_P c^2 = h\nu = h\frac{c}{\lambda}. \tag{166}$$

Next, we consider the reduced $\hbar = h/2\pi$ and presume the wavelength to be $\lambda = 2\pi\ell_P$. The remaining derivation is posed as an exercise for the reader.

Overall, perhaps the simplest derivation of Planck units, based on HUP, mass–energy equivalence, Schwarzschild radius, and $c = \delta R/\delta t$ is presented in [34].

10. Hawking temperature from equipartition theorem

BH mass–energy equivalence can be stated as

$$E_{BH} = M_{BH}c^2. \tag{167}$$

If the BH energy is uniformly divided over N_{BH} bits on the horizon, the temperature can be determined using the equipartition rule for a single degree of freedom (one bit), which is expressed as

$$E_{BH} = \frac{1}{2} k_B N_{BH} T_{BH} , \tag{168}$$

from which relation we can recover the Hawking temperature as

$$\begin{aligned} M_{BH} c^2 &= \frac{1}{2} k_B N_{BH} T_{BH} \\ \frac{D_{BH} c^2}{4G} c^2 &= \frac{1}{2} k_B \frac{\pi D_{BH}^2}{\ell_P^2} T_{BH} \end{aligned} . \tag{169}$$

11. Potential in negative dimensions

If a spherically symmetric fundamental solution (potential) ϕ of Laplace's equation and the general Poisson's equation exists in $\mathbb{R}^n$, it must be in the form of [55, 56]

$$\phi_n = \begin{cases} \dfrac{R^{2-n}}{(2-n) n V_n} & \text{for } n \neq 2 \\ \dfrac{1}{2\pi} \log(R) & \text{for } n = 2 \end{cases} , \tag{170}$$

where $R \neq 0$ and nV_n represents the surface area of the unit radius n-ball (with volume V_n). This can be evaluated in the negative dimensions, as shown in Figure 15 for $R = 1$.

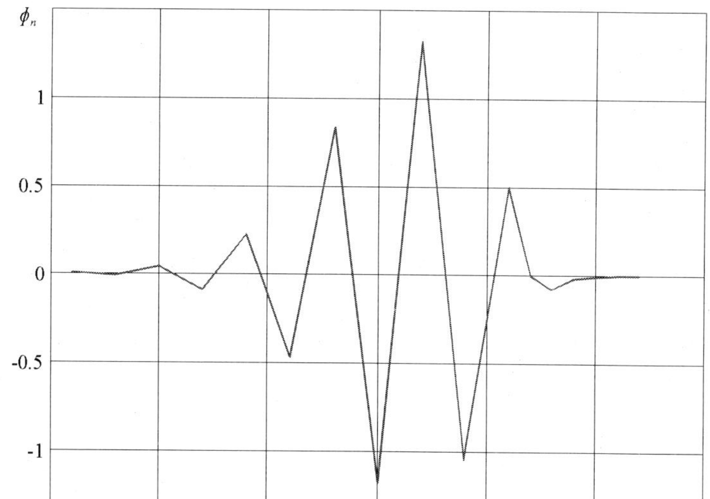

Figure 15. Potential ϕ_n in dimensions n = -20,… 10, R = 1.

Further research is required to evaluate the possible relation of the logarithmic potential (170) in $\mathbb{R}^2$ with Shannon information entropy (60) and Landauer bound (61), (158), or (159).

12. Lorentz contraction and velocity bound

From Equations (90), (104), and (105), the velocity bound (91) is derived as

$$\frac{\delta L^2}{\delta t^2}-\frac{\delta R^2}{\delta t^2}=c^2. \tag{171}$$

Furthermore, the squared-Lorentz contraction (93) can be expressed as

$$\delta L^2=-\frac{\delta L_0^2}{c^2}\frac{\delta R^2}{\delta t^2}, \tag{172}$$

from which equation we substitute $c^2\delta t^2$ into the first equation to reduce it to be time independent.

13. Planck law

We set

$$A_\lambda \doteq \frac{2h}{\ell_P^5}c^2,\quad A_\upsilon \doteq \frac{2h}{\ell_P^3}c^2, \tag{173}$$

in Planck's laws for BH spectral radiance (130), (131). Their derivatives w/r/t l are (derivatives w/r/t d are monotonic)

$$\frac{\partial B_\lambda(l,d)}{\partial l}=\frac{4A_\lambda d\pi^2\exp\left(4\pi^2 d/l\right)}{l^7\left(\exp\left(4\pi^2 d/l\right)-1\right)^2}-\frac{5A_\lambda}{l^6\left(\exp\left(4\pi^2 d/l\right)-1\right)}, \tag{174}$$

$$\frac{\partial B_v(l,d)}{\partial l}=\frac{4A_\upsilon d\pi^2\exp\left(4\pi^2 d/l\right)}{l^5\left(\exp\left(4\pi^2 d/l\right)-1\right)^2}-\frac{3A_\upsilon}{l^4\left(\exp\left(4\pi^2 d/l\right)-1\right)}. \tag{175}$$

From Equations (174) and (175), we can determine the wavelength multiplier l maximizing spectral radiance at a given BH diameter d in terms of the maximum radiation wavelength λ

$$d_\lambda^{\max}=\frac{l}{4\pi^2}\left(W_0\left(-5e^{-5}\right)+5\right)=0.1257678127199937l, \tag{176}$$

$$l_\lambda^{\max}=7.951159986d, \tag{177}$$

or of the maximum radiation frequency ν

$$d_{\nu}^{\max}=\frac{l}{4\pi^{2}}\left(W_{0}\left(-3e^{-3}\right)+3\right)=0.071467894189626l\,, \tag{178}$$

$$l_{\nu}^{\max}=13.99229698d\,, \tag{179}$$

where $W_0(x)$ is the Lambert W function.

It can be seen that the maximum emitted wavelength of BH in the visible radiation range (1 μm $\to l = 6.19E28$) exhibits a diameter of $d_{BH} = 7.78E+27 \to D_{BH} = 1.26E-7$ m, whereas the maximum emitted wavelength of Sagittarius A* ($D_{BH} \approx 2.56E10$ m) amounts $\lambda = 2.04E+11$ m, which is more than 100 times larger than ELF radio waves of 1E8 m (≈3 Hz).

14. Certain properties of the Shannon entropy of holographic spheres

$$H(k)=\log(k)-\frac{k-1}{k}\log(k-1)\,. \tag{180}$$

The 1st derivative is

$$\frac{dH}{dk}=-\frac{\log(k-1)}{k^{2}}\,, \tag{181}$$

yielding the maximum of $H(k)$ at $k = 2$. The 2nd derivative

$$\frac{d^{2}H}{dk^{2}}=\frac{2(k-1)\log(k-1)-k}{k^{3}(k-1)}\,, \tag{182}$$

yields the inflection point of $H(k)$ at

$$k=\frac{1}{2W_{0}\left(\frac{1}{2}e^{-\frac{1}{2}}\right)}+1\approx 3.0935\,. \tag{183}$$

The integral is

$$\begin{aligned}\int H dk &= \left(\log(k-1)+k\right)\log(k)\\ &-(k-1)\log(k-1)+\mathrm{Li}_{2}(1-k)\end{aligned}\,, \tag{184}$$

where $\mathrm{Li}_2(z)$ denotes dilogarithm.

15. The volume of an n-ball in complex dimensions

The volume of an n-ball (B) is known to be

$$V_n(R)_B = \frac{\pi^{n/2}}{\Gamma(n/2+1)} R^n, \tag{185}$$

where Γ ($\mathbb{C}\to\mathbb{C}$) is the Euler's gamma function and R is the n-ball radius. The gamma function is defined for all complex numbers except the non-positive integers. Thus, for $n = a + ib$, where if $a \in \mathbb{Z}$, $a > 0$,

$$\pi^{n/2} = \pi^{(a+ib)/2} = \pi^{a/2}\left[\cos\left(\frac{b}{2}\ln(\pi)\right)+i\sin\left(\frac{b}{2}\ln(\pi)\right)\right], \tag{186}$$

$$R^n = R^{a+ib} = R^a\left[\cos\left(b\ln(R)\right)+i\sin\left(b\ln(R)\right)\right], \tag{187}$$

the volume becomes

$$V_n(R)_B = \frac{\pi^{\frac{a}{2}}R^a\left\{\cos\left[b\ln\left(R\sqrt{\pi}\right)\right]+i\sin\left[b\ln\left(R\sqrt{\pi}\right)\right]\right\}}{\Gamma\left(\frac{a+ib}{2}+1\right)}, \tag{188}$$

and the surface becomes

$$S_n(R)_B = \frac{n}{R}V_n(R)_B = \pi^{\frac{a}{2}}R^{a-1}(a+ib)\frac{\left\{\cos\left[b\ln\left(R\sqrt{\pi}\right)\right]+i\sin\left[b\ln\left(R\sqrt{\pi}\right)\right]\right\}}{\Gamma\left(\frac{a+ib}{2}+1\right)}. \tag{189}$$

The anti-symmetry of the imaginary part of the volume (188) and the surface (189), in a way, establishes the arrow of time and is independent of Re(n) for Im(n) = 0.

INDEX

#

A

B

C

D

I

J

K

L

M

N

T

U

V

W

Z

About the Editors

Valeriy V. Dvoeglazov

Valeriy V. Dvoeglazov, the corresponding editor of the present book, is a professor in the Physics Faculty at University of Zacatecas, México. He holds a degree in physics and education from the Saratov State University and a PhD from the Joint Institute for Nuclear Research at Dubna, both in Russia. A present and past member of several professional associations, he serves as an editor with a number of physical journals, and he was the editor with the "Contemporary Fundamental Physics" and "Relativity, Gravitation, Cosmology" series published by Nova Science Publishers. Other editors are his previous pupils.

M. de G. Caldera Cabral

Unidad Académica de Física
Universidad Autónoma de Zacatecas
Zacatecas, Zac., México

J. A. Cázares Montes

Unidad Académica de Física
Universidad Autónoma de Zacatecas
Zacatecas, Zac., México

J. L. Quintanar González

Unidad Académica de Física
Universidad Autónoma de Zacatecas
Zacatecas, Zac., México